Biotic Interactions and Global Change

BIOTIC INTERACTIONS and GLOBAL CHANGE

Edited by Peter M. Kareiva
Joel G. Kingsolver
Raymond B. Huey
THE UNIVERSITY OF WASHINGTON

Sinauer Associates Inc. • Publishers
Sunderland, Massachusetts

THE COVER

Remains of a crayfish next to a bleached plant on the Cheyenne Bottoms wetland near Great Bend, Kansas. Natural drought as well as draining of water for irrigation resulted in the desiccation of the wetland. Photograph copyright © Jim Richardson, West Light. All rights reserved.

BIOTIC INTERACTIONS AND GLOBAL CHANGE

Library of Congress Cataloging-in-Publication Data

Biotic interactions and global change / edited by Peter M. Kareiva,
Joel G. Kingsolver, Raymond B. Huey.
p. cm.
Includes bibliographical references and index.
ISBN 0-87893-429-4—ISBN 0-87893-430-8 (pbk.)
1. Adaptation (Biology)—Congresses. 2. Man—Influence on nature—Congresses. 3. Climatic changes—Environmental aspects—Congresses. I. Kareiva, Peter M., 1951– . II. Kingsolver, Joel G., 1953– . III. Huey, Raymond B.
QH546.B55 1993
574.5—dc200 92-19162
CIP

Printed in U.S.A.

4 3 2 1

Contents

Contributors

MATTHEW P. AYRES Department of Entomology, Michigan State University, East Lansing, Michigan 48824; and Institute of Arctic Biology, University of Alaska, Fairbanks, Alaska 99775

MARK W. BLOWS Department of Genetics and Human Variation, La Trobe University, Bundoora, Victoria 3083, Australia

STEPHEN R. CARPENTER Center for Limnology, University of Wisconsin, Madison, Wisconsin 53706

JAMES S. CLARK Department of Botany, Duke University, Durham, North Carolina 27706

H. N. COMINS Imperial College, Silwood Park, Ascot, Berkshire SL5 7PY, United Kingdom

TODD E. DAWSON Section of Ecology and Systematics, Cornell University, Ithaca, New York 14853

ARTHUR E. DUNHAM Department of Biology, University of Pennsylvania, Philadelphia, Pennsylvania 19104

THOMAS M. FROST Center for Limnology, University of Wisconsin, Madison, Wisconsin 53706

DOUGLAS J. FUTUYMA Department of Ecology and Evolution, State University of New York at Stony Brook, Stony Brook, New York 11794

MONICA A. GEBER Section of Ecology and Systematics, Cornell University, Ithaca, New York 14853

GEORGE GILCHRIST Department of Zoology, University of Washington, Seattle, Washington 98195

H. C. J. GODFRAY Imperial College, Silwood Park, Ascot, Berkshire SL5 7PY, United Kingdom

NANCY B. GRIMM Department of Zoology, Arizona State University, Tempe, Arizona 85287

MARTHA J. GROOM Department of Zoology, University of Washington, Seattle, Washington 98195

SUSAN HARRISON Division of Environmental Studies, University of California, Davis, California 95616

M. P. HASSELL Imperial College, Silwood Park, Ascot, Berkshire SL5 7PY, United Kingdom

ARY A. HOFFMANN Department of Genetics and Human Variation, La Trobe University, Bundoora, Victoria 3083, Australia

KENT E. HOLSINGER Department of Ecology and Evolutionary Biology, University of Connecticut, Storrs, Connecticut 06269

RAYMOND B. HUEY Department of Zoology, University of Washington, Seattle, Washington 98195

GEORGE C. HURTT Department of Ecology and Evolutionary Biology, Princeton University, Princeton, New Jersey 08544

ANTHONY R. IVES Department of Zoology, University of Wisconsin, Madison, Wisconsin 53706

PETER M. KAREIVA Department of Zoology, University of Washington, Seattle, Washington 98195

JAMES R. KARR Institute for Environmental Studies, University of Washington, Seattle, Washington 98195

JOEL G. KINGSOLVER Department of Zoology, University of Washington, Seattle, Washington 98195

JAMES F. KITCHELL Center for Limnology, University of Wisconsin, Madison, Wisconsin 53706

TIMOTHY K. KRATZ Center for Limnology, University of Wisconsin, Madison, Wisconsin 53706

ROLAND H. LAMBERSON U.S.D.A. Forest Service, Redwood Sciences Laboratory, 1700 Bayview Drive, Arcata, California 95521

RUSSELL LANDE Department of Biology, University of Oregon, Eugene, Oregon 97403

DAVID M. LODGE Department of Biological Sciences, University of Notre Dame, Notre Dame, Indiana 46556

MICHAEL LYNCH Department of Biology, University of Oregon, Eugene, Oregon 97403

DAVID E. MCCAULEY Department of Biology, Vanderbilt University, Nashville, Tennessee 37235

KEVIN MCKELVEY U.S.D.A. Forest Service, Redwood Sciences Laboratory, 1700 Bayview Drive, Arcata, California 95521

WILLIAM W. MURDOCH Department of Biological Sciences, University of California, Santa Barbara, California 93106

BARRY R. NOON U.S.D.A. Forest Service, Redwood Sciences Laboratory, 1700 Bayview Drive, Arcata, California 95521

GORDON H. ORIANS Department of Zoology, University of Washington, Seattle, Washington 98195

STEPHEN W. PACALA Department of Ecology and Evolutionary Biology, Princeton University, Princeton, New Jersey 08544

MICHAEL L. PACE Institute of Ecosystem Studies, Box AB, Millbrook, New York 12545

ROBERT T. PAINE Department of Zoology, University of Washington, Seattle, Washington 98195

JAMES F. QUINN Division of Environmental Studies, University of California, Davis, California 95616

TERRY L. ROOT School of Natural Resources, University of Michigan, Ann Arbor, Michigan 48109

DAVID S. SCHIMEL Climate System Modeling Program, University Corporation for Atmospheric Research, Box 3000, Boulder, Colorado 80307

STEPHEN H. SCHNEIDER National Center for Atmospheric Research, Box 3000, Boulder, Colorado 80307

NATHAN SCHUMAKER Center for Streamside Studies, College of Forest Resources, University of Washington, Seattle, Washington 98195

DAVID TILMAN Department of Ecology, University of Minnesota, Minneapolis, Minnesota 55455

JOSEPH TRAVIS Department of Biological Science, Florida State University, Tallahassee, Florida 32306

Preface

Environmental changes such as climate shifts and landscape modification are accelerating at an unprecedented pace. Not surprisingly, the potential ecological, economic, and social consequences of these changes are increasingly being discussed in scientific, popular, and political arenas. Indeed, terms such as "climate warming," "habitat fragmentation," and "biodiversity" have already joined the vernacular vocabulary. But missing from most of these discussions is an explicit appreciation of the role that evolutionary, population, and community processes inevitably play in global change. As a consequence, predictions of the biotic consequences of global change will remain at best simplistic, and at worst misleading, until analyses of global change consider these basic ecological and evolutionary processes.

To catalyze ecological research and literacy regarding the consequences of global change, we organized a workshop that met at the Friday Harbor Laboratories (San Juan Island, Washington) on 20–23 September 1991. Workshop participants were asked to address aspects of a central question: How do population, community, and evolutionary dynamics influence the way in which the world's terrestrial and freshwater biota respond to and mediate environmental change? In particular, we wanted participants to focus on two aspects of global change: climate change and habitat fragmentation. The resulting discussions ranged over a wide variety of environmental stresses and spanned physiology, evolution, demography, community ecology, and ecosystems analysis. Nevertheless, we made no attempt to be comprehensive, which would have been impossible given the global scale of the issues involved and the narrow focus needed to promote intensive discussion. Accordingly, we did not cover in detail many issues (e.g., direct effects of climate on the physiology and fitness of individuals; ecosystem consequences such as nutrient or energy cycling), except as they related to population, community, or evolutionary interactions. We intentionally discouraged discussions of oceanic systems, which have already been the subject of considerable attention (e.g., GLOBEC). Similarly, practical considerations limited discussions primarily to temperate-zone systems. We recognize, of course, that tropical systems are vital to any complete analysis of these issues.

In selecting participants we attempted to diversify viewpoints. We included several younger workers, whose fresh views are beginning to have an impact on the field. Participants were charged either with giving a presentation on selected topics or with summarizing and evaluating the contributions to one of the main sections. We encouraged them to identify basic population, community, and evolutionary issues that should be included in discussions of global change.

Manuscripts based on the presentations were subsequently evaluated both by the editors and by ad hoc reviewers, and sometimes by other participants. We are extremely grateful to all the reviewers for their constructive suggestions.

The workshop itself began with two "background" talks on patterns of climate and of landscape change. On subsequent days, participants evaluated several research areas (physiology and population responses, evolutionary responses, community responses, landscape change, and habitat fragmentation). Some special feature papers addressed ecological feedbacks and policy implications. Finally, small discussion groups "brainstormed" the key research priorities for each area. A synthesis of these priorities forms the final chapter of this book.

The editors thank the other members of the Steering Committee (Steve Carpenter, Terry Chapin, and Terry Root) for their help at all stages of this project. They were involved with the initial development and organization and the selection of speakers, as well as with the editing. We also acknowledge George Gilchrist, Martha Groom, and Ingrid Parker for their stunning logistic skills. Given that the editors are organizationally impaired, any credit for a well-run meeting is due to the efforts and planning of these graduate students. We also thank Susan Bartroff, Budget Coordinator in the Department of Zoology at the University of Washington, for exercising her usual magic in solving all budgetary problems and in getting all of the workshop participants to an isolated island on time. We especially thank Robert Podolsky, who carried most of the tedious burden of final editorial checking of the manuscripts. Finally, we thank James Reichman (former Program Director of the Ecology Panel of the National Science Foundation) for encouraging us to organize this workshop and for his wise and constructive suggestions at all stages.

The workshop was supported by the National Science Foundation. Any opinions, findings, conclusions, or recommendations expressed in this publication are those of the authors or editors and do not necessarily reflect the views of the National Science Foundation.

Peter Kareiva
Joel Kingsolver
Raymond Huey

CHAPTER 1

INTRODUCTION

Peter M. Kareiva, Joel G. Kingsolver, and Raymond B. Huey

A CHALLENGE FOR ECOLOGISTS AND EVOLUTIONARY BIOLOGISTS

Dozens of books and symposium volumes have been devoted to global climate change; many have even considered ecological facets of this change. Missing, however, has been an examination of how species interactions or evolution might exacerbate or mitigate the dire environmental consequences expected with global warming and massive deforestation. The "ecology" or "evolution" that is typically part of global change discussions can often be summarized in terms of climate performance correlations or actuarial tabulations of losses caused by habitat destruction. Perhaps such a straightforward approach captures all that we need to know about how organisms will respond to global change. However, the intellectual tradition of ecology and evolutionary biology suggests otherwise. We know that the impact of climate warming ought to depend on the genetic flexibility of populations. We know that the effects of environmental change must go beyond a simple adjustment in birth and death schedules. For example, environmental change will surely alter the balance of some competitive and predator–prey interactions. Indeed, it would not be surprising to see entire patterns of community organization jumbled as a result of global change. It cannot be as simple as saying that our summers will be hotter and thus some drought-intolerant plants may disappear locally (or shift northward), or as saying that deforestation will cause a 60 percent reduction in biodiversity. Whatever the outcome of global change, we believe that community and evolutionary dynamics will often play a constraining or amplifying role. If our perception is correct, then expectations regarding global change are likely to be grossly in error whenever they overlook principles of evolution and ecology.

To catalyze input from ecologists and evolutionary biologists into discussions of global change, we organized a workshop and nurtured this book. We focused on two dominant global stresses: climatic change and landscape change (broadly defined to include habitat loss and large-scale ecological invasions). We then challenged leading biologists to *show us why ecology and evolution matters* to the outcome of global environmental change. In other words, we sought ideas from evolutionary and ecological theory that might improve our foresight regarding the implications of global change. This book is the result of that challenge.

ORGANIZATION OF THE BOOK

The book is divided into six sections and includes synthetic literature reviews, critical commentaries, arguments for particular research directions, and the results of original research aimed at the challenge we posed to the authors. Some of this diversity of styles was at our request (we specifically solicited commentaries from several contributors in response to hearing other talks at the workshop); in general, we felt the best way to prompt free-ranging discussion was to avoid imposing any conformity. We did require that the lead-off authors set the stage for the rest of the book by providing baseline data and perspectives. After that, however, the only "order" involves proceeding from individual- and population-level concerns to community- and ecosystem-level concerns.

The first section, "Patterns and Determinants of Climate and Landscape Change," introduces the scale and scope of global change, and provides an overview of the connection between ecological processes and the physical changes emphasized by atmospheric scientists concerned with climate warming. Schneider makes it clear that climate warming is a *fact*; current debates focus on the rate of this change. He summarizes the spatial resolution of current climate models, pointing out that the scale at which biologists study evolutionary or ecological dynamics is far smaller than the scale at which global climate models make predictions. He then reviews some approaches to embedding fine-resolution models of local climate within general circulation models. Groom and Schumaker document the tremendous rates of deforestation that are altering landscapes worldwide and go on to show that crude deforestation rates gloss over much of the ecological detail that is needed for sustainable management. They use a unique data base on old-growth forest fragmentation in the Pacific Northwest to illustrate a level of description regarding landscape changes that proves to be more useful than a simple tabulation of deforestation rates. Schimel closes this stage-setting section by pointing out the role biological feedbacks have both in local climatic variation and in moderating global warming trends. He argues persuasively that the "parameters" determining climate change should be studied as the outcome of species interactions and of evolutionary processes that shape how organisms use carbon.

The next four sections explore the ecological and evolutionary processes that might act as key control points in the response of biotic systems to global change. Contributions to "Physiological and Population Responses to Environmental Change" adopt the individual organism as a focus. Pacala and Hurtt assess forest models, in which the growth and demography of individual trees is used to construct scenarios of large-scale community change. They argue that the failure of present models to distinguish fundamental from realized niches leads to erroneous conclusions about the consequences of climate change for tree distributions. Ayres draws on the theory of resource allocation and plant defenses to illuminate the sometimes dramatic effects of climate on patterns of herbivory. Dunham argues that population consequences of climate change can best be predicted using individual-based, structured population models; he combines biophysical and demographic models to examine the consequences of climate warming. Ives and Gilchrist present an elegant theoretical framework that illustrates how the effect of environmental change on population densities depends on the nature of density-dependent processes operating on the species of concern. After considering all these approaches that build upon individual performance and demography, Murdoch argues that we will be able to anticipate responses to global change if we study a variety of detailed individual-based models of population dynamics. His hope is that such mechanism-rich models will yield simplified abstractions that can cover a wide variety of ecological circumstances.

Ecological and demographic processes are not the only dimension to global change. The next section, "Evolutionary Responses to Environmental Change," explores the many ways in which evolutionary genetics constrain or facilitate the adjustment of species to new environments and new landscapes. Hoffmann and Blows review the evolutionary responses of animal populations subjected to stress, and go on to prescribe the sorts of genetic investigations that are now needed to turn evolutionary genetics into a predictive discipline. They advocate paying particular attention to populations at the margin of species' ranges. Geber and Dawson meld the insights of plant physiology with those of plant genetics to speculate about the direction and rate of evolutionary change expected for plant populations in the face of rapid climate shifts. Holsinger tackles the difficult problem of habitat fragmentation, and asks what are the long-term evolutionary consequences for plants of an increasingly subdivided population structure. Contrary to superficial platitudes about "fragmentation," Holsinger argues that definitive answers regarding the evolutionary implications of subdivided populations are not yet forthcoming. McCauley examines the combined effects of local extinction and recolonization in a fragmented world in order to wrestle with the general consequences of landscape structure for the maintenance of genetic variation. Lynch and Lande use the theory of quantitative genetics to predict critical rates of environmental change that are likely to overwhelm a population's ability to adjust evolutionarily (as opposed to going extinct). The beauty of Lynch and Lande's analysis is that it shows how these critical rates can be estimated using phenotypic standard deviations and a population's intrinsic rate of increase—both eminently measurable parameters. Travis and Futuyma close this section with a provocative commentary that reminds us of the role of systematics, paleontology, and ecological genetics in resolving the evolutionary consequences of rapid global change.

Species do not encounter global change in isolation; consequently, community- or ecosystem-level responses to change must be investigated. The fourth section of the book, "Community Responses to Environmental Change," embraces the perspectives of community and ecosystem ecology. Carpenter et al. begin with an insightful report of whole-lake experiments suggesting that key indicator species give the best warning signal for major ecosystem deterioration, long before chemical or physical changes are evident. If they are generally correct, then Chapter 16 provides a strong argument for a better marriage between population ecology and ecosystem analysis (two disciplines that rarely speak to one another in the current climate, much less share the same bed). Root summarizes distribution patterns in vertebrate, especially bird, communities in concert with environmental gradients. This contribution represents one of the only existing studies that has been conducted at a geographical scale commensurate with the expected scale of global change; her findings suggest that the response of bird communities might be predicted using simple physiological constraints and neglecting species interactions. Grimm looks at stream communities as systems that integrate climatic change across large landscapes, and offers an intriguing "patch analysis" as a mechanism for identifying the controlling factors in a stream's response. One of the unique features of Grimm's contribution is its explicit treatment of "extremes" as opposed to mean values when discussing environmental change. Clark draws on both paleobotanical data and an innovative application of spatially distributed population models to partition the roles of dispersal and reproduction in determining the pattern of forest community variation as a result of climate change. His scrutiny of specific case studies highlights the many "suprises" and indirect effects (such as fire frequency) that can drive major vegetation trends. Tilman uses resource-based competition theory to speculate about the effects of CO_2 enrichment for plant communities; his essay produces fascinating theoretical predictions as well as clear prescriptions for specific field experiments. Paine proves that marine ecology has much to offer in the way of guiding principles; in particular, he counters much of the preceding community discussion with a series of examples meant to dramatize the direct effects of human exploitation of marine communities, which typically dwarf any effects attributable to shifts in water temperature. After considering the many ecosystem and community perspectives that were delivered at the workshop, Pace offers a commentary that promotes the need for large-scale comparative studies, an approach often slighted in the rush to advocate mechanistic models or perturbation experiments.

The fifth section, "Landscape Change and Habitat Fragmentation," addresses a facet of global change that is obvious without any model. Among the most conspicuous elements of landscape change are ecological invasions, whereby non-native species enter new landscapes and are so successful that they become dominant members of the community. Lodge uses studies of aquatic invasions to point out that the effects of these perturbations are much more predictable than one might think if one relied on classic invasion models. Harrison summarizes landscape-level changes in communities by using broad-scale patterns of diversity, particularly between-habitat diversity. Her analyses document the homogenization of landscapes that is well known anecdotally, and suggest that dispersal does not constrain the maintenance of regional diversity. Hassell, Godfray, and Comins use elaborations of classic host–parasitoid models to illustrate how the details of climate (i.e., phenology) and of landscapes (exactly how populations are subdivided) can alter the very character of species interactions. In particular, whether host and parasites engage in stable or wildly fluctuating dynamics, or even spin off to extinction, is found to depend on the very parameters that global change is most likely to affect: emergence from diapause and rates of dispersal between habitat islands. All of this abstract ecological theory is brought down to earth in McKelvey, Noon, and Lamberson's analysis of spotted owl populations in the Pacific Northwest. Here models are called upon to do much more than "offer insight" or "provide guideposts to subsequent research." Models that couple owl dispersal to real landscapes are used to identify those management options most likely to sustain owl populations. Quinn and Karr synthesize ideas from this diverse section by focusing on the processes that control patterns of diversity: dispersal, colonization, extinction, and local uniqueness.

Given all of the specific discussions that make up the bulk of this book, it is natural to ask what they suggest in general for a research agenda, and what their implications are at a policy level. These issues are addressed in the final section of the book, "The Context for Population and Community Research on Global Change." Orians delineates the many implications of global change for biodiversity, paying particular attention to implications at the policy level. He musters a very cogent argument for strong action despite the uncertainties involved, and suggests that we use environmental impact statements as hypotheses that can be tested by setting up ecological monitoring following any large-scale action. The book's editors end the text with a summary of what we feel was a surprisingly strong consensus among workshop participants regarding future research directions.

GLOBAL CHANGE IS THE "TEST" THAT ECOLOGY AND EVOLUTIONARY BIOLOGY MUST PASS

Habitat fragmentation and climate change are only two of many environmental pressures that threaten to degrade our ecological systems. They may not even be the greatest hazards facing us (see Paine's contribution), but by virtue of their dramatic scale and worldwide visibility, they represent crucial test cases for the science of ecology and evolutionary biology. If we as evolutionary biologists cannot anticipate the evolutionary responses of species to fragmented populations and elevated temperatures, then our so-called theory is little more than a retrospective formalism, capable only of describing *a posteriori* what is easily observed in simple laboratory or field environments. Similarly, if we as community ecologists cannot predict the patterns of diversity change associated with these global onslaughts, one has to wonder just what community ecology can predict.

Although the challenges posed by global change for the research scientist are obvious, how to proceed is not obvious. This book presents statements by a range of ecologists and evolutionary biologists regarding what we need to know and how to learn it. It is easy to be daunted by our ignorance in the face of problems as complex as climate change or habitat fragmentation. When asked to address practical problems or controversial environmental issues, academic scientists are often paralzyed by the imperfection of their expertise. But the question is not whether imperfect knowledge is better than perfect knowledge; the question is whether we as ecologists and evolutionary biologists have anything to say about global change, as opposed to being mute. We believe this book is a testimony to how much population biology and community ecology can offer to discussions about the implications, consequences, and directions of global environmental change. We hope it is just a beginning.

PART ONE

PATTERNS AND DETERMINANTS OF CLIMATE AND LANDSCAPE CHANGE

CHAPTER 2

SCENARIOS OF GLOBAL WARMING

Stephen H. Schneider

THE GREENHOUSE EFFECT IS REAL

Despite the public controversy over global warming, the natural "greenhouse effect" is as well established a phenomenon as there is in the atmospheric sciences. It has been known for over a century that certain trace gases in the earth's atmosphere (e.g., H_2O, CO_2, CH_4, N_2O, O_3) transmit a larger fraction of solar radiation (wavelengths of 0.4–1.0 μm) than of terrestrial infrared radiation (wavelengths of 5–20 μm). This is also true of most clouds. These trace constituents of the atmosphere are collectively able to trap enough radiative energy in the lower atmosphere to warm the earth's surface some 33°C above the temperature that would exist if these gases were not present on earth. In other words, the natural greenhouse effect is a noncontroversial and critical component of the coevolutionary history of the earth's climate and biosphere. This natural greenhouse effect is being augmented by human activities, which add trace greenhouse gases to the atmosphere at rates that are typically much faster than previously occurred in geological history. It is only the precise magnitude, timing, and consequences of this anthropogenic component of the greenhouse effect (i.e., global warming) that is widely debated.

THE GLOBAL WARMING DEBATE RAGES ON

In the United States, debate has raged since the intense heat waves in the summer of 1988 over whether a "signal" of global warming has finally been detected against the background "noise" of natural climatic variation. Often this debate has been acerbic and occasionally polemical (e.g., see the Epilogue in Schneider, 1990a). Misinterpreted facts, such as the nonwarming trend in the lower 48 United States (Brookes, 1989) or postulated physical mechanisms that might mitigate any substantial global warming (Ellsaesser, 1984; George C. Marshall Institute, 1989; Lindzen, 1990) have been offered as evidence against global warming. Many scientists (e.g., Raval and Ramanathan, 1989; Hansen and Lacis, 1990; Schneider, 1990b) have challenged these doubts. For the lower atmosphere, 1990 was the warmest year on record in the past century, while it was the coldest in the stratosphere. Taken together, these facts are broadly consistent with the kind of greenhouse effect signal that would be anticipated from the increase in greenhouse gases injected into the atmosphere over the past 150 years: a 25 percent increase in CO_2; a 100 percent increase in CH_4; and the introduction of heat-trapping synthetic chemicals such as chlorofluorocarbons (CFCs) and halons. These emissions are responsible for trapping some 2–3 watts of additional infrared radiative heat per square meter of the earth's surface. This is not a controversial assertion. However, how to translate several additional watts per square meter of heating into *x* degrees of temperature change is controversial, since it involves calculating how many interacting physical, biological, and chemical feedback mechanisms might be amplifying or damping the global warming signal.

Even though the approximately 0.5°C surface warming trend over the twentieth century is "consistent" with the greenhouse gas buildup, the actual cause of the observed warming cannot be determined unambiguously. There is some possibility (perhaps a 10 to 20 percent chance) that the 0.5°C twentieth-century warming trend was largely natural, with little or no contribution from the buildup of greenhouse gases. This scenario, however, is not supported by the extreme differences between the lower atmosphere and stratosphere temperatures found in 1990. Indeed, another possible scenario is that a twentieth-century *cooling* trend, rather than a warming fluctuation, occurred naturally. If this were so, then the world would have warmed more than was observed had there not been a fortuitous natural cooling trend. One could even speculate that the dramatic temperature rise in the 1980s and the record-breaking high temperatures recorded at numerous locations around the world reflect the end of a natural cooling trend combined with rapid establishment of the expected human-caused global warming signal. The lack of detailed experimental control over the processes that force our global climate to vary prevents scientists from knowing precisely the mechanisms of the twentieth century warming trend. Empirical observations of temperature trends, however, are indeed consistent with a CO_2-doubling equilibrium sensitivity of anywhere from 0.5°C to 5.0°C, depending upon various assumptions of how nongreenhouse gases will force the climate to change (e.g., see Wigley and Raper, 1991).

Nongreenhouse gases such as natural and human-emitted SO_2 are photochemically converted in the atmosphere into sulfuric acid aerosol particles that may have intervened to help prevent twentieth-century temperatures from being warmer. Aerosol particles can either directly reflect sunlight back to space in cloudless regions of the atmosphere, or reflect sunlight indirectly through their incorporation into clouds, thereby increasing the number of cloud droplets and consequently making the average cloud brighter. Theory and measurements are not adequate to determine quantitatively the exact mechanisms involved, although some attempts are underway (e.g., Kerr, 1992). Charlson and co-workers (1991) suggest that SO_2 emissions from industrial activities from the 1950s to the 1970s suppressed 2 W/m^2 of extra greenhouse surface heating with some 1.5 W/m^2 of artificial cooling in the Northern Hemisphere. However, pollution controls in the 1970s and 1980s may have reduced SO_2 emissions, which in turn may have led to the 1980s being the warmest decade on record (Figure 1).

In summary, no definitive objection based upon either postulated physical mechanisms or direct empirical evidence contradicts the widespread

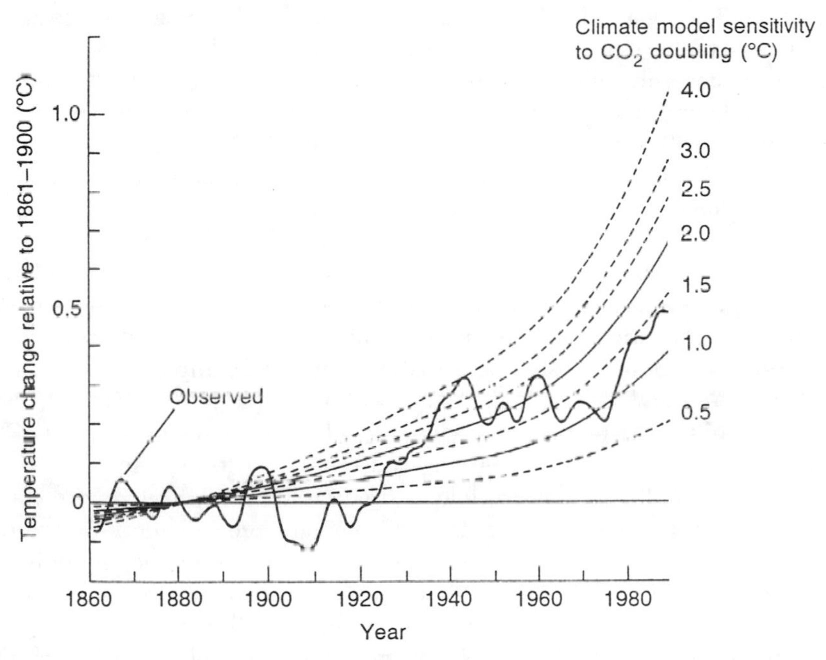

FIGURE 1. Observed global mean temperature changes 1861–1989 compared with predicted values. The observed changes are given in Section 7 of the IPCC report (1990). The data have been smoothed to show decade-averaged and longer time scale trends more clearly. Predictions are based on observed CO_2 concentration changes and on concentration/forcing relationships as given in Section 2 of the IPCC report and have been calculated using the upwelling–diffusion climate model of Wigley and Raper (1990). (After Wigley and Raper, 1991.)

consensus that the world is in a long-term, global-scale warming trend, nor does such evidence contradict the substantial probability that temperatures will continue to go up because of increases in greenhouse gases. Certainly any quantitative connection between the twentieth-century warming trend and greenhouse gas buildup to date cannot yet be unequivocally demonstrated, but it would be folly to assume the connection is not true beyond reasonable doubt (i.e., has only a small probability of being true).

SCENARIOS OF GLOBAL WARMING

The consensus of knowledgeable scientists remains that 1.5–4.5°C additional global warming is likely if CO_2 (or its greenhouse gas equivalent) were to double in the next century. Indeed, the Intergovernmental Panel on Climate Change (IPCC, 1990) recently reviewed vast quantities of evidence, both pro and con, and provided a strongly supported restatement of this consensus. This United Nations-sponsored body used a five-star rating system to describe the probability of various scenarios for global warming (Table 1). They gave the 1.5–4.5°C change in global average surface temperature a ranking of three stars—presumably equivalent to a 60 percent chance. Such probabilities are, of course, based only on well-founded estimates by knowledgeable scientists, and even though the evidence is circumstantial or based on scientific judgments, it is quite persuasive.

Scientists estimating future climatic changes have focused on large-scale models of the climate system—general circulation models, or GCMs—that attempt to represent mathematically the complex physical interactions among the atmosphere, oceans, ice, biota, and land. As these models have evolved, more and more information has become available and more comprehensive simulations have been performed. Nevertheless, the complexities of the real climate system still vastly exceed the comprehensiveness of today's GCMs and the capabilities of today's computers (see IPCC, 1990 for a state-of-the-art review, and Schneider, 1991 for a review of the IPCC process). Many important uncertainties are unlikely to be resolved before significant climatic changes are felt, and certainly not before we are committed to long-term environmental and societal effects. Therefore, we do not have the luxury of working out every detail before we begin applying the models to help avert potentially serious crises involving the degradation of ecosystems. Such preventative applications will actually buy time, thereby allowing us the luxury of continuing interdisciplinary research on the finer details.

What is most needed to evaluate potential ecosystem effects of temperature change is a regional projection of climatic changes that can be applied to ecosystems locally. One response to the need for information on such changes has been the analysis of large prehistoric climatic changes (e.g., Berger et al., 1984; Barron and Hecht, 1985; Budyko et al., 1987; Schneider,

TABLE 1. Major equilibrium changes in climate due to doubling CO_2, as deduced from models.

Degree of confidence[a]	Predicted change
	TEMPERATURE
*****	The lower atmosphere and earth's surface warm
*****	The stratosphere cools
***	Near the earth's surface, the global average warming lies between +1.5°C and +4.5°C, with a "best guess" of 2.5°C
***	The surface warming at high latitudes is greater than the global average in winter but smaller than in summer (In time-dependent simulations with a deep ocean, there is little warming over the high-latitude ocean)
***	The surface warming and its seasonal variation are least in the tropics
	PRECIPITATION
****	The global average increases (as does that of evaporation); the larger the warming, the larger the increase
***	Increases at high latitudes throughout the year
***	Increases globally by 3 to 15 percent (as does evaporation)
**	Increases at mid-latitudes in winter
**	The zonal mean value increases in the tropics, although there are areas of decrease. Shifts in the main tropical rain bands differ from model to model, so there is little consistency between models in simulated regional changes
**	Changes little in subtropical arid areas
	SOIL MOISTURE
***	Increases in high latitudes in winter
**	Decreases over northern mid-latitude continents in summer
	SNOW AND SEA ICE
****	The area of sea ice and seasonal snow cover diminish

(From IPCC, 1990.)
[a]The number of stars indicates the degree of confidence determined subjectively from the amount of agreement between models, our understanding of the model results, and our confidence in the representation of the relevant process in the model. Five stars indicate virtual certainties; one star indicates low confidence in the prediction.

1987). Although such paleoclimatic examples are relevant for estimating future climatic sensitivity to large changes in the radiation balance that forces climate to change, they unfortunately are not precise analogies to the rate and character of present greenhouse gas increases (see the discussion in Section 5.5.3 of IPCC, 1990). Looking at more recent climatic records—the so-called historical analogue method—can provide insights into climatic states and societal vulnerabilities (Pittock and Salinger, 1982; Jager and Kellogg, 1983; Lough et al., 1983). Those records, however, are also based on climatic cause-and-effect processes that probably are different from future greenhouse gas radiative effects (Schneider, 1984; Mearns et al., 1990a). Fortunately, GCMs provide plausible regional climatic change scenarios against which ecologists can judge whether plausible ecosystem responses could be significant. For example, the prediction of summer soil moisture change is highly important for biological assessment (Figure 2). Note the large decrease in summertime soil moisture, especially in midlatitudes, a result common to many models (e.g., Rind et al., 1990). Unfortunately, the smallest scale at which GCMs provide credible information is on the order of 500 km^2 (see Schneider et al., 1990; IPCC, 1990), which is enormously larger than the vast majority of ecological studies (Kareiva and Andersen, 1988). These scale mismatch issues are discussed below.

GCM forecasts of regional or watershed changes in temperature, evaporation, and precipitation are most germane to estimating various consequences of greenhouse warming, such as hydrological implications. But, as Table 2 suggests, such regional forecasts are more uncertain than those at larger scales. Regional temperature changes can be much larger than global changes, and some regions can even show a negative change (Schlesinger

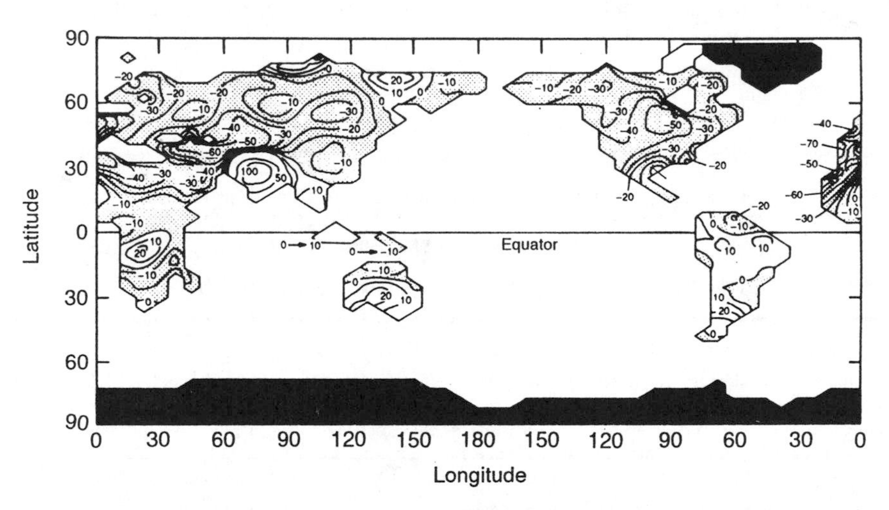

FIGURE 2. Global distribution of CO_2-induced change in summer soil moisture. Soil moisture obtained from a computer model with quadrupled CO_2 is expressed as a percentage of that obtained from a control run with curent CO_2 amounts. Note the nonuniform response of this ecologically important variable to the uniform change in CO_2. (From Manabe and Wetherald, 1986.)

TABLE 2. Scenarios for global warming: A summary.

1. All models show substantial changes in climate when CO_2 concentrations are doubled, even though the changes vary from model to model on a subcontinental scale.

2. The results from models become less reliable at smaller scales, so predictions for smaller than continental regions should be treated with great caution. The continents warm more than the ocean. Temperature increases in southern Europe and central North America are greater than the global mean and are accompanied by reduced precipitation and soil moisture in summer. The Asian summer monsoon intensifies.

3. Changes in the day-to-day variability of weather are uncertain. However, episodes of high temperature will become more frequent in the future simply due to an increase in the mean temperature. There is some evidence of a general increase in convective precipitation.

4. The direct effect of deforestation on global mean climate is small. The indirect effects (through changes in the CO_2 sink) may be more important. However, tropical deforestation may lead to substantial local effects, including a reduction of about 20 percent in precipitation.

5. Improved predictions of global climate change require better treatment of processes affecting the distribution and properties of cloud, ocean–atmosphere interaction, convection, sea ice, and transfer of heat and moisture from the land surface. Increased model resolution will allow more realistic predictions of global-scale changes, and some improvement in the prediction of regional climate change.

(From IPCC, 1990.)

and Mitchell, 1987; IPCC, 1990). For example, surface temperature increases projected for the higher northern latitudes are up to several times larger than the projected global average response, at least at equilibrium. This could imply changes in sea level due to snow balance changes, but estimates (typically 0.5 m ± 0.5 m sea level rise by 2100) are controversial (Schneider, 1992).

Even more uncertain than regional details, but perhaps more important to long-term ecosystem responses, are estimates for such measures of climatic variability as the frequency and magnitude of severe storms, enhanced heat waves, temperature extremes, or reduced frost probabilities (Parry and Carter, 1985; Mearns et al., 1984, 1990b; Wigley, 1985). For example, the physical principle that evaporation increases exponentially with surface water temperature implies that hurricane intensities could increase with warming of the oceans (Emanuel, 1987). This would have ecological implications in those areas where hurricane-induced disturbance is part of ecosystem dynamics (Doyle, 1981; O'Brien et al., in press).

General circulation models, because they typically update their internally-generated weather maps every 30 minutes, are capable of providing estimates of daily, monthly, or interannual variability. Since living things often are more sensitive to changes in such variability measures than in climatic measures, there have been some attempts to analyze the fidelity of GCMs for certain measures of climate variability and even for how such variability might change with global warming (e.g., Rind et al., 1989; Mearns et al., 1990b). One key issue is that contributions of climatic variables often not considered by climate modelers may be of central importance to ecologists (for example, the probability of three weeks of temperatures over 30°C with relative humidity below 25 percent is a major correlate with forest fire likelihood). Thus it is imperative that ecologists make their needs for such nontraditional measures of climatic variability known to climate modelers so that interdisciplinary research collaborations can be built that would not occur if each discipline were to remain intellectually isolated. Such variability issues are just now beginning to be considered and evaluated by climatologists. But these evaluations are based on the results of so-called "equilibrium" climate models, in which CO_2 is doubled and held fixed over time. Because the actual greenhouse gas increases are exponential over time, these issues will, of course, have to be reanalyzed for realistic transient cases to be of maximum value to ecologists or other impact assessors.

NARROWING THE MISMATCH BETWEEN ECOLOGICAL AND CLIMATIC MODEL SCALES

Within the foreseeable future even the highest resolution three-dimensional GCMs suitable for integrations over 50 or more years will not have a grid with nodes much less than 100 km apart; individual clouds and most ecological research (to say nothing of cloud microphysics or microclimatic or microphysiological research) occur on scales far smaller than that. General circulation models will not, therefore, be able to resolve most local biological communities or the important local and mesoscale effects of hills, coastlines, lakes, vegetation boundaries, and heterogeneous soils. (At the typical grid resolution of 4 degrees latitude by 7 degrees longitude, for example, a GCM cannot easily resolve the Sierras and the Rockies as separate mountain chains.) For regions that have relatively uniform surface characteristics, such as a 1000-kilometer savanna or a tropical forest with little elevation change, GCM grid-scale parametric representations of surface albedo, soil type, and evapotranspiration could be used to estimate local changes. Alterations in climate predicted within one grid square would probably apply fairly uniformly across such homogeneous areas. On the other hand, steep topography or lakes smaller than GCM grids can mediate real climate, but are not resolved in the models. Therefore, even if GCM predictions were accurate at grid scale, they would not necessarily be appropriate to local conditions.

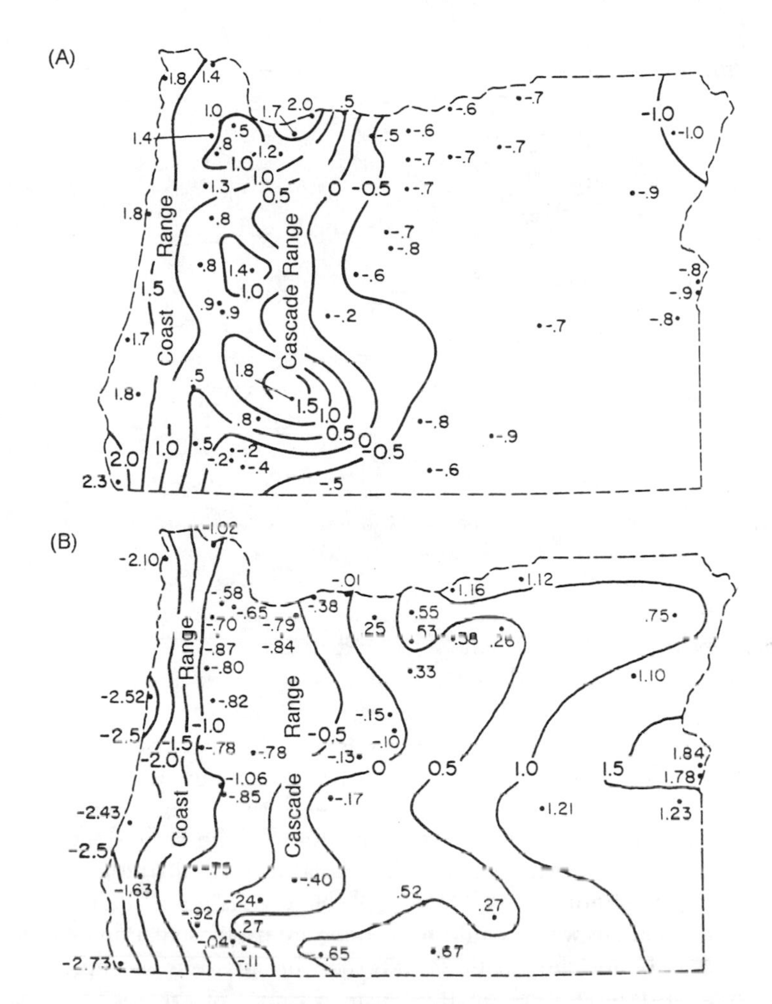

FIGURE 3. The distribution of the relationship between large-scale (area-averaged) and local variations of (A) the monthly mean surface air temperature and (B) precipitation in Oregon, as given by the first empirical orthogonal function determined from 30 years' observational monthly means at 49 stations in comparison with the statewide average. (From Gates, 1985.)

Large-scale observed climatic anomalies can be mapped to local variations. For instance, Gates (1985) analyzed the local climatic variability for the state of Oregon using a technique known as empirical orthogonal functions. A general circulation model run with altered temperatures resolved on the western slope of the Cascades could be applied to the climatic anomaly map of Oregon to determine the climatic effects on local areas on the eastern slopes of the Cascades (Figure 3). Such a map, constructed from variations of climate observed over several years, may seem an ideal way

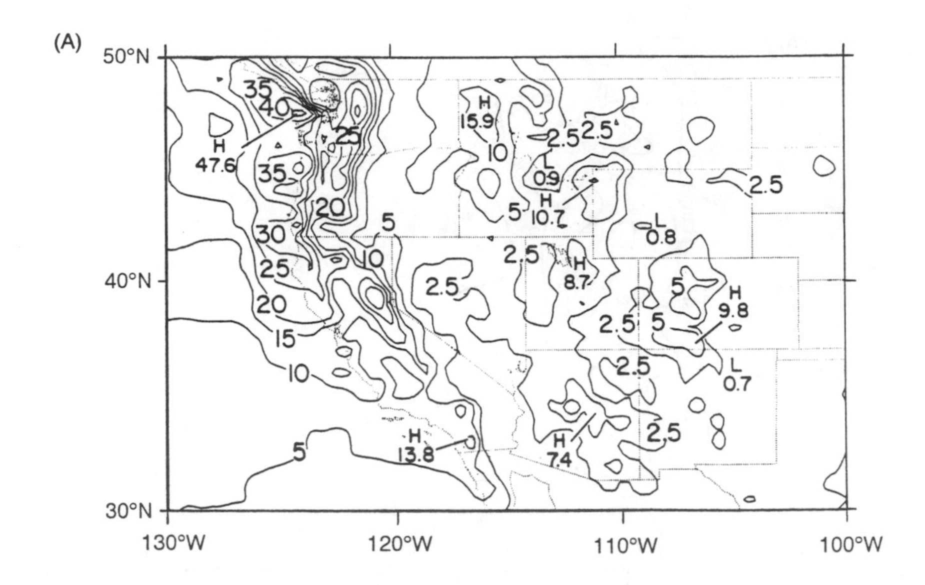

to translate the GCM grid-scale averaged data to the local scale or mesoscale. Because empirical data have been used, however, such a relation would be valid only where the causes of recent climatic variations or oscillations carry forward and include the effect of climatic changes forced by greenhouse gases. The north–south Cascade Mountains translate a simple change in the frequency or intensity of westerly winds into a characteristic climatic signature of either more precipitation on the west slope and less on the east slope, or vice versa. The signature of climatic change from increases in greenhouse gases may not be the same as that from past vacillations, many of which could have been internal oscillations within the climate system, not the result of external forcing. For example, simply increasing the temperature on the western slopes of the Cascades would not necessarily (despite Figure 3) imply a decreased temperature on the eastern slopes if the temperature rise were forced by greenhouse gas heating, whereas eastern slope cooling would likely hold if the western slope warming were a result of weakened westerlies. Thus, other translations of scale need to be considered to map large-scale global change projections to smaller scales that account for the causes of the large-scale changes.

One might embed a high-resolution mesoscale model within a few grid squares of a GCM, using as boundary conditions for the mesoscale model the wind, temperature, and so forth predicted by the GCM at the squares' boundaries (e.g., Dickinson et al., 1989). A mesoscale model with a typical grid-square scale of 50 km × 50 km, could then account for regional topography, soil type, and vegetation cover and could map GCM forecasts to regional topography. Figure 4 shows an example of such a model for the

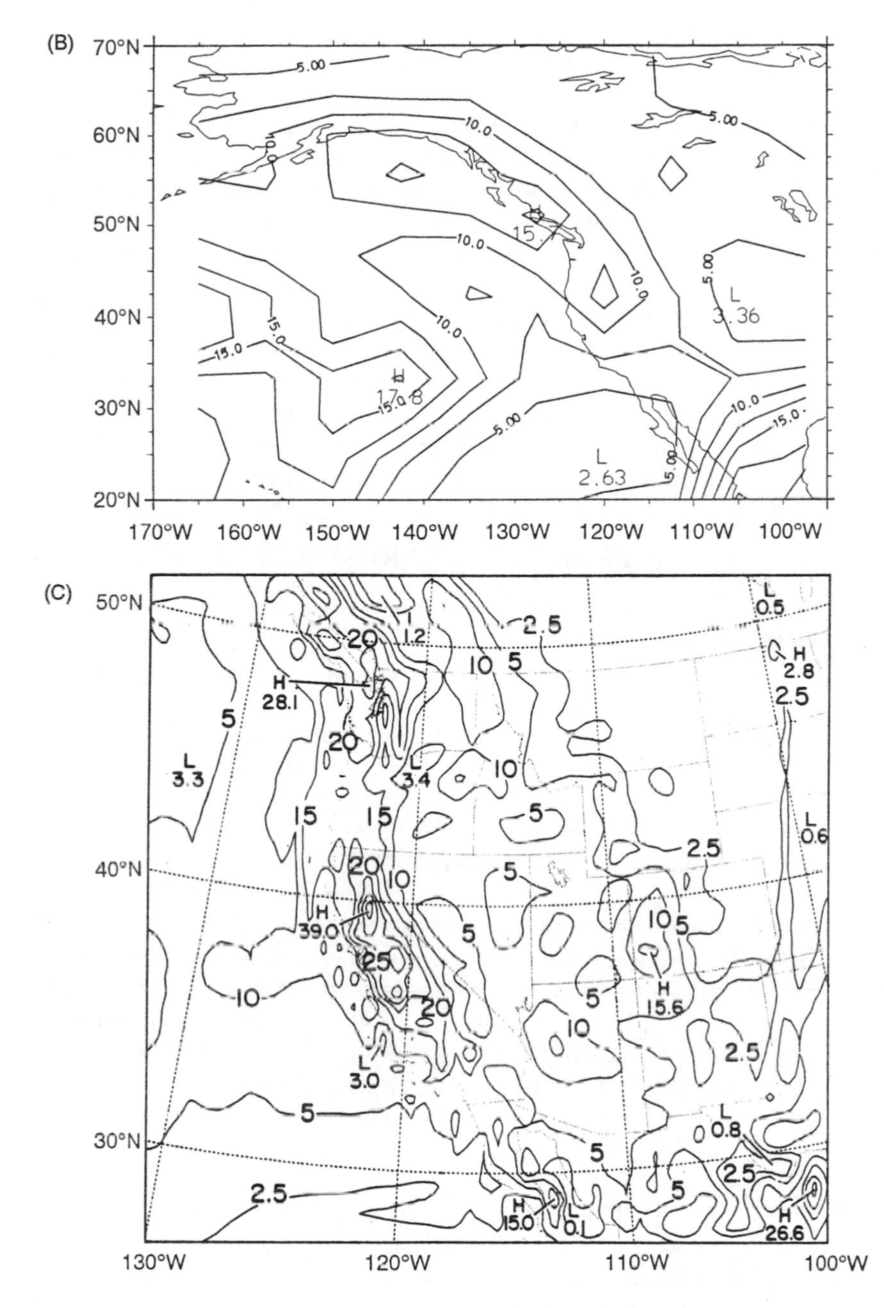

FIGURE 4. Average January total precipitation for the western United States, in centimeters. (A) Observations. (B) R15 general circulation model (4.5° latitude × 7.5° longitude). (C) Mesoscale model driven by output of R15 model. (From Giorgi, 1990.)

western United States. For such a method to have any reasonable hope of success, however, the GCM must produce accurate grid-scale climatic statistics for the special limited grid area. To return to the Oregon case (Figure 3), if the climatic average of the GCM's winds in the unperturbed case (the control case) has the wrong westerly component, the local climate change will probably be misrepresented in a region where topography amplifies any such error in the wind direction. A likely prerequisite for this kind of cross-scale mapping, therefore, is a sufficiently accurate control climate for the important variables. Only then does it makes sense to take the next step of imposing a scenario of greenhouse gas increase on the GCM to estimate how the regional- or local-scale climate might change. Although this technique has shown some promise for wintertime climate situations in mid-latitudes, how embedding would work in the summer or in the tropics (where thunderstorms, smaller in scale than even the mesoscale grid boxes, dominate precipitation systems) is yet to be determined.

REGIONAL CHANGES ARE STILL VERY SPECULATIVE

Neither the IPCC report (1990) nor other assessments (e.g., NAS, 1987; Schneider, 1990a) indicate that a strong consensus exists regarding evolving, regionally specific climatic changes. For example, the world is undergoing a steady increase in greenhouse gas forcing, which is heating the earth reasonably uniformly. Superficially, then, one might expect a uniform global response, but this is far from likely. The centers of continents have relatively low heat-retaining capacity, and thus their new equilibrium climate would be reached more rapidly compared with the centers of oceans, which are thermally more stable. Tropical oceans, however, have a thin (on the order of 50 meters) surface layer of well-mixed waters. The temperature of that mixed layer changes substantially over ten-year time scales, which is much slower than the response time of the centers of continents, but much faster than the rate at which the surface temperatures change in high-latitude oceans, where waters can mix down to the floor of the ocean. Therefore, during the transient phase of climate change, centers of continents, centers of oceans, and high-latitude oceans all approach their new equilibrium temperatures at different rates (Schneider and Thompson, 1981). This means the temperature differences from land to sea and from equator to pole will change over time, which suggests that regional climatic anomalies associated with global warming will not necessarily increase uniformly over time, but could have a transient character very different from the long-term equilibrium character (Thompson and Schneider, 1982; Stouffer et al., 1989; Washington and Meehl, 1989).

The lack of credible predictability over time on a regional scale has implications for evaluating the adaptability of various ecosystems. In agricultural systems, for example, the experience farmers will get with anomalous weather in, say, the 2020s may not help them to prepare for the climate changes of the 2030s. This inhibits "learning by doing," creating a potential lack of adaptability associated with the difficulty of reliably predicting evolving regional climatic consequences. A rapidly changing regional climate would be particularly difficult for natural ecosystems, because the vast majority of habitats cannot be artificially cared for (e.g., by seeding or transplanting hardier species into the area, or by providing needed soil nutrients by applying fertilizers). Moreover, the less predictable regional climatic changes are, the more difficult it will be to adapt farming or water supply systems, or to maintain sound conservation practices. Similarly, the more rapidly the climate is forced to change by, say, greenhouse gas emissions, the less predictable regional changes will become (Thompson and Schneider, 1982).

Despite the difficulties in providing precise regional projections of climate changes, plausible scenarios (such as those given above) can be postulated and used to drive ecosystem response models (e.g., Pastor and Post, 1988). In view of the complexity of both climate and ecosystem models, scenario and sensitivity analyses may be the only assessment tools conservationists have for the foreseeable future—other than observations of possibly dramatic climatic and ecosystem changes in the one uncontrolled "experiment" that is definitive: global change on earth.

IMPLICATIONS FOR POLICY

Climate change is not necessarily a threat to the viability of all climate-sensitive species. However, the transient nature of most projected human-induced climate change is on a time scale of decades, while the adaptability of many species is on a time scale of centuries. As will be illustrated in this volume, substantial disruptions in ecological systems will likely be created owing to maladaptions, significant shifts in species ranges, and inevitable extinctions. The only outcome that can be predicted with virtual certainty is major surprises (Carpenter et al., this volume), and *the more rapidly the climate changes, the higher the probability of substantial disruption and surprise within natural systems.*

Rapid global warming will certainly increase the number of species that are maladapted to their climatically changed habitats. This rapid change, the effects of which may need to be artificially mediated by humans, is created by the increasing use of energy and materials demanded by growing human populations insisting on using the cheapest and most readily available technologies to increase their standards of living (Ehrlich and Holdren, 1971). The likelihood of rapid climate change has led to a debate over whether humans need to intervene as "ecological engineers," deliberately removing soon-to-be-maladapted species and holding them in captive breeding programs for reintroductions later, or translocating wild individuals into potentially more hospitable environments (Orians, this volume; Soulé, 1989; Roberts, 1988; Peters and Lovejoy, 1992).

The most politically acceptable strategy for coping with climate change is to accelerate the implementation of already cost-effective activities that slow down emissions of greenhouse gases into the atmosphere. Such activities include more efficient production and uses of fossil fuel energy, curtailing the production and use of CFCs, and more effective use of nitrogen fertilizer. Although "how to act" always involves value judgments, I believe humans must invest present resources as a hedge against potential change (Orians, this volume). Those actions that are already cost-effective (using and producing energy more efficiently is the most important example) should be vigorously pursued now; political obstacles slowing the penetration of such cost-effective actions should be removed, and incentives to speed such action should be created (e.g. OTA, 1991; NAS, 1991).

Even though the IPCC report (1990) shows that it would take a dramatic 60 percent or so reduction in CO_2 emissions to stabilize CO_2 concentrations in the atmosphere in the decades ahead, less dramatic emissions reductions would reduce the rate at which climatic changes would proceed. As is argued in this volume and elsewhere (e.g., Root, this volume; NAS, 1991), it is the *rate* of change of climate that most threatens to disrupt ecosystems. Therefore, slowing down the rate of global change is a priority item for the world's environment/development policy agenda.

These high-leverage or tie-in strategies (see Schneider, 1990a) are, in my opinion, long overdue for a higher place on the world's action agenda —with or without global warming. Because of the urgency of the need to slow down global climate change, and thereby buy time for humans to assess and the rest of nature to adapt to whatever changes will take place, accelerating the implementation of such actions seems self-evidently the most appropriate immediate policy response to the prospect of unprecedented rapid climate changes.

SUMMARY

Heat trapping by greenhouse gases is a well-established principle. Human additions to global greenhouse gases have added some 2–3 watts per square meter of surface of atmosphere heating over the past century. Translating such heating into precise estimates of global surface temperature increases, however, is controversial, since it involves assumptions about "feedback mechanisms" within the physical, chemical, and biological components of the climate system. Most assessments assign a greater than 50 percent chance to the projection that global warming of 1–5°C is being built into the climate of the twenty-first century, and that this implies rates of change some 10 to 50 times faster than most sustained global average rates of temperature change. The implications of such changes for natural and managed ecosystems are potentially serious.

While a fairly strong consensus of knowledgeable climatologists holds to the view that such a range of potential warming is likely, few such scientists express confidence in the regionally specific projections of evolving climate change so essential to water or agricultural planning, or to ecological impact analyses or biological conservation efforts. Moreover, most climatic projections typically occur over 500 km × 500 km "grid boxes," whereas most field studies or impact assessments occur on scales orders of magnitude smaller. Static and dynamic techniques to bridge this scale mismatch in climatic models are briefly reviewed. The author is cautiously optimistic that some translation of global change projections from grid scale to smaller scales can be made.

Finally, although it is a personal value judgement whether the prospect of unprecedented rapid climatic changes and their potential ecological implications justifies current actions to slow down human activities driving such global changes, the author believes that cost-effective strategies to slow down the forces of global change already exist and should be implemented to buy time to study the issues further and to make it easier for both natural and managed ecosystems to deal with whatever changes do unfold.

ACKNOWLEDGMENTS

The National Center for Atmospheric Research is sponsored by the National Science Foundation. Any opinions, findings, conclusions, or recommendations expressed in this article are those of the author and do not necessarily reflect the views of the National Science Foundation.

CHAPTER 3

EVALUATING LANDSCAPE CHANGE:

Patterns of Worldwide Deforestation and Local Fragmentation

Martha J. Groom and Nathan Schumaker

Human history is a continuing saga of population increase, expanded agricultural activity, and clearing land of forests. Alterations of natural landscapes are among the most visible effects we have on ecosystems, and probably represent the dominant threat to biodiversity. For example, landscape change (habitat loss, degradation, and fragmentation) is the primary cause of endangerment for two-thirds of the world's vertebrates considered at risk of extinction (Prescott-Allen and Prescott-Allen, 1978). While professional ecologists and the general public are well aware of the extensive habitat loss and landscape modification due to human activities, our knowledge of the rates and patterns of such change is remarkably incomplete.

In this chapter we summarize information regarding global deforestation, focusing on this one form of landscape change because it is the best documented. More important, we examine the ways that habitat loss is estimated and suggest improvements in monitoring that would enhance our ability to mitigate habitat destruction. We also examine the analysis of fragmentation at a local scale, where the research and recommendations of biologists can contribute substantially to conservation efforts (e.g., McKelvey et al., this volume). To illustrate some of the problems involved in summarizing fragmentation patterns, we include analyses of a data set on forest fragmentation from the Olympic National Forest in the state of Washington. At both the national and local scales, our ability to summarize landscape change is limited by the availability of accurate information, as well as by biases inherent in the measures, assumptions or definitions used in quantifying habitat change.

Accurate information characterizing habitat loss, degradation, and fragmentation is critical to our efforts to conserve species and resources and to develop methods of sustainable resource use. It is also obviously crucial to investigate the ecological and evolutionary effects of habitat change on biodiversity and ecosystem use. However, it is beyond the scope of our chapter to summarize the ecological effects of landscape change; those effects are examined later in this volume by Quinn and Karr, McKelvey et al., McCauley, Holsinger, Hassell et al., and Harrison. We hope this chapter will increase awareness of the need for better data on landscape change, and for more careful and detailed analysis.

Although our discussion emphasizes forests and forest fragmentation, the issues we raise are pertinent to any case of landscape change and habitat loss. Most of the available data pertain to deforestation and concern losses in tropical forests. However, other habitat types harbor sizable fractions of the earth's biodiversity (Redford et al., 1990), have been modified to appalling extents, and currently are being destroyed at a pace every bit as alarming as that for tropical forests (Table 1).

LANDSCAPE CHANGE ON A GLOBAL SCALE

Deforestation as a long-standing global trend

Tropical deforestation is one of the most important contemporary threats to biodiversity, due to recent increases in the rates of habitat alteration and to the concentration of biodiversity in tropical habitats (Wilson, 1988). It is important to realize, however, that what is now under way in the tropics was completed centuries ago in many temperate ecosystems. For example, Mediterranean forests near Athens were completely denuded by Plato's time (Mather, 1990), France had lost 38 percent of its forests by the year 1 A.D. (Prieur, 1987), and the last virgin forest in England was cut before the end of the twelfth century (Rackham, 1976). At present, forest cover is actually increasing in Europe and in the former U.S.S.R. (WRI, 1991) due to aggressive reforestation programs. Nevertheless, the clearing of virgin forest worldwide has increased exponentially since the 1500s (Mather, 1990). For example, although deforestation has been occurring in China for the past 3000 years, 39 percent of China's original forest cover has been removed within the last 200 years (WRI, 1991). Currently some of the highest rates of forest conversion are found in the humid tropics, having reached 5.2 percent per year in Africa, 4.0 percent per year in Asia, and 3.7 percent

TABLE 1. Examples of original extent and remaining percentage of nonforested habitats.

Country or region	Habitat type	Original extent (km^2)	Percent remaining	Reference
North America	Tallgrass prairie	1,430,000	1	WRI, 1991
Sri Lanka	Thorn scrub	19,800	25	MacKinnon and MacKinnon, 1986b
United Kingdom	Heathland	1,432	27	Nature Conservancy UK, 1984
Nigeria	Mangrove	24,440	50	MacKinnon and MacKinnon, 1986a
Paraguay	Chaco	320,000	57	Redford et al., 1990
South Africa	Fynbos	75,000	67	Mooney, 1988

per year in Latin America (WRI, 1992, Table 19-1, for Côte d'Ivoire, Nepal, and Haiti). Overall, annual rates of deforestation over the past decade have been highest in Central America and Asia and lowest in North America (Figure 1). Forest losses have been most severe in strongly seasonal forests. Whereas 56 percent of evergreen tropical forests remain uncut, only 38 percent of deciduous tropical forests have been spared (Reid and Miller, 1989). Temperate forests have fared the worst: 85 percent have been cleared at least once (Mather, 1990). Before agricultural conversion began, the world's forests covered 5–7 billion hectares. Over 4 billion hectares are estimated to be wooded today (WRI 1991), but 1.7 billion hectares of that total are secondary forests or plantations.

The critical need for accurate data on landscape change

As discussed in detail below, both methodological and political constraints can limit the accuracy of data on the rate and extent of landscape change. Unfortunately, deforestation statistics are often reported and quoted without regard for underlying definitions, assumptions, or potential biases. This can lead to grave errors. To the extent that national and international policy decisions are based on these statistics, even minor errors can have significant consequences for individual nations or for particular development or conservation strategies. Political decisions based on these figures include the allocation of funding for conservation and sustainable development programs, the design of national legislation or multinational agreements to preserve the environment, and the distribution of aid packages. Models of global climate change also rely in part on the accuracy of estimates of land

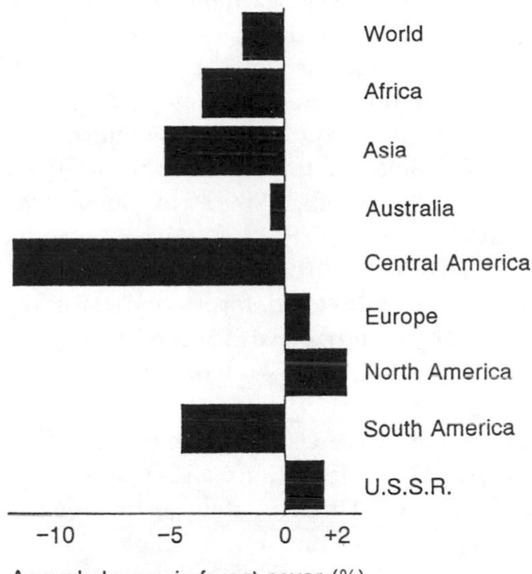

FIGURE 1. Global estimates of annual rates of change in forest cover averaged over the years 1977–1989. Deforestation is estimated as the amount of forest converted to nonforest use. Calculations were based on estimates of forest cover that included natural, undisturbed forests as well as commercially logged areas (regardless of successional state) and tree plantations. (After WRI, 1992.)

use (Schneider, this volume; Schimel, this volume). Finally, simply tracking changes in landscape use or resource depletion is impossible without accurate information. Thus, the way deforestation is evaluated bears heavily on the direction of worldwide development and the fate of natural resources.

Problems in defining and measuring deforestation

Impressive figures on deforestation are easy to generate. Unfortunately, they also are often easy to dispute. The statistics depend on many assumptions that go into quantifying rates of deforestation, and are thus open to bias depending on political agendas or domestic and international pressures. Generally, forest clearing is estimated by a combination of national statistics on wood production and analyses of change in land cover according to maps, aerial photographs, or satellite imagery. The most widely quoted and complete set of statistics regarding deforestation is that for tropical countries published by the Food and Agriculture Organization of the United Nations (FAO). These data have the advantage of being comprehensive and using consistent criteria among countries and among years. However, because some of the data forming these statistics are reported by individual countries in response to FAO questionnaires or queries, the data are not completely uniform. Although FAO requests the same data from all countries, many countries use their own definitions, do not elaborate on their sources, or do not explain the assumptions used in summarizing information on land use (WRI, 1992). Fortunately, FAO has recently put considerable effort into independently verifying deforestation data for tropical countries, and will soon publish new estimates (Singh, 1990). For global data, reports by the World Resources Institute (e.g., WRI, 1992), which compile forest use data from FAO and several other sources, are particularly convenient. These reports include extensive tables of deforestation rates and current forest resources on a country-by-country basis. Much of the information summarized by WRI is updated when independent data on particular countries is available.

In order for data on landscape change to be comparable among countries or regions, it is obvious that definitions and assumptions underlying the statistics must be consistent. Unfortunately, estimates for annual rates of deforestation can vary substantially. For example, Figure 2 displays two deforestation rate estimates for 15 countries; the estimates are particularly

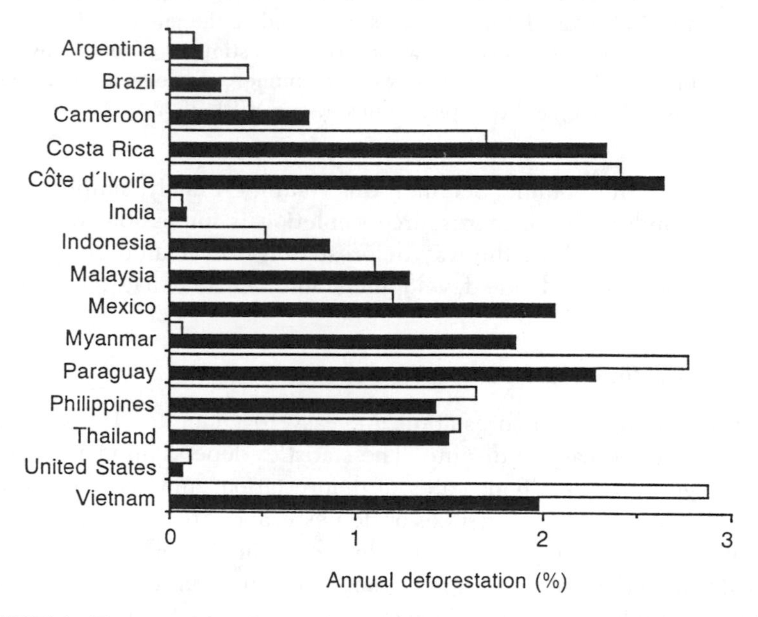

FIGURE 2. Two recent estimates for annual rates of deforestation for 15 selected countries. Dark bars represent data adapted from FAO 1980 and 1990 estimates (WRI, 1992; Table 17.1) and light bars represent data taken from several independent country-based estimates made for various dates between 1980 and 1991 (WRI, 1992; Table 19-1). The methods used for the estimates shown in the light bars are not uniform.

disparate for Cameroon, Myanmar, and Vietnam. Clearly, any conclusion drawn about the rate of resource depletion in Myanmar would be radically different depending on which deforestation measure one used. But obtaining accurate and comparable estimates of total deforestation rates is only part of the battle. To be useful in resource management decisions, the data must also be sufficiently detailed—it is vital to know the rates of change in all major habitats. Primarily, then, there are two sets of decisions that must be made in order to quantify deforestation: What is considered a forest, and what constitutes a loss of forest.

Definitions of forest type The FAO divides deforestation into loss of *closed* vs. *open* forest. Closed forest is defined as "land where trees cover a high proportion of the ground and where grass does not form a continuous layer on the forest floor," whereas open forest is defined as grass–woodland, cerrado, or savanna habitats (FAO, 1981). Open forest constituted about one-fifth of all forested land in 1980 (740 million hectares open forest vs. 2.8 billion hectares closed forest; WRI, 1992). The FAO's first worldwide estimate of forest loss in the tropics (FAO, 1981, 1988) estimated only losses of closed forest, which led to an underestimate of the degradation of forests worldwide. For example, use of only closed forest data for 10 Asian countries underestimated the total losses to broadleaf forest by up to 44.2 percent (Table 2). Recent estimates of national deforestation rates have increased largely because they include open as well as closed forests (Sayer and

TABLE 2. Summary of extent of forests[a] in 10 Asian countries in 1980 according to four broad habitat categories.

Country	Closed broadleaf	Open broadleaf	Coniferous forest	Bamboo forest
India	46,044	5,393	4,357	1,440
Indonesia	113,575	3,000	320	0
Cambodia	7,150	5,100	18	380
Laos	7,560	5,215	250	600
Myanmar	31,193	0	116	632
Nepal	1,610	180	300	1
Pakistan	860	295	1,325	0
Papua New Guinea	33,710	3,945	520	0
Thailand	8,135	6,440	200	900
Vietnam	7,400	1,340	170	1,200

(Adapted from FAO, 1981.)
[a]Extent of forests given in thousands of hectares.

Whitmore, 1991). Comparison of the original FAO statistics and more recent estimates can thus be misleading. In fact, a comparison of FAO data and the most recent estimates listed by WRI (1992, Table 17-1) shows an increase in forest cover for over 40 percent of the countries, when what has actually changed is the definition of "forest."

The reporting of changes in both closed and open forests allows a more complete assessment of national trends in deforestation. However, it does not provide sufficient information to identify specific biological threats, or to develop management options. Fine-scale reporting of changes in different forest types is critical because broad forest classifications mask hundreds of distinct habitat types, each with its own assemblage of species. For example, habitats show extensive altitudinal variation in species composition, and statistics reporting loss of closed forest alone are insufficient to estimate effects on biodiversity. Losses of different forest types can vary considerably; depending on the type of forest, losses range from 0 to 93 percent for Thailand, from 40 to 85 percent for Madagascar, and from 32 to 99 percent for Costa Rica (Table 3). In Costa Rica, where 79 percent of forests have been cleared, the greatest loss (more than 99 percent) has been to dry forest along the Pacific slopes (Janzen, 1988; Table 3). This uncommonly detailed statistic indicates that conservation of this forest type is extremely critical; its loss would have been impossible to detect if statistics on deforestation were lumped for all forest types.

Unfortunately, detailed data on the patterns of deforestation in distinct habitats are scarce. One reason we lack such data is that increasingly we are relying on remote-sensing to quantify landscape change, and neglecting the congruent field studies that are required to identify species assemblages and distinguish especially vulnerable forest types (see "Uses and limitations of satellite imagery," below). Ideally, changes in each habitat type should be monitored separately, with a clear rationale for any lumping of habitats that is performed (e.g., inability to distinguish among certain habitat types; ecological similarity).

Definitions of deforestation Confusion about the "true" rates of tropical deforestation has arisen from disagreement about what constitutes deforestation (Melillo et al., 1985; Fearnside, 1990; Sayer and Whitmore, 1991). For example, estimates of the amount of deforestation in Latin America differ by as much as 30 percent, depending on what types of land alterations are included (Houghton et al., 1991). Most often deforestation rates are calculated from the percentage of forest land converted to nonforest uses, which excludes commercial logging and tree plantations. If our concern is for biodiversity or the functioning of natural ecosystems, such a definition of deforestation obviously underrepresents habitat degradation. At the other extreme, deforestation has been defined as forest cutting for any purpose (Myers, 1989). Because not all forest uses are equally devastating for bio-

TABLE 3. Habitat loss by forest type for Thailand, Madagascar, and Costa Rica.

Forest type	Original extent (km^2)	Percent lost
THAILAND		
Lowland rainforest	12,027	84
Tropical semi-evergreen forest	88,799	51
Tropical pine forest	4,222	93
Montane deciduous forest	144,500	77
Tropical montane evergreen forest	9,331	10
Forest over limestone	200	0
Freshwater swamp	1,250	63
Monsoon forest	6,794	79
Mangrove forest	2,223	69
Dry deciduous forest	219,451	81
Total	507,267	74
MADAGASCAR		
Lowland rainforest	80,729	85
Lowland rainforest/grassland	21,875	80
Moist montane forest	45,312	80
Mixed montane forest	3,646	70
Montane forest/secondary grassland	121,354	70
Dry deciduous forest	51,875	85
Dry deciduous forest/grassland	198,875	70
Deciduous thicket	38,125	85
Thicket/secondary grassland	31,250	70
Mangrove forest and swamp	2,170	40
Total	595,211	75
COSTA RICA		
Dry deciduous forest	3,733	>99
Lowland moist forest	9,903	>99
Lowland wet forest	11,517	78
Premontane moist forest	3,659	>99
Premontane wet forest	12,005	81
Premontane rain forest	4,341	50
Lower montane moist forest	127	>99
Lower montane wet forest	925	86
Lower montane rain forest	3,576	35
Montane wet forest	38	>99
Montane rain forest	1,165	32
Total	50,990	79

(*Sources*: Thailand, MacKinnon and MacKinnon, 1986a; Madagascar, MacKinnon and MacKinnon, 1986b; Costa Rica, Sader and Joyce, 1988.)

diversity, such a definition may overestimate the true impact (Sayer and Whitmore, 1991).

Shifting agriculture may constitute a sustainable use of forests since it creates disturbances that, when carried out at a small enough scale and left to recover for sufficiently long periods, may not lead to long-term degradation of the landscape (Lovejoy and Schubart, 1980; Jordan, 1986). For this reason, such practices have not been included in most estimates of ecologically damaging deforestation (Myers, 1989 is one exception). However, slash-and-burn agricultural practices have been increasing in scale (Jordan, 1986; Uhl et al., 1989), and, in some cases, have paved the way for more intensive and destructive land uses (Uhl and Buschbacher, 1985). Thus, it seems prudent to quantify the amount of forest used in small-scale clearing. Sayer and Whitmore (1991) argue that the mid- to large-scale disturbances created by commercial logging need not be included in deforestation statistics because they have less impact than outright conversion. In some cases, these areas can have greater species diversity than primary forest (e.g., Ewel, 1983), although studies have also documented negative, and sometimes severe effects on wildlife from logging (e.g., Johns, 1985, 1988). These uncertainties about the effects of different types of forest degradation make it clear that greater attention must be paid to tabulating forest disturbances in categories that are relevant to their impact. As a standard practice, at least three categories of deforestation should be used: conversion (permanent change in land use), clearing for wood production, and clearing for rotation or slash-and-burn agriculture. Finally, estimates of forest cover and landscape alteration should specify whether primary or secondary (previously cleared) forests have been lost.

Uses and limitations of satellite imagery

Currently, remote sensing of landscapes through satellite imagery is the most efficient means for tracking habitat conversion on a large scale. Depending on the location, satellite images of much of the globe are available between once every 16 days (Landsat satellites) and once every 6 hours (NOAA AVHRR satellites; see Green et al., 1987). Satellites measure the reflectance of electromagnetic energy from the planet's surface in patterns that can be used to indicate habitat types. The precision of a satellite image is limited by the minimum size unit, or pixel size, of the recorded image. For most nonmilitary satellites, the pixel size is in the range of 30 m^2 to 1 km^2, although the SPOT satellites (the French Système Probatoire d'Observation de la Terre) have a resolution of 5–10 m^2. Spectral properties are averaged over each pixel, resulting in a coarse-grained homogeneity of the landscape at large pixel sizes. The more advanced satellites (SPOT images) can make finer-scale distinctions, such as determining the pattern and extent of damage to selectively logged forests (Myers, 1988). Quattrochi and Pelletier (1991) and Wickland (1991) provide two recent reviews of the ranges of spectral properties measured by current and planned satellite scanners. Satellite paths and the total areas covered are summarized in Asrar (1989).

Although satellite imagery is the most efficient way to monitor forest loss over large scales, the technology has several limitations. Most important, accurate matching of the spectral properties of images and the habitats they represent requires extensive field surveys, or "ground-truthing." These surveys are needed to determine which habitat types can be distinguished by a given satellite's scanner(s) and to determine the effect of spectral averaging on the estimation of land cover for each habitat type. Although field surveys are extremely labor-intensive, they enhance the value of remotely-sensed data because they greatly increase the number of distinguishable forest types (or habitat types) that can be monitored.

Even with extensive field verification, however, some scanners are able to distinguish only grossly different habitat types (Singh, 1987). This problem can be particularly severe in monitoring changes to nonforested habitats (see Table 1), because many habitats look the same in satellite images whether the area is degraded or intact (e.g., rangelands; Makhanya, 1986). In addition, virgin, secondary, and selectively logged forests can all appear the same to some scanners (Stith, 1990; Sayer and Whitmore, 1991). These problems are often exacerbated by physical conditions that limit resolution. For example, some topographic features make distinguishing vegetation types especially difficult (Pastor and Johnston, 1991). Clouds can obscure much of the landscape for a given area, especially in the humid tropics, drastically reducing the number of usable images (Parry, 1986; Singh, 1987; Fearnside, 1990). In fact, in some areas several years have elapsed without obtaining any usable images (e.g., SPOT images of portions of the Brazilian Atlantic coastal forest; Stith, 1990).

Analysis of satellite data is computer-intensive and expensive, particularly at fine scales. Images can cost from $1000 to $4000 per scene, although some data may soon be available in cheaper form on compact disk (Roughgarden et al., 1991). Particularly in developing countries, the high cost of images, as well as the cost and shortage of computer facilities and personnel to interpret them, have limited the use of this type of data (Makhanya, 1986; Green et al., 1987). Thus, although satellite imagery makes large data sets available, these data are expensive and may be inadequate for making crucial ecological distinctions. As the technology improves, and as field verification of images progresses, some of these limitations may be reduced.

However, two other limitations of satellite data are unavoidable. First, only *recent* changes in land use can be measured using remote-sensing technology because satellites have been in use only for the last few decades. Second, and most important, remote-sensing technology can monitor only changes manifested as differences in electromagnetic reflectance (Parry, 1986). The most critical biological effects of landscape change cannot be measured by satellites: what factors cause changes in species abundances and distributions; how forest recovery is influenced by different uses; and how habitat loss, fragment configuration, and surrounding matrix characteristics affect species persistence. These issues must be addressed on the ground by population, community, and ecosystem ecologists. To adequately characterize the responses of species and communities to landscape change, ecologists require detailed data on small-scale patterns of habitat alteration. We turn now to a discussion of how patterns of habitat fragmentation are measured and evaluated on a local scale.

PATTERNS OF FRAGMENTATION AT LOCAL SCALES

Although aggregate data on landscape change are useful in guiding national and international policy, specific management requires more detailed information regarding the state of the remaining habitat. As an area is converted from natural vegetation to new land uses, the remaining habitat is inevitably divided into increasingly smaller fragments. Smaller fragments are more vulnerable to many of the classic threats familiar to conservation biologists, including demographic and genetic risks associated with small population size (Goodman, 1987; Gilpin, 1987), enhanced influences of habitat edges that devalue the interior habitat (Wilcove et al., 1986; Williams-Linera, 1990), and hazards faced by organisms moving between fragments (Fahrig and Merriam, 1985; Harris, 1988; McKelvey et al., this volume). The magnitude of such undesirable effects depends on the details of fragmentation: the number of fragments, their size and shape, the distances between them, and the character of the intervening habitat.

In theory, habitat fragmentation could have a broad range of effects on biodiversity (Wilcove et al., 1986; Quinn and Hastings, 1987; Fahrig and Paloheimo, 1988). Clearly, quantifying the pattern of fragmentation is necessary to predict these effects. Unfortunately, applied work in conservation generally has not emphasized the analysis of detailed quantitative information on habitat change. For example, over the past 5 years (1987–1991) in the journal *Conservation Biology*, 36 of 138 contributed papers dealt with issues of fragmentation, but only 14 reported any quantitative information (usually the size of the fragments studied), and a mere 5 studies related the distribution or abundance of species to the size, shape, or isolation of fragments.

Because the consequences of habitat loss and fragmentation on a local scale may be strongly determined by the arrangement of remaining patches or preserves (Wilcove et al., 1986), the statistics used to summarize landscape change must have ecological meaning. Given the importance of relevant metrics, we now examine several indices designed to quantify aspects of fragmentation, illustrating the usefulness of such indices by applying them to data on old-growth forests in Washington State.

Fragment size

When habitat is lost, the remaining habitat may be left in one large or numerous smaller parcels. The number and size of the remaining fragments will affect the number of species that can maintain sufficiently large populations to persist within each fragment and the number of subpopulations among fragments. The most elementary problem is that the size of a fragment may fall below the size of the required home range for a species. Much recent attention has focused on determining minimum viable population sizes or minimum habitat requirements of vulnerable species (e.g., Soulé, 1987). Beyond accommodating the home range requirements of a species, habitat fragments must also contain the variety of microhabitats required for all life history stages and seasons. That is, some habitat fragments are far larger than an organism's home range, but because they lack a critical resource (e.g., standing dead trees for cavity-nesting birds; salt licks; winter cover), they fail to support that species

The simplest way to summarize the pattern of fragmentation is through a frequency distribution of patch sizes. We will illustrate the visual power of such a simple procedure using data from old-growth temperate rainforests on the Olympic Peninsula of Washington State (Figure 3). Habitat fragmentation has been particularly extensive in the old-growth conifer forests of the northwestern United States. Before the mid-1800s the region was dominated by an unbroken landscape of ancient conifers interspersed with younger successional stages maintained by fire. Now the landscape is a checkerboard of clear-cuts, tree plantations, and remnant old-growth stands (Spies and Franklin, 1988). Recently the Wilderness Society prepared highly accurate and detailed maps of all U.S. Forest Service lands in the Pacific Northwest using a Geographical Information System (GIS) database. We have used their data for the Olympic National Forest in the analyses described below.

In Figure 3, we show the pattern of old growth loss for a 30 × 50-km region of the Olympic National Forest over the period 1940–1988. Clearly, much forest has been lost to logging, but just as striking is the change in the size of patches that remain: between 1940 and 1988, the mean patch size decreased by 93 percent (from 433 to 25 hectares) and its variance decreased by 98 percent. This region now consists of a forest broken up almost entirely into small patches; indeed, during the last 48 years the number of fragments smaller than 100 hectares has increased fivefold. The striking visual impression created by these maps is one of a drastic decrease

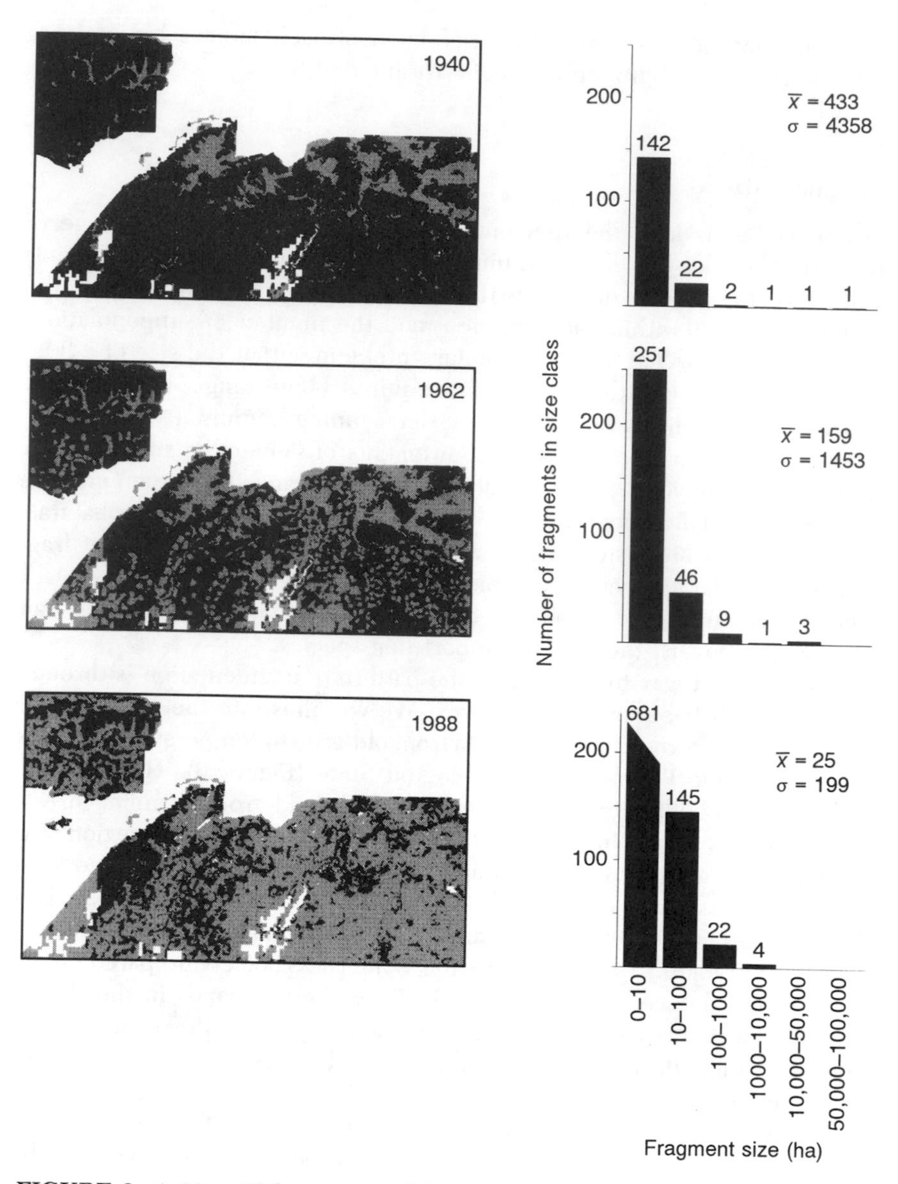

FIGURE 3. A 30 × 50-km region of the Olympic National Forest in 1940, 1962, and 1988. Old growth appears black in all maps; all other habitat types within the national forest are shown in gray. Water, shadows, and areas outside the national forest boundary are shown in white. Beside each map is the frequency distribution of old-growth patch sizes. The pattern of cutting is reflected in the change in the mean and variance of patch sizes (in hectares).

in total old growth and a substantial increase in the proportion that is left in small fragments (Figure 3).

Edge effects

Remaining parcels of a habitat cannot be characterized by their area alone. When habitat is fragmented, the amount of edge habitat increases relative to the interior habitat. Edges and ecotones between habitats are unique as a result of both abiotic and biotic influences. Abiotic influences (e.g., wind, temperature, humidity, light) penetrate into interior habitat and thereby degrade the quality of the habitat island. For example, carrion and dung beetles experience greater desiccation at forest edges and thus may be limited to forest interiors (Klein, 1989), and high light levels at edges can favor pioneer species that outcompete interior tree species (Williams-Linera, 1990). In most cases, these edge effects damage the species composition near the margins of habitat islands (Janzen, 1986). For example, nest predators and avian brood parasites are more common at forest edges than in forest interiors, leading to higher rates of predation and parasitism for birds nesting near edges (Wilcove, 1985; Temple and Cary, 1988). As habitat patches become smaller and are increasingly surrounded by habitats favorable to competitors, predators, and parasites, the quality of the habitat fragment decreases for interior species. The importance of this phenomenon depends on the scale of edge effects: some outside influences penetrate only a few meters from the boundary (e.g., sunlight; see Lovejoy et al., 1986), whereas others infiltrate hundreds of meters into the interior of a fragment (e.g., wind; see Chen, 1991; nest predators; see Wilcove, 1985).

Often our major concern is whether edge effects reduce the amount of interior habitat below the minimum habitat requirement of a species. The simplest way to estimate the amount of interior habitat given an edge diameter, d, is to draw a new perimeter at the distance d in from the fragment's edge, and to calculate the interior habitat according to this new perimeter. Figure 4 provides an example of how rapidly interior area decreases as edge width increases. For an 8 × 8-km section of the Olympic National Forest, using the data described above, we calculated the interior area of old-growth fragments by subtracting edge widths of 57, 114, and 399 m. This range of edge widths was chosen to span the range detected by Chen (1991) for different abiotic variables at the margins of Pacific Northwest old-growth stands (Table 4). Even the smallest edge width (57 m) eliminates the interior area of most fragments (61 small fragments no longer have any interior habitat; Figure 4). If one assumes that edge effects penetrate 400 m into fragments, then there has been a fortyfold decrease in the amount of old-growth interior habitat in the Olympic National Forest since 1940 (Chen, 1991).

Because of the importance of summarizing edge effects, a number of

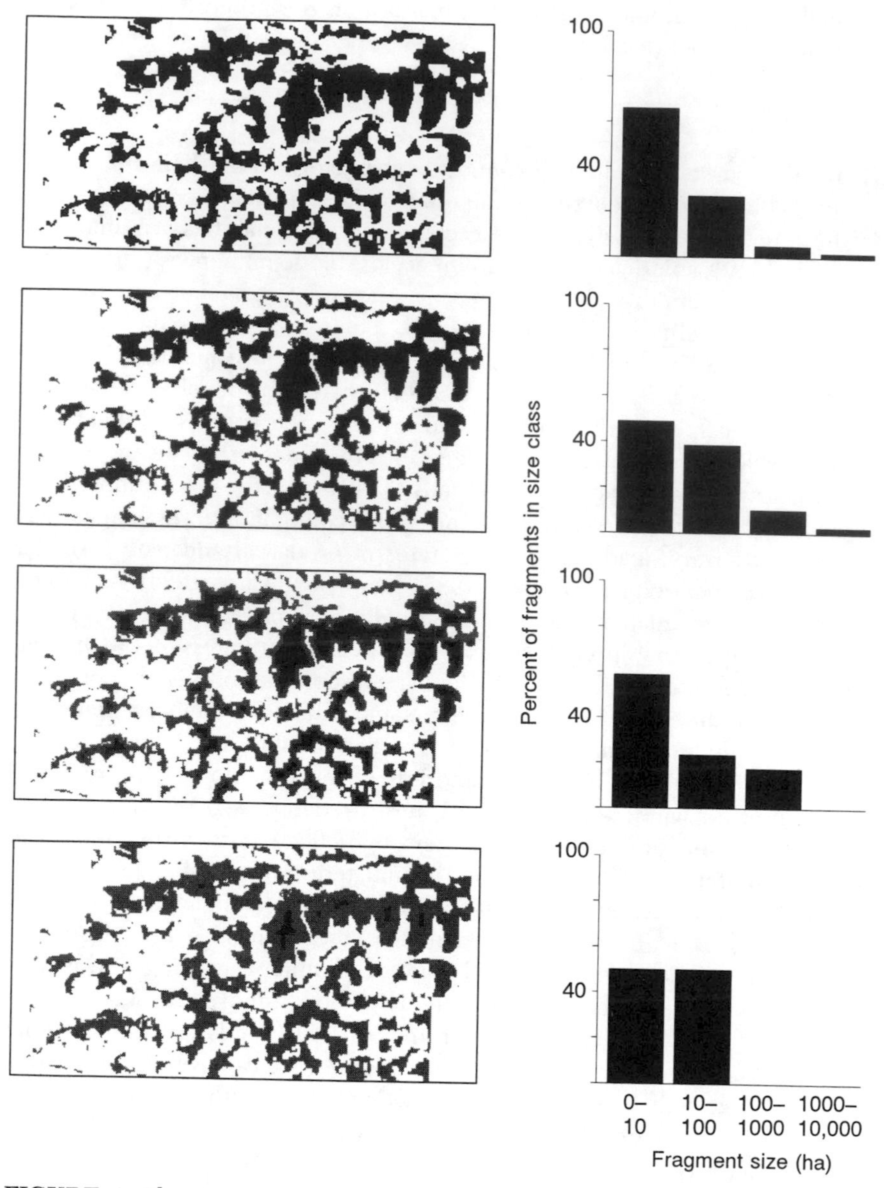

FIGURE 4. The amount of interior area for an 8 × 8-km section of the Olympic National Forest under three edge width assumptions. Patch interiors are shown in black, edge areas in grey. Interior area decreases rapidly with increasing edge width. The site depicted includes some of the largest remaining patches of old growth in the Olympic National Forest.

TABLE 4. Average penetration of edge characteristics into old-growth forest fragments from Washington and Oregon for six abiotic variables.

Variable	Average edge width (m)
Wind speed	180–400+
Relative humidity	180–240
Air temperature	60–120
Soil temperature	60–90
Soil moisture	60–90
Short-wave radiation	60–90

(After Chen, 1991.)

indices have been developed. The simplest of these is the perimeter:area ratio (*P/A*), which indicates the relative edge influence expected for a patch. The relative amount of edge increases with decreasing area; thus, fragments with low *P/A* values should be less edge-influenced. A related measure, the diversity or shape index (*SI*), created by Patton (1975), quantifies edge effects in terms of the deviation of a patch's shape from that of a circle of equal area. Patton's index is expressed as

$$SI = P/[2(\pi A)^{0.5}]$$

where *P* is patch perimeter and *A* is area. For a circle, $SI = 1$, whereas all other shapes have a value greater than 1. This index has the appeal of conveying more information about patch shape than a simple *P/A* ratio. Recently, fractal geometry has been added as a further refinement to quantifying the shape of fragments. The fractal dimension (*D*) of an object (particularly for computer mapped data; Milne, 1991) can be estimated as

$$D = \log A/\log[P/4]$$

where *A* is area and *P* is perimeter. Simple shapes tend to have a fractal dimension near 1, whereas highly complex shapes typically exhibit fractal dimensions approaching a maximum of 2. We have presented the simplest method for calculating *D*; Milne (1991) reviews the diversity of fractal measures with applications to landscape ecology. Fractal dimension can more closely reflect the actual details of patch perimeter, and thus may be a more accurate predictor of edge-to-interior ratios than other indices.

All three edge indices (*P/A*, *SI*, and *D*) are used either to represent the relative amount of edge vs. interior habitat of a fragment, or to reflect the amount of interior habitat in a fragment. We evaluated how well these indices can serve as indicators of interior area by extending the analyses displayed in Figure 4. We calculated the interior area of the forest fragments from the entire Olympic National Forest using a 57-meter edge width. For each fragment, we also calculated *P/A*, *SI*, and *D* and plotted the value of each index against the actual amount of interior habitat (Figure 5). The

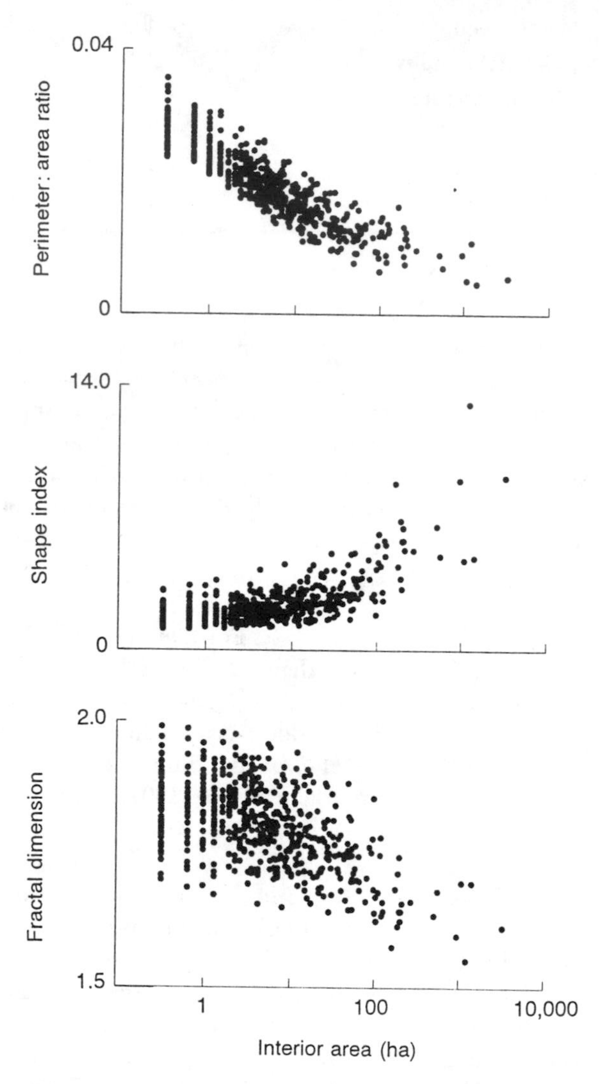

FIGURE 5. Perimeter:area ratio, shape index, and fractal dimension, plotted as functions of interior area, for the Olympic National Forest in 1988. Interior area was computed using a 57-meter edge width. All graphs represent 606 old-growth fragments having non-zero interior areas. An additional 724 fragments had no interior habitat when subjected to a 57-meter edge effect.

indices differed greatly in how well they were correlated with interior area. *P/A* correlated strongly with actual interior area, fractal dimension showed the weakest correlation, and the shape index was intermediate between the other indices. The poor performance of fractal dimension may be an artifact of our use of computer-mapped data where patch sizes are formed by pixels of fixed size. In this case, the fractal dimension may be too constrained, particularly at small patch sizes, to provide an apt description. In summary, although all three indices do reflect trends for interior area, only *P/A* appears to be sufficiently correlated to allow more than a crude estimate of interior area.

Recently, *P/A*, *SI*, and *D* have become popular tools for summarizing the "degree of fragmentation" evident in different landscapes. Although this seems like a good idea in theory, in practice these indices may be more misleading than informative. To illustrate this point, we selected three different portions of the Olympic National Forest, chosen so that each visually represented strikingly different degrees of fragmentation (Figure 6). For each of these distinctive landscapes, we calculated the mean *P/A*, *SI*, and *D* for all old-growth fragments (assuming three different edge widths: 57, 171, and 399 m). When we performed this exercise we found that the three landscapes, which contrasted sharply in fragmentation as judged by the naked eye, exhibited minimal differences in any of our three edge indices. Clearly, none of these indices captured the obvious differences in landscape pattern evident in the maps we present. The only summary statistic that aptly reveals the differences evident in Figure 6 is a direct calculation of interior area. We conclude that shape, perimeter-to area, and fractal indices are not useful as landscape descriptors if the question of interest is the degree of fragmentation.

Isolation effects

An especially robust theoretical prediction from mathematical ecology is that the degree of an organism's dispersal among habitat fragments critically determines its thresholds for extinction (e.g., Levins, 1970; Lande, 1987; Doak, 1989; Hanski, 1991). Populations that are isolated within fragments are vulnerable to extinction through demographic fluctuations (Goodman, 1987; Fahrig and Paloheimo, 1988) and/or via loss of genetic diversity (Gilpin, 1987). Fragments that are too isolated may not receive enough immigrants to maintain genetic diversity or to be rescued from extinction (Brown and Kodric-Brown, 1977). The information necessary for predicting the ease of dispersal among patches, namely some metric of interfragment distances, is regrettably rarely reported. In part, this may be due to the difficulties of measuring dispersal in most organisms. McKelvey et al. (this volume) provide an excellent example of combining data on interfragment

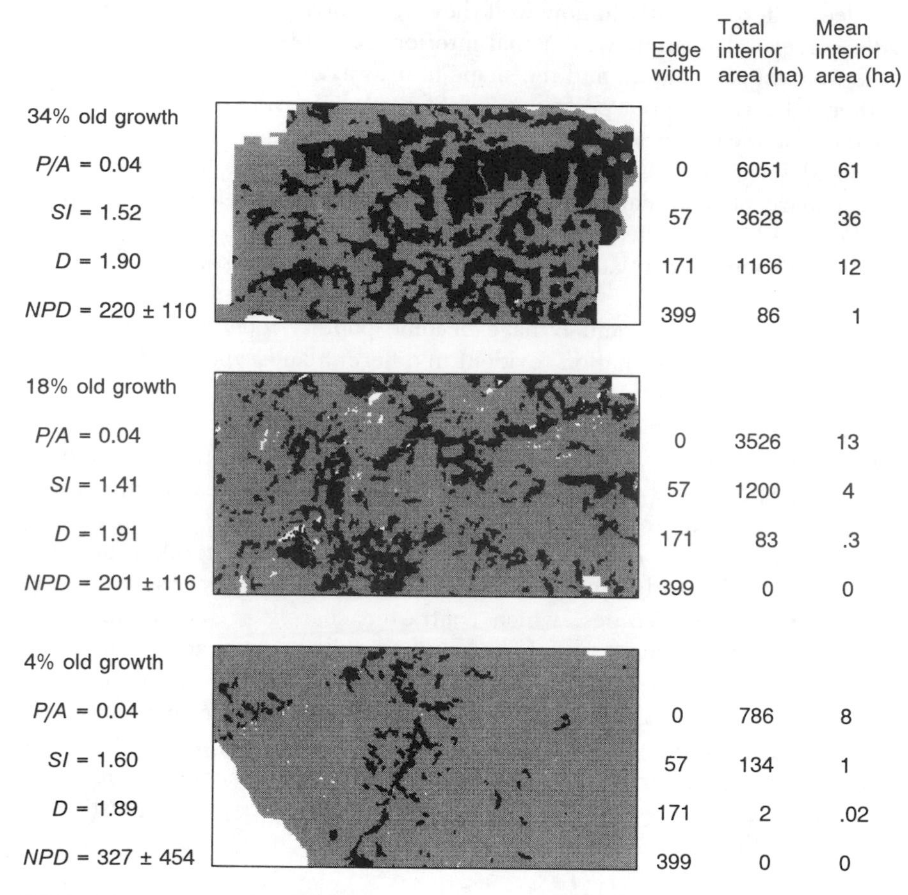

FIGURE 6. Range of variability in patch size and shape of old-growth fragments in the Olympic National Forest. Three different areas of the forest are shown, using data from 1988. Old growth appears in black; all other habitats are shown in gray. Water, shadow, and areas outside the national forest boundary are shown in white. The amount of interior area in each landscape was tabulated for four different edge widths (0, 57, 171, and 399 meters). Mean values (per fragment) for perimeter : area ratio, shape index, fractal dimension, and nearest patch distance (± one standard deviation) are also given.

distances and owl dispersal distances in their analysis of conservation requirements for the northern spotted owl.

One metric that has been used to quantify isolation of habitats is patch dispersion, R:

$$R = 2p^{0.5}r$$

where p is the mean patch density (number of patches /area), and r is the mean nearest-patch distance. However, Ripple et al. (1991) found that patch dispersion did not accurately reflect an increase in fragmentation of several old-growth landscapes in western Oregon. They found that as mean patch density increased, the mean nearest-patch distance simultaneously decreased, yielding similar values for patch dispersion for rather different degrees of fragmentation. An alternative to measuring patch dispersion is to calculate the distribution of nearest-neighbor distances among fragments. Interpatch distances are usually summarized as the arithmetic mean of nearest-patch distances, taken either from edge to edge or center to center (e.g. Opdam et al., 1985; see Figure 6). However, the arithmetic mean may not be a relevant summary of distances between fragments for all species. Most species have a threshold distance above which the likelihood of dispersing among patches becomes virtually zero. Thus, weighted measures, such as the geometric or harmonic mean and variance, may be more appropriate summaries. For example, in both theoretical and empirical studies of pollen movement, the harmonic mean best reflected the distribution of movement distances (Manasse and Kareiva, 1991). Regardless of whether a weighted measure or the arithmetic mean is used, reports of variance should always be included because species may be equally or more influenced by the variance of distances than by the mean.

For species with short dispersal distances, nearest-patch or nearest-neighbor measures are most appropriate because dispersal beyond the nearest fragment is unlikely. However, nearest-neighbor measures may not characterize the effect of fragmentation on more mobile species, which generally will be more sensitive to overall patch distribution. Lynch and Whigham (1984) suggested an alternative that calculated the median distance of the nearest-patch distances in the eight principal compass directions. Another alternative is to take the mean of all possible pairwise distances among fragments adjusted for area.

Incorporating information about an organism's dispersal behavior could lead to much more informative indices of isolation. For example, many organisms set off in a random direction and travel until they find a suitable patch or perish. For such organisms, it makes sense to quantify the degree of isolation of a patch by calculating the average of the nearest patch distances from points distributed around the entire perimeter of the patch. Statistics reflecting other assumptions could also be developed. The key idea is that the best measures of habitat isolation or dispersion will be derived with some particular dispersal behavior in mind.

In addition to the issues already raised here, isolation indices need to take into account patch sizes before they will be biologically useful. In particular, organisms may find very small patches unsuitable, and therefore small patches might be better left out of calculations of isolation. Developing biologically meaningful indices of the isolating effect of fragmentation should be a priority for the emerging field of landscape ecology.

SUMMARY

There are many ways quantification of landscape change can be used to improve conservation planning. For example, concise quantitative descriptors of landscapes should help us identify areas of critical concern, decide among a variety of land use options, and track habitat changes accurately. Population viability models can also benefit from detailed information on the patterns of fragmentation of actual landscapes. These data in turn can be used to devise management strategies for endangered species (e.g., McKelvey et al., this volume). There is a great need to rank landscapes or land use strategies based on specific environmental criteria, and this goal can be advanced by the use of appropriate indices. The challenge is to identify landscape summaries that capture information crucial to the ecological processes under study.

While the purpose of this chapter has been to summarize what we do and do not know about landscape change at global and local scales, it should be understood that quantifying the patterns of change is just one step. To fully understand the magnitude of the problems caused by landscape change, we must devote considerable resources to elucidating species' responses to those changes. Too much attention to description, with too little emphasis on dynamics, would be a sterile exercise of little value to preserving our natural heritage.

ACKNOWLEDGMENTS

We are grateful to Peter Morrison and the Wilderness Society for providing us with GIS maps of the Olympic National Forest. This manuscript was greatly improved by comments from L. Goldwasser, P. Kareiva, R. D. Podolsky and M. Turner. Financial support was provided by a National Science Foundation Graduate Fellowship and a Pew Charitable Trusts Scholarship in Conservation Biology and Sustainable Development to M.J.G., and by U.S. State Department Grant 1753-000574 and USDA/ USFS Grant PNW 90-340 to N.S.

CHAPTER 4

POPULATION AND COMMUNITY PROCESSES IN THE RESPONSE OF TERRESTRIAL ECOSYSTEMS TO GLOBAL CHANGE

David S. Schimel

The biota have significant effects on the chemical and physical processes of the earth system. The present composition of the atmosphere is largely a function of life, both in terms of the gross composition of air and of the burden of trace gases that control radiative balance and photochemistry (Andreae and Schimel, 1989). The climate is influenced by the effect of vegetation on surface energy balance and, in the longer term, by biological control over radiatively active trace gases (Sato et al., 1989; Shukla et al., 1990; Schneider, 1989). While the physical and chemical properties of the ocean determine much of the dynamics of the global carbon cycle over decades to centuries, its dynamics on a shorter (subannual) and much longer (geological) scale are greatly influenced by the biota (Fung et al., 1987). This relatively recent understanding of the strength of atmosphere–biosphere coupling complements the well-accepted role of the biota in controlling geological, geomorphological, and pedogenic processes (Schumm, 1977; Jenny, 1941).

Because of these strong connections, ecology as a discipline should have as strong an identity among the earth sciences as it does within its traditional home among the life sciences. As ecologists become increasingly involved in addressing issues on regional to global scales, we will find as much in common in priorities and methods with geologists, atmospheric scientists, and oceanographers addressing similarly scaled issues as with our traditional colleagues in molecular biology, animal behavior, or developmental biology.

The purpose of this chapter is to briefly present several areas of research where better input from evolutionary, population, and community biology is important to rapid progress. Too often biologists have assumed that because the issues of most global change research are couched in terms of physical processes, they have no role. In fact, many key geophysical processes are significantly mediated by biology, and as climate system models become more sophisticated, so should the level of biology included. Becoming more involved will require biologists to rethink their traditional research priorities and motivating questions and to become more concerned about interactions between biology and the rest of the earth system. This involvement may well dictate new methodologies and approaches but should build on the strong foundations already laid in disciplinary studies.

BIOLOGY AND CLIMATE: LOCAL TO GLOBAL EFFECTS

A number of recent modeling studies have documented the key role that vegetation plays in physical climate. Vegetation influences climate in several ways, though control of latent heat flux via evaporation and transpiration, through albedo and through surface roughness. Much of the study of climate and climate change is done using models of the general circulation of the atmosphere, or GCMs (Schneider, this volume). In GCMs, fluxes of water, energy, and momentum at the land surface must be prescribed or computed. In practice, these are usually simulated using some variant on the Budyko bucket model, which assumes that the land surface is a bucket of specified depth (Dickinson, 1991). Rainfall in excess of that depth is magically transported to the oceans, and evaporation takes place as if from a free water surface until the bucket is empty.

When GCMs are validated against zonal mean climate (e.g., 10–30° bands) these approximations may be acceptable. However, recent simulations using biophysically based simulations within GCMs have shown the difference in climate between bucket models and biophysical models to be appreciable, especially in midcontinental areas (Sato et al., 1989). These differences occur because of the effect of plant control over evaporation on atmospheric moisture, of albedo on boundary layer dynamics, and of evapotranspiration on surface temperature.

Regional climate is likewise affected by vegetation type. A number of climate modelers have simulated the effects of Amazonian deforestation on climate (reviewed by Dickinson, 1991). In these simulations, Amazonian forest was replaced by either desertlike conditions or pasture. The differences were specified in terms of vegetation resistance to transpiration, albedo, and surface roughness. While all of the simulations showed higher temperatures and lowered rainfall, in some models the effects were due mostly to the altered albedo, and in others to altered evapotranspiration. In one of these studies, rainfall was reduced enough to prevent the return of the rainforest, setting up a potential positive feedback, though this was not simulated. The models differed in many ways (spatial resolution, representation of atmospheric dynamics and thermodynamics, biophysics) and so these exercises are of interest mostly because they all demonstrate sensitivity to the behavior of the land surface. No completed study has simulated a likely scenario for deforestation, including the fragmentation of forest and the development of a heterogeneous matrix, although such studies are in progress (P. Martin, pers. comm.).

In addition to the climatic effects of large-scale vegetation change, changes in local vegetation patterns (on scales of tens of kilometers) may also affect climate. A number of modeling studies have shown that alternation of bands or patches of transpiring vegetation with dry soil (oasis effect) can influence atmospheric circulation and mesoscale (10–1000 km) cloudiness, and that these patches may be on the order of tens of kilometers (Pielke et al., in press). Field studies examining irrigated areas in semiarid climates have confirmed these model results (Segal et al., 1988). Recent modeling exercises have shown that differences in vegetation cover as subtle as the boundary between grazed and ungrazed grassland can induce convective winds and influence cloudiness (Pielke et al., in press).

Vegetation heterogeneity, the structure of communities, and disturbance can influence climate if the patches are in a certain range of scales—tens to hundreds of kilometers. The influence of the land surface in these mesoscale simulations is principally through the effects of contrasting surface temperatures and latent heat flux on the three-dimensional temperature–pressure structure of the overlying atmosphere. These contrasts form convective cells with boundaries on or downwind of the land surface boundaries. Where such convergent upward movement occurs, clouds and precipitation may be enhanced. Thus, both natural vegetation contrasts (due to soils or hydrology) and management effects may influence atmospheric dynamics, with possible feedbacks to the vegetation.

POPULATION AND COMMUNITY PROCESSES IN ECOSYSTEM MODELS

The proximate interactions of the atmosphere and biosphere are via the exchange of matter and energy, integrated over fairly large areas. Two primary classes of models have been used in studies of atmosphere–biosphere interactions. First, models of physical processes (momentum transfer, evapotranspiration, albedo) that are influenced by biology (e.g., stomatal resistance, vegetation structure) are used in simulations of climate integrated over relatively short periods of time (Sellers et al., 1988). In these land process models, biological parameters (e.g., vegetation height, vegetation density, stomatal resistance) are static or change in a prescribed way. A second class of models is typically used in simulating longer-term interactions via carbon cycling and trace gas biogeochemistry. These models have been used primarily in identifying potential feedbacks between climate and biogeochemical change (Schimel, 1990; Pastor and Post, 1988). The former class of models are often referred to as *biophysical models* or *land surface parameterizations* and the latter as *ecosystem models* or *terrestrial biogeochemistry simulations*. Several hybrids exist (Running et al., 1989). This chapter will not review biogeographic models based on climate–vegetation type correlations (Prentice and Fung, 1990).

Both of these classes of models are based explicitly on ecophysiology and often rely on evolutionary optimality arguments. In models of physical vegetation–atmosphere coupling, transpiration rates are linked to photosynthesis via stomatal resistance (Collatz et al., 1991) or leaf area (Running et al., 1989). A significant body of theory and experimental data suggests that plants should allocate limiting resources such as investment in photosynthetic capacity as a function of the mean distribution of light to maximize carbon gain (Field, 1991; Hirose and Werger, 1987; Schimel et al., 1991a), based on optimality arguments. The assumption that photosynthetic capacity scales with the mean light profile has been implemented into a satellite data-driven land surface parameterization (Sellers, pers. comm.). The advantages of this assumption are twofold: first, instantaneous CO_2 exchange becomes linear with downwelling PAR (photosynthetically active radiation) flux, since capacity at all heights is scaled to the light environment, and this is in accord with a substantial number of observations. Second, because instantaneous CO_2 exchange is linear with downwelling PAR, averaging over fine-grained variation in light interception should cause minimal error. This latter attribute is consistent with the correspondence of ground- and aircraft-measured fluxes during the First ISLSCP Field Experiment (FIFE; Schimel et al., 1991a) but still requires testing. This improved simulation of photosynthesis translates directly into improved estimates of evapotranspiration and better simulations of climate. The use of evolutionary optimization models in biophysical models demonstrates both the power of optimization arguments in model development and simplification and the close coupling between physical (e.g. evapotranspiration) and biological (allocation, stomatal resistance) processes at the land surface (Schimel et al., 1991b).

The success of the optimization arguments in predicting plant allocation patterns and photosynthesis raises a number of interesting questions at the interface of global change, physiology, and population biology. Clearly, allocation and photosynthesis are closely coupled to prevailing environmental conditions and resource availability (Mooney, 1977; Schimel et al., 1991a,b). Also, clearly, the plasticity of individual organisms and genomes to adjust to changing environments and resource conditions is finite, and this is in fact a major driving force of succession (Pastor and Post, 1986; Tilman, this volume).

As global environmental conditions change via temperature, precipitation and cloudiness (Houghton et al., 1990), and as resource availabilities change with intentional and inadvertent addition of nitrogen and other substances (Melillo et al., 1989), what will control the rate of change of ecosystem response? Certainly the physiology of individual plants is such that some adjustment will occur without population or successional change. If the limits of plasticity in allocation or photosynthesis are reached, that will likely drive changes at population and community levels. If population and community changes cannot keep pace with the rates of change in climate or resources, then currently observed relationships between climate, soils, and the structure and function of vegetation could weaken. Uncoupling of climatic and edaphic constraints on plant adaptations could lead to atmosphere–biosphere interactions that are difficult to predict from current models or data. Uncoupling of climatic and population rates of change should also alter selective pressures, possibly driving evolutionary responses. Changes in the global environment may result in changes in the selective pressures influencing allocation and other aspects of plant physiology that control biophysical feedbacks to the atmosphere (Geber and Dawson, this volume).

Allocation to secondary compounds is a particularly interesting aspect of plant response to global change and atmosphere–biosphere exchange (Ayres, this volume). Biogenic volatile hydrocarbons, produced in many plants in response to herbivory, are highly reactive in the atmosphere. The oxidation of biogenic hydrocarbons is a critical sink for the OH radical in the atmosphere, and a source for ozone in areas with high atmospheric nitrogen oxide concentrations. The population- and community-level processes that control the distribution and abundance of isoprene- and terpene-producing plants have the potential to appreciably influence regional-to-global tropospheric chemistry. For example, Archer (1989) demonstrated large-scale regional succession from grassland to thorn woodland and savanna in the Rio Grande plains of Texas, and that this process was significantly mediated by herbivory. Subsequent work (Archer and Schimel, unpublished data) has shown that this successional process caused a shift from species producing virtually no hydrocarbons to woody species that produce either isoprene or terpenes at appreciable rates. Feedback among insect physiology, population dynamics, and plant allocation to defense compounds (Ayres, this volume) could also lead to complex changes in tropospheric chemistry. The occurrence and abundance of terpenes in plants is susceptible to selection, pointing to a role for evolutionary biologists in global change research. Ecologists studying global change should address the phytochemical traits governing palatability to herbivores in order to understand volatile hydrocarbon biogeochemistry and biology. This subject should provide an exciting area for collaboration between botanists, zoologists, and atmospheric chemists (see Sharkey et al., 1991).

The carbon cycle is a central issue in global change research. Models of ecosystem carbon storage are sensitive to assumptions about plant physiology, rates of herbivory, and variations in physiology and allocation among plant species and populations (Parton et al., 1987; Pastor and Post, 1986; Holland et al., in press; Pastor et al., 1988). Pastor and colleagues (1988) have reviewed a number of interactions between forest succession and biogeochemistry, and between herbivory, succession, and forest biogeochemistry. In grasslands, a model of biogeochemical cycling proved impossible to validate until a simple simulation of herbivory was included; in its absence, the model consistently predicted higher soil carbon levels than observed. Subsequent field studies supported the conclusion that herbivore removal of aboveground net primary production influences soil carbon storage (Bauer et al., 1987).

Holland et al. (in press) and Holland and Detling (1990) showed that the consequences of herbivory for plant physiology (increased allocation to aboveground tissue) had significant effects on nitrogen cycling. As root biomass declined with grazing intensity, immobilization of nitrogen by soil microorganisms decreased due to the reduced supply of carbon. This led to enhanced net nitrogen mineralization, and maintenance of productivity despite grazing pressure. Holland et al. (in press) also examined the consequences of different allocation responses between grazing-tolerant and grazing-intolerant ecotypes. They found that the grazing-tolerant allocation response pattern (i.e., of increased allocation aboveground relative to the intolerant ecotype) resulted in differences in biogeochemical responses to herbivory between the ecotypes. Thus, not only does herbivory affect ecosystem-level processes of carbon storage and nutrient cycling, but population-level physiological traits also modulate ecosystem-level responses. Also, relatively simple models can be used to explore the consequences of population–ecosystem interactions when sensitivity to specific interactions is of interest, despite the complexity of the separate processes.

The significance of trophic interactions for global biogeochemical cycles is just beginning to be appreciated. For example, herbivores contribute significantly to global greenhouse gas emissions. Domestic ungulates are one of the largest sources of methane, and increases in human holdings and forage management are likely one of the significant reasons for global methane concentration increases (Fung et al., 1991). Nitrogen in animal urine is the largest source of the basic gas ammonia to the atmosphere, significantly affecting acid–base chemistry in aerosols and rainwater (Langford et al., unpublished data). In a fascinating paper, Huntley et al (1991) invoked mammalian and bird respiration as a significant control over the carbon budget of the southern ocean. They argued that the southern ocean food web is very inefficient, releasing 25 percent of the carbon fixed through top predator (bird and mammal) respiration. In another article, Broecker (1991) suggested that the only way in which biotic change could affect oceanic carbon uptake was by reducing respiration through overfishing. That would presumably increase the efficiency of food webs, although Broecker may have presented this idea to illustrate how absurd he considered any non-geochemical model of marine CO_2 uptake.

As the importance of herbivory and other trophic interactions in the earth system becomes clearer, improved global explanations for the dynamics and abundance of relevant organisms will be required. This presents a clear opportunity for population ecologists to work in an arena with direct relevance to biotic feedback on the carbon cycle and other biogeochemical processes.

PLANT PHYSIOLOGY, SOIL CARBON STORAGE, AND GLOBAL CHANGE

The soil temperature increases that accompany global warming should result in increased decomposition rates, leading to a loss of soil carbon (Schimel, 1990). Several recent papers have explored the uncoupled response of soil carbon to temperature and predicted significant losses (Jenkinson et al., 1991; Townsend et al., in press). The countervailing process of enhanced primary production due to CO_2 fertilization may lead to enhanced carbon storage and several recent papers have suggested that the effects of CO_2 fertilization are already evident in the atmospheric CO_2 burden (Tans et al., 1989; Enting and Mansbridge, 1991). Despite theoretical arguments and global concentration data, a number of authors have suggested that ecosystem sensitivity to CO_2 fertilization should be greatly curtailed by other limiting resources to plant growth (e.g., water and nutrients) (Pastor and Post, 1988; Schimel, 1990).

We recently completed a series of analyses of the coupled response of the plant–soil system to increased CO_2 and warming using the Century ecosystem model (Parton et al., 1987; Schimel et al., 1990; Kittel et al., unpublished data). The simulations showed that warming increased NPP by releasing soil nutrients through higher decomposition rates, and that CO_2 fertilization caused larger increases in dead than in biomass carbon storage because reduced nitrogen content in litter lowered decomposition rates. This exercise suggested that (1) nutrients significantly modulate biophysical responses to climate change, and (2) the behavior of the coupled plant–soil system is often different from that predicted from the separate components.

Our simulations also emphasize the importance of understanding physiological responses to environmental change. We know that physiological responses are conditioned by evolutionary pressures and that they vary among species and populations. The development of simplified but sophisticated parameterizations of ecosystem response to global change requires the inclusion of appropriate generalizations derived from more detailed population-level studies. Contrary to the view of many biologists, the process of including such biological sophistication in ecosystem models is well advanced, and mechanisms for doing so are increasingly evident. Mechanisms for capturing the dynamics of functional groups (e.g., C_3 vs C_4 plants) and other taxonomic groupings are less evident but are absolutely essential to realistic prediction of rates of change.

DISCUSSION

The so-called ecosystem level is a natural level for the integration of atmospheric and biospheric processes, since the atmosphere is most influenced by the spatially aggregated flux of matter and energy at the land surface and by the variability of those fluxes at the mesoscale. Yet ecosystem processes are the collective result of the activities of many species—processes carried out by individual species (e.g., nitrogen fixers) that affect many others, or processes that involve the interaction of biotic with physical forces (Schimel et al., 1991b). Predictive modeling of ecosystem processes in a changing environment requires a strong underpinning from understanding of physiological and population processes, organized by consistency with evolutionary theory.

The goals and intellectual center of evolutionary and population biology have been in understanding how the unique attributes of species and populations arise and function, although there are significant counterexamples (e.g., Mooney, 1977). This is an important agenda and perhaps crucial to the development of a theoretical basis for conservation biology. However, another agenda—that of discovering regularities that can be used in models of large-scale processes in the immense variety of morphological, physiological, and life history traits—is crucial both to progress in biophysics and biogeochemistry and to regional and global extrapolation of population processes.

Key species attributes reflect resource availability (water, nutrients, light), and influence resource availability (via decomposition rates, water depletion, light use) in a predictable way (Tilman, this volume; Schimel et al., 1991a; Running and Gower, 1991; Holland et al., in press). These facts may serve as a conceptual framework for incorporating optimization and control responses based on evolutionary predictions into models of biogeochemistry and biophysics. Trophic interactions, especially herbivory, may also be susceptible to simplified analysis within this same paradigm (Holland et al., in press; Pastor et al., 1988). Considerable effort will be required to extend other aspects of population and evolutionary ecology to regional and global scales.

Considerable effort will be needed to develop and apply techniques for the validation of such models at appropriate scales (Schimel et al., 1991b). The validation process may require the understanding of population processes influencing ecosystem properties (properties such as canopy chemistry and its spectral correlates, or regional atmospheric composition) that are inherently more measurable at global scales than the abundance of individuals. On the other hand, as Carpenter et al. (this volume) point out, the local abundance of sensitive species may provide an early warning of ecosystem change, and appropriate models should predict this type of change as well. The development and testing of regional-to-global population models will provide both a more rigorous foundation for the biotic component of earth system modeling and a significant benefit in the understanding and management of biotic diversity.

Finally, ecologists must view their work as having a place in understanding the earth as a system, in which strong reciprocal interactions take place among physical, chemical, and biological processes. Ecology is a key discipline in the earth sciences, and to the extent that ecologists are unaware of this status, scientists in atmospheric, geochemical, and oceanographic sciences have proceeded independently in the study of inherently biological processes. While this can, and has, led to creative new approaches to problems, it can also lead to error or duplicative efforts. The generally low representation of ecologists in global change science is due less to exclusion by others than to ecologists' general lack of awareness or interest in the key scientific issues, and to a lack of common language. This volume hopes to significantly alleviate these problems.

SUMMARY

The biota have significant effects on chemical and physical processes in the earth system by influencing energy exchange at the land surface and by control of atmospheric composition. Predictive modeling of the role of the biosphere in a changing environment requires linking understanding from biophysics, physiology, and ecosystem ecology. In order to predict realistic rates of change in the coupled atmosphere–biosphere system, a strong, underpinning understanding of physiological and population processes is needed.

ACKNOWLEDGMENTS

This essay was written while the author was Project Scientist for the University Corporation for Atmospheric Research's Climate System Modeling Program. Chris Field, Terry Chapin, Tim Kittel, Steve Archer, Beth Holland, Bill Parton and Joe Berry aided in developing the ideas presented in this paper. Model results on CO_2 fertilization and preliminary observations of hydrocarbon emissions were supported by NASA's Earth Observing System program, and were developed in collaboration with Tim Kittel, Buck Sanford, Bill Hunt, Steve Archer, Alex Guenther, Beth Holland and Mary Wildermuth.

PART TWO

PHYSIOLOGY AND POPULATION RESPONSES TO ENVIRONMENTAL CHANGE

CHAPTER 5

TERRESTRIAL VEGETATION AND CLIMATE CHANGE: Integrating Models and Experiments

Stephen W. Pacala and George C. Hurtt

One difference between plant and animal ecology is that plant ecologists have never, as a group, seriously doubted that density-dependent interactions play a significant role in governing population dynamics. Anyone connected with agriculture or forestry understands the importance of competition within and between species, and knows that mutualists and natural enemies can affect dynamics in at least some systems. Moreover, it is increasingly apparent that to understand natural vegetation, one must understand the interplay among plant performance, physical factors, and the consumption and renewal of abiotic resources (Tilman, 1982). Changes in community composition with changes in soils or climate are simply too common and abrupt to believe otherwise. For these reasons and because natural vegetation is a primary sink and source of atmospheric carbon, plant community ecologists inevitably must play a role in the debate about global climate change.

This chapter is about the nature of that role. In what follows we first discuss paleoecological studies that illustrate the profound effects of climate on plant distribution and abundance. These studies also show that current associations between climate and vegetation cannot be used to forecast reliably the consequences of future climate change. Instead, we must understand the relevant population dynamic processes.

Second, we examine some experimental studies designed to determine the effects of future climates on plant population dynamics. We argue that, because of the complexity and time scale of climatic effects, experiments will not by themselves lead to an ability to forecast future changes. Rather, we will be forced to rely on experimentally calibrated population dynamic models.

Third, we examine the JABOWA–FORET family of forest simulators, which dominate the modeling literature on the response of vegetation to climate change. This literature focuses on forest models, apparently because of the overwhelming role that forests play in the carbon cycle. Forests store 80 percent and annually fix 65 percent of global biotic carbon (Solomon, 1986). Sedjo and Solomon (1989) estimate that the global addition of forest to an area of 465 million hectares (the size of the continental United States west of the Mississippi River, minus the states of Montana, Idaho, and North Dakota) could halt the current increase in atmospheric CO_2.

Although the JABOWA–FORET models are a useful first step, they do not represent a healthy integration of modeling and experiments. Published models are not calibrated primarily with experiments. Rather, individual plants are assumed to be able to grow only between climatic extremes estimated from the latitudinal limits of each species' range. We refer to this practice as *the confusion of the fundamental and realized niches*. Also, published studies rely on nonspatial models with open recruitment (recruits are not produced by the modeled trees). Species diversity is artificially maintained by a random number generator that introduces new recruits, and there is no possibility of population dynamic time lags caused by limited dispersal. We refer to this practice as *the problem of unlimited dispersal.*

Fourth, we study a simple numerical model to examine the effects of the confusion of the fundamental and realized niches and the problem of unlimited dispersal. We discuss the interaction of these two problems: The fundamental and realized niches are approximately equal only if dispersal is unlimited. We also show that these problems can seriously distort forecasts of the response of vegetation to climate change in either a falsely alarmist or falsely reassuring direction.

Finally, we discuss work in progress indicating that the functions in forest simulators can be estimated from experimental data. The desired coupling between models and experiments can thus be achieved now. Mooney et al. (1991) have recently proposed a national network of experimental sites to determine the responses of terrestrial ecosystems to climate change. We close with a call to add to this kind of network the experiments necessary to calibrate population dynamic models of vegetation.

LESSONS FROM PALEOECOLOGY

Climates of the late Pleistocene and early Holocene were unlike any on earth today. Temperatures ranged from several degrees cooler than the current climate during glacial maxima to 2–3°C warmer during interglacial periods (Schneider and Londer, 1984; Dorf, 1976). The concentration of atmospheric CO_2 was as low as 200 parts per million 18,000 years ago (compared to 280 ppm in the preindustrial late Holocene and 350 ppm today; Davis, 1989). Patterns of precipitation and disturbance caused by fire (Grimm, 1983, 1984; Clark, 1989, 1990) and windstorms (Davis, 1985; Spear et al., 1989) changed markedly over the past 10,000 years.

The 5°C warming forecast by some global climate models would produce a climate warmer than any in the past 100,000 years (Woodwell, 1989). This change is predicted to occur at a rate approximately ten times faster than past changes (Davis, 1989). Each 1°C increase in temperature corresponds to approximately 100 km of latitude or 150 m of elevation (MacArthur, 1972).

Studies of the pollen record and of ancient packrat middens lead to two generalizations about the response of vegetation to past climate change. First, whole communities do not remain intact and simply move north or south as the climate changes (Davis, 1989; Webb, 1987; Van Devender and Spaulding, 1979). For example, 18,000 years ago there were extensive areas containing species present today but grouped in communities rarely found today (e.g., spruce–oak woodlands and black ash tundra; Wright, 1987). Between 3000 and 8000 years ago, large areas of Minnesota were dominated by oak savanna, which is found today only in a narrow ecotone (McAndrews, 1967; Davis, 1989). Spruce was a dominant species in the central portion of the North American continent 18,000 years ago and is dominant in the boreal forest today, but was rare for several thousand years approximately 8000 years ago (Kutzbach and Wright, 1985; Wright, 1987). The depauperate forests of northern Europe are the result of extinctions of species such as tulip tree, hemlock, and white cedar during Pleistocene cooling and of extinctions of other species following postglacial warming (Seddon, 1971; Tralau, 1973).

Second, changes in vegetation sometimes lag behind changes in climate. For example, although some species responded rapidly to postglacial warming, others attained their current distributions only within the last 2000 years (Davis, 1981). Some time lags may have been caused by limited dispersal and others by slow local dynamics (Davis, 1989). For example, the postglacial northward expansion of *Tsuga* and *Fagus* halted at the Great Lakes dispersal barrier for 1000 years (Davis et al., 1986), whereas the lagged response of birch in northern Europe is thought to have been caused by the slow accumulation of soil organic matter (Pennington, 1986).

One approach to forecasting the response of modern vegetation to climate change is to use the correspondence between modern plant communities and climate together with the predictions of global climate models to produce maps of the future ranges of species (see, for example, Emmanuel et al., 1985; Sedjo and Solomon, 1989). Although this is a useful first step, the lesson from paleoecology is that it is not sufficient (Davis, 1989). The novel climates of the future are expected to produce novel communities, just as in the past, and time lags lasting longer than the life spans of political systems are expected to occur for some species but not for others, just as in the past. This latter problem is particularly important because future climate change is predicted to be ten times faster than any changes that occurred over the past 100,000 years (Davis, 1989). Many species will be unable to respond at this rate, especially given the barriers to dispersal caused by modern habitat fragmentation (Lovejoy, 1980). Apparently, our only recourse is to understand the population dynamics and movements of interacting species as a function of climate change (Davis, 1989).

LESSONS FROM EXPERIMENTAL STUDIES

The enormous literature on experimental studies of the physiological and population dynamic responses of natural vegetation to climate change has been reviewed recently by Bazzaz (1990) and Mooney et al. (1991). We make no attempt to repeat these reviews here. Rather, we cite examples of experimental studies to argue for the development and experimental calibration of mechanistic and spatial population dynamic models. Such models would be constructed from functions that summarize: (1) the performance of each species of plant as a function of temperature and of local resource availability (light, water, CO_2, and soil nutrients); and (2) local resource availability as a function of the local plant community, site attributes, and climate.

The reason that experiments alone will not allow us to forecast the response of vegetation to climate change is simply that the experiments would take too long. If the predictions of global climate models are correct, the global climate will change drastically long before experiments in systems dominated by long-lived perennials could be completed. Even so, a purely experimental approach might be sufficient to answer some questions. For example, Oechel and co-workers (Oechel and Strain, 1985; Prudhomme et al., 1984; Oechel and Riechers, 1986) have studied the response of tundra vegetation to changes in temperature and CO_2. They found that the net uptake of carbon increased in response to increased length of growing season, thaw of permafrost, and CO_2 fertilization. Although the net carbon flux apparently equilibrated in three years, the possibility remains that long-term population dynamic changes ultimately will further alter the net uptake of carbon (Mooney et al., 1991). In the majority of systems, the response of community composition to changing climate will take place over decades to centuries and will involve range extensions and contractions of hundreds of kilometers. Clearly, spatial models will be essential to extrapolate the results of comparatively short-term and small-scale experiments.

Because experiments have shown striking and unpredictable interspecific variation in responses to higher temperature and CO_2 levels, separate functions for each species may need to be estimated. For example, the growth rate increases in response to CO_2 fertilization in some species, but not in others (Bazzaz et al., 1989; Sionit et al., 1985; Tolley and Strain, 1984a,b, 1985). Generalizations—such as the prediction that C_3 plants will respond to increased CO_2 but C_4 plants will not—are often true (Bazzaz, 1990; Williams et al., 1986). However, root biomass of *Spartina patens* (a C_4 plant) increased 85 percent in response to elevated CO_2 levels (Curtis et al., 1989a,b; Drake, 1989). Climate change may tip the competitive balance in favor of shade-intolerant tree species (Tolley and Strain, 1984b; Bazzaz 1979) or shade-tolerant ones (Bazzaz, 1990). Flowering time may be earlier or later, and seed set may increase or decrease (Carter and Peterson, 1983; Garbutt and Bazzaz, 1984; Bazazz, 1990).

Three reasons suggest that the most useful population dynamic models will be mechanistic. First, the number of parameters that would have to be estimated increases as the square of the number of species in phenomenological models of competition, but only linearly in mechanistic models (Pacala, 1989). Second, experiments have shown that elevated CO_2 and temperature can alter nutrient cycling and availability. For example, changes in root growth and exudation caused by increases in atmospheric CO_2 may result in increased mycorrhizal colonization (Strain and Bazzaz, 1983) and nitrogen fixation (Lamborg et al., 1983; Luxmoore 1981). Similarly, CO_2 fertilization may increase carbon:nitrogen ratios in leaf litter (Melillo, 1983) and thus reduce rates of decomposition and nitrogen mineralization (Pastor and Post, 1988; Strain and Bazzaz, 1983; Williams et al., 1986). Third, we need models not only to predict the responses of vegetation to climate change, but also to understand the effects of vegetation on climate. Changes in plant community composition will alter surface albedo, evapotranspiration, and whole-ecosystem rates of photosynthesis and respiration, and will thus affect the global exchange of energy and matter between the biosphere and atmosphere (Bazzaz, 1990; Mooney et al., 1991; Davis, 1989). To understand how population dynamics affect climate, we must have population dynamic models that deal in the same currency as climate models.

To summarize, we need calibrated models to extrapolate the results of experiments that must be completed before the effects of climate change occur. We need models formulated at the species level because paleoecological and experimental studies demonstrate that the responses of species are individualistic. We need mechanistic models because phenomenological models would be prohibitively difficult to calibrate and because the exchange of matter and energy between plants, soils, and the atmosphere and hydrosphere is the essence of the problem.

Although the experimental literature demonstrates the possibility of estimating many of the building blocks required for population dynamic models of the response of plants to climate change, very little information is available about possible effects on their animal and microbial mutualists and natural enemies (but see Ayres, this volume). One reason to expect discordant changes in the distribution and abundance of plants and of their mutualists and natural enemies is that green plants are affected by CO_2 fertilization, whereas other species are not. Some studies have examined the effects of climate change-induced alterations in the nutritive value of leaves. For example, insect herbivores may compensate for increased C:N ratios in leaf tissue by elevating their feeding rates (Fajer, 1989; Lincoln and Couvet, 1989; Lincoln et al., 1986), but may still suffer increased mortality (Akey and Kimball, 1989; Fajer et al., 1989) and reduced growth and fecundity (Fajer et al., 1989). Clearly, more work in this area is vital.

LESSONS FROM MODELING STUDIES

JABOWA (Botkin et al., 1972), FORET (Shugart and West, 1977) and related forest simulation models (see the review in Shugart, 1984) have been used extensively to predict the response of North American forests to future climate change (Solomon, 1986; Pastor and Post, 1988; Hanson et al., 1989, 1990; Solomon and West, 1987; Botkin and Nisbet, 1992; Shugart, in press) and to analyze responses to past climate change (Solomon et al., 1981; Solomon and Webb, 1985; Davis and Botkin, 1985). These simulators model the growth, mortality, and recruitment of individual trees located in a single spatial cell (usually approximately 0.1 ha). Individual trees have sigmoid growth in diameter under ideal conditions, with a species-specific asymptotic size. An individual's actual growth rate is determined by multiplying its ideal growth rate by a series of species-specific *growth modifying functions* (GMF) that describe the effects of resource availability and climate on growth. Mechanistic submodels determine how each individual modifies the resources available to others. For example, each plant reduces the light available to all smaller plants according to Beer's Law. Mortality is stochastic and may depend on growth rate. Typically, the probability of mortality increases as growth rate decreases, and the same dependence of mortality on growth rate is used for all species. Recruits are not produced by the trees in the model. Rather, in each iteration, recruits are drawn from a fixed list of species.

Note that these simulators have several of the attributes called for in the previous section. For example, the models are mechanistic and defined at the species level. The models have, however, two particularly debilitating problems (discussed in the two sections that follow).

Confusion of the fundamental and realized niches

The majority of functional forms and parameter values in published forest simulators are *not* estimates obtained by regression from experimental data.

Instead, one of two practices is used to determine how plant performance is affected by resources and climate. Consider the response of growth to temperature. All published studies of the response of vegetation to climate change use the practice started by Botkin et al. (1972) to "estimate" the temperature GMFs. Botkin et al. (1972) assumed that the temperature GMF is a parabola. Growth is positive only between the parabola's two roots and is thus highest at the temperature midway between the roots (where the GMF equals one). Here, "temperature" is in terms of the number of degree days greater than 5°C. The latitudinal or elevational limits of a tree species' distribution typically correspond to degree-day isopleths. Botkin et al. (1972) set the lower root of a species' temperature GMF equal to the temperature of the species' northern range limit and the upper root equal to the temperature of its southern range limit. Thus, all published simulators of the response of forests to climate change assume that fundamental and realized thermal niches are equal. This same practice is used to calibrate the other GMFs that summarize effects of climate (for example, the function that determines the response to drought; Solomon, 1986).

By definition, a species' fundamental niche is greater than or equal to its realized niche. However, in the few cases where they have been measured experimentally, fundamental niches of plants typically were found to be much larger than realized niches (reviewed in Austin and Austin, 1980). A remark attributed to Robert MacArthur is that were it not for competition with tropical species, most temperate species would grow in the tropics. The implication is that the fundamental climatic niches in forest simulators are much too small. This should produce an overly rapid local elimination of species whose fundamental limits to growth are (falsely) exceeded by the rapidly changing climate.

The second practice used in "calibrating" forest simulators is simply to assign "reasonable" functions to each species. For example, in all published models that concern climate change, each species is assigned to one of two or three shade tolerance classes, and each class is assigned a different light GMF. The light GMF for shade intolerants increases slowly with light level to a high asymptote, and the function for shade tolerants increases rapidly to a low asymptote. Similarly, the function relating growth rate and probability of mortality is not estimated but rather simply chosen. In Solomon (1986), for example, the probability of mortality increases to an admittedly arbitrary 0.368 if an individual's growth rate is less than 10 percent of its ideal growth rate.

Unlimited dispersal

Because recruits are drawn from an unchanging list of species, all species are always present to respond instantly to simulated climate change. A time lag caused by limited dispersal is not possible. Thus, in addition to the overly rapid local elimination of species caused by the confusion of the fundamental and realized niches, an overly rapid rebound of local diversity as warm-adapted species swarm northward should be caused by the problem of unlimited dispersal.

Experimental studies indicate that short dispersal of tree seeds is the rule. For example, Seddon (1971) found that fewer than 3 percent of Engelmann spruce seeds disperse farther than 200 meters downwind (see also Figure 4C). The problem of unlimited dispersal is widely recognized in the modeling literature, and several attempts have been made to modify the random selection of recruits to mimic limited dispersal (see Solomon and Webb, 1985) or to make the model spatial (Hanson et al., 1989, 1990).

Additional problems

Two additional problems with these models are caused by the lack of feedback between adults and recruits. First, at some spatial scale, changes in the abundance of adults must affect the number of recruits. When we have added internal recruitment and space to similar forest simulators, the predicted diversity has collapsed (Pacala, unpublished). This raises serious questions about the forecasted effects of climate change on forest diversity, over and above the absence of time lags caused by limited dispersal. Second, because adult abundance does not affect recruitment, species can never really go extinct. The worst that can happen is that a species will fail to grow and mature if climatic conditions are outside the species' fundamental niche (e.g., outside the lower or upper temperature for growth in the temperature GMF). Because fundamental climatic niches are assumed to be equal to realized climatic niches, adults of a species inevitably are present in climates that occur within the observed range of the species and are never present in climates outside this range. In other words, the simulators are simply a fancy way to do what studies in paleoecology warn against: To use associations between current ranges and climate to predict future ranges (as in Emanuel et al., 1985; Sedjo and Solomon, 1989).

LESSONS FROM A SIMPLE MODEL

We have seen that the confusion of fundamental and realized niches and the problem of unlimited dispersal should interact to amplify the pace of population dynamics following climate change in forest simulation models. In this section, we examine a simple spatial model of forest dynamics to demonstrate other possible consequences of these problems.

Lottery models (Chesson and Warner, 1981; Shmida and Ellner, 1984) are among the simplest and most widely studied models of competition for space. Consider a collection of spatial cells, each of which contains at most one adult, and let $N_i(t)$ be the fraction of cells occupied by species i at time

t. We assume that the number of cells is large enough to ignore demographic stochasticity. Competition occurs only among the juveniles that occupy a cell made vacant by the death of an adult. The juvenile that succeeds in capturing a vacant cell is chosen by a lottery—a biased random selection of a single individual from the population of juveniles in the cell. If d is the adult death rate, dispersing seeds are uniformly distributed across all cells, F is the per capita fecundity of adults, and α_i is a measure of the competitive ability of a species i juvenile (the bias in the lottery), then

$$\frac{dN_i(t)}{dt} = -dN_i(t) + d\,\frac{\alpha_i F N_i(t)}{\sum_{j=1}^{Q} \alpha_j F N_j(t)} \tag{1}$$

where Q is the number of tree species.

Equation 1 states that the change in abundance of species i $[dN_i(t)/dt]$ is equal to the fraction of cells lost by species i $[-dN_i(t)]$ plus the fraction of cells captured (the fraction on the extreme right-hand side). The fraction of cells captured is simply the fraction that are vacant (d) times the probability that a species i juvenile is chosen in the random lottery that determines the cell's new occupant. The lottery term is given as the density of species i juveniles in the cell $[FN_i(t)]$, weighted by the competitive ability α_i and divided by the total weighted density of juveniles of all species. It is easy to show that the species with the largest α will competitively exclude all others.

Now suppose that we greatly expand the spatial scale so that an entire collection of cells governed by Equation 1 occupies a single point along a much larger transect. One can think of a 1000-kilometer transect containing 1000 different collections of 10,000 100-square meter cells (each collection of cells occupies 1 km^2). Let $N_i(x,t)$ be the fraction of cells occupied by species i in the collection located at position x along the transect. Further, assume that there is a monotonic gradient in some physical factor (such as temperature) along the transect, and suppose that the competitive abilities of the species change with this physical factor. Let $\alpha_i(x)$ be the value of the competitive weight for species i at location x. A species' fundamental niche is the range of environmental conditions for which its α is greater than zero, or, equivalently, the range of values of x for which its α is greater than zero. Finally, let $p(y)$ be the probability that a seed disperses a distance y along the transect. We assume that the mean of $p(y)$ is zero and so the direction of dispersal is unbiased. Then the model for the entire transect is:

$$\frac{\partial N_i(x,t)}{\partial t} = -dN_i(x,t) + d\,\frac{\alpha_i(x)\int_{-\infty}^{\infty} p(y) F N_i(y + x,t)\,dy}{\sum_{j=1}^{Q} \alpha_j(x)\int_{-\infty}^{\infty} p(y) F N_j(x + y,t)\,dy} \tag{2}$$

Although it looks complicated, this model is very similar to the simple lottery model (Equation 1). The primary difference is the presence of the integrals in the lottery term. These integrals are simply the densities of seeds that disperse into the collection of cells at transect location x. If there is no dispersal along the transect [the variance of $p(y) = 0$), then Equation 2 specifies a separate model, identical to Equation 1, at each position along the transect.

To make Equation 2 numerically tractable, we approximate the $N_j(x + y,t)(j = 1, 2, \ldots, Q)$ in the integrals by second-order Taylor expansions about $y = 0$. This produces the diffusion equation:

$$\frac{\partial N_i(x,t)}{\partial t} = -dN_i(x,t) + d\,\frac{\alpha(x)\left[N_i(x,t) + D\,\dfrac{\partial^2 N_i(x,t)}{\partial x^2}\right]}{\sum_{j=1}^{Q} \alpha_j(x)\left[N_j(x,t) + D\,\dfrac{\partial^2 N_j(x,t)}{\partial x^2}\right]} \tag{3}$$

where D is a diffusion coefficient ($D = \sigma^2/2$) and σ^2 is the variance of $p(y)$. The rate of movement along the transect increases with D.

We studied model (3) numerically. In all cases, we assumed absorbing boundaries (seeds that dispersed off either end of the finite transect were "lost"), initial abundances equal to $1/Q$ for all species in all locations, and parabolic functions for the α's. The function for species i was:

$$a_i(x) = -(x - M_i - W/2)(x - M_i + W/2)$$

if $(M_i - W/2) < x < (M_i + W/2)$ and $\alpha_i = 0$; otherwise, where M_i is the center of species fundamental niche (the value of x at which α_i is greatest) and W is the width of the fundamental niche. We spaced the centers of the Q different fundamental niches evenly across the transect. Finally, we examined both cases in which the d's were constant across the transect as well as cases in which adults died instantly outside their fundamental niches ($d\Delta t = 1$ outside the fundamental niche).

Results

In the model, the fundamental niche of each species is larger than the realized niche at equilibrium because species are competitively displaced from parts of their fundamental ranges. The magnitude of this effect depends on the dispersal rate (Figure 1). If there is no dispersal, then each species occurs at equilibrium only where it is the dominant competitor, and hence its realized niche is relatively small. In contrast, rapid dispersal continually moves each species from source areas in which it is competitively dominant to sink areas in which it is competitively inferior. The assumption of published forest simulators that realized and fundamental niches are equal is generally true only in the large dispersal limit.

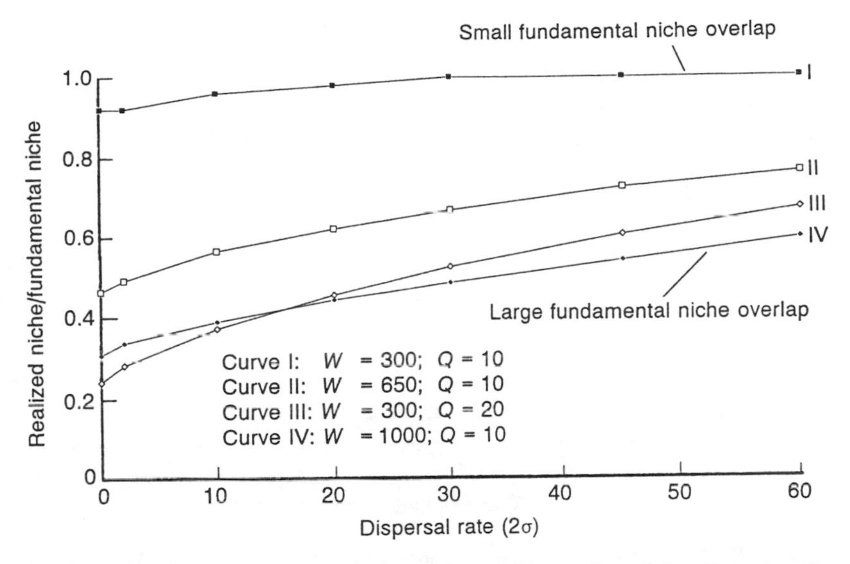

FIGURE 1. The ratio of the realized niche width predicted by Equation 3 at equilibrium to the fundamental niche width (W) as a function of the dispersal rate. The realized width for each species was calculated as the length of habitat in which the species occupied $\geq 1\%$ of the spatial cells. The mean of these widths for all species was used to calculate the ratios shown. In all runs, there were 100 spatial cells in the discretized approximation of the model. Step sizes were $\Delta t = 1.0$, $\Delta x = 30.0$. The equilibrium abundances are independent of the adult death rate (d). Q is the number of species.

The dispersal rates in Figure 1 must be evaluated relative to the range limits given by the width of the fundamental niche (W). Values of W from 300 to 1000 were used to produce the curves shown. We think of these units as kilometers, so two standard deviations of the dispersal function range from 0 to 60 km. This latter limit is unrealistically large. For a Gaussian dispersal function, it implies that 5 percent of seeds disperse more than 60 km. Thus, the most biologically meaningful part of the figure is the extreme left side.

In curve I of Figure 1, the ratio of the equilibrium realized and fundamental niches is nearly one for all dispersal distances because there is only a 9 percent overlap between the fundamental niches of species with adjacent niches, and thus little space along the transect in which species could displace one another. Realized niches are substantially smaller than fundamental niches for the remaining curves because there is more overlap between the fundamental niches (58% in curve II, 78% in III, and 73% in IV) and thus more potential for competition to restrict the realized ranges.

Figure 2 graphs the width of the realized niche at equilibrium as a

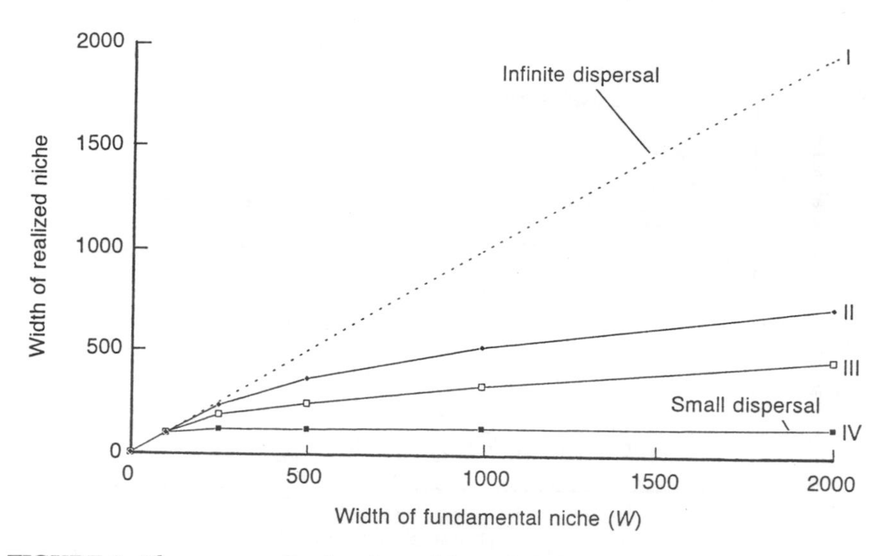

FIGURE 2. The mean realized niche width predicted by Equation 3 at equilibrium as a function of the fundamental niche width for selected dispersal distances. See the legend to Figure 1 for a description of how the mean realized niche width was calculated. Parameters were identical to those in Figure 1 (see legend), except: Curve II, $\sigma = 30$; curve III, $\sigma = 10$; curve IV, $\sigma = 0.01$. Curve I is the theoretical prediction for the large dispersal limit.

function of the width of the fundamental niche for four different dispersal rates ($2\sigma = \infty$ in curve I; $2\sigma = 60$ in II; $2\sigma = 20$ in III; and $2\sigma = 0.02$ in IV). This figure shows that *a small fundamental niche with a large dispersal rate gives the same realized niche width as a large fundamental niche with a small dispersal rate.* For example, suppose that we wish to estimate the fundamental niche width (W) and observe a realized niche (range) of 150 km. We could assume that curve I applies and set $W = 150$ and $2\sigma = \infty$ (as in published simulators of the response to climate change). However, we also could set $W = 170$ and $2\sigma = 60$ (curve II), $W = 190$ and $2\sigma = 20$ (curve III), or $W = 1500$ (!) and $2\sigma = 0.02$, because these values also yield a realized niche of 150.

To examine the joint effects of the confusion of fundamental and realized niches and of the problem of unlimited dispersal, we performed series of runs with simulated climate change. Runs within a series were identical in every respect except one: *Values of W and σ differed among the runs within a series, but were chosen to yield identical realized niches prior to climate change.* To simulate climate change, we first iterated the model to equilibrium. We then added a constant (the abrupt climate shift, S) to the mean of each species' fundamental niche (M_i). After the climate shift, each species was best adapted to conditions a distance S from the center of its prior range. We then iterated the model to determine the reassortment of ranges following climate change.

We observed two qualitatively distinct consequences of the problem of unlimited dispersal and the confusion of fundamental and realized niches (illustrated in Figure 3). First, forecasts of models (e.g., published forest simulators) that assume that W equals the realized niche and that dispersal is very large are sometimes overly alarmist. In Figure 3a, we plot the mean local species richness (number of species) following climate change for four different combinations of W and σ, ranging from high W and low σ in curve I to low W and high σ in curve IV. Again, these four runs had the same mean realized niche width and the same mean local species richness before the climate changed. The species in curve IV had fundamental niches approximately the same size as realized niches, as in JABOWA–FORET simulators. Although extinction of all species occurred in run IV because the climate shifted outside of each species' small fundamental niche, local diversity actually increased in run I. This increase was caused by the transient mixing of "northern" and "southern" floras during the reassortment that followed climate change. Species were not driven extinct, as in curve IV, because their fundamental niches were large enough to contain the climate shift.

Second, the assumption that W equals the realized niche and that dispersal is large can also produce falsely reassuring forecasts. In Figure 3b, we show two runs, again with equal realized niches, in which species responded to a continuously shifting climate (an amount S added to the M_i in each iteration). Rapidly dispersing species ($\sigma = 14.8$, curve I) were able to keep ahead of the pace of climate change despite their small fundamental niche ($W = 150$), and thus species richness was relatively insensitive to climate change. In contrast, the species richness of the slowly dispersing species ($\sigma = 3.2$, $W = 450$, curve II) decreased. As in published JABOWA–FORET simulators, the fundamental niches of the rapidly dispersing species (curve I) were approximately the same size as the realized niches.

The implication of Figure 3 is that forecasts of published models could be seriously distorted in a direction that depends on the actual sizes and shapes of fundamental niches and dispersal curves. The only way to avoid the problem of unlimited dispersal and the confusion of the fundamental and realized niches is to measure directly the components (i.e., GMFs, dispersal functions, and mortality functions) from which models are constructed.

A PLEA FOR COUPLED EXPERIMENTS AND MODELS

How is one to predict the response of vegetation to climate change? Studies in paleoecology make it clear that we will not succeed solely by associating current distributions with climate. Experimental studies demonstrate that

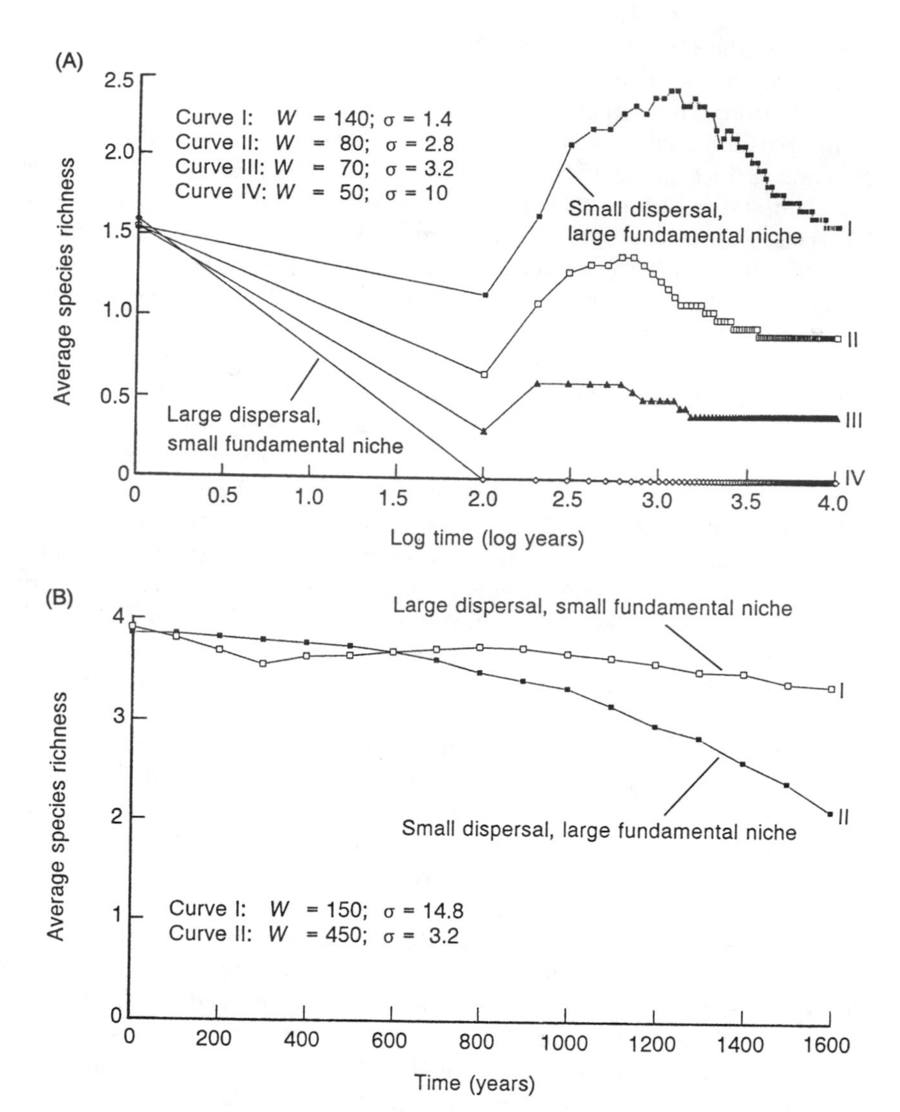

FIGURE 3. Average species richness following climate change as predicted by Equation 3. The average species richness is the mean of the number of species with abundances greater than 0.01 per spatial location. During all runs, abundances were set equal to zero if they fell to less than 0.0001. At the beginning of each run, abundances were at equilibrium under the conditions prior to the climate change. (A) Trajectories for equal realized niches. We used 20 spatial locations; $Q = 20$, $\Delta t = 1.0$. $\Delta x = 30$, $d = 0.01$ within a species' fundamental niche, and $d\Delta t = 1.0$ otherwise. The climate shift ($S = 60$) occurred instantaneously at the start of the trajectories shown (year 0). (B) Trajectories for different dispersal rates. We used 75 spatial locations; $Q = 35$, $\Delta t = 1$, $\Delta x = 15$, and $d = 0.01$. The climate shift occurred gradually. In each iteration shown, the climate shifted by an amount $S = 1.5$.

we can measure short-term responses of species to the novel climates of the future, but leave us guessing about the all important long-term and large-scale population dynamic effects and feedbacks between the biosphere and atmosphere. The population dynamic models published to date make the required predictions, but from an inadequate foundation.

Even so, we are optimistic about the future. Although time-consuming and expensive, estimates of all of the functions in forest simulation models are now possible. For example, in Figure 4, we show light GMFs, mortality functions, and dispersal functions for several species of trees. These functions were estimated by regression from data rather than simply assigned. The light responses in Figure 4A show that the light GMFs in published forest simulators do not capture the responses in nature (Pacala et al., in preparation). Note that the most shade-tolerant species (hemlock and beech) have low-light growth rates that are no higher than those of the most shade-intolerant species (yellow birch and red maple). The mortality functions in Figure 4B show that species differ enormously in the response of mortality to growth rate (Kobe et al., in preparation). Recall that published simulators assign identical mortality functions to all species. Calibrated simulators demonstrate that the observed interspecific differences among mortality functions play an overwhelming role in maintaining local diversity and largely determine the pace of population dynamics (Pacala et al., in preparation). Also, Figure 4C shows just how short dispersal is as well as substantial interspecific variation among species in dispersal rates (Ribbens et al., in preparation). Calibrated simulators demonstrate that the short dispersal of shade-tolerant species (e.g., hemlock) substantially slows the rate of succession and promotes long-term coexistence of shade-tolerant species (Pacala et al., in preparation).

To avoid the confusion of fundamental and realized niches in models of the response of plant communities to climate change, one would need to transplant species along gradients of temperature and rainfall as well as to manipulate the concentration of CO_2 experimentally (see Mooney et al., 1991, and references therein for a discussion of the necessary methods). These experiments would provide the data required for regression estimates of the functions in simulation models of the response to climate change. We believe that successful models of the response to climate change are likely to be calibrated extensions of JABOWA–FORET. Thus our criticism of JABOWA–FORET simulators is not intended as a general condemnation of these models, but rather a call for their further development.

We recognize that experimental calibration does not by itself ensure that a model will be a useful reflection of nature. We must begin using estimated sampling distributions to produce confidence limits on the predictions of simulation models (as in Pacala and Silander, 1990). The required methods are straightforward and readily accessible. This practice will serve to govern the growth in complexity of models, to ensure that predictive

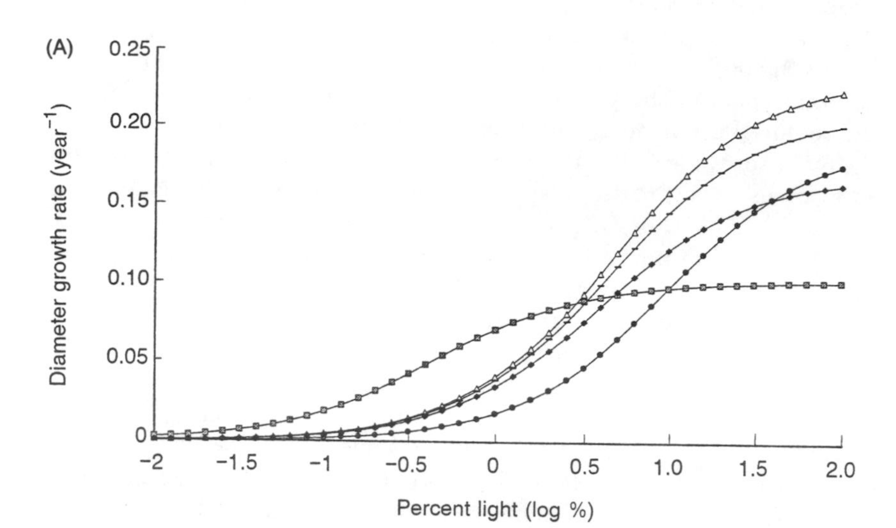

power does not evaporate in a statistical fog, and to eliminate the dependence on unestimated parameters. We must also work to simplify simulation models and to produce analytically tractable substitutes that lead to understanding. Finally, we must test models against all available data.

Recently, Mooney et al. (1991) have proposed a national network of experimental sites dedicated to assessing the effects of climate change on ecosystem processes. Community ecologists should endorse this plan, and add to it the experiments necessary to calibrate population dynamic models of the feedback between climate change and vegetation dynamics.

SUMMARY

We argue for the development and experimental calibration of population dynamic models that forecast the response of vegetation to climate change. This argument consists of three parts. First, studies in paleoecology imply that simple associations between current vegetation and climate are probably insufficient to predict how vegetation will respond to future climate change. Instead, we must understand population dynamic processes well enough to construct reliable models of the feedback between climate and vegetation. Second, experimental studies show that it is currently possible to estimate the components of mechanistic models of the response of vegetation to climate change. Third, we critique the way in which the JABOWA–FORET simulators of forest dynamics have been empirically calibrated. We show that forecasts of the response to climate change by these models could be seriously distorted in either an alarmist or reassuring

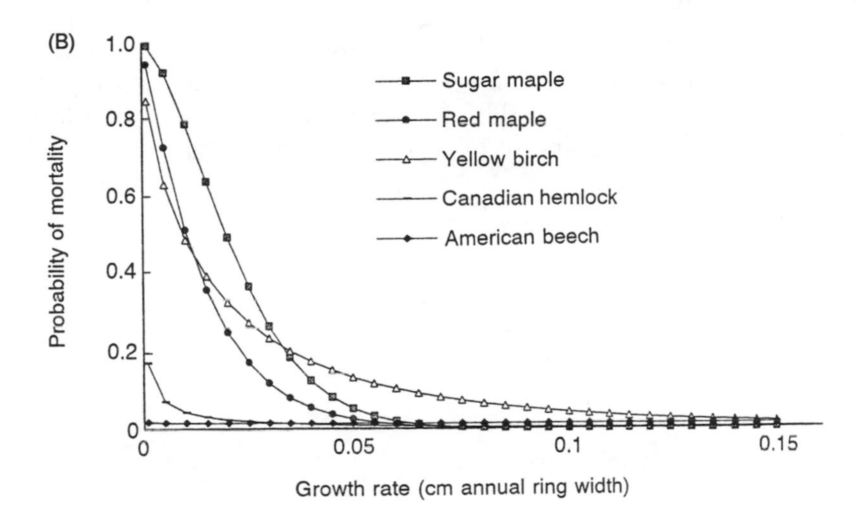

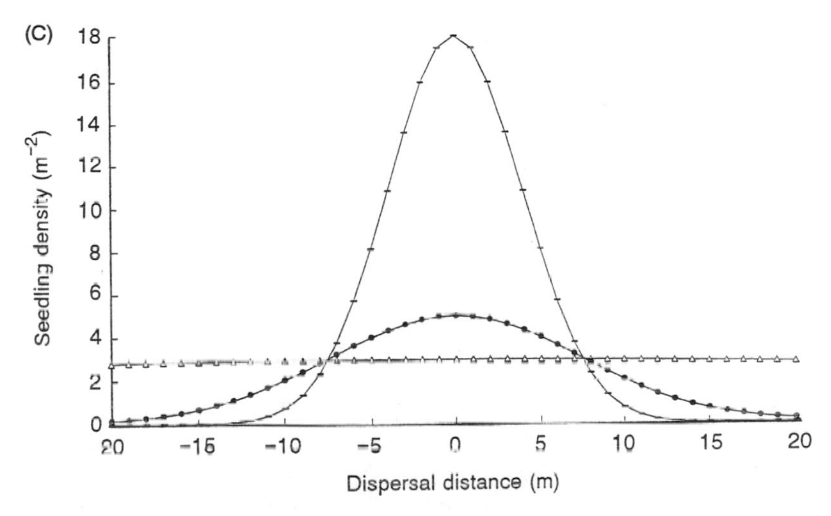

FIGURE 4. (A) Growth predictors. Estimated diameter growth rate as a function of light (percentage of total radiation during the growing season, obtained by fish-eye photography; Canham, 1988). (B) Mortality predictors. Understory mortality as a function of annual ring width. (C) Recruitment predictors. Recruitment as a function of dispersal for some northeastern forest tree species.

direction. Finally, we close with a call for the development of new JABOWA–FORET simulators that reflect a more healthy integration of models and experiments.

ACKNOWLEDGMENTS

We thank J. Silander and K. Holsinger for comments on the manuscript, and R. Huey for his thorough and professional editing. The support of the National Science Foundation (BSR-8811884), the Department of Energy (DE-FG02-90ER60933), and NASA is gratefully acknowledged.

CHAPTER 6

PLANT DEFENSE, HERBIVORY, AND CLIMATE CHANGE

Matthew P. Ayres

Anticipated patterns of climate change (Schneider, this volume) will have multiple direct effects on the physiology of both plants and herbivores. However, the community-level consequences cannot be predicted solely from these direct effects because climate change can also have surprisingly large effects on the interactions between plants and their herbivores. For example, a minor increase in average temperature (only 1°C) altered the interactions between mountain birch and a principal insect herbivore enough to triple the potential rate of population growth of the insect (see "Increased temperature," p. 88). Consequently, any attempt to predict how changing climates will affect plant–herbivore communities must address the complexities of such biotic interactions.

Climate change can influence interactions between plants and herbivores in several ways. For example, expected environmental changes in carbon dioxide levels, temperature, cloud cover, and water and nutrient availability will affect plant susceptibility to herbivores (Larsson et al., 1986; Waterman and Mole, 1989; Johnson and Lincoln, 1991). Climate change will also directly affect herbivore nutritional requirements (Robbins, 1983; Scriber and Slansky, 1981), development time (Taylor, 1981; Solbreck, 1991), and overwinter survival (Pease et al., 1979; Bale, 1991). In addition, herbivores will experience the indirect effects of changes in food quality (Bryant and Reichardt, 1992). Thus global change will certainly alter the nature and strength of many plant–herbivore interactions and will sometimes yield dramatic consequences at the level of populations, communities, and ecosystems.

Considerable empirical evidence attests to the impact of herbivores on plant population dynamics (Louda, 1983; McAuliffe, 1986; Rice, 1987; Crawley, 1989; Swank and Oechel, 1991). Exclosure studies, apparently without exception, indicate that plant communities are very sensitive to vertebrate herbivory (Watt, 1960; Ross et al., 1970; Pease et al., 1979; Cargill and Jefferies, 1984). Even modest levels of herbivory can modify plant communities by altering competitive interactions (Cottam et al., 1986; Kakes, 1989) and by changing the trajectory of plant succession (Launchbaugh 1960; Brown et al., 1987; Brandner et al., 1990; De Steven, 1991). Herbivory may also affect ecosystem attributes such as nutrient flux, primary productivity, insolation, fire frequency and severity, and gas and water exchange (Miller et al., 1976; Grant et al., 1985; Bazely and Jefferies, 1986; McNaughton and Georgiadis, 1986; Dyer et al., 1991; Harrington, 1991; Hobbs et al., 1991; Veblen et al., 1991; Schimel, this volume).

There are empirical and theoretical bases for expecting global climate change to affect the physiology of plants and herbivores in ways that impact plant–herbivore interactions and ripple through ecosystems. For example, when a recent irruption of the autumnal moth (*Epirrita autumnata*, Geometridae) killed hundreds of square kilometers of birch forest in Fennoscandia (Tenow, 1972; Kallio and Lehtonen, 1975), the abrupt edge of the outbreak zone was apparently delimited by winter temperatures lethal to *Epirrita* (Tenow and Holmgren, 1987). Similarly, the population dynamics of *Epirrita* may be influenced by the effects of summer temperature (Niemelä, 1980) and nutrient availability (Tuomi et al., 1984) on birch tree resistance. Thus, changes in winter temperatures, summer temperatures, or nutrient availability could alter the probability and extent of future outbreaks in this system. Agricultural losses to the European corn borer (*Ostrinia nubilalis*, Noctuidae) are clearly related to climatic effects on the life history and population dynamics of the pest (Showers, 1981; Showers et al., 1989). In fact, temperature-driven models that relate insect phenology to stages of plant susceptibility have become a vital tool in the management of many agricultural and horticultural pests (Hatfield and Thomason, 1982; Herms, 1990; Taylor and Shields, 1990). Plant stress resulting from unusual weather patterns may underlie catastrophic herbivore outbreaks in ecosystems throughout the world (White, 1984; Mattson and Haack, 1987a; Larsson 1989).

The diversity of plausible mechanisms by which global change could alter plant–herbivore interactions [Mattson and Haack (1987a) suggested dozens relevant to drought alone] indicates a need for identifying where the effects are most likely to be dramatic. In this chapter, I summarize relevant models of plant and herbivore physiology and evaluate their predictions regarding the response of plants, herbivores, and plant–herbivore interactions to expected patterns of global change. This chapter provides a framework for integrating existing data and also serves as a vehicle for directing future research towards critical or inadequately understood mechanisms.

RESOURCE ACQUISITION AND ALLOCATION IN PLANTS

Plant growth and reproduction are dependent upon the acquisition and use of three classes of resources: carbon, mineral nutrients, and water. Plants vary tremendously in their allocation of these resources depending upon the genotype, ontogeny, and environment of the plant (Bloom et al., 1985). In general, plants may allocate photosynthates and mineral nutrients to reproduction (providing the most immediate fitness benefit), to growth (maximizing the potential for future reproduction), to defense (protecting resources already acquired), or to storage (Figure 1). In addition, plants expend substantial carbon in respiration (Ryan, 1991). The high phenotypic plasticity that characterizes plants is a joint function of environmental variation in resource availability and of phenotypic plasticity in patterns of resource acquisition and allocation. By changing resource availability and allocation, global change is likely to alter plant characteristics that are relevant to herbivores, to affect herbivore population dynamics, and to cause herbivore-mediated changes in plant communities.

MODELS OF PHENOTYPIC RESPONSE IN PLANTS

At least three theoretical models have been proposed to explain phenotypic responses in plants: (1) the carbon/nutrient balance hypothesis, (2) the growth/differentiation balance hypothesis, and (3) the optimal allocation hypothesis. Hypotheses based on carbon/nutrient balance and growth/dif-

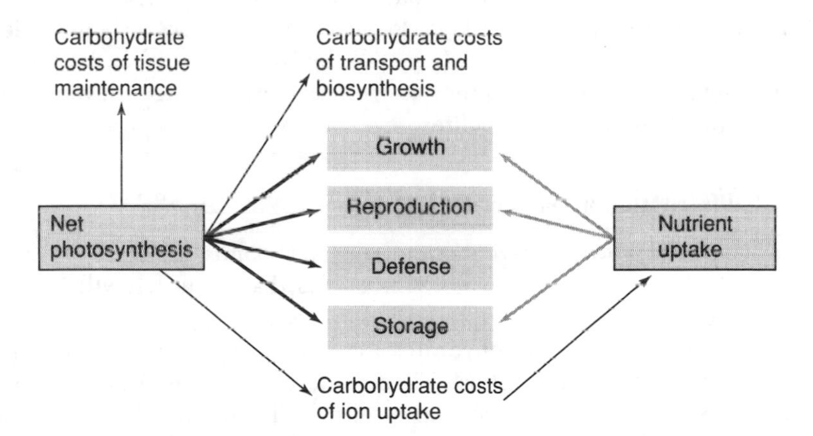

FIGURE 1. A model of resource allocation in plants. Carbohydrate products of photosynthesis that are not expended in respiration may be invested in growth, reproduction, defense, or storage. Plant growth and especially reproduction require substantial nitrogen and are therefore dependent on nutrient uptake, but many plant defenses are purely carbon based. Plant respiration includes the costs of tissue maintenance, ion uptake, and transport and biosynthesis.

ferentiation balance follow from considerations of plant form and function; they emphasize physiological constraints and predict phenotypic responses that may or may not be evolutionarily adaptive. Conversely, the principle of optimal allocation assumes that natural selection has resulted in plant responses that maximize growth (a fitness surrogate) and predicts phenotypic responses that may or may not be possible within the physiological milieu of the plant. Each of these models has well-established explanatory powers, and collectively they provide a solid foundation for predicting the effects of climate change on plant allocation patterns and resistance to herbivory. Where their domains overlap, predictions are frequently (but not always) congruent.

Carbon/nutrient balance

Plant growth requires both carbon and nutrients (see Figure 1). Nutrient limitations, which are ubiquitous among wild plants (Chapin, 1980), typically limit growth more than they do photosynthesis. Consequently, nutrient-limited plants tend to accumulate more carbohydrates than can be immediately invested in growth. If the ratio of carbon to nutrients increases, excess carbohydrates may be shunted into secondary metabolism and the production of carbon-based compounds like condensed tannins may increase (Bryant et al., 1983). Such secondary metabolites can deter herbivory and are often described as "defensive" (Rhoades and Cates, 1976; Haukioja, 1980; Mooney and Gulmon, 1982; Coley, 1986). However, under the carbon/nutrient balance hypothesis, secondary metabolites may alternatively be thought of as "storage" or "waste" products that only have incidental effects on herbivores (Tuomi et al., 1988). Predictions based on carbon/nutrient balance differ for nitrogen-based (e.g., alkaloids) versus carbon-based secondary metabolites. This chapter, as well as most of the relevant data, focuses on carbon-based metabolites.

Growth/differentiation balance

The principle of growth/differentiation balance (Loomis, 1932, 1953; Lorio, 1986, 1988; Herms and Mattson, 1992) asserts that tissue growth is negatively associated with tissue differentiation because both processes compete for the same pool of carbohydrates and because differentiation requires a mature intracellular architecture. Tissue differentiation in plants is defined to include processes that typically occur after cell expansion has been completed: lignification of cell walls, thickening of leaf cuticle, production of thorns and trichomes, and secondary metabolism leading to products such as tannins, terpenes, and alkaloids. Plants with a surplus of carbohydrates beyond that which can be invested in growth (because of other limitations such as water or nutrients) are said to be "sink-limited" (Wareing and

Patrick, 1975; Wardlaw, 1990; Luxmoore, 1991), and are predicted to invest proportionally more carbon in differentiation. Conversely, "source-limited" plants (whose growth is limited by carbon supply) are predicted to invest proportionally more carbon in growth. Plants with a high allocation to differentiation should have relatively low concentrations of water and protein, but relatively high concentrations of secondary metabolites (Herms and Mattson, 1992). Any kind of environmental stress that retards growth more than it does photosynthesis (e.g., nutrient stress or mild water stress) should shift allocation of carbohydrates towards differentiation, and this shift should generally be detrimental to herbivores. Predictions based on growth/differentiation balance and carbon/nutrient balance overlap broadly, but not completely (see "Changes in water balance," below). The growth/differentiation balance hypothesis is unique in specifying the importance of cellular development.

Optimal allocation

Because plants deal in so many resources and currencies, plant fitness is strongly influenced by strategies of allocation (Mooney, 1972; Bazzaz et al., 1987). Applying the tools of resource economics, Bloom et al. (1985) and Chapin et al. (1990) predicted optimal phenotypic responses of plants to various environmental scenarios. Optimally allocating plants should adjust to changes in resource availability in ways that stabilize exchange ratios (the relative amounts of two resources that can be acquired for a given expenditure of a single resource). A corollary is that plants should balance their allocation such that growth is equally limited by all three resource classes (i.e., nitrogen, carbon, and water). Empirically based models provide a powerful means of refining and testing these predictions (Hirose, 1988; Pons et al., 1989; Sinclair and Horie, 1989; Hilbert et al., 1991).

HERBIVORE NUTRITION AND ENERGETICS

The growth and reproduction of herbivores, like that of plants, depends on the acquisition of carbon (chiefly as digestible carbohydrates) and nutrients (e.g., protein nitrogen) (Figure 2). However, because animals require more nitrogen than plants do (tissue concentrations of 7–14 percent versus 0.5–4 percent), herbivore growth may commonly be limited by dietary nitrogen (Mattson, 1980; White, 1984). The upper limits of diet quality for both vertebrate and insect herbivores are thought to be a joint function of protein content and of digestible carbohydrates (Robbins, 1983; Slansky and Scriber, 1985). If either or both are low, herbivores can sustain only low productivity (growth, fattening, fecundity, and lactation). Moreover, plant secondary chemicals can constrain herbivore growth below these limits by deterring consumption (Bernays and Simpson, 1982), interfering with digestive effi-

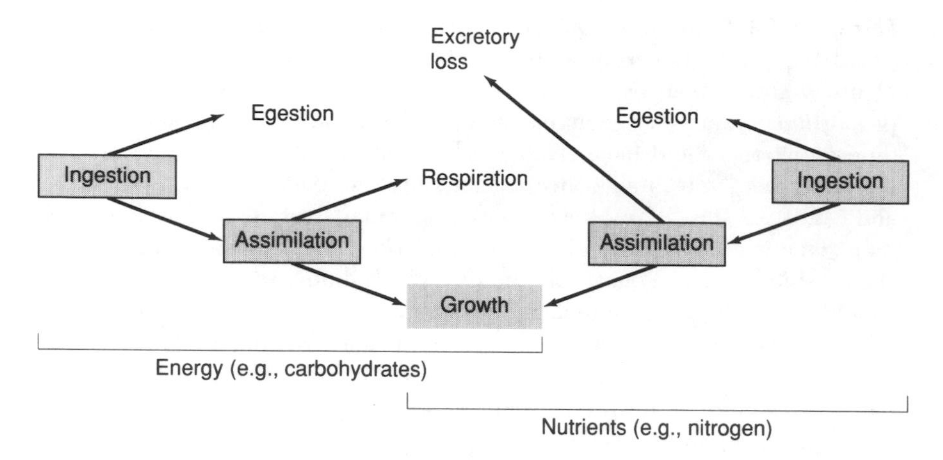

FIGURE 2. A model of herbivore energetics and nutrition. Herbivore growth is dependent upon the acquisition of both energy and nutrients. Of that which is consumed, some fraction of energy and nutrients is assimilated. Some fraction of assimilated energy is expended in respiration, and some fraction of assimilated nutrients is lost as excretory waste products (e.g., urea). The remainder supports herbivore growth. Plant secondary metabolites may deter consumption, interfere with assimilation, elevate respiratory expenses, or increase excretory losses.

ciency (Robbins et al., 1987), or elevating respiratory and/or excretory losses (Thomas et al., 1988).

In this chapter, I characterize changes in plant palatability based in part on tissue concentrations of nitrogen and secondary metabolites, but I emphasize that host "quality" is also a function of the particular herbivore. Different taxa of herbivores, even when fed the same diet, can differ dramatically in growth and survival (Lederhouse et al., 1992), usually because they are differentially sensitive to secondary metabolites. Other plant attributes, such as toughness, water content, fiber content, and micronutrients, are also of demonstrable relevance to herbivores, but these typically covary with nitrogen (Coley, 1983; Scriber, 1984) and are less often measured.

EFFECTS OF GLOBAL CHANGE ON HERBIVORY

Increased carbon dioxide

Under the carbon/nutrient balance hypothesis, an increased supply of carbon will increase the ratio of carbon to nutrients and should therefore lead to reduced tissue concentrations of nitrogen, but to increased tissue concentrations of carbon-based secondary metabolites. As predicted, leaf nitrogen concentration consistently declines about 18 percent under CO_2 enhancement. However, no study has detected significant changes in secondary metabolite concentrations (Table 1). For example, the concentration of cardenolides in *Digitalis lanata* changed by less than 1 percent (7510 versus 7450 nmol/g) when CO_2 was tripled, even though plant productivity was increased by 63 percent (Stuhlfauth et al., 1987). The number of studies measuring secondary metabolite concentrations under CO_2 manipulation is presently limited and is skewed toward herbaceous plants. Also, some of the measured compounds may be metabolic intermediates with such high turnover rates that static concentration is a poor indicator of defensive investment (Reichardt et al., 1991). Nonetheless, this stability of secondary metabolism seems counter to predictions based on carbon/nutrient balance.

The consistent depression of leaf nitrogen concentrations under elevated CO_2 levels (Table 1) could either be a nonadaptive perturbation or an optimizing allocation response that maximizes whole-plant carbon gain by producing more leaf area of lower nitrogen concentration. These alternatives could be tested by evaluating the joint response of photosynthesis to leaf nitrogen and atmospheric CO_2 concentration. Photosynthetic rate generally increases with leaf nitrogen content (Field and Mooney, 1986; Hirose, 1988), presumably because of increased concentrations of a rate-limiting enzyme (rubisco, RuBP carboxylase). Enhanced CO_2 immediately increases photosynthetic rates (average of 62 percent increase in 18 experiments with 9 tree species; Eamus and Jarvis, 1989; Bazzaz, 1990), although photosyn-

TABLE 1. Effects of enhanced CO_2 on plant tissue concentrations of nitrogen and secondary metabolites.

Percent change in tissue concentration[a]			
Nitrogen	Secondary metabolites[b]	Plant	Reference
−24*	−19	*Plantago*	Fajer et al., 1989
−17*	9	*Plantago*	Fajer, 1989
−26	—	Lima bean	Osbrink et al., 1987
−11*	—	Soybean	Lincoln et al., 1984
−7	—	Soybean	Lincoln et al., 1986
−25*	7	Peppermint	Lincoln and Cuovet, 1989
−18*	−20	Sagebrush	Johnson and Lincoln, 1990
−18*	4	Sagebrush	Johnson and Lincoln, 1991

[a]Effects are expressed as percent change relative to control.
[b]All secondary metabolites are carbon-based (i.e., contain no N).
*$P < 0.05$

thetic rates may return to near their original levels following plant acclimation (perhaps involving reallocation of nitrogen; Tissue and Oechel, 1987; Eamus and Jarvis, 1989). Whole-plant carbon gain per unit of nitrogen (μmol $CO_2 \cdot$mol $N^{-1} \cdot s^{-1}$) could be maximized at a lower concentration of leaf nitrogen in an atmosphere with elevated CO_2 concentration (Hilbert et al., 1991).

Depending on assumptions, the principle of optimal allocation could predict increases, decreases, or no change in concentrations of carbon-based antiherbivore defenses under enhanced CO_2. If the benefits of defense increase linearly with investment in secondary chemistry, and if the cost function declines under enhanced CO_2, then optimal allocation predicts that increased CO_2 will favor increased concentrations of defensive metabolites. However, if antiherbivore defenses are maximally effective at relatively low concentrations (the "qualitative" defense scenario; Rhoades and Cates, 1976), then the optimal investment in defense would probably not change under enhanced CO_2. Clearly, these predictions are so all-encompassing as to be useless without additional information.

Increasing atmospheric CO_2 should not have any direct physiological effects on herbivores (Fajer et al., 1991). However, CO_2-induced changes in host plants may sometimes be detrimental to herbivores. In 4 of 9 studies, insect herbivores raised on plants grown with enhanced CO_2 had reduced survival, reduced growth rate, increased development time, or reduced pupal mass (Table 2). Insects feeding on enhanced-CO_2 plants appeared to compensate for low nutritive quality (low nitrogen content) by increasing food consumption (average of 48 percent higher, significant in 6 of 6 studies). In 3 of 6 cases of increased consumption, growth rates did not change, indicating successful compensation. The effects of enhanced CO_2 on insect growth, survival, and pupal mass were usually less than those of some other sources of variation in host quality (e.g., long-term inducible resistance, leaf maturation, and among-tree variance; Table 2).

Soil nutrient availability

Global climate change should alter soil moisture, soil temperature, and litter quality in ways that affect decomposition processes and thus nutrient availability for plants (Anderson, 1991). Many scenarios are possible (Pastor and Post, 1988; Mooney et al., 1991), and patterns are likely to vary regionally, but the general effect may be to reduce mineral nitrogen availability because of increased C:N ratios in the litter (Bazzaz, 1990). Nitrogen fertilization typically leads to plant tissue with higher concentrations of nitrogen and lower concentrations of secondary metabolites (Table 3; note that the one evergreen—Scots pine—was an exception in that resin acids actually increased by 20 percent). Thus, plants that become increasingly nitrogen-limited due to climate-induced changes in mineralization rates are

TABLE 2. Effects on herbivores of growing plants in an enhanced CO_2 environment.[a]

Effect on herbivore (percent change)					
Survival	Relative growth rate	Relative consumption rate	Development time (T_{Dev})	Pupal mass (M_{Pupa})	Reference
			ENHANCED CO_2		
−15*	—	—	7*	3	Fajer et al., 1989
−2	−13*	34*	—	—	Fajer, 1989
—	—	—	0	−1	Akey et al., 1988
—	—	—	−1	—	Osbrink et al., 1987
—	−12	81*	—	—	Lincoln et al., 1984
—	−5*	20*	—	—	Lincoln et al., 1986
—	10	23*	—	—	Lincoln and Cuovet, 1989
—	26	74*	—	—	Johnson and Lincoln, 1990
—	−40*	53*	—	—	Johnson and Lincoln, 1991
			LONG-TERM INDUCIBLE RESISTANCE		
−25*	—	—	6*	−18*	Haukioja et al., 1985
—	−32*	—	—	—	Neuvonen and Haukioja, 1984
			LEAF MATURATION		
—	−40*	−18*	—	—	Ayres and MacLean, 1987a
			AMONG-TREE VARIANCE		
—	−32*	—	—	—	Ayres et al., 1987

[a]Effects (expressed as percent change relative to control) due to some other sources of intraspecific variation (long-term inducible resistance, leaf maturation, and among-tree variance) are included for comparison.
*$P < 0.05$

likely to decline in quality for herbivores. The phenotypic response of plants to altered nutrient availability may be substantial (average of 38 percent increase in tissue nitrogen and 26 percent decrease in secondary metabolites in fertilized plants; Table 3), and the secondary effects on herbivores may also be considerable. Compared to foliage of control trees, foliage of fertilized quaking aspen had lower concentrations of condensed tannin (−83 percent), tremuloiden (−31 percent), and salicin (−50 percent), but higher concentrations of nitrogen (+100 percent). Apparently as a result, the large aspen tortrix (Lepidoptera: Geometridae), a defoliator of aspen, produced 25 percent larger pupae on fertilized trees (Bryant et al., 1987b). Snowshoe

TABLE 3. Effects of fertilization on plant tissue concentrations of nitrogen and secondary metabolites.

Percent change in tissue concentration[a]			
Nitrogen	Secondary metabolites[b]	Tissue	Reference
56*	−24*	Leaf	Larsson et al., 1986
100*	−55*	Leaf	Bryant et al., 1987a
26*	−29*	Leaf	Johnson and Lincoln, 1991
26*	20*	Needle	Bjorkman et al., 1991
31*	−38*	Shoot	Bryant et al., 1987b
−11	−29*	Shoot	Bryant, 1987

[a]Effects are expressed as percent change relative to control.
[b]All secondary metabolites are carbon based.
$*P < 0.05$

hares were nearly perfect at recognizing and preferentially consuming shoots from fertilized birch trees (which contained 38 percent less papyriferic acid than control trees; Bryant et al., 1987a), and moose browsed preferentially on Scots pine growing in high-nutrient sites (Danell et al., 1991).

The typical effects of nutrient availability on plant tissue composition and palatability to herbivores conform well with all three theories. Under the carbon/nutrient balance hypothesis, fertilization yields increased nitrogen, therefore reduced C:N, higher tissue nitrogen, and lower secondary metabolism. Under the growth/differentiation balance hypothesis, fertilizing a nutrient-limited plant promotes growth, and the resulting growth-dominated plant contains higher nitrogen and lower secondary metabolites. Under the principle of optimal allocation, plants in a high-nutrient environment have less difficulty replacing nutrients lost to herbivory, absorb a greater opportunity cost if they do not grow maximally, and therefore invest less in antiherbivore defense (Bryant et al., 1983).

Changes in cloud cover

Reduced radiation, such as that resulting from increased cloud cover, tends to depress plant carbon budgets and to lower secondary metabolism (average of 52 percent reduction in tissue concentrations of secondary metabolites; Table 4). Shading appears to have a larger effect on secondary metabolism than does either fertilization or enhanced CO_2, but a lesser effect on tissue nitrogen concentration. Although studies are few, shading can clearly have

TABLE 4. Effects of shading on plant tissue concentrations of nitrogen and secondary metabolites.

Percent change in tissue concentration[a]			
Nitrogen	Secondary metabolites[b]	Tissue	Reference
−4	−70*	Leaf	Larsson et al., 1986
4	−44*	Leaf	Mole et al., 1988
0	−62*	Shoot	Bryant et al., 1987b
7	−33*	Shoot	Bryant, 1987

[a]Effects are expressed as percent change relative to control.
[b]All secondary metabolites are carbon based.
*$P < 0.05$

enormous effects on the palatability of plants to herbivores. For example, snowshoe hares consumed nearly all the birch shoots from shaded trees while consuming virtually none of the control shoots (Bryant et al., 1987a). Similarly, leaf beetles consumed five times as much foliage from willows grown in low light compared with control willows (Larsson et al., 1986).

Carbon/nutrient balance readily explains the reduced concentrations of secondary metabolites in shaded trees (reduced carbon yields reduced C:N, yields reduced allocation of carbon to secondary metabolism). However, reduced secondary metabolism can also be interpreted as a reduced commitment to differentiation processes in response to source limitations. Both the carbon/nutrient balance and the growth/differentiation balance hypotheses appear to predict increased concentrations of tissue nitrogen, which does not occur. Reduced allocation to secondary metabolism in response to shading may be optimal if carbon stress is an early symptom of being overgrown by competitors (maximal allocation to growth under these circumstances may increase the probability of surviving to reproduction, even though defense is compromised); but if the increased risk of catastrophic herbivory outweighs the competitive benefits, then the response is maladaptive.

Changes in cloud cover will affect some poikilothermic herbivores directly by altering their operative temperature (Rawlins and Lederhouse, 1981; Grossmueller and Lederhouse, 1985; Casey et al., 1988; Kukal et al., 1988; Weiss et al., 1988; Dunham, this volume). General circulation models (GCMs) do not appear to be as explicit or accurate with regard to changes in radiation as would be desirable (Schneider, this volume) given the likely biological effects.

Changes in water balance

Correlative and experimental evidence suggests that changes in plant water balance can alter palatability to herbivores and influence herbivore population dynamics (White, 1984). Mattson and Haack (1987a) highlighted 10 insect families in 5 orders "that historically have reached outbreak proportions following drought."

Moderate water stress tends to retard growth more than it does photosynthesis (Kozlowski, 1982; Allen et al., 1987; Wardlaw, 1990). Thus, carbohydrates accumulate, and plant allocation to differentiation processes (e.g., secondary metabolism) should increase. Under severe water stress, however, photosynthesis declines (Ehleringer and Cook, 1984; Gollan et al., 1985). Plants then become carbon-limited, and their secondary metabolite content should decline. This reasoning, which predicts a curvilinear response of secondary metabolism to water stress, has been supported by studies of tannin in Douglas fir (Horner, 1990) and of rubber in guayule (Reddy and Das, 1988). In many other studies, however, the concentration of secondary metabolites simply increases following drought stress (Table 5; 27 of 27 studies reviewed by Mattson and Haack, 1987b). Presumably plant species differ in the water deficit (MPa of leaf water potential) at which secondary metabolism is maximized. Many studies to date involve only two water treatments, which are obviously too few to describe a curvilinear function. Considerations of carbon/nutrient balance alone seem inadequate

TABLE 5. Effects of water stress on plant tissue concentrations of secondary metabolites.

Percent change in secondary metabolites[a]	Tissue	Leaf water potential (MPa)	Reference[b]
61*	Leaf	−0.05 to −0.6	Charles et al., 1990
−14	Leaf	−0.7 to −1.3	Stuhlfauth et al., 1987
7	Leaf	−0.35 to −0.65	Horner, 1990
−17*	Leaf	−0.65 to −1.00	Horner, 1990
71*	Stem	−0.5 to −2.5	Reddy and Das, 1988
−5	Stem	−2.5 to −3.0	Reddy and Das, 1988
163*	Stem	−1.8 to −3.0	Allen et al., 1987

[a]All secondary metabolites are carbon-based. Effects are expressed as percent change relative to control.

[b]In two studies where a nonlinear response was suggested, the regions of positive and negative slope were evaluated separately. Twenty-seven of 27 other studies reviewed by Mattson and Haack (1987b) showed an increase in secondary metabolites following drought stress.

*$P < 0.05$

to predict changes in secondary metabolism based on changes in water balance. Water stress is thought to retard photosynthesis more than it does nutrient uptake (Bloom et al., 1985; Mattson and Haack, 1987a), which should reduce the C:N ratio and lead to reduced secondary metabolism. This prediction of the carbon/nutrient balance hypothesis seems inconsistent with the data.

The consequences for herbivores of drought stress in their host plants can be beneficial (bark beetles attacking white fir; Ferrell, 1978), negligible (gypsy moths feeding on birch trees; Herms, 1991), mildly detrimental (sawflies feeding on ponderosa pine; McCullough and Wagner, 1987), or extremely detrimental (six times higher neonatal mortality in shoot-galling sawfly larvae attacking willows; Preszler and Price, 1988). Water stress made 12 of 41 plant species more palatable to locusts, but 5 of 41 less palatable (Bernays and Lewis, 1986). Spider mite populations were highest on either well-watered or severely stressed bean plants, but low on moderately stressed plants (English-Loeb, 1990); this pattern is predicted if secondary metabolism rises, then falls, with increasing water stress. The overall diversity of effects is likely due to differences among plants in their response to drought stress as well as to differences among herbivores in their sensitivity to plant responses (Larsson, 1989). There is no substitute for more empirical studies with ecologically relevant plants, herbivores, and treatments.

Bark beetles, which have great potential for altering ecosystems (Rykiel et al., 1988; Veblen et al., 1991), may be a critical agent of change in forests subjected to drought stress (Balch and Prebble, 1940; Christiansen et al., 1987; Lorio, 1986). Seasonal changes in the resistance of pine to southern pine beetles may be explained in terms of water deficits affecting plant allocation to growth versus differentiation processes (Lorio et al., 1990). During early summer, tree growth is rapid, and the production of oleoresin defenses in attacked trees is relatively low (Lorio and Sommers, 1986). This is the time of greatest population growth in the bark beetles. With the onset of moderate water deficits in midsummer, tree growth declines, oleoresin production increases, and resistance to bark beetles also increases. This change corresponds to the ontogenetic transition from the production of earlywood to the production of latewood (Lorio, 1988). [Latewood is characterized by highly differentiated cells (forming vertical resin ducts) that function in the production and transport of oleoresin.] Under this scenario, reduced precipitation would shorten the seasonal period of rapid tree growth when vulnerability to bark beetles is greatest. However, extreme water stress, which limits photosynthesis as well as growth, should reverse the effect by limiting oleoresin production and increasing vulnerability to beetle attack. Within a region, reduced precipitation could thus lead to increased beetle damage in dry sites, but reduced beetle damage in mesic sites.

High commitments of drought-stressed trees to reproduction may exacerbate their vulnerability to wood-boring beetles. Birch trees defend themselves against the bronze birch borer (Coleoptera: Buprestidae) by rapidly growing a callus around beetle larvae feeding in the cambium (Herms, 1991). The larvae die if encapsulated by the callus, but the tree dies if more than a few larvae escape their callus. Growth of the birch callus is driven by carbohydrates transported from the leaf canopy. The photosynthetic rate of ten-year-old birch trees was reduced under a low irrigation treatment; consequently, these drought-stressed trees exhibited slower callus growth and sustained higher mortality due to birch borers (Herms, 1991). This effect was accentuated because the slow-growing trees invested heavily in female reproduction, which must have reduced the carbon available to support callus growth.

Increased temperature

General circulation models consistently predict increases of about 2–4°C in average air temperatures, with larger increases at high latitudes (Mitchell et al., 1989; Schneider, this volume). However, surprisingly little is known about the effects of such temperature changes on plant allocation patterns or resistance to herbivory. In warm years compared with cold years in Fennoscandia (thermal sums above a 5°C base = 540 to 300 degree-days from 1979 to 1984), mature foliage of *Vaccinium myrtillus* tended to have lower leaf nitrogen (1.4 percent versus 2.4 percent dry mass) and higher Folin-Denis phenolics (20 percent versus 10 percent) (Laine and Henttonen, 1987). Greenhouses that elevated average daily summer temperatures by about 3°C produced a comparable effect (13 percent reduction in leaf nitrogen compared with control plots, and a 38 percent increase in phenolics). Figure 3 suggests a preliminary model to explain these patterns in terms of temperature effects on net photosynthesis, maintenance respiration, and growth rate. The model assumes that the summer climate in Finnish Lapland includes low temperatures at which *Vaccinium* can photosynthesize but not grow. The model predicts that populations of the same plant in different regions could respond differently to the same temperature elevation, with southern populations becoming carbon-stressed but with northern populations accruing a carbon surplus. Based on results from other carbon/nutrient manipulations, I would expect increased herbivory and improved herbivore performance on plants with a temperature-induced carbon deficiency.

Many herbivores have evolved life histories in which their maximum nutritional demand coincides with the season when developing (as opposed to mature) plant tissue is maximally available. These herbivores include ungulates (Klein, 1965; Cooper et al., 1988), primates (Milton, 1979), rodents (Batzli et al., 1980), geese (Sedinger and Raveling, 1986), gypsy moths (Hough and Pimentel, 1978; Raupp et al., 1988), spruce budworm (Law-

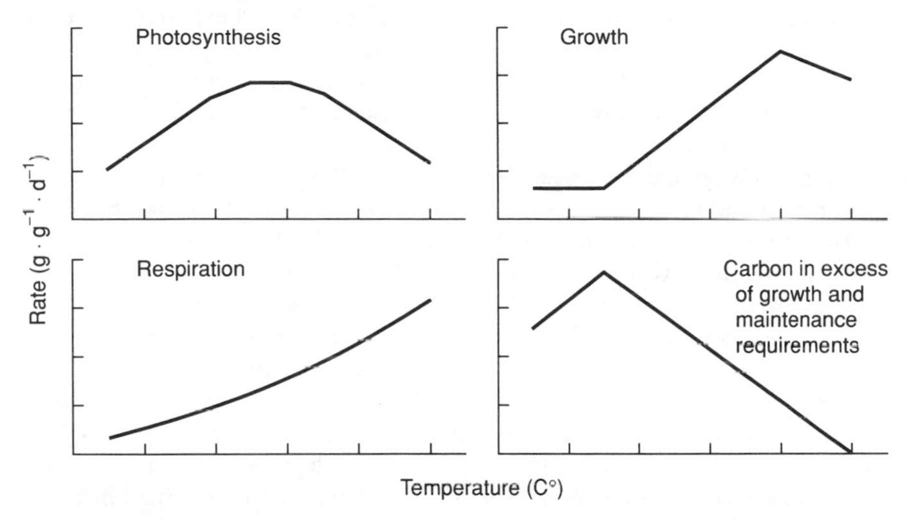

FIGURE 3. Generalized response of plant carbon balance to temperature. Photosynthesis tends to have a broad temperature optimum, while respiration rate increases across the full range of temperatures (Berry and Bjorkman, 1980; Sveinbjornsson, 1983; Patterson and Graham, 1987). There exists a range of low temperatures at which photosynthesis proceeds but growth does not (Pollack et al., 1983; Pollack, 1990). Beyond this lower threshold, plant growth rate tends to increase with temperature (Scott, 1970; Watts, 1974; Parsons and Robson, 1980). Consequently, carbon available for secondary metabolism (i.e., carbon in excess of growth and maintenance requirements) is predicted to increase, then decrease, with increasing temperature.

rence, 1990), and leafhoppers (Townsend, 1989). About half of forest insect pest species (27 of 54 in Martineau, 1984) appear to be early-season specialists on immature plant tissue. Developing plant tissue is typically low in fiber (thus highly digestible), high in nitrogen, and low in secondary metabolites (Mattson, 1980; Coley, 1983). Maturation of this plant tissue imposes phenological barriers to herbivores (Feeny, 1970; Schweitzer, 1979; Schroeder, 1986; Turgeon, 1986); hence herbivores of immature tissue engage in a developmental race with their host plants (MacLean, 1983). Larvae of *Epirrita autumnata* (Lepidoptera: Geometridae) grew very quickly (doubling their mass every 1 to 2 days) on young birch leaves, but insect growth rate declined dramatically (−32 percent) over a period of just 3 to 6 days following the completion of leaf expansion (Ayres and MacLean, 1987a). Global warming should reduce the time during which high-quality immature foliage is available because leaf development rate increases with temperature (Scott, 1970; Ong and Baker, 1985; Watts, 1974; Pollack et al., 1983; Graves et al., 1989; Pollack 1990). This seems unavoidably detrimental to homeothermic herbivores, and may result, for example, in chronically lower breeding success in geese and caribou (Sedinger and Flint, 1991; Kuropat and Bryant, 1979). However, insect herbivores, like the plants on which they feed, tend to grow faster at higher temperatures (Figure 4; Taylor, 1981; Scriber and Lederhouse, 1983; Ayres and MacLean, 1987b), so predictions are less clear.

Increasing global temperatures can affect the developmental race between poikilothermic herbivores and their host tissue if the temperature sensitivity of the herbivore differs from that of the host (MacLean, 1983). We tested this in the *Epirrita*–mountain birch system by building greenhouses around in situ trees (S. F. MacLean and M. P. Ayres, in preparation). Leaf maturation, and a concomitant decline in food quality for *Epirrita*, occurred more rapidly in greenhouse than in control trees (25 percent lower larval growth rates 3 weeks after budburst). However, *Epirrita* also increased development rates in response to the temperature elevation; larvae grown on greenhouse trees pupated 3–6 days earlier than did those on control trees (8–15 percent shorter development time). Furthermore, greenhouse larvae had better survival rates and produced larger pupae

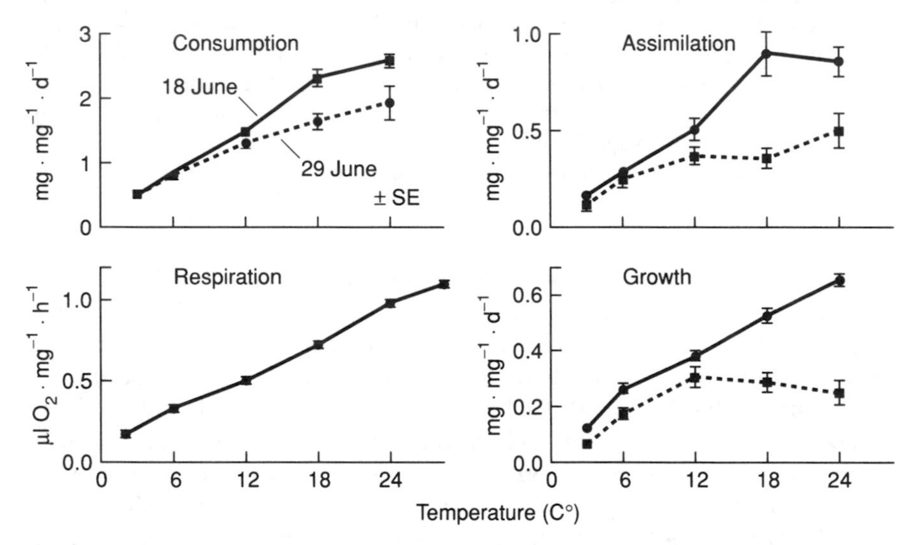

FIGURE 4. The interacting effects of temperature and host quality on the growth of a birch herbivore, *Epirrita autumnata.* On young expanding leaves (18 June), larval consumption rate, assimilation rate, respiration rate, and growth rate all increased with increasing temperature from 3°C to 24°C. Following the completion of leaf expansion (29 June), assimilation rate declined due to reduced digestive efficiency and reduced consumption rate. Consequently, herbivore growth rate declined markedly, especially at high temperatures. On mature leaves, respiration rate increased as much as assimilation rate over the range from 12°C to 24°C, and the difference between them, growth rate, did not change (S. F. MacLean and M. P. Ayres, unpublished data).

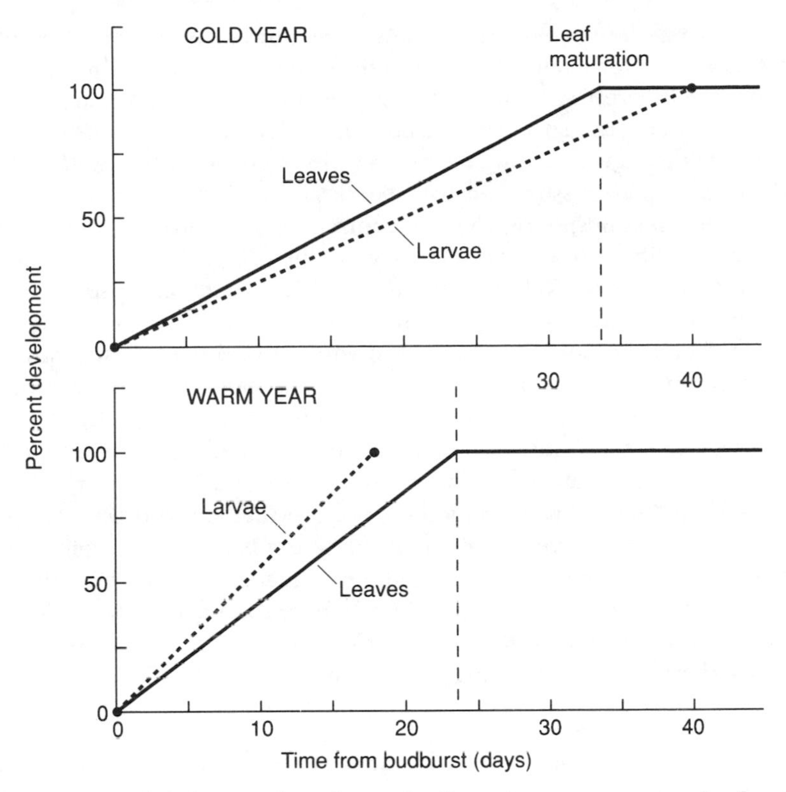

FIGURE 5. Model showing hypothesized effect of temperature on the developmental race between maturing leaves and insect herbivores that consume them. Both leaf maturation and larval development are temperature-sensitive processes, but larval development is accelerated relatively more in a warm year. Leaf quality for the herbivores declines dramatically at leaf maturation, so warm years favor the insects by allowing them to complete development before the leaves do.

(yielding more fecund adults) than control larvae did. As a result, the rate of population increase of moths in the greenhouse treatment increased up to 2.9 times, suggesting that temperature increases in this system will favor the insect. Remarkably, these effects were the result of only a 1°C increase in daily mean temperature inside the greenhouses. If insects are generally more temperature-sensitive than their host plants, global warming of 2–4°C may lead (through the mechanism suggested in Figure 5) to outbreaks of many insect herbivores.

SUMMARY

Climate change could affect plant–herbivore interactions in many ways. The challenge is to identify the mechanisms that will most directly transduce climate change into ecological impacts. This process is expedited by physiological models of plant allocation and herbivore nutrition, but the most appropriate theoretical framework seems to vary depending upon the herbivore guild and the specific environmental change (Larsson 1989). I find it useful to distinguish among herbivores of mature plant tissue, herbivores of developing plant tissue, and bark beetles.

Changes in atmospheric CO_2 concentration, soil nutrients, cloud cover, water availability, and temperature all affect the composition and palatability of mature plant tissue (Table 6). Increased CO_2, reductions in soil nitrogen, and sometimes increased temperature, tend to reduce tissue nitrogen; these changes should be detrimental to herbivores. Reductions in soil nitrogen, reductions in cloud cover, and sometimes increases in temperature and drought stress tend to increase concentrations of plant secondary metabolites; these changes should also be detrimental to herbivores. However, we cannot generalize that global change will be detrimental to herbivores of mature plant tissue. Such predictions require that we forecast changes in photosynthetically active radiation, plant water balance, and soil nutrient availability. Yet changes in each will presumably vary in direction and magnitude from region to region. Both temperature and CO_2 are likely to increase on a global scale (Schneider, this volume), but the consequences for plant–herbivore interactions of doubling CO_2 could easily be reversed

TABLE 6. Effects on plant tissue composition of five environmental factors associated with global change.

Environmental change (≈ amount)	Percent change in concentration[a]		Tissue
	Nitrogen	Secondary metabolites	
+ CO_2 (2×)	−18*	−4	Leaf
+ Soil nitrogen (50 g/m^2)	61*	−36*	Leaf
	26*	20*	Needle
	10	−34*	Shoot
+ Shade (50%)	0	−57*	Leaf
	3	−48*	Shoot
− Water potential (1 MPa)	—	9*	Leaf
	—	76*	Stem
+ Temperature (3°C)	−13*	38*	Leaf

[a]Values for CO_2, soil nitrogen, shade, and water potential indicate average percent change from Tables 1, 3, 4, and 5, respectively; temperature effect is from Laine and Henttonen (1987). Effects are expressed as percent change relative to control.

*Average includes at least one significant effect ($P < 0.05$).

by changes in one or more other factors (the effects of CO_2 on host plant quality may be the least among five aspects of global change; Table 6). We expect nonlinear effects of temperature and water potential on plant defense. Thus, meaningful predictions require that we know the initial state of the plant, the direction and magnitude of the environmental change, and the precise form of the plant's physiological responses. This knowledge is lacking for virtually all plants. Moreover, we need a better understanding of the ways by which various environmental perturbations interact. Can the effect of simultaneously increasing temperature and reducing precipitation be predicted from the additive effects of each?

Bark beetles deserve special recognition because of their extraordinary economic and ecological impact, and because their outbreaks have been so often linked to climatic anomalies. Lorio et al. (1990) are accumulating support for a model of tree defense that relies on water balance to predict the timing of an ontogenetic switch from poorly defended earlywood to well-defended latewood.

Numerous herbivores, including perhaps half of the economically important outbreak insects, specialize on developing plant tissue that is of high quality but only ephemerally available. For these herbivores, a phenological mismatch of a few days, perhaps caused by differential effects of temperature on plants and herbivores, can halve fecundity (Ayres and MacLean, 1987a; Keese and Wood, 1991) and thus result in huge impacts on plant–herbivore communities. Plants can reduce damage through slight phenological displacement from the herbivore (Eidt and Little, 1970; Aide, 1988; Crawley and Akhteruzzaman, 1988; Townsend, 1989; Tuomi et al., 1989). The availability of high-quality food (or, from the plant's perspective, the period of high vulnerability) is strongly temperature sensitive. Warmer temperatures will reduce the temporal availability of food, but will also increase the rate of consumption and development in poikilothermic herbivores. Unless the temperature sensitivity of plant development and insect development are identical, changes in temperature will favor either the plant or the insect (see Figure 5). Virtually nothing is known about the relative temperature sensitivity of insects and their host plants, but greenhouse experiments (above) in Finnish Lapland indicate that an increase of 1°C can potentially triple population growth of the herbivore.

Physiological models of plant allocation and herbivore nutrition already allow general predictions regarding the effects of climate change. The theory of carbon/nutrient balance provides a parsimonious explanation for many phenotypic responses to alterations of CO_2, light, and nutrient availability. Its failures suggest the role of natural selection in adapting plant responses to the ecological challenges of particular environments (optimal allocation hypothesis). The theory of growth/differentiation balance explains the general importance of immature plant tissue for herbivore nutrition and provides a robust framework for evaluating the effects of water balance on plant resistance. I anticipate that changes in temperature and water availability will have the largest impact on plant–herbivore interactions. These are the same areas where our theoretical and empirical understanding seems weakest. Nonlinear responses to temperature and moisture imply that changes in climatic variability (even without a change in the mean) can alter plant–herbivore interactions. Assessing the ecological and economic risks that accompany global change demands experiment-based research that builds on existing physiological models to predict responses of populations, communities, and ecosystems. Progress will be expedited through the selection of experimental systems that involve naturally interacting plants and herbivores.

ACKNOWLEDGMENTS

I thank B. D. Ayres, J. P. Bryant, F. S. Chapin, C. Field, M. A. Geber, D. A. Herms, K. S. Johnson, R. C. Lederhouse, P. L. Lorio, S. F. MacLean, W. J. Mattson, P. Niemelä, and J. M. Scriber for helpful discussions and comments on the manuscript. D. A. Herms generously shared his literature collection. Financial support was provided by USDA Grant 90-37153-5263, NSF Grant BSR 88-01184, and a Fulbright research fellowship.

CHAPTER 7

POPULATION RESPONSES TO ENVIRONMENTAL CHANGE: Operative Environments, Physiologically Structured Models, and Population Dynamics

Arthur E. Dunham

Major and rapid change in global climates is expected to occur during the coming century (Schneider, this volume). Environmental change of the magnitude and rate predicted by current models will doubtless have profound ecological effects. However, the nature of those effects is still poorly understood. Consequently, ecologists now face a challenge to develop testable models that accurately predict how changing and even novel environments will affect populations, local diversity, and communities.

One promising approach for modeling the ecological consequences of environmental change involves developing a mechanistic theory that links the dynamics of environmental variation, organismal physiological ecology, and populations. Such an approach is built on the undeniable premise that environmental factors (e.g., microclimates) influence key population-level rates (birth, mortality, and migration rates), initially by affecting the physiological ecology of individual organisms, and then by affecting the growth and reproduction of those individuals. Consequently, developing a predictive theory of ecological responses to environmental change requires an understanding of the dynamics of the mechanisms that link the biophysical and physiological ecology of individuals with population-level rates and processes.

This mechanistic approach begins by developing a biophysical model of how variation in a given environmental factor such as air temperature influences the physiology (e.g., rates of exchange of mass and energy), activities, and in turn the life history phenotypes (e.g., growth rate, age of reproduction, clutch size, age of death) of individuals. The next step involves integrating these patterns across individuals and thus modeling how variation in the environment influences population-level patterns (e.g., birth and death rates). Finally, by incorporating information on feedback dynamics (e.g., density dependence, interspecific interactions; see Ives and Gilchrist, this volume; Murdoch, this volume), a predictive model will be completed. Such a comprehensive, mechanistic approach holds considerable promise not only for predicting the responses of populations to many forms of environmental change—not just climate change—but also for understanding why particular responses occur.

Many of the tools necessary to predict the ecological consequences of climate change are already available. For example, given data on local climates, biophysical ecologists already can predict—quite accurately—the effects of a increase in air temperature on body temperature and on potential activity times of many ectotherms. Similarly, physiological ecologists are beginning to model how such changes in body temperature and activity times will in turn alter an individual's energetics, growth, and reproduction; and population biologists can now interpret such "individual" data in the context of individual-based, structured-population models (below). Thus we now have in essence all the fundamental analytical tools required to predict how environmental change will influence population dynamics.

This biophysical, mechanistic, structured-population approach, though powerful, is both complex and data-voracious. Accordingly, it may not always be practical on a large scale (but see below and Murdoch, this volume). Nevertheless, it has distinct advantages over two alternative and simpler modeling approaches. *Compartment models* typically characterize the responses of communities or ecosystems to environmental change through broad generalizations about the responses of constituent populations (e.g., Pacala and Hurtt, this volume). In contrast, *statistical models* "explain" system dynamics by empirically analyzing past environmental and population data (e.g., key factor analysis; see Ives and Gilchrist, this volume). Both types of models, though relatively simple to develop and apply, have serious limitations. For example, both assume that populations are functionally homogeneous. In other words, they assume that all individuals in a population (or in a given developmental stage; see Ives and Gilchrist, this volume) respond identically to environmental change. However, individuals even of the same age can differ dramatically in physiological state and size (Bennett, 1987; Huey et al., 1990), and they can also differ dramatically in their life histories (e.g., age of maturity, reproductive success). Consequently, individuals are unlikely to respond identically, and thus any model that assumes individual homogeneity may thus generate a misleading portrait of population dynamics. In addition, statistical models are descriptive, not mechanistic, and thus provide only limited insight into the mechanisms underlying the patterns of interest. They are also unsuited for predicting responses to novel environmental changes (for which per force no response data are available), for they could do so only by extrapolating beyond the range of available data, thus violating a primary rule in statistical modeling and rendering the prediction logically indefensible.

This chapter provides an heuristic basis for understanding how environmental variation can be transduced into population-level rates, characteristics, and processes. First I briefly describe some ways in which biophysical and physiological models can be used to describe how environmental variation influences the physiological ecology and population dynamics of individual organisms. I then describe several promising and related models for estimating population-level rates from such individual-level data. Finally, I provide a simple mechanistic model of how climate warming might influence the population dynamics of a lizard (*Sceloporus merriami*). This example illustrates not only how an individual-based approach can predict the population effects of climate change, but also some of the problems and limitations currently associated with this approach.

"OPERATIVE ENVIRONMENTS" AND BIOPHYSICAL MODELS

Environmental factors influence population dynamics only by influencing population rates of birth, mortality, immigration, and emigration. Consequently, any attempt to develop a mechanistic model of the consequences of environmental change should begin by identifying and modeling the factors ("operative" environments, cf. Spomer, 1973; generalized by Dunham et al., 1989a) that influence one or more of these population-level rates. A variety of environmental factors—resource, biophysical, social and demographic, and exploitative (predation, parasitism, and disease)—are known to influence population levels. For present purposes, however, I will focus only on microclimate factors.

Much of contemporary biophysical and physiological ecology is involved with measuring and modeling how variation in two operative environments, the biophysical environment and energy resources, influences organismal physiological properties. A key goal of these fields is to produce mechanistic models that (1) explain processes such as heat-transfer dynamics and mass flow (e.g., water economy; Porter et al., 1973) and (2) predict the proximate physiological (Muth, 1980; Tracy and Christian, 1986; Huey, 1991) and ecological consequences (Porter et al., 1973; Bakken and Gates, 1975; Christian and Tracy, 1981; Grant and Dunham, 1988, 1990; Kingsolver, 1983, 1989) of those dynamics.

These fields have been remarkably successful at meeting these goals. Indeed, by integrating small-scale microclimate data (e.g., direct solar radiation, thermal radiation, air temperature) and organismal data (e.g., size, shape, reflectivity), biophysical models can readily and accurately predict the equilibrium body temperature of an ectotherm (or the heat loads on an endotherm) anywhere in its environment. (Porter et al., 1973; Bakken, in press; see below). Consequently, once climate modelers develop the tools to predict how climates will change on a local scale (see below and Schneider, this volume), then biophysical ecologists will be able to predict the proximate impact of those changes on organisms.

Biophysical ecology in general, and biophysical modeling in particular, have also furthered our understanding of the rate of mass and energy exchange between organisms and their environments (Gates, 1980). We now realize that a change in microclimates can have complex and interacting physiological effects on organisms. For example, if ambient air temperature increases because of climate warming, ectotherms might experience higher body temperature, which would increase their metabolism and their need for food, water, and oxygen as well as altering their "discretionary" energy available for growth, storage, and reproduction. Endotherms might balance any increased convective heat gain by increasing evaporative cooling, which would of course affect their water balance and perhaps mass balance.

Because mass and energy must be conserved, these multiple interactions, though complex, can be conveniently conceptualized in a "coupled-equation" model (Figure 1). Along the main diagonal are the biophysical equations balancing heat gain and heat loss. Crossing the diagonal are mass-

Heat in (Radiation + Convection + Conduction)
+
Mass in = Metabolism + Mass out + Mass stored
\\
Heat out (Radiation + Convection + Conduction)
+
Water in = Evaporation + Water out + Water stored
+
Heat stored

FIGURE 1. A simplified version of the coupled heat balance and mass balance equations relating variation in operative environments to thermoregulatory biology and net allocatable resources (Modified from Dunham et al., 1989b, after Porter and Tracy, 1983.)

balance equations that balance intake of food (and water) with "loss" from metabolism (evaporation), excretion, growth, reproduction, and storage.

Biophysical constraints and trade-offs

The coupled equations in Figure 1 enable one to predict how a specified climate change will influence the mass and energy balance of individuals. However, before one can use that "individual" information to predict how climate change will influence population dynamics, one must first develop a detailed understanding of how operative environments constrain the options of individual organisms as well as how they affect physiological trade-offs. The nature and magnitude of environmental constraints depend on the particular operative environments experienced by an individual, on the size, structure, and physiological state of the individual, and on the physiological constraints imposed by biochemical pathways and related processes. Trade-offs result from the requirement that individual heat, mass, and time budgets must balance. These constraints and trade-offs are discussed below.

The nature of the constraints imposed by the physical environment, and their effects on organismal fitness, have been addressed in many studies. Biophysically imposed constraints have been shown to affect time available for activity (e.g., Porter et al., 1973; Chappell and Bartholomew, 1981; Grant and Dunham, 1988, 1990), microhabitat availability and habitat selection (Christian et al., 1983; Grant and Dunham, 1988, 1990; Huey, 1991), opportunities for mating (Kingsolver 1983, 1989; Kingsolver and Watt, 1984), and risk of predation (Christian and Tracy, 1981)

The coupled equations in Figure 1 also demonstrate that one must understand how an environmentally induced change in an individual's energy budget will affect its partitioning of available energy into maintenance, activity, synthesis, etc., or how an induced change in its potential activity time will affect its allocation of time to foraging, social interactions, rest, etc. Trade-offs inevitably occur here because assimilated energy (or nutrients) and potential activity time must be allocated among functions or activities that are often mutually exclusive.

Despite the power of biophysical approaches, one caveat is required. Any application of these biophysical models to predicting the responses of a particular system to climate change currently faces a serious limitation: Current climate models (i.e., general circulation models) are large in scale and provide credible information only on a scale of about 500 × 500 km (Schneider et al., 1990; Schneider, this volume). This scale is obviously much larger than that on which most ecological processes occur (Kareiva and Andersen, 1988). Consequently, before we can realize our goal of forecasting ecological consequences of climatic change in any particular system, considerable attention must be given to the development of methods for translating large-scale climate projections onto a scale appropriate for organisms.

INDIVIDUAL-BASED POPULATION MODELS

Individual-based models explicitly view a population as an assemblage of potentially interacting individual organisms, and they attempt to predict population dynamics only after the dynamics of individuals are specified. In such models the survival, growth, reproduction, and, in some cases, behavior of individual organisms (or sets of individuals) that make up each population of interest are followed simultaneously. Individual-based population models include: (1) physiologically structured models such as those described by De Roos et al. (1992), Gurney et al. (1983), Nisbet et al. (1989), Oster (1977), Sinko and Streiffer (1967, 1969), and numerous models cited in Metz and Diekmann (1986); and (2) individual allocation models of the type described by Adams and DeAngelis (1987), DeAngelis et al. (1991), Dunham et al. (1989a), Huston et al. (1988), Kingsolver (1983), and Ayres (this volume). Physiologically structured models of the first type are based on systems of differential equations and generally follow sets of individuals (e.g., size or age classes) rather than following each individual per se. In such models all members of a given set of individuals are assumed to have identical values for each state variable. In contrast, individual allocation models attempt to specify the internal dynamics of the state variables for each individual separately. This second class of models is potentially more intensive computationally, especially if the number of individuals in the population is large. Both types of models base the specification of the set of state variables on the physiological characteristics of the individuals in the populations being modeled (see below).

For several reasons, the individual organism is the appropriate level for modeling the transduction of environmental variation into population-level rates of reproduction, mortality, and migration. First, each individual is ecologically unique, not only because of inheritance but also because each individual has had a different sequence of experiences and environments during its life. Although this assertion may appear trivial, its ecological consequences are significant. Indeed, one consequence is that the "life history rules" (Dunham et al., 1989a, see below) by which organisms allocate energy (to growth, maintenance, storage, and reproduction; see Figure 3) will be both individual- and context-specific. In turn, then, the dynamics of the population, as well as the rates of energy and mass transfer to other levels of ecological organization, must be sensitive to individual variation. Second, because environments (like individuals) are heterogeneous, small-scale environmental heterogeneity can affect individual performance and thus have a major impact on population dynamics. An individual-level approach lends itself to an explicit consideration of such environmental variability. Third, it follows from the above two points that not all individuals in a population contribute equally to the key population-level rates (e.g., birth, death). Indeed, only a few individuals in a population often dominate recruitment, so the "average" individual may make (in comparison) a minor contribution to population-level rates and processes.

Modeling the individual organism

All of the modeling approaches discussed here require information on several characteristics of individual organisms [for example, sex, age, size, and physiological state (e.g., energy reserves, water content, nutrient pools available for biosynthesis)]. These characteristics make up the "state space" of the individual. Examples of such individual specification include Adams and DeAngelis (1987), DeAngelis et al. (1991), and Roff (1983) for fish populations; Dunham et al. (1989a) for lizards; Oster and his colleagues (Oster, 1977) for sheep blowflies; and Gurney et al. (1990), and Nisbet et al. (1989) for *Daphnia* (see also Murdoch, this volume).

A critical consideration in the development of individual-based and physiologically structured models is adequate specification of the set of variables necessary to characterize an individual. The particular set of required variables is likely to vary with the organism in question. In general, however, that set should include any mechanism whereby environmental variation has a significant impact on the growth, reproduction, death, or migration of individuals. Unfortunately, this class of models is generally analytically intractable, and numerical solutions must be obtained. Frequently the computation time necessary for such algorithms increases exponentially as a function of the number of variables used to characterize individuals. Discovering the minimum number of such variables that retain the important dynamics for a given system is an important practical, as well as philosophical, question (see below and Murdoch, this volume).

Dunham et al. (1989a) provide a general model for terrestrial ectotherms that conceptually transduces operative environmental variation into population-level rates, characteristics, and processes. They begin by defining a "life history" as the result of interactions whereby variation in the operative environments of individuals is transduced into population-level estimates of age- or size-specific birth rates and probabilities of mortality and migration (Figure 2; Dunham et al., 1989a). To understand such transductions, one must understand three fundamental sets of allocation decisions that individuals make. First, how do individuals allocate "available time" to activities such as mate and resource acquisition? This allocation may, of course, be a prime determinant of social status and of net resources available for future allocation. Second, how do individuals allocate assimilated resources (water, energy, amino acids, trace elements, etc.) into the competing functions of growth, maintenance, activity, storage, and reproduction (Figure 3; see

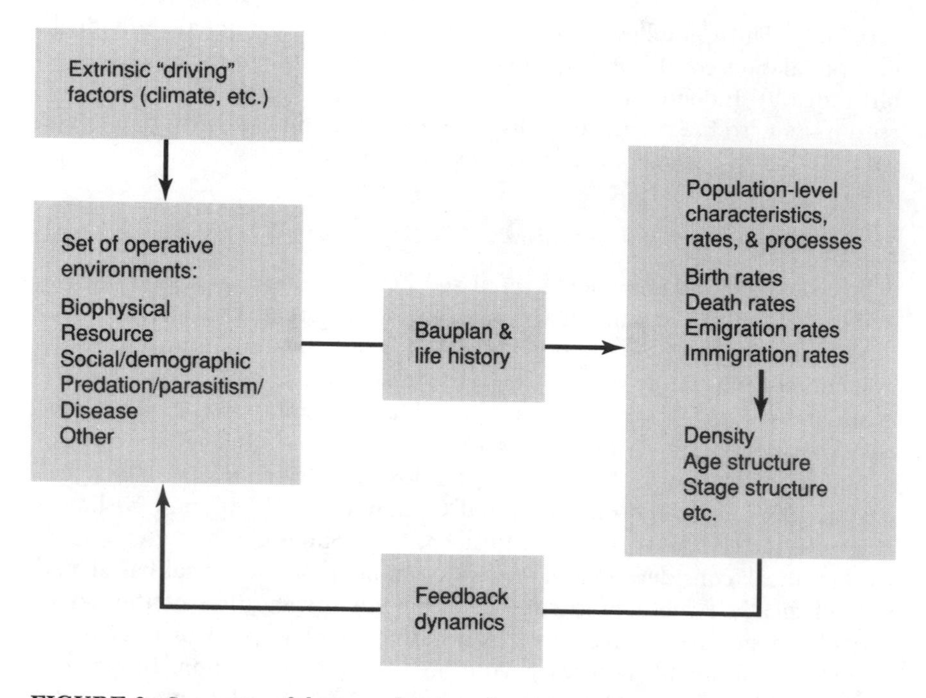

FIGURE 2. Summary of the transduction of operative environmental variation into population-level rates, characteristics, and processes. This transduction is the result both of the sequential set of allocation "decisions" characterizing each individual and of the phenotypic consequences of those decisions. The relevant allocation "decisions" and their attendant conditional risks of mortality are described in the text and illustrated in Figure 3. Birth, death, and migration rates result from averaging over all individuals and determine population density. Population density may influence variation in any operative environment by a number of different mechanisms (e.g., direct density-dependent effects on resource availability and foraging success).

FIGURE 3. Summary of factors influencing daily activity budgets of ectotherms. ➤ Four operative environment types that potentially influence time available for allocation are indicated. This allocation results in a daily time–activity budget and incurs a risk of mortality. Foraging success and digestive physiology determine net assimilated resources available for allocation to the competing functions of growth (G), maintenance (M), storage (S), and reproduction (R). This allocation determines the seasonal energy–mass budget, incurs a risk of mortality, and determines individual growth rates and reproductive output. (Modified from Dunham et al., 1989a.)

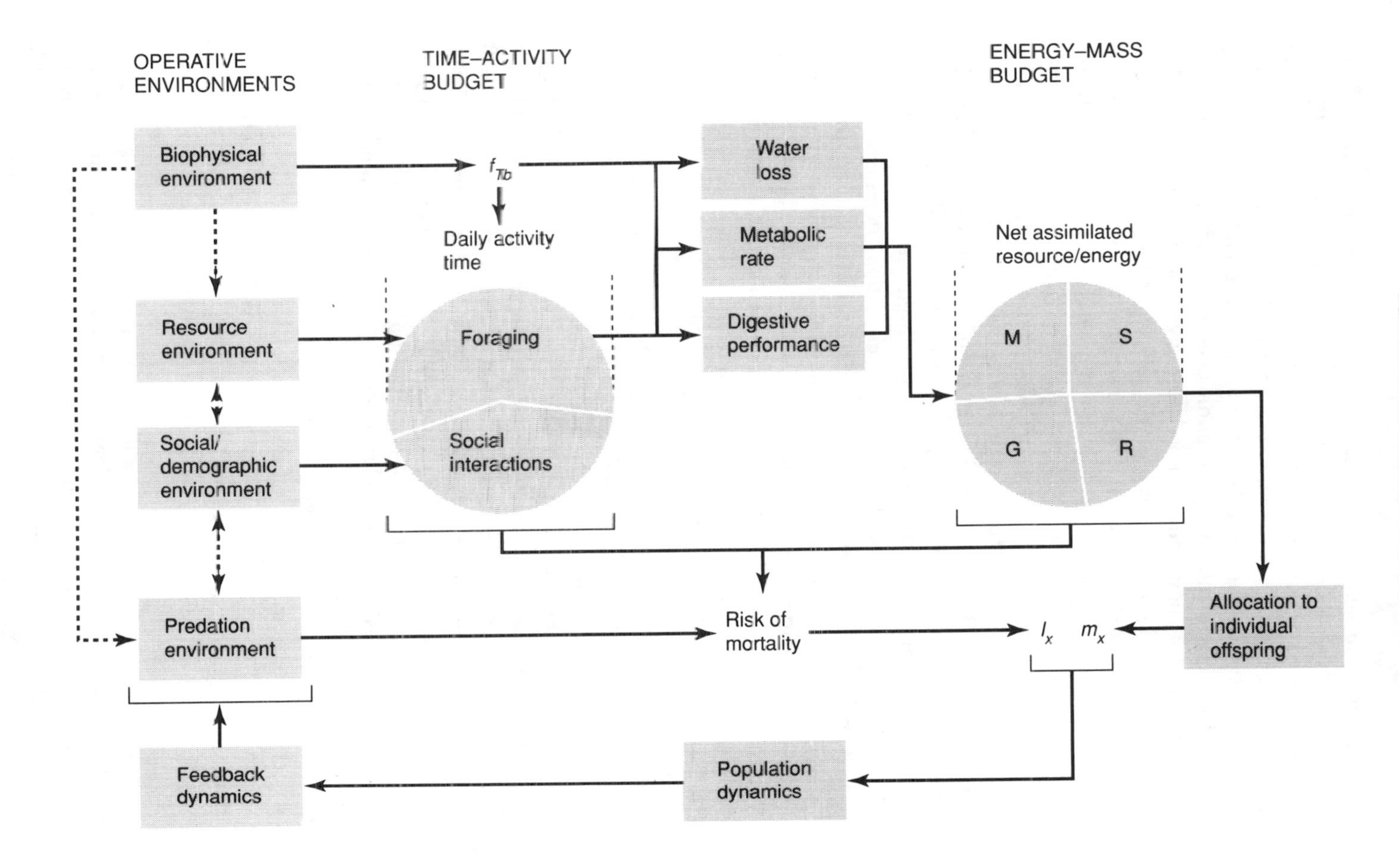
OPERATIVE ENVIRONMENTS
TIME–ACTIVITY BUDGET
ENERGY–MASS BUDGET
Biophysical environment
f_{Tb}
Daily activity time
Water loss
Metabolic rate
Digestive performance
Net assimilated resource/energy
M
S
G
R
Resource environment
Foraging
Social interactions
Social/ demographic environment
Predation environment
Risk of mortality
l_x m_x
Allocation to individual offspring
Feedback dynamics
Population dynamics

Congdon, Dunham, and Tinkle 1982)? Third, how (and how frequently) do individuals package their reproductive allocation into individual offspring? The time-ordered sequence of these allocation decisions determines an individual's life history phenotype—that is, the timing of its maturity, the number, provisioning, and scheduling of its offspring, and so on (Figure 3).

Many factors will influence allocation decisions. Individuals may well differ genetically in how they respond to a given environmental context. Moreover, their allocation decisions will almost certainly be sensitive to several factors discussed above: (1) the constraints imposed by the environment, (2) the organism's physiological state, and (3) the trade-offs and constraints imposed by its physiology. Thus, each life-history "genotype" could potentially represent a range of "allocation phenotypes." This range of phenotypes constitutes the "norm of reaction" of a given life history (Schmalhausen, 1949; Stearns and Koella, 1986; Via and Lande, 1985), and the life history trajectory followed by a given individual will depend on the sequence of operative environmental variation it encounters.

Of course, a risk is associated with each allocation "decision." For example, a decision to increase the amount of energy devoted to reproduction (at a cost to maintenance) could increase the risk of mortality due to environmental stress (Hoffmann and Blows, this volume) or to predation. Formally, we can view the risk associated with any allocation decision as the conditional probability of mortality of an individual that makes allocation decision x, given both a specified individual "state space" (above) and a specified set of operative environments.

Attempting to specify the relevant set of allocation rules and associated risks is, perhaps, the most difficult problem in applying physiologically structured models to natural populations. The number of functionally distinct life-history genotypes is potentially large, and the number of life-history phenotypes is necessarily even larger. Moreover, it will be difficult to anticipate how the population distribution of life history genotypes will shift with novel environmental change. Nevertheless, the range of interindividual variation in those allocation rules could be determined directly by manipulative experiments (e.g., by manipulating food levels or environmental temperatures). Alternatively, likely allocation rules could be predicted on the assumption that selection has favored heritable sets of allocation rules that result in the highest expected lifetime reproductive success in a specified environment. Specifically, optimal allocation rules could be estimated by using a structured population formulation and the methods of optimal control theory (Intrilligator, 1971; Bertsekas, 1987). Some progress has been made using this approach (Chiarello and Roughgarden, 1984; Hom, 1987, 1988; Ayres, this volume).

Once one determines the allocation rules (and associated risks) that a given individual will follow in a particular environment, one can then predict that individual's life history. In other words, one can predict its probable growth rate and body size trajectory, its age-specific fecundities, offspring sizes, and age at death. As shown in the next section, the demographic characteristics of a population can be determined by integrating these life histories across individuals and over a biologically meaningful time period.

Population dynamics

Appropriate methods for dynamically linking individual state to population dynamics were described as early as 1967 (Sinko and Streiffer, 1967, 1969) but have been largely unused until recently. These methods constitute a class of "transport" models that describe changes in population density due to growth, reproduction, and mortality as a function of physiological state. The formal models involve the solution of systems of first-order hyperbolic partial differential equations, and the general model in one dimension is:

$$\frac{\partial N}{\partial t} + \frac{\partial[g(x,t) \cdot n(x,t)]}{\partial x} + \mu(x,t) \cdot n(x,t) = 0 \tag{1}$$

with boundary condition

$$n(0,t) = \int \beta(x,t) \cdot n(x,t)dx \tag{2}$$

where N is the total number of individuals in the population; $n(x,t)$ is the number of individuals with a given value of physiological state variable x at time t; $g(x,t)$ describes the growth rate of physiological variable x; $\mu(x,t)$ is the time and physiological state-dependent mortality rate; and $\beta(x,t)$, the physiological state- and time-dependent birth rate, determines the boundary condition at the next time step. This formulation is easily modified to include multiple physiological state variables (e.g., Oster 1977; Cushing, 1990).

This general model (Equation 1) represents, however, changes in population density in a single environment. To be useful in the present context, it requires two major modifications. First, it must be expanded to include the dynamics of all important sources of environmental variation (e.g., climatic forcing of operative environmental variation). (As indicated above, adequate methods for specification of these extrinsic sources of environmental variation on an appropriate scale do not currently exist.) Second, it must specify how component growth rates and mortality rates (Equation 1) are determined by these patterns of environmental variation. Third, it must incorporate a mechanistic model of the relationship among operative environmental variation, individual allocation performance, and the resultant population demography. Finally, it must incorporate an understanding of the mechanisms of density dependence operating within the system (Ives and Gilchrist, this volume; Murdoch, this volume).

Relatively simple structured models have already been applied to a few populations. For example, Oster and colleagues (Oster, 1977) produced a structured population model of the dynamic behavior of Nicholson's (1957) experimental populations of sheep blowflies. They modeled the state space of the individual blowfly as its protein content (considered not to vary among individuals), and then modeled individual growth and birth rates as explicit functions of individual protein content in a given experimental environment. DeAngelis et al. (in press) developed a set of individual-based models of smallmouth bass (*Micropterus dolomieui*) populations. They used size as the relevant physiological state variable, and they developed submodels of the bioenergetics of growth and reproduction, of foraging, and of density feedback dynamics. They could then predict the density-dependent dynamics of populations under different environmental regimes of food availability and size-specific predation. Gurney et al. (1990), McCauley et al. (1990), and Nisbet et al. (1989) have developed a detailed stage-structured, physiologically based model for population dynamics of *Daphnia*. Murdoch (this volume) provides a detailed discussion of this and several other structured models.

These structured, individual-based models will generally be very complex as well as very labor-intensive to apply. Model simplification will be essential if this method is to be widely applied. Sequential hypothesis testing (comparing the performance of alternative simplified models) and model simplification following sensitivity analysis should allow discovery of comparatively simple models that retain the essential dynamics for many ecological systems (see also Murdoch, this volume).

Structured models must, of course, be well validated and tested against alternative models (e.g., compartment or statistical models; see above). All of these models can be thought of as competing hypotheses, and their relative merits can be tested in several ways. First, one can determine empirically how well each model explains the dynamic behavior of a population in response to an observed (past or present) pattern of environmental variation. Second, one can compare how well the models do in predicting population responses to experiments in which specific operative environments are manipulated. My expectation is that structured, individual models will generally yield the most accurate predictions, especially for "novel" environmental conditions. Moreover, structured models are the only models that provide an understanding of the mechanistic causes of population responses.

EFFECTS OF CLIMATE WARMING ON A LIZARD

To illustrate how a structured approach can be used to predict how a change in climate (specifically, an increase in average air temperature) might affect a population of lizards, I have developed some simple simulations for a small iguanid lizard (*Sceloporus merriami*) that lives in the Chihuahuan Desert of northern Mexico and west Texas. This lizard is a suitable subject for an individual-based approach because much is known about its long-term demography, physiology, and behavior as well as its biophysical environment (e.g., Dunham, 1978, 1981; Dunham et al., 1989a; Grant and Dunham, 1988, 1990; Huey et al., 1990). I present a model for a population that lives in the Grapevine Hills, Big Bend National Park, Texas.

A complete individual-based model requires diverse information and analyses. For example, it needs detailed microclimate data for the biophysical components of the analyses. It also needs population-specific data on ecological energetics, the thermal and size dependence of digestive physiology and metabolic rates, the energetics of individual growth, allometric relationships, social structure, and mating system, as well as data on the dependence of mortality rates on the age, size, and social status of individuals. Fortunately, these data are available or can be readily estimated for *S. merriami*. The details of this model, the input data, and the numerical methods will, however, be presented elsewhere.

The modeling exercise begins by estimating how climate warming will affect the thermal environment of the lizards. I then examine how such changes will affect potential times of activity of the lizards and how in turn this will affect their growth and reproduction.

Effects of climate warming on equilibrium body temperatures

The first step involves the development of a biophysical submodel that accurately estimates the equilibrium body temperatures (operative environmental temperatures, T_e) that an individual lizard will have in any particular environment (Bakken, 1976). With this model we can easily predict how a given change in climate might affect the distribution of T_e of a given ectotherm at a given locality.

The particular mechanistic model used here is based on Porter and Tracy (1983) and Dunham et al. (1989b). This model requires information on the lizards [specifically, body size as well as the lizard's surface heat-transfer properties (e.g., absorptivity) and kinetics] and on the thermal environment (e.g., direct solar radiation, air temperature, etc.). Microclimate data used in this analysis come from the Grapevine Hills site (Big Bend National Park, Texas) described in Dunham et al. (1989a) and Grant and Dunham (1988, 1990). The model then calculates T_e for a lizard at a specified site and with a specified orientation (see below) by estimating the magnitude of heat loss and gain from long- and short-wave electromagnetic radiation, conduction, convection, and evaporation (A. Dunham, W. Porter, and K. Overall, unpublished; Grant and Dunham, 1988; 1990). Despite their inherent complexity, such biophysical models work remarkably well (Porter et al., 1973). Estimated T_e for this implementation are accurate to within 1°C or so (Dunham et al., 1989b; Grant and Dunham, 1988); that is, calculated T_e (below) agree very closely with field measurements of T_e (Grant and Dunham, 1988; 1990).

Body size can have a large impact on T_e (Stevenson, 1985), and consequently an explicit consideration of body size may be crucial to attempts to model the consequences of climate change. To explore the importance of body size for the Grapevine Hills simulations, I estimated T_e for individual *S. merriami* of 3-, 4-, and 5-g body mass and also for two hypothetical lizards of the same shape and surface heat-transfer properties as *S. merriami*, but with body masses of 10 g and 50 g, respectively. (Individual *S. merriami* rarely exceed 6 g in body mass.)

Incorporating these lizard and environmental data into the biophysical model, I estimated T_e for the different sized lizards perched in full sun, oriented tangent to the sun, oriented oblique to the sun, or perched in full shade. These four "classes" of solar radiation environments bracket the range of available thermal environments in Big Bend, at least above ground (see Grant and Dunham, 1988, 1990).

To explore how climate change would affect T_e, I also estimated T_e under two scenarios of climate warming: I increased average ambient air temperatures (T_{alh}, measured at the height of the lizard) by 2°C or by 5°C. No other environmental factors (e.g., radiation loads) were changed.

Results The biophysical simulations for *S. merriami* in the Grapevine Hills show that T_e is obviously sensitive to air temperature (Figure 4) For example, for a lizard in a particular solar radiation environment (e.g., in full sun), a 2°C increase in ambient air temperature should increase T_e by an average of 1.6°C (±0.08, SD), whereas a 5°C increase in air temperature should increase T_e by 4.1°C (±0.10, SD). Thus, air temperature directly affects T_e, but the relationship is clearly not 1:1.

Equilibrium body temperatures are also strongly influenced by the solar radiation environment (e.g., full sun vs. shade; Figure 4). Thus, air temperature and the solar radiation environment interactively determine T_e and thus the range of thermal microclimates available to these lizards (Figure 4). Even so, the radiative environment clearly has a much greater effect on T_e than does air temperature, at least for *S. merriami* over the range of operative environments simulated here.

Equilibrium body temperatures are also dependent on body size (Figure 4); for example, T_e of large lizards (50 g) in sun will generally be about 2°C warmer than that of small lizards (3 g). Body size effects are minor, however, for lizards in shade. (An increase in T_e with body size for ectotherms in the sun is well known—see, e.g., Dunham et al., 1989b; Porter and Tracy, 1983; Stevenson, 1985—and is a consequence of the fact that heat flux in large ectotherms is strongly influenced by the radiative environment, whereas that in small ectotherms, with their thin boundary layers, is relatively more sensitive to convective heat loss.)

An important lesson emerges from the observation that body size can influence T_e. To the extent that individuals in a population (or different species in a locality) differ in body size, those individuals (or species) may well perceive a given thermal environment quite differently (Stevenson, 1985; Huey, 1991). As a consequence, the constraints imposed by the

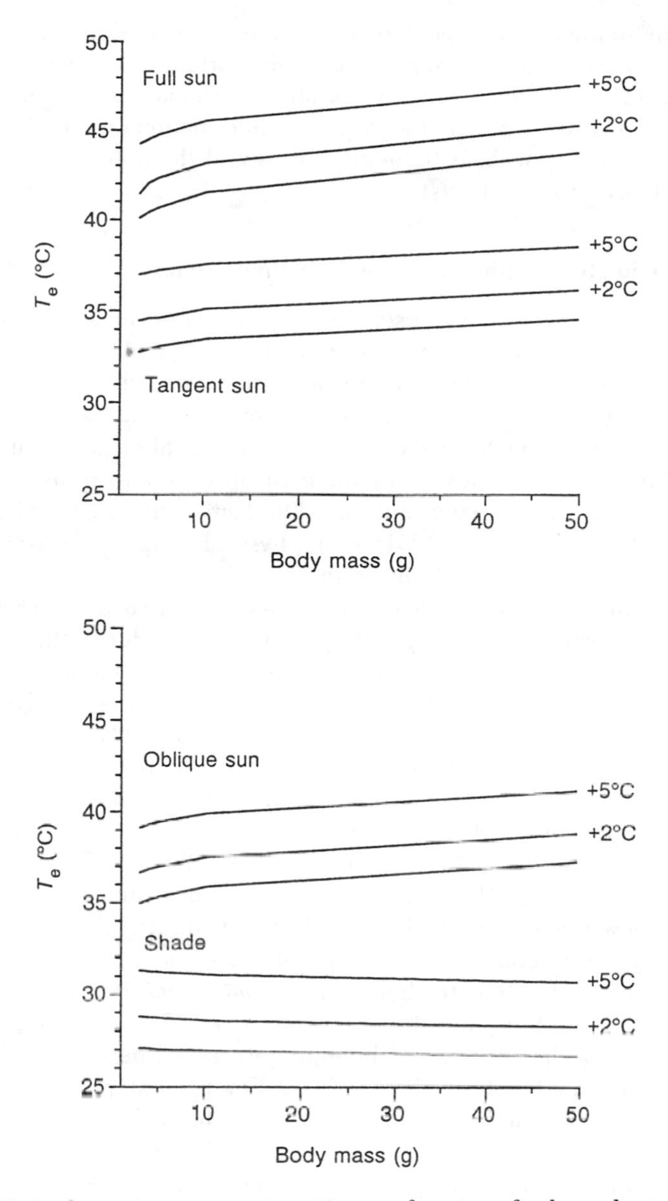

FIGURE 4. Operative temperature T_e as a function of solar radiation class, air temperature at lizard height, and body mass for lizards with the shape and surface heat transfer properties of *S. merriami*. Simulation conditions for this graph are those typical of mid-June in the Grapevine Hills population at 10:30 CDT. Biophysical conditions for standard simulation: elevation, 1036 m; T_{a2m}, 27°C; T_{alh}, 30°C; $T_{substrate}$, 31°C; wind velocity, 0.5 m/s; relative humidity, 50 percent; lizard absorptivity to short-wave solar radiation, 0.85; direct-beam solar irradiance, 750 Wm^2. (Data are from unpublished measurements by A. Dunham, W. Porter, B. Grant, and K. Overall.)

thermal environment will depend on an organism's size. Moreover, because body size can sometimes differ dramatically within or among species, a given change in climate will clearly not affect T_e equally for all individuals or for all species. By explicitly incorporating information on individual size, an individual-based mechanistic model can reveal the potential magnitude and importance of such effects.

Effects of climate warming on potential activity times

Using the above model, I can estimate how climate change (2°C and 5°C increases in ambient air temperature) will affect the daily pattern of availability of suitable microhabitats for *S. merriami* during the activity season. Then, knowing the range of thermal microhabitats acceptable to the lizards for activity, I can estimate how climate change will alter potential activity times for these lizards. These heuristic estimates are an extension of previous studies by my colleagues and me (Dunham et al., 1989a; Grant and Dunham, 1988; 1990; Grant, 1991) of biophysically imposed constraints on time and space utilization in *S. merriami*.

To estimate how climate change will affect potential times of activity, one must first estimate the range of T_e available at different times of day and also the range of body temperatures the lizards will find acceptable for activity. One can then readily determine the times during the day during which acceptable T_e are available (Porter et al., 1973). Data on the distribution of T_e vs. time of day for the Grapevine Hills population are taken from Grant and Dunham (1988). To incorporate the effects of climate warming, I then added to each T_e measurement the expected increase in T_e resulting from climate change scenarios involving 2°C and 5°C increases in average air temperature (1.6°C and 4.1°C, from above simulation). Thus we generate a new range of available T_e at different times of day (Figure 5).

The range of acceptable body temperatures—and hence the potential activity times—depends on the behavior of *S. merriami*. Body temperatures of *S. merriami* from this population averaged 32.2 ± 2.1°C (±SD; n = 232), and 95 percent of the observed body temperatures fell between 26.9°C and 37.8°C (Grant and Dunham, 1988; 1990). For present purposes, I have considered two extreme behavioral scenarios: (1) individuals are active only when they can achieve (anywhere above ground) body temperatures (T_b) equivalent to the modal temperature measured for the population (32.2°C); and (2) individuals are active whenever they can achieve T_b between 32.2°C and 37.8°C. (Some *S. merriami* will accept high body temperatures when exposed to elevated ambient temperatures in summer and thereby increase their potential activity times during the hot times of day; Grant and Dunham, 1990; Grant, 1990.)

Existing climates dramatically constrain activity times of Grapevine Hills lizards: they can be active for a maximum of only 2.5 or 5.25 hours a day, depending on whether they accept activity at high T_b. Not surprisingly,

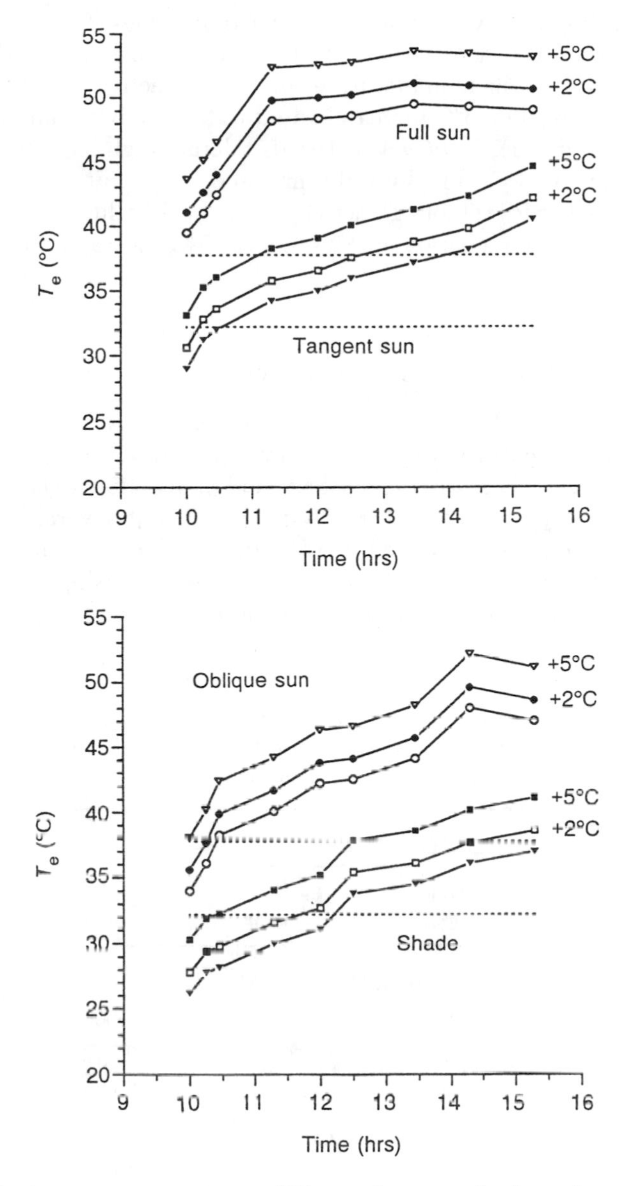

FIGURE 5. Operative temperature (T_e) as a function of solar radiation class (see text), air temperature at lizard height, and time of day for lizards with the shape and surface heat transfer properties of *S. merriami*. Biophysical data for each solar radiation class are measured values and are typical of conditions at this locality (Grant and Dunham, 1988). Curves for +2°C and +5°C indicate expected T_e that result from adding (to the measured T_e trajectory) the average increase in T_e (1.6°C and 4.1°C, respectively) resulting from increases in air temperature of 2°C and 5°C at lizard height in the T_e simulation described in Figure 4. The dotted lines at 32.2°C and 37.8°C are body temperatures used to calculate time available for activity.

climate warming will even further restrict activity times. If lizards are active only when they can achieve a body temperature of 32.2°C, then a 2°C increase in average air temperature would reduce activity time to less than 2 hours a day, while a 5°C increase would reduce activity time to only 0.6 hours a day (Table 1). If, however, lizards tolerate high body temperatures (T_b as high as 37.8°C), biophysically imposed constraints on activity time will be somewhat less (Table 1): activity times will be limited to 4.1 or 2.6 hours a day under scenarios of +2°C and +5°C increases in average air temperature, respectively.

Consequences of reductions in activity time

Thus far I have used a biophysical model to predict how climate change will constrain activity times of *S. merriami*. Next I translate those time constraints into effects on the ecology and population dynamics of this population. Such effects are currently unknown, but several likely consequences are probable. Reduced activity time will necessarily reduce the time available for foraging and social interaction (courtship, territorial defense, etc.). As a consequence, foraging success will probably be reduced,

TABLE 1. Changes in time available for activity, relative age-specific fecundity, and estimated intrinsic rate of increase (r) in the Grapevine Hills population of *S. merriami* under two climate change scenarios.

Climate change scenario	Time for activity: Available (hours)	Time for activity: Percent reduction	Relative r	Fecundity
	MAXIMUM T_b 32.2°C			
Measured	2.50	—	1.000	0.19
+2°C	1.92	23.2	0.572	−0.16
+5°C	0.58	76.8	0.000	—
	MAXIMUM T_b 37.8°C			
Measured	5.25	—	1.000	0.19
+2°C	4.08	22.2	0.296	−0.52
+5°C	2.58	50.9	0.000	—

Note: Measured values for time available for activity are from Grant and Dunham (1988, 1990). The intrinsic rate of increase, r, is the solution of $1 = \int e^{-rx} \cdot l_x \cdot m_x dx$, where l_x = survival from oviposition to age x, m_x = age-specific fecundity, and x = age. Measured values for r are calculated from the life table for the Grapevine Hills population over the time period 1974–1978 presented by Dunham (1981). Values are calculated from data in Figure 5. The climate change scenarios of +2°C and +5°C are defined in Figure 4.

and it could be reduced even further if climate change also lowers food availability.

As stated above, lizards could limit climate-induced constraints on their activity time by accepting higher than normal body temperatures (Table 1). However, that behavioral "decision" would have complex—and not necessarily positive—consequences. For example, hot lizards will incur additional maintenance costs because metabolic rates increase with temperature (Beaupre, Dunham, and Overall, in preparation). In addition, their evaporative water (respiratory and cutaneous) loss rates will also increase. The net consequences of elevated T_b will depend on the trade-offs between increased resource acquisition due to increased time available for foraging and the increased rate of dissipation of allocatable energy and water resources due to higher T_b.

I next used a physiologically structured model to estimate the total egg production of a typical female in her first reproductive season under the two climate change scenarios (+2°C and +5°C air temperature) relative to that of an identical female under current climate conditions. I assumed that food resource availability was directly proportional to the amount of time available for activity and standardized by the time available to the female not experiencing the climate change. (I do not know whether resource availability will in fact be affected in this manner; I make this assumption purely for illustrative purposes.) I repeated the simulation twice. In the first simulation, all females were assigned 24-hour body-temperature profiles typical of lizards at the Grapevine Hills site (Grant and Dunham, 1988, 1990; Dunham, unpublished). In the second simulation, females experiencing climate warming were assigned body temperatures between 32.2°C and 37.8°C during the active part of the day but "normal" T_b profiles at other times.

Under current climatic conditions, female *S. merriami* can reach maturity and reproduce as yearlings (Dunham, 1978). However, the model suggests that climate change will have a major and negative impact on their reproduction. For example, if air temperature increases by 5°C, yearling females will *not* be able to reproduce at all in either simulation, and the population will go extinct quickly (Table 1). If, however, air temperature increases by only 2°C, females will be able to reproduce, but their fecundities will be much lower than those of females under current conditions (−43 percent, −70 percent, respectively). Interestingly, females that accept elevated T_b in response to climate warming can roughly double their activity times and hence their food intake rates, but such females will nevertheless have very low relative fecundities (Table 1). Fecundities are presumably reduced because of the increased maintenance "cost" of activity at high body temperatures (Beaupre, Dunham, and Overall, in preparation).

These simulations demonstrate that accurate prediction of the reproductive consequences of climate change will depend on knowing which temperature regulation "decision" individual females make. A shift in acceptable body temperatures has a profound effect on reproduction (Table 1), and an effect of that magnitude will surely influence population dynamics greatly (see below). Individual "decisions" can potentially have profound effects on population-level responses to climatic change; and, to the extent that individuals vary in these decisions, individual-based mechanistic approaches can make explicit the consequences of such differences.

Reduced reproduction caused by a climate change of +2°C will likely have serious demographic consequences for the Grapevine Hills population. To estimate the general pattern of these demographic trajectories, we must determine how climate change will affect age-specific fecundities (m_x) and survival rates (l_x). Age-specific fecundities for female *S. merriami* under current conditions are well characterized from long-term studies (Dunham, 1981 and unpublished data). Female *S. merriami* from this population generally mature at 11–12 months, thereafter generally produce two clutches of eggs per year, and rarely live beyond 3 or 4 years of age. Body size and clutch size increase somewhat with age (Dunham, 1981).

To simulate the impact of climate change on fecundity, I reduced the current fecundity of each age class by either 29.6 percent or 57.2 percent (representing the two behavioral scenarios in Table 1). However, I assumed that survival schedules (l_x) of the various age classes would unaffected by climate change (but see below). I then computed life tables for populations facing climate warming and compared them to the average life table characterizing this population under current climate conditions (Dunham, 1981).

If lizards attempt to regulate T_b at 32.2°C in the face of a 2°C increase in average air temperature, their intrinsic rate of growth (r) will become negative ($r = -0.160$), and population density will be cut in half in 4.3 years. If, however, lizards allow T_b to rise to 37.8°C, their intrinsic rate of growth will be even more negative ($r = -0.52$), and population density will decline by one-half in 1.3 years. Under either behavioral scenario, the population will go extinct quickly (thought not at fast as a population exposed to a +5°C increase).

The above simulation assumes that survival schedules are unaffected by climate change. However, the reductions in activity time may well reduce mortality rates. Accordingly, we can determine how much age-specific survival would have to increase to maintain a stable population. This estimation can be readily accomplished using the known demographic and reproductive schedules from the Grapevine Hills population (Dunham, 1981).

Figure 6 shows the values for survival to first reproduction and for average adult survival that permit population persistence. Values are shown assuming age-specific fecundities characteristic of the present Grapevine Hills population (curve C) or for climate-change populations (+2°C) using one of the two behavioral scenarios (curve A for T_{bmax} = 32.2°C; curve B for T_{bmax} = 37.8°C). For the Grapevine Hills population to persist during a +2°C climate warming, either survival to first reproduction or average annual adult survival (or both) must increase dramatically, regardless of the

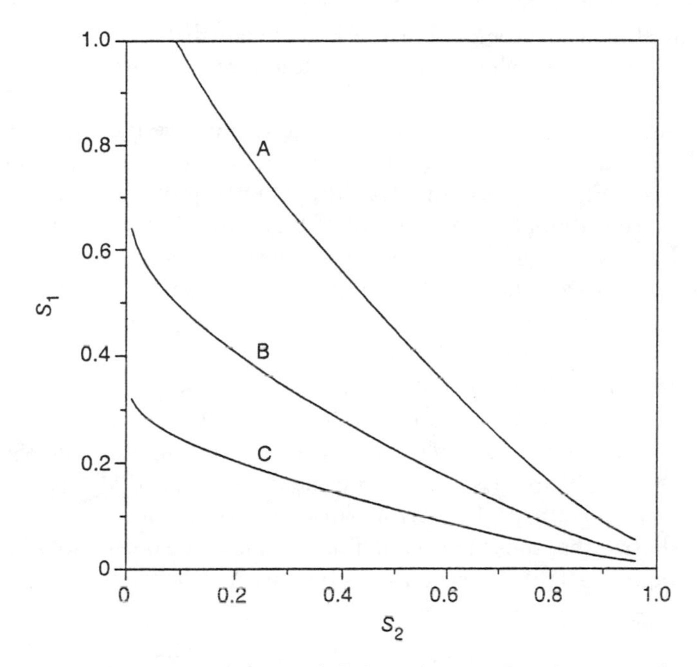

FIGURE 6. Combinations of the probability of survival to first reproduction (S_1) and of average annual survival of adults after first reproduction (S_2) that are necessary to allow persistence of the Grapevine Hills population of *S. merriami*, as a function of relative age-specific fecundity (RAF). Each curve represents the set of combinations of S_1 and S_2 for which $r = 0$ for a given level of relative age-specific fecundity. Combinations of S_1 and S_2 that lie on or above the $r = 0$ isopleth will allow population persistence. Age-specific fecundity is relative to the long term average for the Grapevine Hills population. For each level of RAF, r is the intrinsic rate of increase (assuming stationary fecundity and mortality schedules) and is computed as in Table 1. Curve A corresponds to RAF = 0.269, curve B corresponds to RAF = 0.572, and curve C corresponds to the long-term average age-specific fecundity.

thermoregulatory "decision" made by individuals. For example, if adult survival averages 0.4 (a reasonable figure for this population), then survival from egg to the age of first reproduction would have to increase from about 0.15 (current level) to about 0.3 or 0.6, depending on the thermoregulatory strategy. However, hatchling survivorships this high have never been observed in 18 years of demographic observations on this population (Dunham and Overall, unpublished). Therefore, the Grapevine Hills population will almost certainly go extinct if faced with the +2°C climate change scenario.

The climate change scenarios presented above assume that air temperature is increased only during the active season of the lizard. However, the simulations can be repeated for a scenario in which air temperature is increased over the entire year. Effects of a +2°C or a +5°C increase in average air temperature throughout the year can be estimated from the long-term climate data and from the knowledge that *S. merriami* become active in the spring, when average air temperatures exceed 18°C (Grant and Dunham, 1990).

The annual activity season would probably increase by slightly less than 1 month for a 2°C increase in air temperature and by about 2 months for a 5°C increase (Figure 7). Estimating the population-level effects of these changes is considerably more complex than the simple simulations presented above and will not be presented here. In any case, this present exercise emphasizes that rather detailed (i.e., month by month) climate projections will be required if we are to attempt to predict the population-level effects of global climate change.

The impact of climate change on predicted population dynamics of the Grapevine Hills population depends ultimately, of course, on the nature of density dependence in resource acquisition, mortality rates, and the rate of climate change. These issues are complex and beyond the scope of this chapter, but they are vital to any attempts at prediction.

I emphasize that these population projections are specific to *S. merriami* in the Grapevine Hills. Sympatric species with higher operating tempera-

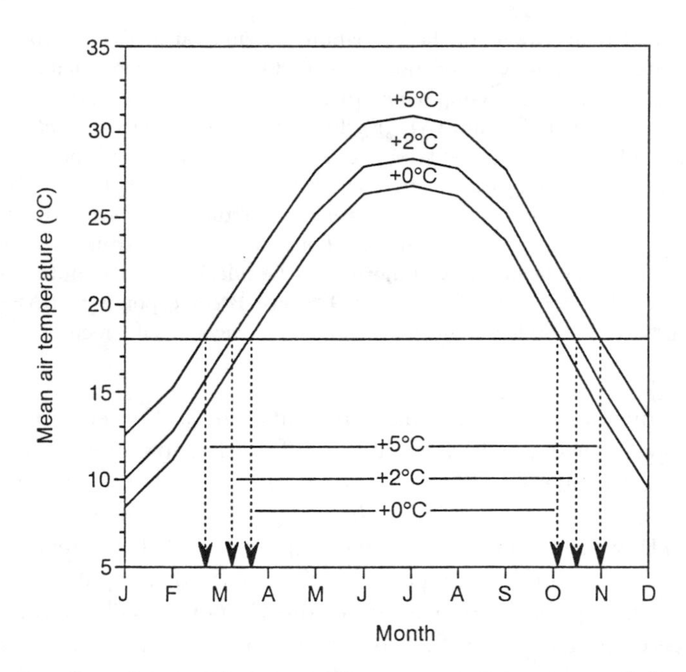

FIGURE 7. Effect of 2°C and 5°C increases in air temperature on daily mean air temperature in different months in the Grapevine Hills. The horizontal line represents the minimum daily air temperature (18°C) necessary for activity. Climate warming should slightly expand the potential activity season of *S. merriami*.

tures and different physiologies might show positive, not negative, responses to climate warming. Moreover, high-altitude populations of *S. merriami* itself might also benefit from climate warming (see Dunham et al., 1989a).

DIFFICULTIES IN IMPLEMENTATION

Several limitations must be faced in applying physiologically structured, individual-based models to understanding population responses to environmental change. The most serious limitation is that these models are inherently complex, and they will only become more so when the relevant feedback dynamics (e.g., density dependence) and interspecific interactions are incorporated. Clearly, this approach will be very labor-intensive to apply, and its wide-scale application is quite problematical (see also Murdoch, this volume). Even so, several groups of investigators have made considerable progress toward the development and validation of these models.

A further difficulty is the necessity of knowing allocation rules and relevant behavioral decisions. We saw, for example, that reproductive output of a yearling female can vary by as much as 52 percent, depending on her thermoregulatory strategy (Table 1). These rules and decisions are likely to depend strongly on the particular populations being investigated, but they can be elucidated (at least in principle) through well-planned experiments.

The various complexities, plus those imposed by computational algorithms (numerical methods) and by the need for data on variation in operative environments, combine to make model simplification essential. The discovery of the simplest individual-based model that retains the fundamental dynamics of a system is a crucial but relatively unexplored area (Murdoch, this volume). Sensitivity analyses probably represent an appropriate first step in model simplification. To date, however, published sensitivity analyses are few (see DeAngelis et al., in press).

No individual-based model has yet produced unpredictable or chaotic population dynamics. The possibility of chaotic population (and higher-level) dynamics arising from realistic implementations of these models does exist (May and Oster, 1976), and the recent results of Hastings and Powell (1991) on simple three-species food webs suggest that chaotic dynamics may frequently occur in multiple interacting population models. Well validated individual-based models should be investigated for such behavior. In addition, stochastic versions of these individual-based models remain to be developed.

None of the models reviewed here has yet investigated the consequences of spatial variability (e.g., metapopulation structure *sensu* Howe and Davis, 1991; Pulliam, 1988; Pulliam and Danielson, 1991) for the formulation and implementation of the models. These approaches may require considerable modification to include populations with high migration rates, with very coarse-grained distributions, or with complex spatial distributions of sink and source subpopulations.

Finally, we need to develop alternative approaches to structured individual-based models for situations in which they are insufficient or impractical for predicting population-level responses to environmental change. Experiments in which selected operative environments are manipulated may provide insight into the population-level effects of such variation. For example, Ayres (this volume) describes the population- and community-level consequences of minor increases in air temperature achieved in portable greenhouse experiments. Where such manipulations are impractical (e.g., placing a greenhouse over the Grapevine Hills), transplant experiments may be a useful approach (Pacala and Hurtt, this volume). To the extent that climate modelers can specify likely scenarios of change in particular operative environments (e.g., air temperature, moisture levels), experiments can be designed to mimic those scenarios, and likely system dynamics can be inferred from the outcome of such perturbation experiments.

At present any experimental approach suffers from two limitations. First, as noted above (see also Schneider, this volume), no methodology currently exists for generating climate change scenarios on a micro scale that is relevant to populations in nature; this problem affects modeling approaches as well. Second, even if we knew, for instance, that air temperatures would increase by 2°C, experimentally increasing air temperature by that amount might be difficult—to put it mildly—for organisms, like *S. merriami*, with large home ranges.

SUMMARY

This chapter outlines a research program that is designed to predict how climate change will affect responses at all levels of ecological organization (individual, population, community, ecosystem, landscape). The program presented here is fundamentally mechanistic and hierarchical. The initial step involves developing a mechanistic, individual-based, physiologically structured model that predicts the response of population dynamics to environmental change. Starting from a mechanistic, individual-level base is critical and appropriate because the ecological effects of environmental change (human-induced or otherwise) are functionally transduced through individual organisms, and because all individuals do not respond equivalently. Consequently, any attempt to develop an ecological theory that ignores either mechanistic processes or individual variation necessarily runs the risk of yielding predictions that are superficial and potentially inaccurate. These individual-to-population models can then be integrated as critical components of higher-level, composite models that are designed to predict the community, ecosystem, and landscape responses to climate change. Understanding the dynamics of populations will be critical to such predictions because the rate at which mass and energy is transferred among trophic levels will very likely depend on the dynamics of constituent populations.

Models of how variation in the environment affects the dynamics of individuals, populations, and communities are clearly complex and require diverse kinds of data. The development of such models will be difficult, but hardly impossible. Indeed, many of the key links are now very well understood. For example, current biophysical models accurately predict how changes in microclimate variables (e.g., air temperature) affect equilibrium body temperatures of ectotherms. Thus, the program I advocate here (see also Murdoch, this volume) is practical, and much of it is practical now.

To provide an example of a mechanistic approach, I present a simple set of biophysical and structured-model simulations that explore the likely consequences of climate change to a population of the lizard *Sceloporus merriami* in Big Bend, Texas. Projected increases in air temperature of +2°C or +5°C should increase average equilibrium body temperatures and thereby reduce (sometimes drastically) potential time periods during which these lizards can be active during summer. With less time available for feeding, reproduction by yearlings will probably be reduced or perhaps even eliminated. Moreover, even if survival rates are increased, reduced reproduction will almost certainly drive the population to extinction, even under the most favorable of scenarios.

These simulations, though simplistic, demonstrate the potential utility and power of a mechanistic, individual-based approach to predicting the consequences of environmental change. Much needs to be done. Future work will involve incorporating, for example, feedback considerations (e.g., density dependence) and stochasticity. Moreover, because these models will necessarily be too complex for widespread application, we need to encourage sensitivity tests as well as parallel experimental manipulations to guide model simplification and application.

CHAPTER 8

CLIMATE CHANGE AND ECOLOGICAL INTERACTIONS

Anthony R. Ives and George Gilchrist

Anticipation of large-scale changes in the world's climate has led to growing concern about the effects of climate change on the world's ecosystems. Research into the ecological effects of climate change has focused primarily on three areas: paleobiology (Clark, this volume), physiological ecology (Ayres, this volume; Dunham, this volume; Root, this volume), and ecosystem ecology (Carpenter et al., this volume). These areas of research are critical for predicting the response of species to climate change, but an understanding of the ecological interactions within and among species is no less critical (Carpenter et al., this volume). Even though species interactions will certainly play a role in the response of ecological communities to climate change, as yet there are no theoretical predictions about what this role will be. Theoretical ecologists thus need to enter the research arena of climate change. In this chapter we develop two theoretical predictions concerning how species interactions determine the sensitivity of an ecological community to climate change. First, the change in the mean population density of a species in response to climate change is inversely proportional to the strength of intraspecific density-dependent population regulation. Second, ecological interactions alter the sensitivity of mean population densities to climate change according to the ecological roles occupied by species. Species occupying unique roles are much less sensitive than those competing with other species.

Historically, many populations have responded to environmental shifts by altering their geographic ranges and thus tracking the changing climate (Peters 1991, and references therein). With the advent of industrial society, however, many of the potential corridors available for these biogeographic shifts have been destroyed or fragmented (Groom and Schumaker, this volume; Harrison, this volume; McKelvey et al., this volume). As a result, climate tracking will be limited or impossible for many species, and many populations will be confined to their existing geographic ranges. It is within this context that our predictions are made. We ask explicitly how climate change will affect ecological interactions, potentially altering biological diversity within a given refuge.

In this chapter, we present the theoretical foundation upon which our two predictions are based, then discuss how current information on species interactions can be related to the predictions. To develop the theory, we begin by considering a single species, then proceed to a multispecies community. Our analysis divides the problem of predicting the effects of climate change into two distinct components, one associated with the direct effects of climatic factors on the per capita population growth rate of a species, and the other determined by indirect effects operating through interactions within and among species. We then describe two different empirical approaches that can be used to obtain the information necessary to predict the responses of species to climate change. The first approach uses long-term data on population dynamics in nature, and the second involves detailed "microcosm experiments" conducted under controlled conditions. Our general aim in presenting these examples is to show what information is needed to predict the consequences of climate change, and to encourage further research.

THEORY

Single species

We begin with a hypothetical example illustrating the fundamental theoretical question at issue: How important is density dependence in mediating the influence of climate change? Climate changes will affect the mean population density of a species by altering its reproductive and mortality rates (Dunham, this volume), which together are summarized by the per capita population growth rate of the species. Figure 1A depicts the direct, positive effect of a climatic trend on the per capita population growth rate in the absence of any density dependence. Over the course of 200 generations, the mean of the log per capita population growth rate is increased by 25 percent; this is equivalent to increasing the average number of offspring produced per individual by 30 percent. Figure 1B shows the response in the log population density of two species subjected to the same climatic trend. One species shows an increase in log density by a factor of 2.5 over 200 generations, while the other increases by a factor of only 1.5. The difference between these species is that the former experiences only weak intraspecific density dependence, while the latter is subject to strong density

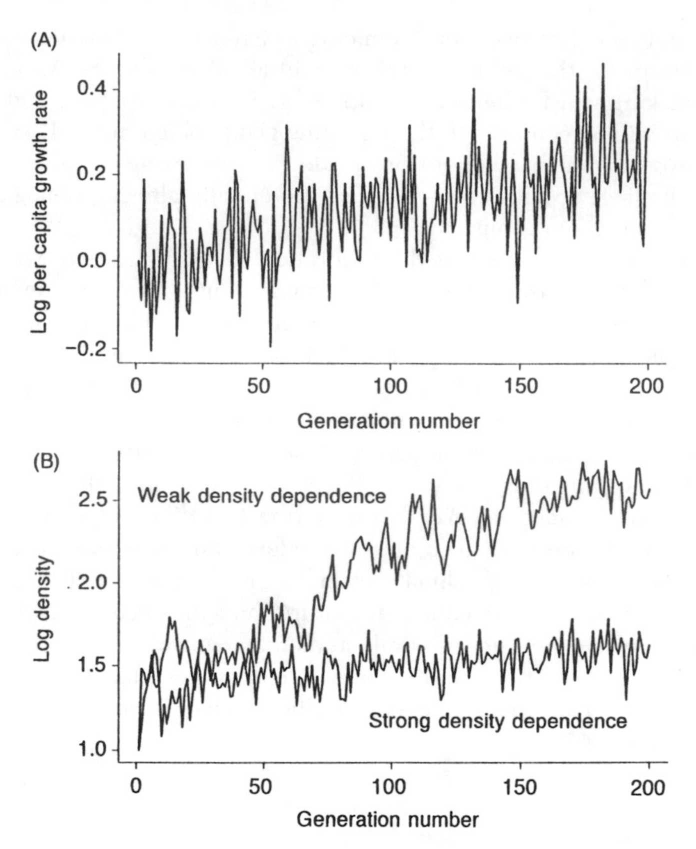

FIGURE 1. Response of species to a directional trend in the environment that enhances population growth. (A) The log per capita population growth rate (scaled to equal zero at time zero) in the absence of intraspecific density dependence. (B) The log population density of two noninteracting species through time. Although the climatic effect on reproductive rates is the same in each species, the species differ in the degree of density dependence; the lower population growth rate is shown by the species with greater intraspecific density dependence. For both species, the population dynamics are given by the equation $n(t + 1) = \lambda\, e^{cT}\, n(t)\, e^{-rn(t)}$. T is normally distributed with variance 0.01 and mean ranging from 0 to 0.25. Parameter values differing between species are $\ln \lambda = 0.1, 1.0$; $r = 0.025, 0.25$; and $c = 1.0$.

dependence. With strong intraspecific density dependence, increases in the per capita population growth rate are counteracted by strong competitive interactions among individuals. Although more offspring are produced each generation, density-dependent interactions reduce the number that survive to reproduction.

We can proceed to quantify the predicted change in population density for a species subject to an environmental trend by defining $F(N,T)$ as the per capita population growth rate of a species over the course of a year. $F(N,T)$ depends on N, the population density of the species, and T, a parameter summarizing the effects of the climate on the per capita population growth rate. Because population growth is an exponential process, it is easiest from the onset to describe population density on a logarithmic scale, so assume that N is the natural log of the population density, and F is the natural log of the per capita population growth rate. Thus, the population density of a species in year $t + 1$ depends on the density in year t according to the equation

$$N(t + 1) = F(N(t),T) + N(t) \qquad (1)$$

We leave the definition of T loose for the mathematical analysis; for different species, climatic factors influencing the population growth rate could be very different (e.g., number of frost-free days in a year, average monthly summer temperature, or minimum daily winter temperature). The units of T will depend upon what the environmental parameter actually measures. Furthermore, the model makes the simplifying assumption that climatic effects on the population growth rate can be summarized by a single parameter. This assumption implies that a single environmental variable largely explains the year-to-year fluctuations in population density (or that multiple variables are highly correlated). Although this assumption probably will not hold for any species in nature (Schneider, this volume), multiple climatic parameters can be included in the model in the same way as a single climatic parameter; we postpone this complication until dealing with actual data in a later section.

This model is phrased in terms of year-to-year fluctuations in population density, despite the fact that many environmental factors operate over a shorter time interval than a year. For example, a cold snap during a vulnerable insect stage may cause high mortality. However, because the model is designed to predict long-term changes in population densities, we have chosen an annual time step. Even though an environmental factor may act only over a short time, it will nonetheless be included in the model if it affects year-to-year fluctuations in population densities. This procedure has the advantage of allowing one to average over seasonal changes of species densities that may contribute little to year-to-year population fluctuations.

Because T is a climatic parameter, it is best described by a random variable with mean, variance, and other statistical moments. Because the per capita population growth rate $F(N,T)$ depends on T, the population density N must also be described by a random variable. The distribution of N is determined by the distribution of T and the per capita population growth rate according to Equation 1. If the distribution of T does not change, the distribution of N will approach a stationary distribution N^* that is analogous to a stable equilibrium for deterministic equations. In Appendix

A we demonstrate how to derive an exact solution for this stationary distribution.

The objective in analyzing this model is to predict how changes in the distribution of the environmental parameter T affect the mean species abundance, as illustrated in Figure 1. Specifically, if the mean of T changes by a small amount $\Delta\overline{T}$, then the change in the mean of N^* is approximately given by

$$\Delta\overline{N}^* = -\Delta\overline{T}\,\frac{E[\delta F/\delta T]}{E[\delta F/\delta N]} \tag{2}$$

(see Appendix B), where $\delta F/\delta T$ denotes the change in F with respect to changes in T, $\delta F/\delta N$ denotes the change in F with changes in population density N, and $E[\cdot]$ denotes the expectation taken over the distributions T and N^*. Thus, the change in the average population density of a species is directly proportional to the change in the per capita population growth rate caused by climatic changes, but is inversely proportional to the strength of intraspecific density dependence in the population growth rate.

To demonstrate the importance of the density-dependent term $E[\delta F/\delta N]$ in Equation 2, we give an example using the specific equation

$$n(t+1) = \lambda\, e^{cT}\, n(t)\, e^{-rn(t)} \tag{3}$$

To present this equation in a familiar form, we have written it in terms of actual density rather than logarithms; in the earlier notation, $N(t) = \ln n(t)$, and $F(N,T) = [\ln \lambda + cT - re^N]$. The parameter r controls the degree of intraspecific density dependence in the per capita population growth rate, λ scales the density-independent component of the per capita population growth rate, and c scales the environmental parameter T. Also, we assume that T has a normal distribution, so e^{cT} is lognormally distributed. Figure 2 shows that increasing the degree of density dependence in the per capita population growth rate decreases the sensitivity of the mean population density to climate change. Figure 2A shows the relationship between

FIGURE 2. The effects of climate change on species abundances with varying degrees of intraspecific density dependence. (A) The change in population densities in successive years for three different degrees of density dependence, increasing from lines B to D, with each line corresponding to the results shown in Figures 2B–2D. The dashed line indicates no density dependence. (B–D) The corresponding changes in the distribution of the species abundances; in all cases, increasing the environmental parameter increases the mean population density. With increasing density dependence, the change in the mean abundance decreases for the same change in the environmental parameter T. For all graphs, $F(N,T) = [\ln \lambda + cT - re^N]$ and T follows a normal distribution with variance 0.1 and mean 0, 0.05, and 0.1 for the three lines in each graph. Values for $\ln \lambda$ and r are: (B), 0.1, 0.025; (C), 0.3, 0.075; and (D) 0.6, 0.15; and $c = 1.0$. ►

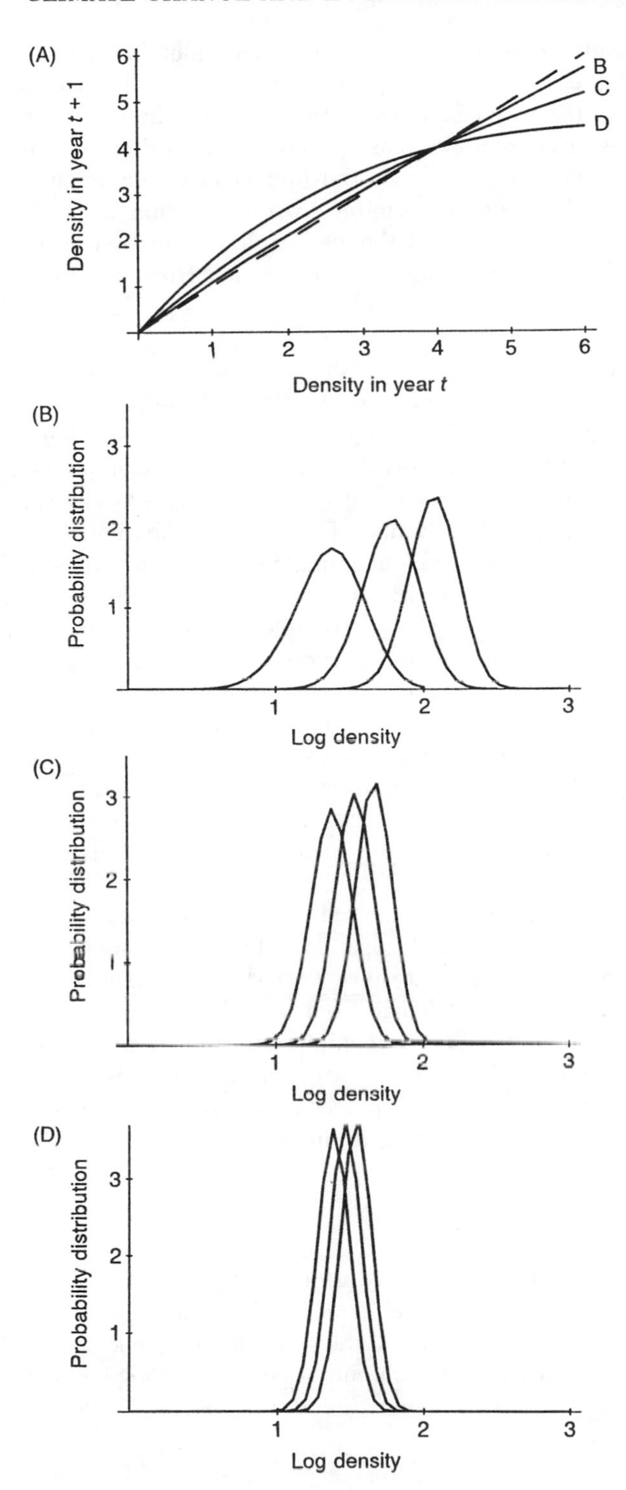
(A)
Density in year t + 1
Density in year t
B
C
D
(B)
Probability distribution
Log density
(C)
Probability distribution
Log density
(D)
Probability distribution
Log density

$n(t + 1)$ and $n(t)$ in the absence of environmental variability for three different degrees of density dependence, increasing from B to D. Figures 2B–2D show the response of the population density to climate change for each of these examples. Increasing density dependence in the per capita population growth rate reduces the impact of changing the value of the climatic factor T. For these examples, we have assumed that increasing the climatic parameter increases the per capita population growth rate, and consequently increases the population density. However, the same general patterns hold for the case when climatic changes produce lower population densities.

There is an interesting relationship between the variance in the stationary distribution of population density, N^*, and the response to changes in the climate parameter T. In Figure 2, increasing the degree of density dependence not only decreases the change in the mean population density in response to climate change, but it also gives rise to low variance in the stationary population distribution. This might make one expect that populations exhibiting low variance in annual population fluctuations will also be the least susceptible to climatic changes. However, this expectation is not always correct. Figure 3 shows examples of population density for two additional species with density dependence greater than those in Figure 2. The lowest variance in the population density occurs when the curve of $n(t + 1)$ against $n(t)$ is flat, as it is for the curve labeled B in Figure 3A. For curve C in Figure 3A, populations tend to oscillate between years, and this produces greater variance in the population density distribution (Figure 3C). However, the reason for these oscillations is very strong density dependence, so the change in the mean population density with climatic change is small.

These observations lead to the following generalization: Populations that fluctuate greatly from year to year but do not exhibit overcompensating population dynamics will likely have weak density dependence and respond more strongly to climatic changes than populations exhibiting either small year-to-year fluctuations or overcompensatory population dynamics. However, this generalization must be viewed cautiously with respect to real populations; it only applies to species that experience the same degree of environmentally induced variability in population growth rates.

Multiple species

Examining the responses of many interacting species to climate change can be done in very much the same way as for single species, although the problem is complicated by the interactions among species. Thus, for a given species, say species 1, in a community containing S species, the equation governing the year-to-year population dynamics is

$$N_1(t + 1) = F_1(N_1(t), N_2(t), \ldots, N_S(t), T) + N_1(t) \tag{4}$$

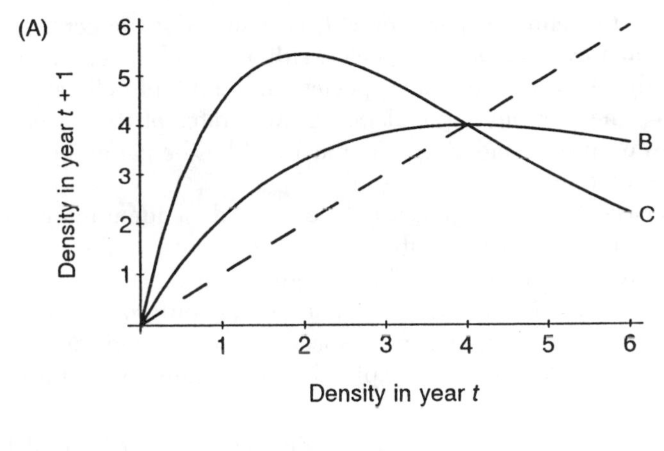

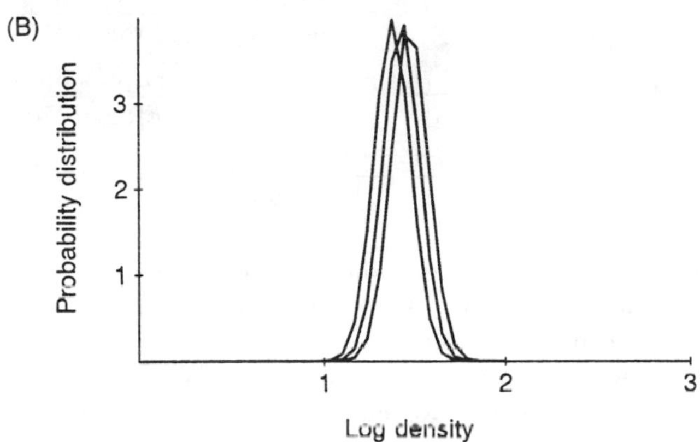

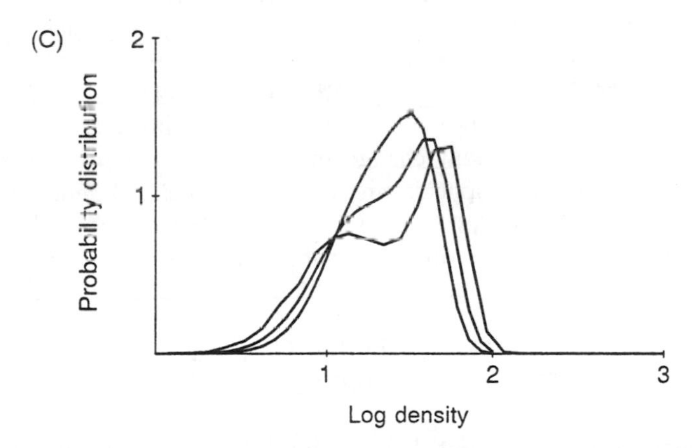

FIGURE 3. The effects of climate change on species abundances for greater intraspecific density dependence than shown in Figure 2. The equation used for population dynamics is the same as in Figure 2, but values for the parameters ln λ and r are: (B), 1.0, 0.25; and (C), 2.0, 0.5.

Here, the climatic parameter T is assumed to affect the per capita population growth rates of all species, although each species may be affected differently. As with the single-species case, multiple climatic factors may influence the per capita population growth rates of the species, but here we focus on just a single factor summarized by the parameter T to simplify our presentation.

The easiest way to summarize the density-dependent interactions among species is to construct a matrix **C** with elements in the ith row and jth column given by the ratio $-E[\delta F_i/\delta N_j]/E[\delta F_i/\delta N_i]$. Here, $E[\delta F_i/\delta N_j]$ is the expected change in the log per capita population growth rate of species i with respect to the log density of species j; expectations are taken over all values of T and N^*. As an example, for a community with three species

$$\mathbf{C} = \begin{bmatrix} -1 & \dfrac{-E[\delta F_1/\delta N_2]}{E[\delta F_1/\delta N_1]} & \dfrac{-E[\delta F_1/\delta N_3]}{E[\delta F_1/\delta N_1]} \\ \dfrac{-E[\delta F_2/\delta N_1]}{E[\delta F_2/\delta N_2]} & -1 & \dfrac{-E[\delta F_2/\delta N_3]}{E[\delta F_2/\delta N_2]} \\ \dfrac{-E[\delta F_3/\delta N_1]}{E[\delta F_3/\delta N_3]} & \dfrac{-E[\delta F_3/\delta N_2]}{E[\delta F_3/\delta N_3]} & -1 \end{bmatrix} \quad (5)$$

Thus, the rows of the matrix give the average changes in population density for each of the species with respect to the population density of all other species in the community, scaled by the effect of the species on itself. We have constructed the matrix to have values of -1 along the diagonal, because the majority of species are expected to be self-regulated and therefore to have a negative effect on their own population density. However, this does not exclude the possibility that individuals of the same species interact in a mutualistic fashion or do not interact at all.

The change in the mean population density of species 1, $\Delta\overline{N_1}^*$, in response to changes in the mean value of the climatic parameter, $\Delta\overline{T}$, depends on the consequent change in the per capita population growth rates both of the focal species (in this case, species 1) and of all other species. To simplify the presentation, we will first consider the effects of climate change operating through changes in the per capita population growth rate of the focal species alone, which is approximated by the equation

$$\Delta\overline{N_1}^* = -\Delta\overline{T}\,\frac{E[\delta F_1/\delta T]}{E[\delta F_1/\delta N]}\,(\det\mathbf{C}_1/\det\mathbf{C}) \quad (6)$$

(see Appendix B). Here, det denotes the determinant of a matrix, and $\mathbf{C}_1$ is the matrix produced by replacing all elements of the first row and column of matrix **C** with zeros, except the element in the top left-hand corner. For the three-species case above,

$$\mathbf{C}_1 = \begin{bmatrix} -1 & 0 & 0 \\ 0 & -1 & \dfrac{-E[\delta F_2/\delta N_3]}{E[\delta F_2/\delta N_2]} \\ 0 & \dfrac{-E[\delta F_3/\delta N_2]}{E[\delta F_3/\delta N_3]} & -1 \end{bmatrix} \qquad (7)$$

Thus, matrix $\mathbf{C}_1$ corresponds to the matrix of species interactions in the case in which species 1 does not interact with any other species. Note that Equation 6 is the same as found for the single-species case (Equation 2), but modified by the term det$\mathbf{C}_1$/det$\mathbf{C}$. Therefore, the ratio det$\mathbf{C}_1$/det$\mathbf{C}$ summarizes the effects of interspecific interactions on the change in mean species density due to climate change.

The critical information needed to interpret Equation 6 is that the value of the determinant of a matrix is greatest when its rows and columns contain dissimilar values. Each row i in matrix $\mathbf{C}$ summarizes the effects that other species in the community have on the population growth rate of species i. Thus, if two rows are similar, the species corresponding to those rows have similar patterns of interactions with other species. To interpret det$\mathbf{C}_1$/det$\mathbf{C}$, notice that $\mathbf{C}_1$ corresponds to the case in which species 1 does not interact with other species in the community. This gives the neutral case for the interactions of species 1 with other species. If the patterns of interactions of species 1 are similar to those of other species, then the first row in matrix $\mathbf{C}$ will be similar to other rows. In this case, det$\mathbf{C}$ will be smaller than det$\mathbf{C}_1$, and det$\mathbf{C}_1$/det$\mathbf{C}$ will be greater than one. Thus, from Equation 6, the effect of species interactions is to increase the change in population density of species 1, $\Delta\overline{N}_1{}^*$. The opposite occurs if species 1 exhibits patterns of interactions dissimilar to those of other species. In this case, det$\mathbf{C}_1$ will be relatively large, and the ratio det$\mathbf{C}_1$/det$\mathbf{C}$ will be less than one. Therefore, species interactions moderate the changes in population density of species 1. This heuristic discussion of determinants of matrices and of Equation 6 has a useful graphic interpretation, presented in Appendix C.

The biological reason for the results of Equation 6 can be explained as follows. Consider first two species that hold the same ecological role, and hence are competitors. If they are strong competitors, then small climate-driven changes in the per capita population growth rate of one may be sufficient to make it dominant over the other species. Thus, the densities of competitors should be relatively sensitive to climatically driven changes in per capita population growth rates. Now, consider two or more species, with one holding a unique ecological role, such as a unique predator in a community of prey. If, for example, the overwintering survival of the predator increased due to a warming climatic trend, its density would immediately increase, driving a subsequent decrease in the density of its prey. However, low prey density might then limit the increase in predator population density. Thus, the predator–prey interactions mitigate the direct effects of climate change on the predator's population growth rate.

To illustrate these predictions with numerical examples, we will calculate the response of species to climatic change for the case of two competitors and for the case of a prey species and its predator. For the competition example, let the population dynamics of the two species be governed by the equations

$$n(t + 1) = \lambda_1 e^{cT} n(t) e^{-r_1[n(t)+\alpha m(t)]} \tag{8a}$$

$$m(t + 1) = \lambda_2 m(t) e^{-r_2[m(t)+\beta n(t)]} \tag{8b}$$

To make these equations more familiar, we have expressed them in terms of the absolute population densities $n(t)$ and $m(t)$ rather than the natural logs of population density. These equations are simply the discrete-time analogues of the Lotka–Volterra competition equations. The parameters λ_1 and λ_2 determine the magnitudes of the density-independent components of the per capita growth rates, r_1 and r_2 measure the strength of density dependence, and α and β are the competition coefficients measuring the relative strength of interspecific to intraspecific competition. We assume for the sake of simplicity that the climatic parameter T only influences the population growth rate of species 1, and that T is distributed according to a normal distribution and scaled by the parameter c. Figure 4A shows the subsequent changes in the density of competitor 1 with increases in the mean value of T.

For the predator–prey example, let the population dynamics for the two species be governed by the equations

$$n(t + 1) = \lambda_1 e^{cT} n(t) e^{-r_1 n(t)} e^{-ap(t)} \tag{9a}$$

$$p(t + 1) = \lambda_2 n(t)[1 - e^{-ap(t)}] \tag{9b}$$

Again we have written these equations in terms of absolute densities, rather than the natural logs of densities. The equation for the prey (Equation 9a) is the same as the equation for the competitor in Equation 8a, except only a fraction $e^{-ap(t)}$ of the prey escape from the predators; in this expression, a gives the searching efficiency of the predators (Holling, 1959; Hassell, 1978). The number of prey consumed by predators, $n(t)[1 - e^{-ap(t)}]$, leads to the production of λ_2 predators in the next generation. This direct translation of consumed prey into new predators is an appropriate assumption for host–parasitoid interactions, and Equation 9 is a modified version of the well-known Nicholson–Bailey host–parasitoid model (Hassell, 1978). Figure 4B shows the changes in population density of the prey species with increases in the mean value of T. Comparing these results with those obtained

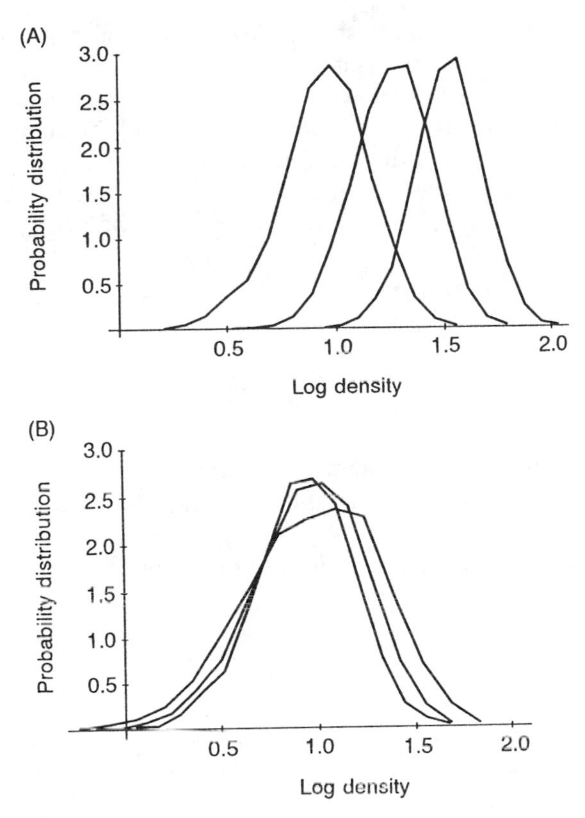

FIGURE 4. The effects of climate change on population density in communities containing two species. In (A) the focal species competes with the other species, while in (B) the focal species is a prey of the other, predatory, species. In each graph, the environmental parameter T has variance 0.1 and mean values 0, 0.05 and 0.1. It is assumed that increasing values of T increase the population growth rate of the focal species, and therefore result in increases in mean density. (A) The equations governing the population dynamics are given as Equations 8a and 8b. The parameter values are $\lambda_1 = \lambda_2 = 1.22$, $r_1 = r_2 = 0.05$, $\alpha = \beta = 0.5$, and $c = 1.0$. (B) The equations are given as Equations 9a and 9b, where $n(t)$ is the density of prey and $p(t)$ is the density of predators. The parameter values are $\lambda_1 = 1.22$, $\lambda_2 = 0.81$, $r_1 = 0.05$, $a = 0.5$, and $c = 1.0$. Note that the equations for the focal competitor in (A) and the prey in (B) depend in the same way on T and have the same intraspecific density dependence.

for the case of two competitors (Figure 4A) demonstrates that the population density of the competitor is more sensitive to changes in climate than the population density of the prey.

The discussion so far has centered on the change in population density resulting when climatic changes directly alter the per capita population growth rate of the focal species only. To consider also the effects from changes in the per capita population growth rate of other species, a new matrix, $\mathbf{M}_1$, must be constructed by replacing the first column of $\mathbf{C}$ with the vector whose elements are $-E[\delta F_i/\delta T]/E[\delta F_1/\delta T]$. For example, for the three-species case in which the focal species is species 1,

$$\mathbf{M}_1 = \begin{bmatrix} -1 & \dfrac{-E[\delta F_1/\delta N_2]}{E[\delta F_1/\delta N_1]} & \dfrac{-E[\delta F_1/\delta N_3]}{E[\delta F_1/\delta N_1]} \\ \dfrac{-E[\delta F_2/\delta T]}{E[\delta F_1/\delta T]} & -1 & \dfrac{-E[\delta F_2/\delta N_3]}{E[\delta F_2/\delta N_2]} \\ \dfrac{-E[\delta F_3/\delta T]}{E[\delta F_1/\delta T]} & \dfrac{-E[\delta F_3/\delta N_2]}{E[\delta F_3/\delta N_3]} & -1 \end{bmatrix} \quad (10)$$

Note here that the environmental parameter T may affect each species differently, since $\delta F_i/\delta T$ may have different magnitudes and signs for the different species.

The equation for the change in mean population density of species 1 is now

$$\Delta\overline{N_1}^* = -\Delta\overline{T}\,\frac{E[\delta F_1/\delta T]}{E[\delta F_1/\delta N]}\,(\det\mathbf{M}_1/\det\mathbf{C}). \quad (11)$$

Although this equation is more complicated than its predecessor (Equation 6), the analysis above reveals the same general expectation for the role played by species interactions in accentuating or mitigating the effects of climate change. If the vector given by the first column in $\mathbf{M}_1$ is more similar to the other columns than is the first column in $\mathbf{C}$, then $\det\mathbf{M}_1/\det\mathbf{C}$ will be less than one. This will decrease $\Delta\overline{N_1}^*$, and species interactions will buffer the effects of climate change. The opposite is true if the first column in $\mathbf{M}_1$ is more dissimilar to the other columns than is the first column in $\mathbf{C}$. Thus, when many species in the community experience direct effects from climatic changes, the influence of species interactions on the change in density of a focal species depends on the pattern of interactions among species and the pattern of climatic effects on population growth rates of the interacting species.

We end this section with two comments about this theoretical approach. First, our focus has been on predicting relatively small changes in population densities resulting from climate change. In particular, equations for $\Delta\overline{N}^*$ (Equations 2, 6, and 11) assume small changes in the climatic parameter $\Delta\overline{T}$. Predictions of the consequences of large environmental perturbations, such as may occur with climatic change, would require knowing the strength of species interactions at densities far from their current values and under environmental conditions very different from those at present. We, however, have focused on small environmental changes, because predictions of the consequences of relatively small environmental changes are more likely to be accurate.

Second, our approach is very different from those used to address the stability properties of communities (May, 1974; Pimm, 1982, 1984). Because our models are stochastic, the stability or instability of the dynamics is a moot question. Instead, we focus on perturbations imposed upon the community by external environmental changes. Although questions have been asked about external perturbations from evolutionary (Levins, 1975) and ecological (Lane, 1986) perspectives, we believe ours is the first attempt to use a stochastic approach to address this question.

From a practical point of view, the most important idea from our models is that the direct effects of climate change on population dynamics should be separated from the indirect effects operating through interactions within and among species. This simple mathematical result bridges studies in two disparate areas of ecology: physiological ecology, which addresses the effect of the environment on individual and population growth rates (Dunham, this volume), and population ecology, which examines the strength of ecological interactions among species. Because the response of population densities depends on both of these, predicting the consequences of climate change requires research in both fields simultaneously for the same organisms.

DATA

We have two separate objectives in coupling theory with data. The first is to validate the theory by testing its predictions. The most direct way to test our theory is to experimentally manipulate the environment of communities in controlled "microcosms" supporting organisms with very rapid generation times. Independent studies on population growth rates and interactions among species can produce predictions for the changes in population densities that will occur when an environmental factor in the microcosm is experimentally manipulated. The predictions can be tested directly using long-term measurements of population dynamics. These methods will be discussed in greater detail below. A less direct alternative is to use a comparative approach. For communities that have experienced known climatic trends, the theory predicts the relative response of species' densities. Therefore, if the communities differ in the type or strength of species interactions, the theory will predict qualitatively different responses of species densities to climate change. This approach has the advantage of being more natural than controlled microcosm experiments. Unfortunately, it requires far more information about species interactions and the response of population growth rates to climatic changes than is known for any community of which we are aware.

The second objective in coupling theory with data is to predict the consequences of climate change for communities at some time in the future. This procedure tacitly assumes that the theory is applicable, or has been validated using other species. Confidence in the applicability of the model may be built by constructing more detailed models that capture greater realism of the species in question, but ultimately the theory's applicability depends on how closely its assumptions match reality.

In the sections that follow, we first present examples of the type of experiments needed to validate the theory. We then apply the theory to published data to predict the potential response of two insect populations to climatic change. Throughout this discussion we focus primarily on single-species interactions relevant to our first theoretical prediction because experiments with single species are more tractable.

Testing the theory and its predictions: What is needed?

Validation of the model requires (1) an assessment of the effect of the environmental variable(s) of interest on population growth rates, (2) an estimate of the effect of the density-dependent factor(s) on population growth rates, and (3) experiments run over several generations in which both factors are manipulated. Ideally, we would like to present a single study that satisfies all three of these requirements. Unfortunately, we know of no such study, and so instead resort to describing three separate examples, each of which illustrates one of the three components required to test our predictions.

Howe's 1962 study of the grain beetle *Tribolium* typifies the physiological ecologist's assessment of the temperature sensitivity of population growth rate. Howe cultured *T. castaneum* under a variety of controlled temperature and humidity regimes, and recorded the development rate of larvae and the oviposition rate of females. From these data, he calculated the intrinsic rate of increase (r) under different environmental conditions (Figure 5). The left-skewed curve, with a sharp drop-off in fitness beyond the optimum point, is typical of the insects for which these data are available. This figure reveals a caution that must be applied to ecological studies. Many ecological models that include temperature, such as the degree-day models often used in applied ecology (Pruess, 1983; Wang, 1960), assume that there is a linear relationship between temperature and various demographic rates (such as development time or fecundity) that directly affects per capita population growth rates. Although degree-day models often predict year-to-year dynamics with considerable accuracy (reviewed in Higley et al., 1986), the inherent nonlinearity of species responses to temperature (and other environmental factors) may make these models less useful for longer-term forecasts or under rapidly changing conditions such as may arise due to global warming. If this is the case, a more complex, nonlinear

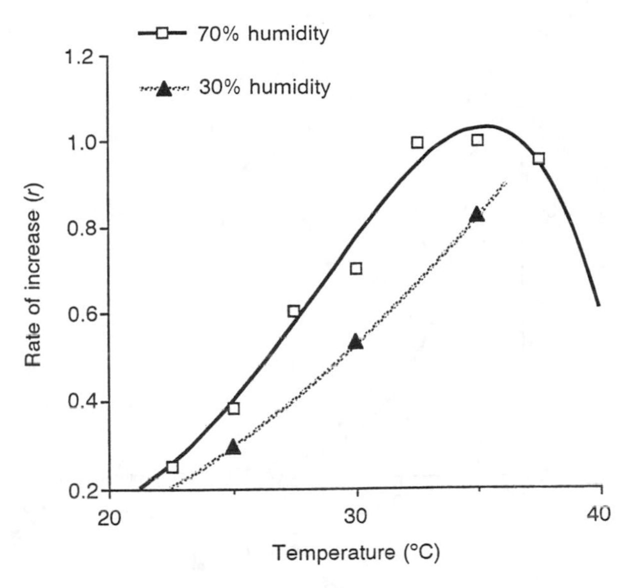

FIGURE 5. The intrinsic rate of increase (r) of *Tribolium castaneum* as a function of temperature under two humidity regimes. (Data from Howe, 1962.)

function (e.g., Logan et al., 1976; Sharpe and DeMichele, 1977) must be used to describe the response of species to environmental changes.

Many population ecologists have studied the role of density-related factors, such as competition and predation, on population growth rates. Frank's 1960 studies of intraspecific competition in *Daphnia pulex* cultures illustrate the typical methodology. In these studies, the density of organisms per unit of food resource was varied, and the population(s) grown for several generations under controlled conditions. Throughout the experiments, age at maturity, survivorship, fecundity, or other demographic parameters were measured. The typical response of per capita population growth rate to increasing density is nonlinear, as illustrated by *D. pulex* at densities above 4 per ml (Figure 6).

The third requirement to validate our models is to conduct experiments that vary environmental and density-dependent effects on population growth rates. The experiment we found that came closest to satisfying this requirement is Nicholson's (1958) classic work on the sheep blowfly, *Lucilia cuprina*. Rather than change an environmental factor in the experiments, however, he imposed density-independent mortality by removing a fixed proportion of individuals from cages at fixed time intervals. Because many environmental sources of mortality act in a density-independent fashion, we will equate density-independent mortality with environmental effects

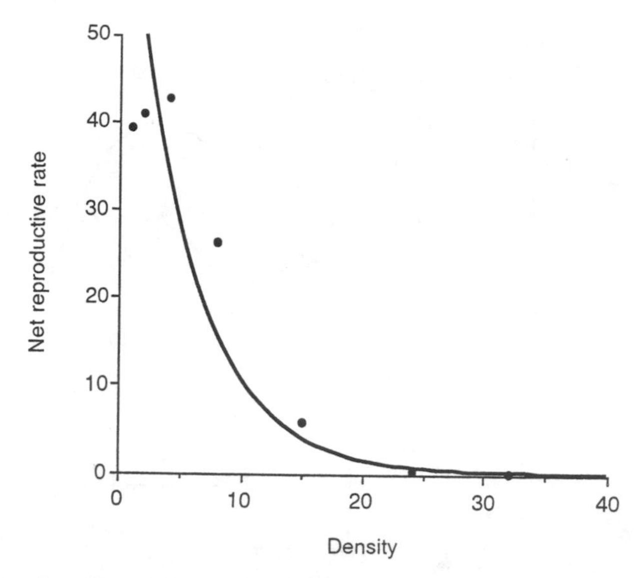

FIGURE 6. The effect of density in *Daphnia pulex* cultures on the net reproductive rate per generation (R_0). (Data from Frank, 1960.)

on population growth rates. Nicholson determined that both larval and adult flies compete for food resources, and he manipulated density dependence by varying the food supply. As the degree of density-independent mortality increased, the density of adults decreased in both food-limited and unlimited cultures, but the effect was less dramatic in the food-limited treatments. In fact, the mean adult population size in cultures experiencing adult and larval competition with 50 percent of the adults destroyed every second day was approximately the same as that in control cultures where there was no density-independent mortality. Nicholson's (1958) experimental results are consistent with our theoretical predictions: density-dependent factors can ameliorate the effect of density-independent mortality.

The three examples we presented above show that the types of experiments necessary to test our predictions are possible; what is needed is a series of experiments like those we describe, all conducted on the same organisms. These examples also point to the complications caused by nonlinearities in both environmental and density-dependent effects on parameters governing the population growth rates of species. Although we may ignore nonlinear effects for small climate changes, we might not properly do so for large changes. This underscores the need to perform detailed experimental studies on the long-term response of species densities to a changing environment.

Predicting the consequences of environmental change: Life table data with key factor analysis

Life table compilations record the dynamics of a population over several generations and, when coupled with data on environmental changes over the course of the census periods, can be used to identify density-dependent and density-independent factors that may regulate the population. This procedure takes advantage of naturally occurring fluctuations in weather and density. One statistical method used to analyze life table data is key factor analysis (Varley and Gradwell, 1970; Varley et al., 1974; Southwood, 1982). In this section, we will use published key factor analyses to examine two different, noninteracting insects. We then estimate the two parameters needed to predict the response of each species to climate change: (1) the direct effects of climate on population growth rates and (2) the strength of density dependence in population dynamics. The two insects we discuss are a plant bug, *Leptopterna dolabrata* (McNeill, 1973), and the Douglas fir tussock moth, *Orgyia pseudotsugata* (Mason and Torgersen, 1987).

The analysis of these data has two parts. First, environmental factors that influence population dynamics must be identified. The most sensitive procedure is to analyze sources of mortality in individual developmental stages, rather than analyzing the effect of environmental change on yearly fluctuations in population abundance; examining between-stage effects factors out variability in mortality in other life stages. Second, the environmental and density-dependent effects of the relevant environmental factors on annual population fluctuations must be determined. Even though mortality in a particular stage of an insect might be strongly correlated with an environmental factor, it will have a significant impact on long-term changes in population densities only if mortality in this stage is correlated with annual fluctuations in population density.

Table 1 shows life-table data for *L. dolabrata* (McNeill, 1973) and for *O. pseudotsugata* (Mason and Torgersen, 1987) for single representative years; for key factor analysis, data over many consecutive years were used. For each developmental stage, a k-value is calculated as the difference between the logarithms of the number alive in the previous and in the present stage. If specific sources of mortality can be identified, then these values can be partitioned into several classes. For example, in the 1965–1966 season of McNeill's study, an average density of 5.12 eggs/m^2 successfully hatched, while the density of nymphs entering instar 3 was 1.55/m^2. Thus the k-value for young nymphs is $\ln 5.12 - \ln 1.55 = 0.518$. The sum of the k-values, $K = k_1 + k_2 + \ldots + k_i$, gives the total generation mortality. Density dependence due to intraspecific competition can be detected by checking for a significant regression coefficient between the k-values and the log of population densities obtained for different years (Varley

TABLE 1. Life stages used in key factor analysis of *L. dolabrata* and *O. pseudotsugata*.

Age class	Number per m^2	k-Factor	k-Value
		L. DOLABRATA	
Fecundity	73.6	Maximum possible (k_{0a})	0.289
	37.8	potential (k_{0b})	0.360
Eggs	16.5	Parasitism (k_1)	0.104
		Infertility (k_2)	0.069
		Other (k_3)	0.101
Hatched	5.12	Loss of young nymphs (k_4)	0.518
Instar 3	1.55	Loss of late instar (k_5)	0.031
		Parasitism (k_6)	0.111
Adults	1.12		
		O. PSEUDOTSUGATA	
Eggs	4.537	Predation (k_1)	0.136
		Parasitism (k_2)	0.258
Early larvae	1.831	Parasitism (k_{3a})	0.006
		Disappearance (k_4)	0.584
Mid-larvae	0.471	Parasitism (k_{3b})	0.016
		Disappearance (k_{5a})	0.628
Late larvae	0.107	Parasitism (k_{3c})	0.014
		Disappearance (k_{5b})	0.107
Pupae	0.081	Parasitism (k_6)	0.135
		Predation (k_7)	0.194
Adults	0.038	Residual mortality (k_8)	0.165

Note: Data from McNeill, 1973 for *L. dolabrata* in the years 1965–1966; and from Mason and Torgerson, 1987, for *O. pseudotsugata* in the years 1974–1975.

et al., 1974; Hassell, 1985). Similarly, the potential effects of different climatic factors can be estimated by regressing the k-values in different years against weather data. Key factors are defined as the k-values that are correlated with the pattern of total mortality K across generations. Podoler and Rogers (1975) suggest regressing the k-values against K; the factors having the highest slopes (perfect correspondence would yield a slope of 1.0) are judged the key factors.

Once the environmental factors that influence annual population dynamics have been identified, the estimates needed to predict changes in mean population densities following climatic changes can be obtained using multiple regression. Specifically, assume that H environmental factors affecting population dynamics have been identified, and denote these by T_i (i = 1 to H). The effects of these environmental factors and population density on population fluctuations can be estimated from the regression equation

$$N(t + 1) = b_N N(t) + \Sigma b_{T_i} T_i + C \tag{12}$$

From this regression and Equation 1, we obtain $E[\delta F/\delta T_i] = b_{T_i}$ and $E[\delta F/\delta N] = (b_N - 1)$; subtracting 1 from b_N is required because F is measured on a per capita basis. This allows reformulating the expression for changes in the mean population density in response to climatic changes as

$$\Delta \bar{N}^* = -\Delta \bar{T}_i b_{T_i}/(b_N - 1) \tag{13}$$

If more than one environmental factor affects the per capita population growth rate, the effect of each factor on mean population density can be assessed separately.

The key factors affecting the plant bug *L. dolabrata* are nymphal mortality and variation in potential fecundity from year to year (McNeill, 1973). Two climatic factors were correlated with the key factors. Mortality during the young nymph stage is correlated with heavy rains, which cause drowning (b = 0.3996), whereas mortality in older nymphs is correlated with high June temperatures (b = 0.195) (McNeill, 1973). Density dependence is suggested by significant regressions of two k-values on log density: reduced realized fecundity (b = 0. 514) and increased old-nymph mortality (b = 1.093) were observed at high population densities (McNeill, 1973). We used Equation 12 to obtain the multiple regression coefficients of $N(t)$, June temperature, and rainfall on $N(t + 1)$ (R^2 = 0.9996, $F_{3,1}$ = 760.924, P = 0.0266). The high portion of variance explained is a direct consequence of McNeill's identification of the key density-independent factors affecting the population and the relatively strong density dependence. Using Equation 13, these results predict decreases in mean population density with either increasing May rainfall or June temperatures (Table 2).

The data provided by Mason and Torgersen (1987) for the Douglas fir tussock moth consisted of k-values for the six stages listed in Table 1. None of the k-values was judged to reflect key factors (Mason and Torgersen, 1987); however, k_5 (disappearance of mid to late stage larvae) was the largest measured source of mortality. We obtained weather data (maximum and minimum daily temperature) from the ClimateData database for the Crater Lake Ranger Station, approximately 25 kilometers north of the area in which the tussock moth was studied. From data in Brookes et al. (1978), we approximated the dates at which each stage was present in the field and computed the mean minimum and maximum daily temperatures, the extremes, and the standard deviations for the time spanning the mid- to late-larval stage in each of the years covered by the study. We asked if factor k_5 was correlated with any of the temperature statistics during that stage. Of the six regressions, only the regression on absolute maximum temperature

TABLE 2. Predicted changes in mean density with climate change.

Demographic stage	Factor	Regression factor affected[a]	$\Delta\bar{N}^*$ effect[b]
		L. DOLABRATA	
Female fecundity and mortality of old nymphs	Log density	−1.34 ± 0.11	—
Mortality of young nymphs	Heavy rainfall	−0.062 ± 0.029	−0.046 (log density/mm rainfall/h)
Mortality of old nymphs	Mean June temperature	−0.160 ± 0.017	−0.12 (log density/°C)
		O. PSEUDOTSUGATA	
Parasitism of eggs, larvae, and pupae?[c]	Log egg density	−1.13 ± 0.45	—
Mid-larval mortality	Maximum temperature	−0.02 ± 0.18	−0.018 (log density/°C)

[a]Values of b_N or b_T from Equation 12. Key factor analysis was used to identify statistically significant environmental and density-dependent effects in different stages.
[b]Predicted change in mean log population density in response to changes in the environmental factors.
[c]No key factor was identified in this study; however, parasitization was the major cause of density-dependent mortality.

had a slope significantly different from zero (b = 0.1583, $P < 0.05$, R^2 = 0.5866). Biologically, this suggests that mid- to late instar larvae experience increased mortality at higher temperatures, as has been found in the lab (Brookes et al., 1978). The only density dependence affecting the k-values was the combined parasitization of all stages, which was correlated with the density of eggs in the present (b = 0.144) and the previous (b = 0.334) generations, suggesting a weak, delayed density dependence (Mason and Torgersen, 1987). Berryman (1978) has suggested that the periodic outbreaks of the Douglas fir tussock moth are driven by a lagged density dependence; however, the population in this study was at sub-outbreak density throughout the census period. Neither the egg density in year t nor that in year $t - 1$ explained a significant amount of variation in adult population size, so we have ignored the lagged effect and present only the weak, direct density dependence to illustrate the analysis in Table 2. We again used Equation 12 to obtain multiple regression coefficients of $N(t)$ and maximum temperature during the mid- to late larval stage (R^2 = 0.0283,

$F_{2,3} = 0.0436$, $P > 0.9$). No key factors were identified that might explain the year-to-year variation in population dynamics (Mason and Torgersen, 1987). This results in a regression model that explains virtually none of the between-generation variance in population size. Thus, although weather and parasitism had significant effects on mortality at some stages, Equation 13 predicts virtually no change in the long-term mean population density with temperature.

The changes in mean population density with climate change as predicted by these examples illustrate an important point: density dependence must be considered in anticipating the responses of populations to environmental change. The changes in k-values with respect to an environmental parameter were used to identify potential shifts in population density associated with climate change. For example, neither egg density nor temperature explain year-to-year variation in the number of adult tussock moths in the study population, so the model predicts little response of the population to temperature change despite increased larval mortality at higher temperatures. In contrast, high June temperatures that increase larval mortality reduce the adult population density of *L. dolabrata*. If this were the only factor affecting population growth, the potential change in the log population density from one year to the next would be −0.16 log density/°C. However, the predicted change (Table 2) in mean equilibrium population density, $\Delta\overline{N}^*$, is only −0.12 log density/°C. Thus, density dependence operating after the old-nymph stage partially compensates for the annual changes in population density driven by the effect of temperature on old-nymph mortality.

We end with three cautions. Key factor analysis suffers from several statistical and methodological problems, especially in the detection of density dependence (Varley et al., 1974; Hassell, 1985; Hassell, 1987; Caston and Lawton, 1987). Some recent reviews of key factor studies in insect populations (Dempster, 1983; Stiling, 1988) have failed to find evidence for density dependence in about half of the studies examined. Hassell (1985, 1987; Hassell et al., 1989) suggests several reasons why these reviews underestimate the prevalence of density dependence. First, the likelihood of detecting density dependence is positively correlated with the duration of the study (Hassell et al., 1989); because density dependence is identified via regression methods, short-term studies simply do not provide enough data. Secondly, studies of a single population based on a single census per generation may miss density dependence that acts only over limited portions of the life cycle (Hassell, 1985); collecting census data at several stages within each generation minimizes this problem. Finally, spatial variation in density dependence is often obscured by averaging population values within generations (Hassell, 1987). Again, attention should be paid to within-generation variation in density dependence.

Our second caution is that lagged, or delayed, density dependence is often overlooked in life table studies (Varley et al., 1974; Turchin, 1990). Although methods such as time series analysis (Turchin, 1990) can detect delayed density dependence, these methods demand more generations of data than are normally available. When significant delayed density dependence is overlooked, our model will underestimate density-dependent effects on population growth, resulting in an overestimate of the effects due to environmental change. (Our examples above did not include lagged density dependence because of the relatively limited data available.)

Our final caution is that although increasing the number of censuses within generations allows one to identify more sources of density dependence, it also increases the number of variables that enter the analysis. Similarly, one can look for correlations between mortality at any stage and as many environmental factors as one chooses to analyze. However, in searches for correlations among a large set of variables, some relationships will be significant purely by chance. In applying these models, we have used the statistical significance level only to identity factors that *might* be influenced by climatic factors and population densities, rather than to provide statistical confidence in our estimates; our predictions are best regarded as hypotheses. The only solution to this problem is to take these results into the laboratory or field for manipulative experiments to quantify the relationship between the mortality factor and the environment or density. Longer-term studies with more intensive sampling are needed to provide the basic data on potentially susceptible species and life stages, but the problem of controlling experiment-wide errors in analyzing those relationships remains.

SUMMARY

Our main conclusion is that density dependence and species interactions can play a major role in determining the changes in densities of species that will result from climate change. Because population ecologists have long been interested in interactions within and between species, they should be well positioned to help predict the long-term changes in population densities that will result from either natural or anthropogenic changes in climate. We also provide some predictions and tools to direct further research in this area.

From our theoretical model, we predict that (1) density-dependent factors limiting per capita population growth rates can ameliorate the direct effects of climate change on individual reproductive rates, and (2) for species holding unique ecological roles, and therefore having few competitors, interactions with other species in the community act to buffer population densities against climate change.

The above two predictions can potentially be validated using laboratory experiments. Examples of studies done on *Tribolium*, *Daphnia*, and blowflies illustrate the types of experiments needed to validate the theory. Although we know of no population for which all of the necessary data are currently available, the methodology to conduct the tests is well established. Testing this theory will also provide us with important information on the roles of density-dependent and density-independent factors in regulating populations.

Long-term data sets on population dynamics and annual weather fluctuations can provide the necessary information to predict the response of mean population density to long-term climatic trends. We demonstrated this by applying key factor analysis to two species of insects.

APPENDIX A

The stationary population density distribution, N^*, is obtained from the identity

$$N'^* = F(N^*,T) + N^* \quad \text{(A1)}$$

where N'^* denotes the random variable with the same distribution as N^*. This equation may be solved explicitly for the case when $F(N,T)$ can be written in the form $F(N,T) = F_1(N) + F_2(T)$, as will result when the environmental parameter T affects only the density-independent component of the per capita population growth rate. For functions of $F(N,T)$ with this form, no generality is lost in rescaling the environmental parameter such that $F(N,T) = F_1(N) + T$. If $p(n)$ and $q(t)$ denote the probability density functions of N^* and T, then $p(n)$ must satisfy the convolution

$$p(n) = \int q[(n - u - F_1(u)] \, p(u) \, du \quad \text{(A2)}$$

This equation is a Fredholm integral equation of the second kind, which can be solved using standard numerical methods (Mayers, 1974).

APPENDIX B

To calculate the effects of changing the distribution of T on the distribution of N^*, assume that T is changed by adding the distribution τ which has small mean, variance, and other moments. This will result in a change to a new stationary distribution. The new stationary distribution can be approximated to at least the first two moments with the distribution $N^* + \nu$, where ν is a distribution independent from N^*. By definition, the expected value of the per capita population growth rate calculated for the new stationary distribution and the new distribution of T must equal zero, so

$$0 = E[F(N^* + \nu, T + \tau)] \quad \text{(B1)}$$

This expected value is calculated over four random variables, N^*, ν, T, and τ. Taking the expected values over ν and τ first,

$$0 = E_{N^*,T}[E_{\nu,\tau}[F(N^* + \nu,\ T + \tau)]]$$
$$\cong E_{N^*,T}[E_{\nu,\tau}[F(N^*,T) + \{\delta F(N^*,T)/\delta N\}\nu + \{\delta F(N^*,T)/\delta T\}\tau]]$$
$$= E_{N^*,T}[\delta F(N^*,T)/\delta N]\Delta\overline{N} + E_{N^*,T}[\delta F(N^*,T)/\delta T]\Delta\overline{T} \quad \text{(B2)}$$

where $\Delta\overline{N} = E[\nu]$ and $\Delta\overline{T} = E[\tau]$. Equation 2 follows immediately. This approximation assumes that for any fixed values of N and T, $F(N + \nu, T + \tau)$ can be written as a linear function of ν and τ. However, even when this assumption is violated, density dependence operating through $E[\delta F(N,T)/\delta N]$ still acts to decrease the change in population density ΔN^* as found in Equation 2.

The multispecies case proceeds in the same manner. Let $\mathbf{N} = \{N_1, N_2, \ldots, N_S\}$ and $\phi(\mathbf{N},T) = \{F_1, F_2, \ldots, F_S\}$, the vector containing the per capita population growth rates of all species. For any change in the distribution of the environmental parameter, τ, there is a resulting change in species densities, $\nu = [\nu_1, \nu_2, \ldots, \nu_S]$. The new stationary distribution is approximated by $\mathbf{N}^* + \nu$, which must satisfy

$$0 = E[\phi(\mathbf{N}^* + \nu,\ T + \tau)]$$
$$= E_{\mathbf{N}^*,T}[E_{\nu,\tau}[\phi(\mathbf{N}^* + \nu,\ T + \tau)]]$$
$$\cong E_{\mathbf{N}^*,T}[E_{\nu,\tau}[\phi(\mathbf{N}^*,T) + \{\delta\phi(\mathbf{N}^*,T)/\delta\mathbf{N}\}\nu + \{\delta\phi(\mathbf{N}^*,T)/\delta T\}\tau]]$$
$$= E_{\mathbf{N}^*,T}[\delta\phi(\mathbf{N}^*,T)/\delta\mathbf{N}]\ \Delta\overline{\mathbf{N}} + E_{\mathbf{N}^*,T}[\delta\phi(\mathbf{N}^*,T)/\delta T]\ \Delta\overline{T} \quad \text{(B3)}$$

Here, $\Delta\overline{T} = E[\tau]$, $\Delta\overline{\mathbf{N}} = \{E[\nu_1], E[\nu_2], \ldots, E[\nu_S]\}$, and $E_{\mathbf{N}^*,T}[\delta\phi(\mathbf{N}^*,T)/\delta\mathbf{N}]$ is a matrix with elements in the ith row and jth column $E_{\mathbf{N}^*,T}[\delta F_i(\mathbf{N}^*,T)/\delta N_j]$. Applying Cramer's rule (Apostol 1969, p. 93), the value of $\Delta\overline{N_1^*}$ is given by Equation 11 in the text.

To examine the case in which the environmental parameter T affects only the per capita population growth rate of the focal species 1, assume that $E_{\mathbf{N}^*,T}[\delta F_i(\mathbf{N}^*,T)/\delta N_j] = 0$ for all $i \neq 1$. In this case, $\det\mathbf{M}_1 = \det\mathbf{C}_1$, which leads to Equation 6.

APPENDIX C

This appendix presents a geometrical interpretation of determinants and Equation 6. To give a concrete example, we first consider the following matrix $\mathbf{C}$ for a community containing three competitors:

$$\mathbf{C} = \begin{bmatrix} -1 & -0.6 & -0.6 \\ -0.6 & -1 & -0.6 \\ -0.6 & -0.6 & -1 \end{bmatrix} \quad \text{(C1)}$$

Here, the values -0.6 of the off-diagonal elements imply that the competition coefficients between species are 0.6. Each row i in the matrix represents a vector of the effects of species i on the other species in the community. The determinant of this matrix corresponds to the volume of the parallelepiped constructed by adding together all of the row vectors. For example, Figure 7A shows the vectors defined by matrix **C**; because the species are competitors, all vectors point in the negative direction. The parallelepiped defined by these vectors in shown in Figure 7B, and the volume of this polygon equals the determinant of **C**.

Analysis of the matrix $\mathbf{C}_1$ is much the same. For this example,

$$\mathbf{C}_1 = \begin{bmatrix} -1 & 0 & 0 \\ 0 & -1 & -0.6 \\ 0 & -0.6 & -1 \end{bmatrix} \tag{C2}$$

The determinant of $\mathbf{C}_1$ is similarly given by the volume defined by its row vectors. However, because the first row contains zeros except at the diagonal, the magnitude of the volume given by det$\mathbf{C}_1$ equals the magnitude of the area of the two-dimensional trapezoid lying in the horizontal x–y plane of Figure 7B; this is given by the matrix

$$\begin{bmatrix} -1 & -0.6 \\ -0.6 & -1 \end{bmatrix} \tag{C3}$$

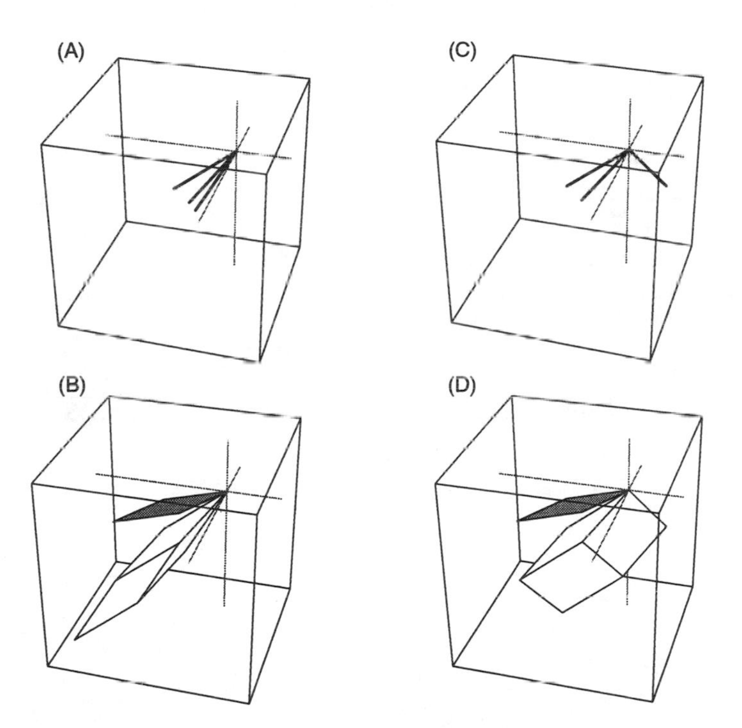

FIGURE 7. Vectors defined by matrices presented in Appendix C.

This trapezoid is simply the projection of the parallelepiped onto the plane caused when the dimension corresponding to species 1 is removed. Thus, the value of $\det\mathbf{C}_1/\det\mathbf{C}$ is the area of the trapezoid divided by the volume of the parallelepiped. In this particular case, $\det\mathbf{C}_1/\det\mathbf{C} = 1.81$.

To compare with the example of three competitors, we also present an example of a focal predator and two prey species, with the interaction matrix given by

$$\mathbf{C} = \begin{bmatrix} -1 & 0.6 & 0.6 \\ -0.6 & -1 & -0.6 \\ -0.6 & -0.6 & -1 \end{bmatrix} \qquad \text{(C4)}$$

Because species 1 is a predator of the other two species, the two nondiagonal elements in the top row are positive. Figure 7C shows the vectors defined by this matrix, and Figure 7D shows the parallelepiped defined by the matrix $\mathbf{C}$ and the trapezoid defined by matrix $\mathbf{C}_1$. The vector showing the effect of other species on species 1 (the first row in matrix $\mathbf{C}$) now lies in a different quadrant, because increases in the density of either of the prey species will increase the population growth rate of their predator. This increases the value of $\det\mathbf{C}$, leading to a value of $\det\mathbf{C}_1/\det\mathbf{C}$ equal to 0.69. Thus, the interactions among predator and prey act to mitigate the effects of climate change relative to interactions among competitors.

CHAPTER 9

INDIVIDUAL-BASED MODELS FOR PREDICTING EFFECTS OF GLOBAL CHANGE

William W. Murdoch

Global environmental change, and especially changes in climate, will initially affect ecological communities by altering the physiology and behavior of individual organisms. Such alterations will directly change the dynamics of populations, primarily by affecting vital rates (e.g., birth, death, and emigration rates) and will also indirectly alter vital rates of other interacting populations (e.g., prey species). The links between individuals and communities are forged by the effects of the environment on the dynamics and interactions of populations. A key prerequisite for predicting the effects of global change therefore, is models that describe how environmental change—via its effects on individuals—affects the dynamics of populations and communities.

Individual-based models of population and community dynamics provide an obvious framework within which to investigate and forecast the likely effects of global change. The key feature of such models is their grounding in the mechanisms determining vital rates, and hence dynamics. Such a mechanistic understanding will in many cases be necessary to predict events under qualitatively different circumstances from those now existing.

Individual-based models usually trade off generality for a detailed account of mechanisms in a *particular* population or system. This feature renders them in principle highly testable. However, their particularity, coupled with their inherent complexity, may render them nongeneral.

Individual-based models constitute one of two approaches suggested by contributors to this volume (see chapters by Dunham and by Pacala and Hurtt). The second approach seen in this volume involves relatively simple models that aim at generality (chapters by Hassell et al., Ives and Gilchrist, and Pacala and Hurtt). This approach concentrates on, and isolates, an ecological phenomenon of interest and evaluates its dynamic effects in a model otherwise containing little biological detail. The approach is well illustrated by Hassell et al.'s chapter exploring the possible consequences of an early-season mismatch in the emergence times of a parasitoid insect and its host.

Simple models have, however, a major disadvantage: They aim at generality not so much by generalizing biology as by omitting it. For example, the logistic model ignores the fact that birth and death rates are determined by different processes. Simple models are therefore difficult to test rigorously because their structure does not match the processes and mechanisms that drive particular real ecological systems. The crucial question is whether a simple model has omitted essential biology rather than discardable "detail," and that question is not easy to answer empirically.

In what follows I address three main issues: (1) the need for models that are tested in as broad a range of circumstances as possible, (2) the need for simplification in the face of complexity, and (3) the need for generality in a highly particular world. I propose not to choose between the two modeling approaches outlined above, but rather to suggest that a successful strategy needs to link them. Indeed, although an individual-based approach can lead to a well-tested predictive model of a particular system, ultimately its main use may be to serve as the basis from which simpler and more general models can be developed. That is, I view detailed individual-based models as a step in a process rather than an end result.

A strict interpretation might restrict the term "individual-based model" to simulations of collections of nonidentical individuals (Metz and Diekmann, 1986; Huston et al., 1988; Hallam et al., 1990; Dunham, this volume). Here I use the term more broadly to include, for example, models that track the dynamics of different groups of individuals, the members of each group being assumed to be the same (Metz and Diekmann, 1986). "Stage-structured" models of insects provide a good example because individuals within an instar often are very similar, but the properties of the various instars may be very different (Gurney et al., 1983; Murdoch et al., 1987; Godfray and Hassell, 1989; Murdoch et al., in press b).

INDIVIDUAL-BASED MODELS FOR POPULATIONS

Although our concern here is with global change, individual-based models can play a central role in developing a broad understanding of ecological systems. It will therefore usually be appropriate to develop the model with a broader purpose than predicting the effects of human-induced changes.

In any case, a central issue is whether the model can predict dynamics under the range of existing conditions. The question of testability is therefore central.

The need to understand mechanisms leads to a focus at the individual level. However, to provide insight into *population* dynamics, an individual-based model needs to be based on an explicit hypothesis about those dynamics. For example, forest models are based on the central population-level hypothesis that competition among individuals for nutrients, light, or other resources determines the outcome (Pacala and Hurtt, this volume). This leads to a focus on the acquisition and allocation of resources. Because tree size has a dominant effect on those processes, an individual-based model needs to distinguish individuals of different sizes. Size-dependent acquisition and allocation of energy is also thought to be the key to dynamics in lake fish populations (e.g., Adams and DeAngelis, 1987; Madenjian and Carpenter, 1991) and in the zooplankter *Daphnia* (McCauley and Murdoch, 1987). Consequently, a model describing the acquisition and allocation of energy by individual lake fish or *Daphnia* also requires information on size (McCauley et al., 1990; Gurney et al., 1990).

Indeed, energy needs (or nutrients in general—I will refer henceforth only to energy) are likely to be central in the dynamics of most systems. Thus predicting the effects of global change involves, at a minimum, predicting how such change will alter energetic processes in individuals. In the case of increasing temperature, for example, some of the effects on the relevant rates are already known for some species (Dunham, this volume).

The starting point is thus a model of individual physiology and behavior. The model consists of rules determining (for an individual of a given size, age, or stage) properties such as intake rate, assimilation efficiency, maintenance costs, and allocation of assimilate to growth and reproduction, all as a function of the environment. For plants the rules will cover, for example, allocation to above ground versus below-ground structure. Where predation is an important part of population dynamics, the model may also need to consider allocation to defensive features, such as antipredator spines in zooplankton or chemical defenses in plants (Ayres, this volume). Behavior may also be an important component of the model. For example, an individual may adjust the time it spends foraging in a high-food but also high-risk area in response to changes in the individual's hunger state, its size, and the density of its predators.

For many systems, such a model of individual physiology and behavior is intrinsically subjectable to a rich array of experimental tests based on predictions derived from the model (Step 1 in Figure 1). For example, models of individual lake fish and *Daphnia* predict many different aspects of individual performance, such as size-specific growth rate and fecundity, time to maturity, size at maturity or at the onset of winter, and ultimate size—all as a function of the rate of supply of food.

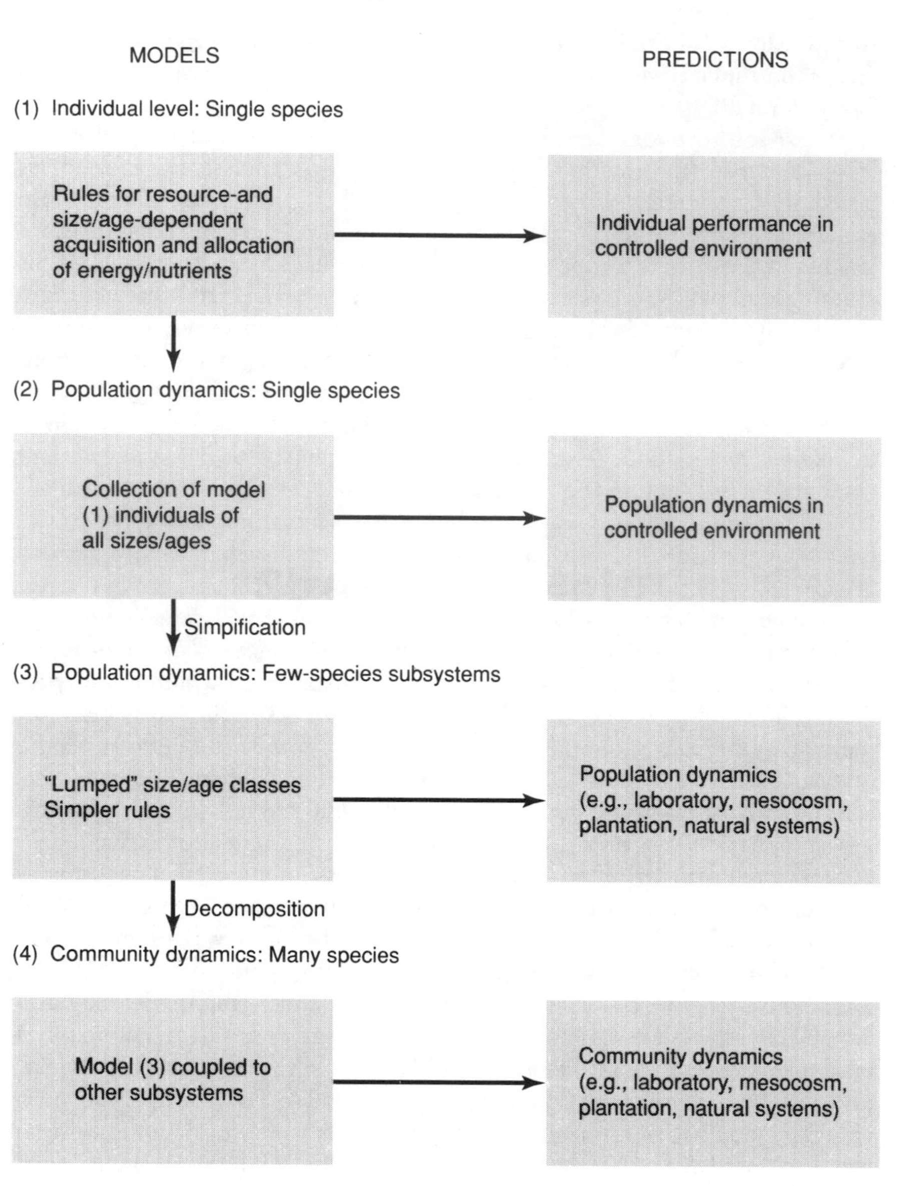

FIGURE 1. Outline of a process for developing predictive models for the response of populations and communities to human-induced changes in the environment.

The key feature of this process is the clear distinction between information that defines the model and information needed to test it. In *Daphnia* and lake fish, for example, the processes in the models are described by functions relating energy uptake, and its allocation to growth and reproduction, to such variables as the individual's length, weight, and the density of food particles. These functions are derived from, for example, observations on feeding and measures of individual respiration in different circumstances. Tests of the models' predictions then require wholly different experiments. For example, individuals are raised at a range of food supplies, or are followed under varying food abundances in the field, and their age-specific fecundity schedules or ultimate sizes are compared with those predicted by the models (Gurney et al., 1990; DeAngelis et al., 1991). Where predation is important, predictions concerning individual performance might also include, for example, the size-specific probability of death from fish predation in zooplankton, or the percentage of defoliation at a given level of insect attack in plants.

Figure 1 suggests that tests of the individual model might be most effectively done in controlled environments, though this is certainly not intended as a proscription. In *Daphnia* the process has been relatively straightforward and has led to quite rigorous laboratory tests of the model (Gurney et al., 1990). However, fair success was also achieved in field tests of a model of first-year growth of walleye perch (Madenjian and Carpenter, 1991) and of first-year growth and survival of smallmouth bass (DeAngelis et al., 1991), though the tests of assumptions and some of the parameter estimation were less direct than could be done in a more controlled environment. The challenge to test the model independently of the data that parameterized it seems much more difficult for models of forest trees, another group in which extensive individual-based modelling has been done (Pacala and Hurtt, this volume); experimental tests may be restricted to the smallest size classes.

The next logical step in the process is to develop a population model that represents a collection of individuals, each obtaining energy according to the rules of the individual model (Step 2 in Figure 1). Indeed, this step is sometimes the final aim as, for example, in the JABOWA-derived models for forest communities and in current models of lake fish populations (DeAngelis and Rose, in press).

In principle, a population model consisting of a collection of model individuals should also be highly testable because it can be asked to predict detailed population dynamics in simple environments (Figure 1). For example, a *Daphnia* model of this sort should be able to predict, for populations raised in the laboratory, such details as mean density, mean size structure, presence of cycles, cycle period and amplitude, and changes in demography through time. Successful prediction of this rich set of details would provide confidence in the individual-based population model. The model could possibly also be used to predict detailed dynamics in somewhat more complex situations, for example, in "mesocosm" (stock tank) experiments in which the algal population is also dynamic and a genuine predator–prey interaction exists between the *Daphnia* and algae (McCauley and

Murdoch, 1987). Equivalent tests in plant species could be made in uniform plantations.

However, population-level individual-based models may not be testable in natural situations for several reasons (Murdoch et al., in press a). First, because many species usually interact in real systems (see below), a full model would require many (perhaps hundreds) of parameters. Thus such models are extremely data-hungry (but see DeAngelis and Rose, in press). Second, there is a temptation, and perhaps a need, to curve-fit rather than test because of the model's complexity and the virtual impossibility of estimating all of its parameters independently in a fluctuating and uncontrolled environment. Third, even though the reason for building such models is to specify and test hypotheses about mechanisms, they become hard to understand, so we may lose the ability to see why a given change in the model has a particular dynamic effect.

The problem of testing complex models raises the second issue mentioned above: the need to simplify individual-based population models. Little experience exists, however, to tell us when or how best to simplify individual-based models. A priori, the obvious place to simplify is after the individual-based model has succeeded in predicting details of population dynamics in well-controlled environments with one or very few species (Step 3 in Figure 1). Only then would one have confidence in the model.

The crucial feature of the individual-based modeling approach is that it allows simplification to be a hypothesis-testing process. Several approaches are possible. For example, Nisbet et al. (1989) lumped size classes to produce a *Daphnia* model that recognizes only a limited number of stages (eggs, juveniles, "adolescents," and adults) and assumes that individuals in a given stage all have the same properties. Whether this simplification is too crude can again be tested by comparing its predictions (e.g., mean density, stage structure) with the observed laboratory population dynamics.

Notice that the population model resulting from such a process of simplification may ultimately be as simple as that generated by the second approach discussed in the Introduction, in which we begin with a simple model that aims at generality by isolating the phenomenon of interest. Thus the process of deriving individual-based models is in principle able to effect a smooth and continuous linkage between individual-based models and simple models of the second type.

Two key differences exist, however, between these two approaches. First, only the individual-based modeling approach forces us to be explicit about the biological mechanisms and details that are omitted from the model. Second, this approach allows us to test the effects of these omissions on our ability to predict dynamics; i. e., we are forced to *test* the claim that we have omitted only "discardable detail." Indeed, the process should allow us to choose which aspects of the dynamics we are willing not to predict.

INDIVIDUAL-BASED MODELS FOR COMMUNITIES: ARE COMMUNITIES DECOMPOSABLE INTO WEAKLY COUPLED SUBSYSTEMS?

The need for simplification, which is pressing in one-species or few-species systems, is inescapable for multispecies communities. Here two aspects of complexity must be considered. The first is that discussed above, namely, the difficulty of incorporating all of the individual-level details for all the species in the interaction. For a model that needs to be testable in natural situations, this is neither feasible nor desirable. The second is that, even with the simplest of representations at the population level, not all of the species in the system can be included explicitly. In this section I will consider this second aspect.

In fact, no model (individual-based or otherwise) includes all the interacting species in the community being studied. All community models contain, whether explicitly or not, the hypothesis that some species can be ignored. Forest models (Pacala and Hurtt, this volume), for example, mostly assume without discussion that herbivores (and the herbivores' predators, etc.) are not important determinants of the distribution and abundance of forest tree species and thus omit these groups. As a consequence, allocation of resources to defense against herbivores, for example, plays no role in the dynamics (but see Ayres, this volume).

Forest models thus implicitly assume that communities are decomposable into subsystems whose member species interact strongly with one another but interact weakly—or not at all—with the rest of the community. The second class of models discussed in the Introduction, namely simple models aiming at generality, typically also make this implicit assumption; indeed, they usually involve only two or a few species. A predictive theory for human-induced change at the community level needs to make explicit this assumption about decomposability. Moreover, it needs to develop ways to distinguish those groups of interacting populations that can be treated as subsystems, and to subject that hypothesis itself to experimental test (Figure 1).

Murdoch and Walde (1989) argue that a system is decomposable into subsystems if these subsystems are only weakly coupled. They present formal definitions for weak and strong coupling between species: From the point of view of species 1, species 2 is weakly coupled (or even uncoupled) if we can successfully describe the dynamics of species 1 without an explicit equation describing the dynamics of species 2. Suppose the interaction between two species can be specified as follows:

$$dN_1/dt = N_1 f_1(N_1) + a_1 N_1 g_1(N_2)$$

$$dN_2/dt = N_2 f_2(N_2) + a_2 N_2 g_2(N_1)$$

where the functions f_i and g_i are, respectively, the intraspecific and interspecific interactions, and the a_i are numbers that determine whether the interspecific effects are large or small. Suppose further that a_1 is large (i.e., species 2 has a strong effect on the rate of change of species 1). If a_2 is also large, there is a complete feedback loop from species 1, through species 2, and back to species 1. Thus both equations are needed and the two populations are considered strongly coupled. This system cannot be decomposed into two subsystems of one species each. However, if a_2 is small (species 1 has a weak effect on species 2), the feedback from species 1 through species 2 is incomplete, because the dynamics of species 2 is largely independent of (i.e., uncoupled from) the density of species 1. Now the system is decomposable: An equation for species 2 is not needed to predict the dynamics of species 1, and species 2 can be treated merely as part of species 1's environment. However, because species 2 has a strong effect on species 1 (recall that a_1 is large), it needs to be incorporated into the description of species 1's environment via some function(s) and parameter(s). If a_1 and a_2 are both small, the two species are largely irrelevant to each other's dynamics; they do not even need to be incorporated as parts of each others' environment.

These ideas have implications for studies that try to define community dynamics in terms of food webs (see also Hastings, 1990). They suggest that we need to know not only the direction and strength of trophic interactions (the links or arrows in food webs), but also whether feedback loops are complete. The benefit of such hard-earned information, however, is a major simplification of the task of modeling community dynamics.

Freshwater systems seem particularly well suited to tests of decomposability because they are often physically isolated and lend themselves to manipulative experiments. Figure 2 illustrates a freshwater community, albeit an artificial one, in which decomposability has been demonstrated experimentally (Murdoch and Walde, 1989; Orr et al., 1990). In this case, the *Daphnia*/algal interaction is tightly coupled. However, the dynamics of *Daphnia* and the backswimming bug *Notonecta* are only weakly coupled, for the following reasons. Although small instars of *Notonecta* feed on *Daphnia*, adult *Notonecta* eat prey that come largely from outside the system (arthropods trapped on the water surface); moreover, this food source is the factor regulating the *Notonecta* population, via adult fecundity and cannibalism rate. *Daphnia* and *Notonecta* are therefore only weakly coupled because the effects of *Notonecta* on the *Daphnia* population do not feed back into *Notonecta* dynamics. In addition, because *Notonecta* cannot affect the rate at which terrestrial prey land on the water, it can be treated as a single-species system with environmental forcing. The *Daphnia*/algae and *Notonecta* subsystems can therefore be modeled separately; though, for example, the effect of *Notonecta* would need to appear in *Daphnia*'s death rate.

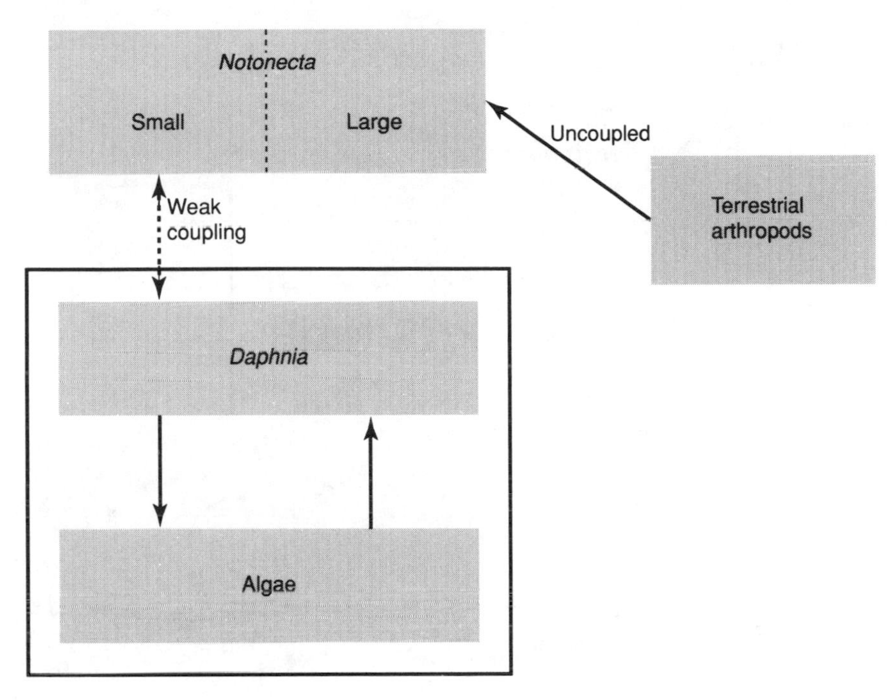

FIGURE 2. Diagrammatic representation of the results of experiments in stock tanks demonstrating that the stock tank community is decomposable into subsystems within which interactions are strong but between which they are weak. (Modified from Murdoch and Walde, 1989. Source, Orr et al., 1990, and Murdoch, unpublished data.)

Whether decomposition into small subsystems is possible may depend on the part of the community in which we are interested (Figure 3). For example, if we want to understand the short-term dynamics of *Daphnia* (e.g., whether or not they cycle over the summer; Murdoch and McCauley, 1985) in lakes with planktivorous fish that attack *Daphnia*, we may be able to treat the *Daphnia*–fish interaction as a weak coupling because of differences in time scale. Thus, although the fish may help determine *Daphnia*'s average density, representing a strong effect, *short-term* changes in *Daphnia* density should not influence fish dynamics, which operate on a seasonal time scale. The effect of *Daphnia* on short-term fish dynamics is therefore weak. In other words, feedback from *Daphnia* through fish and back to *Daphnia* is incomplete on a short time scale. Hence, the fish population can be treated adequately as part of *Daphnia*'s environment (Figure 3A).

The effect of fish in this instance can be incorporated via *Daphnia*'s functional response. Thus, if individual fish respond to changes in *Daphnia*

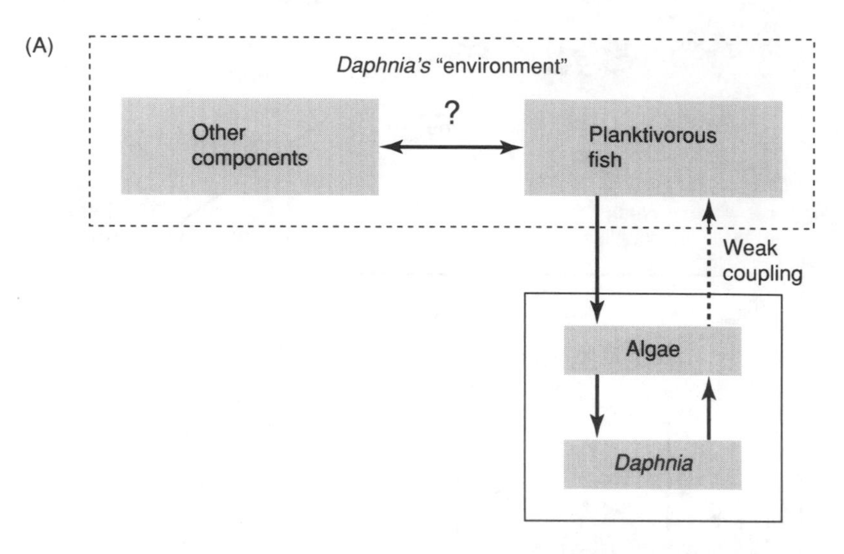
(A)
Daphnia's "environment"
Other components
?
Planktivorous fish
Weak coupling
Algae
Daphnia

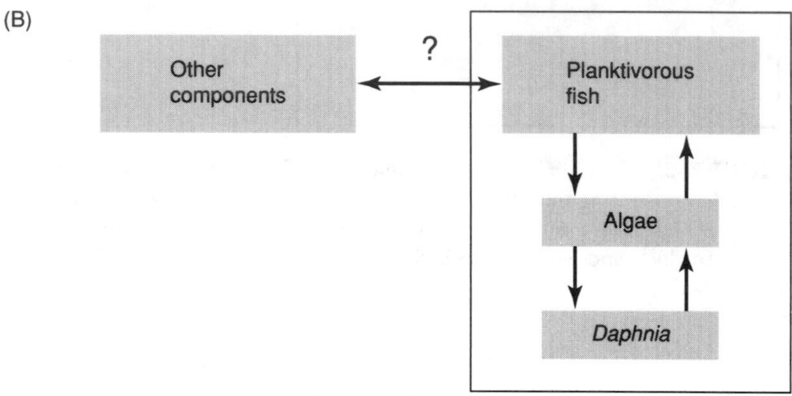
(B)
Other components
?
Planktivorous fish
Algae
Daphnia

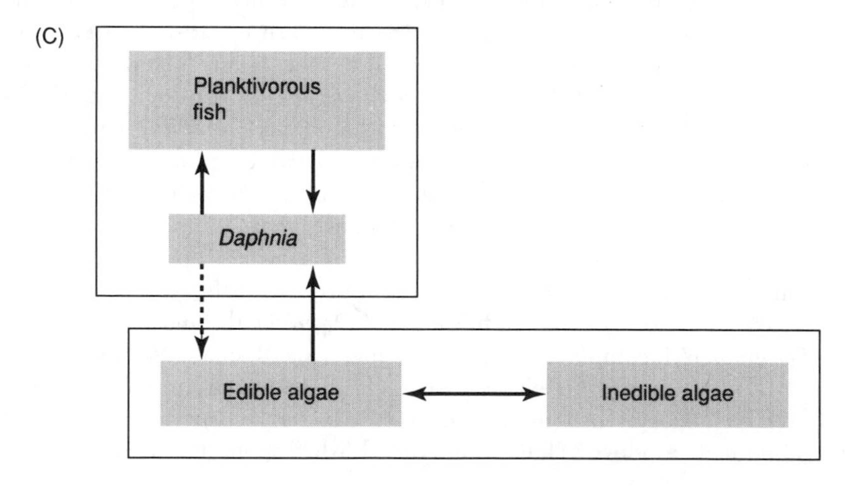
(C)
Planktivorous fish
Daphnia
Edible algae
Inedible algae

◀**FIGURE 3.** Possible decompositions of the major interacting species in a lake community. (A) Only the short time scale of planktonic dynamics is of interest here. Populations of fish responding on a longer time scale are part of *Daphnia*'s environment and do not need to be modeled explicitly in the plankton model. (B) Fish dynamics over the long time scale are of interest here. The planktivorous fish affect *Daphnia* dynamics, but the *Daphnia* population is still on average sufficiently dense to affect and be affected by algal dynamics. Thus all three species need to be modeled explicitly. (C) Planktivorous fish suppress the *Daphnia* population below the level at which it can strongly influence algal dynamics, and the system decomposes into two weakly coupled subsystems. Solid arrows represent strong coupling; dotted arrows represent weak coupling. Boxes with solid lines delineate strongly coupled species constituting a subsystem.

density with a type 2 functional response, the number of *Daphnia* being eaten per unit time is:

$$g(D) = \alpha D/(1 + \alpha b D)$$

where D is *Daphnia* density, α is the instantaneous attack rate exerted by the fish population, and b is the usual handling time. Here, however, the parameter α is aF, where a is the per-head attack rate of the average fish and F is fish density. F might be the average fish density over the season or, if seasonal variation in fish abundance is marked, $F = F(t)$, where $F(t)$ is now a seasonal forcing function.

The key here is that F does not respond to short-term changes in *Daphnia* density. Consequently the *Daphnia*/algal interaction can be described by a set of predator–prey equations that incorporates fish density but does not include fish dynamics explicitly (Figure 3A).

If we are interested in fish dynamics, however, such a time-scale argument may not be valid. Here the appropriate time scale is seasonal and year-to-year. Now, yearly changes in average *Daphnia* density will certainly influence changes in fish density over time, unless *Daphnia* is only a small component of the fish's diet. Furthermore, changes in fish density from one year to the next will influence annual changes in *Daphnia* density. In this case, fish, *Daphnia*, and algae will represent a strongly coupled system, which would require a tritrophic-level model (Figure 3B).

Even here, however, a different decomposition may be possible. If the planktivorous fish population strongly suppresses *Daphnia* density, and if the *Daphnia* population then has only a weak effect on algal dynamics, the *Daphnia* and algal dynamics will be only weakly coupled. In this case the system can be modeled as a fish/*Daphnia* subsystem weakly coupled to an algal subsystem (Figure 3C).

Other species in the lake may also be strongly coupled with each subset, but again these other species need not be modeled explicitly unless they form part of a complete feedback loop. This is true even when a species has strong effects that reverberate, or cascade, through the system. Consider, for example, an experiment in which a top predator is introduced into the lake, thus reducing the density of planktivorous fish. Suppose this leads to an increase in the density of *Daphnia*, which in turn suppresses algal density (for a real example see Carpenter, 1987). Now *Daphnia* density will be determined mainly by the rate of production of algae, and the planktivorous fish will have little effect on *Daphnia* density. The planktonic subsystem and the fish system will therefore be only weakly coupled, as in Figure 3A. Irrespective of whether we are interested in short- or long-term plankton dynamics, we now do not need to include fish explicitly in our plankton model; we simply need to change the parameter α, induced by the reduction in F. That new parameter value is generated by the model of the tightly coupled planktivorous fish–predator fish interaction.

The above examples illustrate two factors that tend to make communities decomposable: differences between species in time scale, and a tendency for one of the processes (e.g., predation or food limitation) to be the major regulator of density. Many herbivorous insect populations in forests probably also exemplify these two factors because they usually operate on a much shorter time scale than their host trees and are frequently suppressed by their natural enemies. In many cases, therefore, the host tree can probably be treated simply as part of the insect's environment.

Two other factors that may enhance decomposability are breadth of diet and interactions that impinge on only a part of the life history. With regard to the first, the dynamics of very generalist predators may be only weakly coupled to the fluctuations in abundance of any one prey species. In fact, communities with many species, many of which are characterized by broad diets (or by prey that are attacked by many enemies), paradoxically may be quite decomposable. If the species in which we are interested is influenced by many other species, most of them will necessarily have a relatively weak effect, in which case we may be able to consider the sum of these effects as part of the stochastic environment of the population in question. With regard to the second factor, size-selective predation is common, as is stage-specific parasitism.

By this process of decomposition, we may be able to develop a program for investigating the dynamics of "the community." The process involves considering various "nodal" populations, or interactions of interest, and then determining in how many directions, and how far, along the web of interactions we need to go before we have defined the strongly interacting populations that make up the subsystem in which our nodal species is central. Other subsystems that articulate with the subsystem of interest then constitute part of its environment and provide one or more of the necessary parameter values determining its dynamics. The behavior of the "community" is then described by the collection of those subsystems that correspond with species or interactions that interest us. The process will inevitably be more complicated than I have suggested because communities are more complicated than my examples. Nevertheless, the reliability of the process can be tested experimentally, for example, by manipulating major components in the community and testing the ability of the models to predict the effects.

PREDICTING THE EFFECTS OF GLOBAL CHANGE

Testing the model

Global change will likely bring conditions different from those in which the model has been developed and tested. Some such changes might be dealt with simply. For example, if the temperature is expected to change, a model of individual physiology and behavior could be reparameterized for higher temperatures on the basis of laboratory studies. Such a model possibly could be tested under limited field conditions in some communities. Ayres (this volume) shows that we are likely to be able to predict the effects of increased carbon dioxide levels on resource acquisition and allocation in different types of plants, and Dunham (this volume) has predicted the consequences of environmental temperature change on the time budgets and reproduction of a lizard.

However, global change may also alter many other aspects of the physical environment. Consequently, directly testing the model's ability to predict in these new circumstances in advance may be difficult. One way to increase confidence in the ability of the model to predict in novel circumstances is to test it under a wide range of *present* circumstances. For example, forest models that are developed from truly independent individual-level observations could be asked to predict community composition along various contemporary environmental gradients, or to predict the community effects of invading species. Gradients of increasing nutrients have been used by freshwater ecologists to test dynamic models of plankton (McCauley et al., 1988; Leibold, 1989).

Generality

The individual-based modeling approach is very time consuming and cannot be done for a very large number of systems. Furthermore, one of the likely effects of global climate change is that some species will be lost from, and others will be added to, the community of interest, thus changing the species in the system. Both of these considerations emphasize that we somehow need to generalize our models, and to test how well such generalization works.

The evidence is encouraging for generality at the level of models of *individual* performance. For example, Ross and Nisbet (1990) have modified the energetics model for individual *Daphnia*, reparameterized it for mussels, and successfully predicted individual performance. This is a useful result because energetic considerations are likely to be key in the dynamics of many systems and will therefore lie at the heart of any individual-based model. Ayres's chapter (this volume) makes it seem likely that we will be able to predict the effects of increased carbon dioxide levels on broad classes of plants. So far, however, there appear to be no published efforts to modify an individual-based *population* model to predict the behavior of a system other than the one for which it was developed.

In developing models with broad applicability, modeling again needs to be combined with experimental testing. Recent experiments manipulating whole communities offer promising settings for the testing of community models. Testing could be done both in mesocosms and in natural communities. Carpenter et al.'s whole-lake experiments (this volume) reinforce the message delivered above that freshwater systems seem to be particularly amenable environments in which to develop and test predictive models. Additional opportunities for testing models could follow the release of new agents of biological control or even the accidental invasion of species new to an area. Taking advantage of any of these situations, however, requires the existence of a range of well-tested models, and there is a pressing need for ecologists to increase the number of systems for which mechanistically based models have been developed.

DISCUSSION

The difficulties posed by the need to predict the effects of global change are common to the analysis of ecological systems in general: With great effort, we might understand a few systems in great detail, but every ecological situation is in some way unique. How do we predict the behavior of the systems we have not studied well? The answer must surely be, by extrapolation from the systems we do know well. This can succeed only if there are generalities—i.e., strong similarities—among different members of a particular class of ecological system. That in turn requires us to recognize the classes and their distinctive features. Once again, the processes of defining such classes, and of identifying the key features of their members, are subjects for hypotheses and testing.

One approach might be to develop the type of individual-based models described above for a number of systems that are thought to be dynamically similar, and then to determine whether they do in fact simplify to the same or at least very similar simple models. But some communities will probably have different dynamic processes. For example, communities in which generation time increases markedly up the food chain (e.g., lakes) may differ fundamentally from those in which the opposite is true (e.g., trees attacked by insects that are attacked by microorganisms). The process of developing individual-based models for different systems may well speed up as we do it more frequently, so carrying out such studies for a number of different kinds of systems may be feasible.

None of this is meant to suggest, however, that an individual-based, or even a mechanism-oriented, approach is the only useful one. For example, "what if?" models (such as that explored by Hassell et al., this volume) already provide insight into how different mechanisms might work, and this is needed to understand more realistic and hence more complex models. Furthermore, "what if?" models can point to the types of ecological structures that are likely to yield similar dynamics and to those that will not. Even models with very little structure, such as that of Ives and Gilchrist (this volume), can raise new questions and suggest features (in this case, the pervasiveness of density dependence in the system) that might be crucial in defining different classes of ecological systems.

There is, however, no way to avoid the hard task of developing models that convincingly predict the behavior of some particular real systems. Individual-based models give promise of being a useful way to approach this goal.

SUMMARY

Global environmental change will affect communities and populations through changes in the physiology and behavior of individual organisms. Individual-based models provide a framework for testing hypotheses about the consequences of such changes and the mechanisms that produce them. I discuss how these models can be tested, simplified, and ultimately serve as the basis for general theory. Results from programs on *Daphnia,* freshwater fish, and trees illustrate how this can be done and also show the difficulties involved.

It should be possible to create models of communities, consisting of individual-based models of interacting populations, by decomposing the community into subsystems, each of which contains strongly coupled populations but is only weakly coupled to other subsystems. The criterion for weak coupling is that the interaction is most strong in only one direction; i.e., the feedback from one species, through the other and back to its own dynamics, is weak. This allows the dynamics of the first species to be modeled without coupling it to the dynamics of the second species, which can be treated like a part of the first species' environment. The community can then be viewed from various "nodes" (species of special interest) and we need to go along various directions in the food web radiating out from the node only as far as is needed to include those populations strongly coupled to the nodal species. This process of simplification is also testable experimentally. I provide an experimental example and discuss features of communities that enhance their decomposability.

ACKNOWLEDGMENTS

I am grateful to Ed McCauley, Roger Nisbet, and Craig Osenberg for discussions on the ideas presented here, and to Roger Nisbet for commenting on a draft. Several years ago Peter Chesson provided crucial discussions on decomposability. Ray Huey and Joel Kingsolver provided invaluable editorial advice. The research was supported by NSF Grant BSR89-17774 and Grant DE-FG03-89ER-60885 from the Ecological Research Division, Office of Health and Environmental Research, U.S. Department of Energy; the latter support does not constitute an endorsement of the views expressed here.

PART THREE

EVOLUTIONARY RESPONSES TO ENVIRONMENTAL CHANGE

CHAPTER 10

EVOLUTIONARY GENETICS AND CLIMATE CHANGE: Will Animals Adapt to Global Warming?

Ary A. Hoffmann and Mark W. Blows

The greenhouse scenario is likely to increase mean temperature by several degrees during the next century (see Schneider, this volume). In addition, it may change environments in other ways. For example, it may alter rainfall patterns, sea level, and even the intensity of other stresses, including pollutants.

Predictions of the biological impact of these changes have been based mainly on the association between present distributions and climatic factors and on the anticipated expansion and contraction of species ranges as these factors change, although the ability of species to track environments via migration has also been considered. Evolution is usually assumed to be irrelevant because it is thought that most species cannot evolve significantly or rapidly enough to counter climate changes (Peters and Darling, 1985).

This view is partly justified on the basis of fossil data (Travis and Futuyma, this volume), which suggest that many lineages display evolutionary stasis during periods of climate change (e.g., Coope, 1977; Sheldon, 1987; Cronin and Schneider, 1990). Fossil studies, however, can detect only morphological change; they have a very restricted capacity to detect change in physiological traits, which are likely to be more important in climatic adaptation. Physiological evolution can occur in the absence of much morphological change, as indicated by the large physiological differences between some sibling species that are almost indistinguishable on morphological grounds. For example, species of *Drosophila* from the *melanogaster* group are morphologically similar but differ strikingly in stress resistance. Widespread species such as *D. melanogaster* and *D. simulans* are much more resistant to climatic extremes than are their sibling species that are restricted to the tropics (Parsons and Stanley, 1981; David et al., 1983).

There is little doubt that evolutionary responses to global warming are a possibility in some animals (see Hoffmann and Parsons, 1991; Travis and Futuyma, this volume) and we need to address the likelihood of such responses. In particular, we need to identify features of populations and environments that facilitate adaptation, as well as factors constraining evolutionary responses. This is particularly important for animals with restricted distributions conserved in habitat reserves because those species may be able to cope with global warming stresses only by adaptation (Lynch and Lande, this volume).

We begin by briefly reviewing genetic responses to the climatic stresses that are likely to increase under global warming. Although the evidence suggests that at least some animal species may have the necessary genetic variation to evolve, this evolutionary potential may not be the same for all populations or for all species. Possible mechanisms leading to general stress resistance are therefore discussed, since understanding such mechanisms may aid in predicting whether or not populations or species are able to evolve resistance. Genetic correlations and tradeoffs may complicate these predictions. We also outline the importance of marginal populations as an experimental tool for understanding what limits species distributions and how distributions may change under global warming. Finally, management steps to maximize the chance of conserving the "right" genes in threatened species are suggested, along with future directions for research in evolutionary genetics relevant to climate change.

POTENTIAL FOR ADAPTIVE RESPONSES TO CLIMATE CHANGE

Climatic variables are important in the distribution and abundance of animals. As well as being associated with mean temperature or rainfall, distribution limits of many species seem to be correlated with climatic extremes such as the temperature of the hottest or coldest month (e.g., Caughley et al., 1988; Kohlmann et al., 1988; Root, 1988). Such extreme conditions are likely to increase in frequency and severity under a global warming scenario (see Schneider, this volume).

Adaptation to unfavorable climatic conditions can occur either (1) by evasion of these conditions or (2) by the evolution of enhanced resistance. Evasion occurs when animals migrate, switch resources, or alter behavior to avoid stressful conditions. Evasion may also be achieved by dormancy during some life cycle stages or by altering life history patterns. Evolutionary changes increasing resistance may involve traits preventing mortality as well as traits enhancing fecundity and fertility under stress. Both evasion and resistance traits may, of course, interact with biotic factors. For example, an evolutionary increase in stress resistance of one species might enable it to displace a competitor.

Evolutionary responses involving resistance and evasion following an environmental change are well documented (Hoffmann and Parsons, 1991). A striking case is the evolution of increased resistance in many animals (especially insects) to chemicals introduced into the environment. But what about stresses likely to arise from the greenhouse scenario? Can populations evolve to cope with increased temperatures and drought conditions that might, without evolution, be lethal? There is little doubt that most animal populations contain high levels of genetic variation, and attempts to demonstrate genetic variance in quantitative traits under laboratory conditions are usually successful (Mousseau and Roff, 1987). Unfortunately, little information is available for traits likely to be involved in global warming responses.

Resistance to climatic stress

A prerequisite for evolutionary change is genetic variation, and selection experiments are often used to evaluate levels of genetic variation within populations. In artificial selection experiments in the laboratory, resistant (or nonresistant) individuals are directly selected and their progeny are used to establish the next generation. In experiments using natural selection in the laboratory, a population is held in stressful conditions for many generations to select indirectly for stress resistance. There are many experiments on altered resistance to environmental factors (particularly chemical stresses), although resistance to heat and desiccation have been considered less frequently (Hoffmann and Parsons, 1991). In *Drosophila melanogaster*, desiccation resistance increased rapidly under artificial selection, and resistance to other environmental stresses including heat also increased as a correlated response (Hoffmann and Parsons, 1989a,b). Selection for altered adult resistance to heat has been successful in experiments with insects (e.g., White et al., 1970; Stephanou and Alahiotis, 1983; Quintana and Prevosti, 1990; Huey et al., 1991), although the response to selection for increased resistance has often been small. Selection has also successfully increased heat resistance in agricultural animals such as beef cattle (Frisch, 1981).

Genetic variance in heat resistance has been documented with techniques other than selection. For example, Bradley (1978) compared families of copepods and Bowen and Washburn (1984) compared siblings of Japanese quail to demonstrate heritable variation in heat resistance. However, heritability estimates from such experiments may only be useful in predicting selection responses for a few generations (Falconer, 1981), and maternal effects often cannot be evaluated in family studies.

Field studies on heat resistance have involved sampling individuals from two or more environments and testing their progeny when they are reared under controlled environmental conditions. Heritable variation is implicated when there are differences in resistance between the progeny of individuals from different environments. This approach has demonstrated heritable variation in heat resistance in aquatic animals inhabiting water bodies heated by effluent. For example, progeny of copepods collected near a power station had higher heat resistance than those from other sites (Bradley, 1978, 1981). Similarly, Holland et al. (1974) found that bluegill fish from an effluent area were more resistant to heat than those from other sites acclimated in the same environment, although a genetic basis was not confirmed by progeny testing.

Apart from studying stress resistance directly, heritable variation in resistance has also been established by associating specific genotypes or phenotypes with resistance in the field. The frequencies of shell banding patterns and colors in the snail *Cepaea nemoralis* were changed when a dune population was exposed to a heat stress (Richardson, 1974), and wing color affects body temperatures in *Colias* butterflies and can reduce the likelihood of overheating (Kingsolver and Watt, 1983). Body color may also indirectly influence heat resistance in endotherms such as indigenous African cattle (Finch and Western, 1977). Variation at enzyme loci has often been correlated with environmental factors including temperature (Nevo, 1988), but how allozyme polymorphisms are influenced by climate is not usually clear. Exceptions include the elegant studies discussed in Watt (1985), Powers (1987), Koehn (1987), and Powers et al. (in press).

The possibility that resistance may evolve via the selection of increased levels of plasticity has not received much attention apart from the work by Levins (1968). Animals can often counter the effects of climatic change by plastic changes when they become acclimated after exposure to nonlethal conditions, and highly plastic genotypes (i.e., genotypes that can produce a wide range of phenotypes in different environments) may be selected under stressful conditions. Plastic and nonplastic changes may represent alternative evolutionary pathways; indeed, species with low resistance to a stress and a high level of plasticity may coexist with a more resistant but less plastic species (e.g., Yamamoto and Ohba, 1982). We do not know which environmental factors favor plastic changes. However, there are a number of possibilities, such as the frequency and speed of an environmental change (Levins, 1968; Hoffmann and Parsons, 1991), and these could be evaluated by selecting lines under different conditions.

Evasion of climatic stress

Most research on stress evasion focuses on the avoidance of unfavorable seasonal periods of low temperature and reduced food availability; only a few studies examine evasion of stresses likely to be associated with global warming. Drought stress may be evaded by switching to other resources, as in Darwin's finches (*Geospiza fortis*) in the Galapagos. Drought greatly increased finch mortality, and survivors tended to have bigger beaks and body sizes than did nonsurvivors. These traits are known to have a high heritability (Boag and Grant, 1978, 1981). Large birds seemingly survived the drought stress because they were able to feed on the large and hard seeds that were the only types of seeds available during the drought. Climate change might thus induce evolutionary changes indirectly by affecting food availability.

Life history changes may also mediate the evasion of drought stress, as in selection on the *abnormal abdomen* genotype in *Drosophila mercatorum* (Templeton and Johnston, 1988). Flies with *abnormal abdomen* reach reproductive maturity earlier than other flies but do not live as long. This altered life history pattern was favored when dry conditions drastically reduced adult longevity. Life history changes may also mediate the evasion of high temperature stress. For example, Meffe (1991) found that mosquitofish collected from the heated outflow of a nuclear reactor reproduced all year, produced larger clutches, and had smaller offspring than those from an ambient environment. However, the genetic basis of these differences was not determined, and plasticity may have been involved.

Behavioral evasion of climatic stresses has been largely ignored in evolutionary genetics even though many stress responses are likely to be behavioral as argued by Davenport (1985). Behavioral evasion may often act in addition to physiological mechanisms. For example, while bluegill fish adapt physiologically to heated effluent (see above), they are also able to evade high water temperatures during reactor discharges by retreating to refugia where temperatures are cooler (Block et al., 1984). Unfortunately, we know little about adaptive behavioral responses because behavior is difficult to study genetically, which is why behavior geneticists have tended to focus on simple laboratory traits, often with little relevance in the field. One trait that could be easily studied from this perspective is migratory ability in insects, which is often under some degree of genetic control (Harrison, 1980). Flight may enable insects to evade locally unfavorable conditions arising from climatic changes.

IN SEARCH OF GENERALIZATIONS

The above laboratory and field studies suggest that animal populations have the potential to evolve in diverse ways when faced with stresses likely to arise from global warming. However, it is not clear whether all species or populations will show adaptive responses. Recent selection experiments with *Drosophila*, using chemical stresses and desiccation, suggest that populations often show differences in the degree and rate of evolutionary response (Figure 1) (Cohan and Hoffmann, 1989; M. W. Blows, unpublished data). Correlated selection responses in these experiments indicated that different mechanisms are involved in different populations. This suggests that not all populations or species will respond similarly to greenhouse stresses and that some may show little response.

Predicting evolutionary responses to climate change becomes an onerous task if each animal population responds in a different way. For this reason, it is of interest to test for general responses to stress. Similar genes or physiological mechanisms may be involved in different animals. Variation in these genes or mechanisms could be assessed to evaluate the potential of populations to adapt to environmental stresses.

An obvious candidate for a common mechanism underlying stress resistance is the heat shock protein system. These proteins are inducible by a number of stresses and are universally present in animals (Ananthan et al., 1986; Lindquist, 1986). Variation in heat shock proteins might indicate a population's ability to respond to stress if these proteins form a major component of climatic adaptation. Unfortunately, not many data relate variation in the heat shock response to adaptation. The temperatures at which heat shock proteins are induced and maximally synthesized are correlated with the thermal environment of species (Huey and Bennett, 1990; Powers, in press), but whether or not this extends to the intraspecific level is unknown. Relevant data could be collected by examining heat shock proteins in lines selected for altered stress resistance (Alahiotis and Stephanou, 1982) or in populations subject to different climates.

Several other mechanisms may be generally important in temperature and desiccation adaptation. In particular, many animals maintain approximate membrane fluidity during temperature perturbations by altering the percentage of unsaturated lipids in their membranes (Hochachka and Somero, 1984). Individual differences in temperature resistance might generally be associated with membrane lipid composition and this needs to be addressed at the genetic level. Lipids are also important components of cuticles, and differences among arthropods in cuticle composition may influence stress resistance. For example, Gibbs et al. (1991) found genetic variation in the melting point of epicuticular lipids within and between populations of a grasshopper species. Melting points tended to be higher in populations from locations where ground temperatures were warmer in summer months.

Another response mechanism that may be common to many animals involves a change in metabolic rate. There is some controversy about whether high or low rates lead to increased resistance to stressful environ-

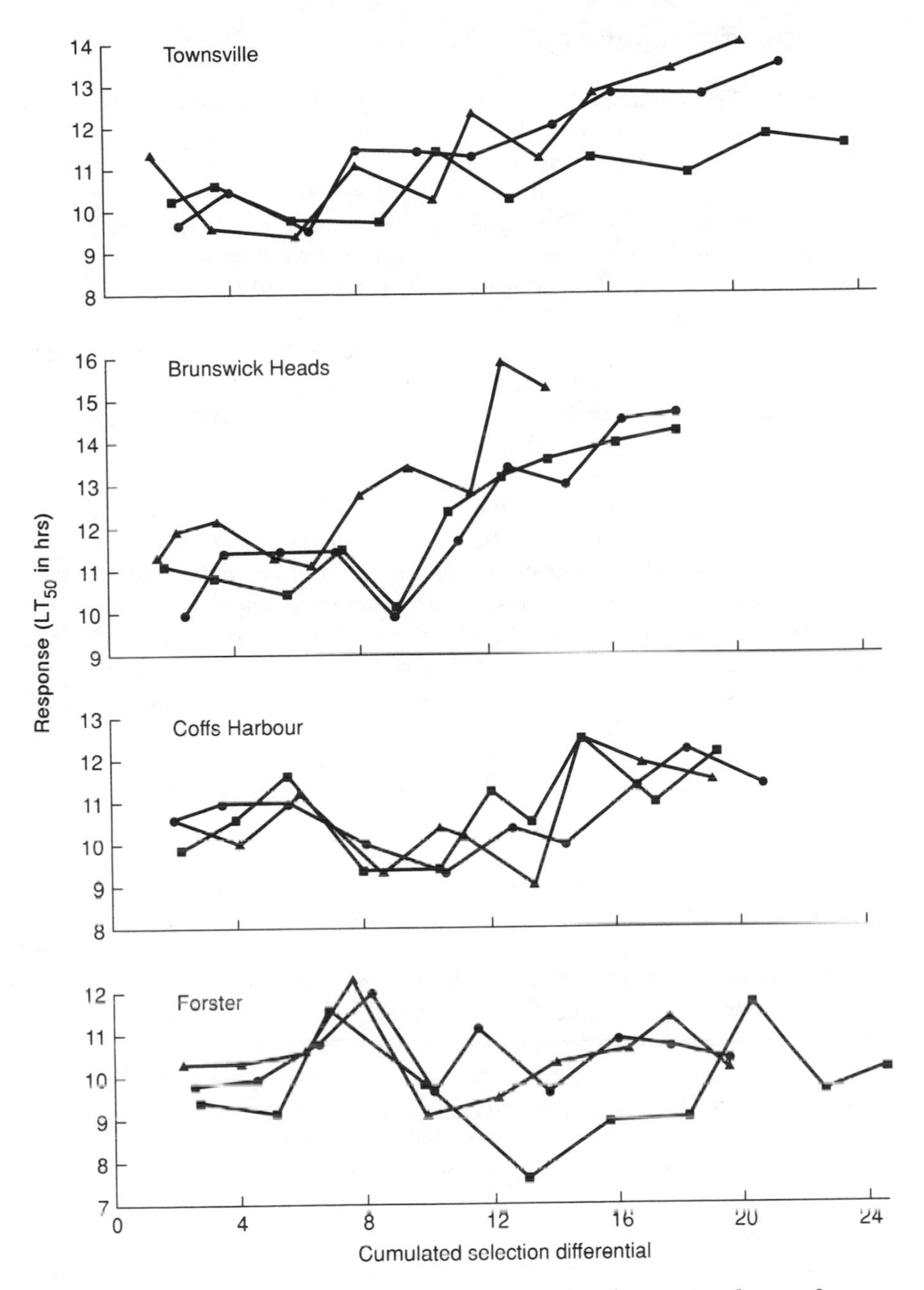

FIGURE 1. Response of four *Drosophila serrata* populations to selection for increased resistance to desiccation. Three replicate lines were independently selected for each population for 10 generations. The populations are in order of increasing proximity to the species' southern limit. Townsville is the most northern of the populations and is at the center of the distribution of *D. serrata*, whereas Forster is the most southerly point at which this species is found. (M. W. Blows, unpublished data.)

ments. A high metabolic rate may enable animals to counter the effects of many stresses because metabolic energy is required to combat stresses (see Calow, 1991). However, there is evidence that resistant animals often have low levels of metabolism, as illustrated at the intraspecific level for mussels (Koehn and Bayne, 1989), *Drosophila* (Hoffmann and Parsons, 1989a; Blows, unpublished) and cattle (Frisch, 1981) and at the interspecific level for a number of other groups (Hoffmann and Parsons, 1991; Parsons, in press). Moreover, plastic responses to climatic stresses may involve a reduction in metabolic rate, as in the case of dormant life cycle stages. Having a low metabolic rate enables animals to conserve energetic resources. This can be beneficial when animals encounter stresses such as starvation, desiccation, and high temperatures that restrict the acquisition of food or increase the rate of energy wastage.

The relative advantage of high or low metabolic rates in stress resistance may depend on the type of stress. High metabolic rate may promote the active exclusion and detoxification of chemicals, but low metabolic rate may promote survival of climatic stresses (Hoffmann and Parsons, 1991). In any case, more research is needed to assess genetic variation in metabolic rate under field conditions and relate this variation to stress resistance. Laboratory studies should determine the proportion of the variance in stress resistance attributable to metabolic rate or any other common physiological mechanism. The question of whether changes in metabolic rate reflect other factors such as size rather than being directly responsible for increased resistance also needs to be addressed.

Generalizations have been sought about the ways in which genetic parameters change when populations encounter climatic stress and whether populations evolve more rapidly under stressful or benign conditions (Langridge, 1963; Parsons, 1983). Heritabilities often depend on the environment and it has been argued that they will decrease when organisms encounter a more stressful environment (Blum, 1988). The environmental variance may increase as stress increases because small environmental changes will have much larger phenotypic effects under stressful conditions; for example, the impact on survival and growth of encountering a food item by chance will be greater when food is scarce than when it is abundant. However, there are also reasons why the genetic variance might increase under climatic stress (Hoffmann and Parsons, 1991). These include a reduction in the canalization of a trait and the expression of genes that have not been under directional selection. An increase in the genetic variance under stress could increase heritability if the environmental variance does not change much. The limited empirical data on this issue do not clearly support either view (Hoffmann and Parsons, 1991), and the question of whether stressful conditions generally increase or decrease evolutionary responses remains to be resolved.

GENETIC CORRELATIONS AND ENVIRONMENTAL STRESS

Interactions among traits have been emphasized in many recent evolutionary studies, particularly in those on life history traits. They are measured at the genetic level by looking for genetic correlations between traits. Large negative correlations are particularly important because these may indicate tradeoffs between traits. Tradeoffs may constrain the evolution of an adaptive trait and account for the persistence of genetic variance within populations (Rose, 1982; Arnold, 1987).

Most work on genetic correlations has focused on life history tradeoffs measured in relatively optimal environments and there have been few attempts to study tradeoffs in changing or stressful environments (Sibly and Calow, 1989). Interactions between life history traits will probably change as the environment changes. For example, genetic correlations between these traits may tend to become more positive under conditions of environmental stress because genes that allow an organism to survive in a new environment are likely to affect a range of fitness components (Bell and Koufopanou, 1986; Holloway et al., 1990). Whether this is generally true is not yet clear, although the positive correlations found in many life history studies could be interpreted from this perspective (Hoffmann and Parsons, 1991). Genetic correlations between fitness traits under optimal conditions may therefore have little relevance to evolutionary constraints under stressful conditions.

Genetic correlations are likely between traits important in the resistance of climatic stress and life history traits measured under benign conditions. In *Drosophila melanogaster*, resistance to starvation is correlated with postponed senescence and increased early fecundity (Service et al., 1985; Rose et al., in press), and increased desiccation resistance is associated with decreased early fecundity and increased longevity (Hoffmann and Parsons, 1989b; Graves et al., in press). In a parasitic hymenopteran, stress resistance and longevity are positively correlated (White et al., 1970), but heat resistance and growth rate are negatively correlated in endotherms (Jonsson et al., 1988; Wilson et al., 1975). Negative correlations between stress resistance and some fitness traits measured under benign conditions are predicted by the association between a low metabolic rate and stress resistance (see above) because a high active metabolic rate has been positively correlated with high fitness under benign conditions in many animals (Pough, 1989; Parsons, 1990; Hoffmann and Parsons, 1991). Body size or color in mammals may form the basis for negative correlations between heat-stressed and unstressed environments (Finch and Western, 1977; Jonsson et al., 1988). Such correlations may constrain evolutionary responses to environmental stress if they are large enough.

Genetic interactions may also occur between stress evasion and fitness in benign conditions. The ability to evade a stress is often correlated with a reduced reproductive output, as illustrated by the lower fecundity of winged insect morphs compared to wingless morphs (Harrison, 1980; Kaitala, 1988). Stress evasion via a rapid development time can decrease reproductive output by causing animals to metamorphose at a smaller size, as found in toads and frogs (Berven and Gill, 1983; Newman, 1988).

These interactions suggest that genes favored during times of climatic stress will often not be favored in benign conditions. This complicates predictions about evolutionary responses in animal populations. For example, increased resistance to heat or drought may lead to a decreased reproductive output during favorable conditions, and may thus reduce competitive ability. Consequently, ample genetic variance for stress resistance is no guarantee that an animal will maintain its distribution and abundance under climatic change.

MARGINAL POPULATIONS AND CONSERVATION

Empirically, the best area in which to seek answers to many of the above problems would seem to be in populations at the ecological margins of species. Here the effects of climate change will first become obvious. Moreover, marginal populations provide an opportunity to study what limits adaptation. If we understand the processes that currently determine margins, then we may be able to predict how distributions will shift when environments change. Studies on marginal populations have tended to be ecological rather than evolutionary. Abundance in marginal populations often seems to be determined by density-independent factors (Whittaker, 1971; Randall, 1982; Brown, 1984) but little is known about which traits actually limit species ranges (Caughley et al., 1988; Gaston, 1990). The identification of relevant traits is only the first step in an evolutionary explanation of species margins because we need to understand why such traits do not evolve. Although repeated calls have been made for evolutionary studies on this problem (e.g., Mayr, 1963; Krebs, 1972; Antonovics, 1976), little has been accomplished, apart from the construction of several hypotheses (Table 1). Margins are often assumed to simply reflect low levels of genetic variation for traits limiting distributions either because of a small population size, physiological limits, or a large environmental variance under marginal conditions. Alternatively, gene flow from central to marginal populations may prevent adaptive differentiation at the margins (Mayr, 1963; Antonovics, 1976), or interactions among traits or environments may limit adaptation at margins (Hoffmann and Parsons, 1991).

Information on evolutionary factors limiting species borders may help to predict distribution changes under global warming. If alleles allowing for range expansion simply do not arise, then changes in species distributions may be predicted by considering climatic changes. However, predictions

TABLE 1. Hypotheses about evolutionary limits to range expansions when ranges are determined by environmental stresses.

1. Levels of genetic variance for stress response traits are low as a consequence of directional selection and physiological limits.
2. Levels of genetic variation are generally low because of a persistent small population size.
3. Levels of genetic variance are low for stress response traits when these are tested under stressful conditions but not under benign conditions.
4. Changes in several independent characters are required for range expansion to occur.
5. Negative genetic correlations exist between performance traits under stressful and optimal conditions.
6. Negative genetic correlations exist among fitness traits and these correlations are expressed under stressful conditions.
7. Favored marginal genotypes are swamped by gene flow from central populations.
8. Inbreeding occurs as a result of small marginal population size.
9. Heterozygotes are favored under extreme conditions.

(From Hoffmann and Parsons, 1991.)

become more difficult if borders are associated with other factors. Environmental changes leading to habitat fragmentation may suppress gene flow between central and marginal populations, facilitating an evolutionary response if gene flow from central populations normally restricts adaptation (see McCauley, this volume). Evolutionary responses are more likely in large outbred populations with high levels of genetic variation, so any effect of climate change on population size may have evolutionary consequences if margins are determined by inbreeding or small size. Predictions become particularly difficult if tradeoffs exist between stress resistance/evasion and fitness traits influencing biotic interactions. For example, animals may adapt to counter a climatic stress but still undergo a contraction of their range if increased resistance leads to decreased competitive ability.

Most genetic studies have examined ways in which marginal and central populations differ based on allozyme variation and inversion polymorphisms. In *Drosophila*, allozyme heterozygosity tends to be similar in central and marginal populations (Brussard, 1984), but other taxa display reductions at margins (Soulé, 1973; see also Sjogren, 1991 and references therein). In any case, allozyme heterozygosity is a poor predictor of the ability of insects to adapt to new environments (Myers and Sabath, 1980 and references therein), and substantial differences between populations for ecological traits can occur in the absence of substantial allozyme differentiation. Electrophoretic screening of species may therefore be of little use when predicting responses to climatic change. Inversion polymorphism tends to be lower in some marginal drosophilid populations (Brussard, 1984), but not in others (Soulé, 1973). Moreover, inversion polymorphism may be rare in most other animals (Remington, 1968). Consequently, information on inversions is also unlikely to yield a general answer to the species border problem.

To understand why species margins exist, genetic studies on traits important in limiting species distributions are needed (Parsons, 1990). Given that climatic extremes are at least in part responsible for species distributions, assessing the relative amounts of quantitative genetic variation for climatic stress responses in central and marginal populations may provide useful information. One of the few attempts to compare levels of genetic variation for resistance to a climatic stress in central and marginal populations of an animal (Figure 1) showed that a marginal population of *Drosophila serrata* (Forster) was less able to respond to desiccation stress than more central populations were. However, desiccation may not be responsible for the species border, and comparisons should ideally be carried out with ecological traits known to limit the distribution of a species.

IMPLICATIONS FOR MANAGEMENT AND FUTURE RESEARCH

Threatened species restricted to reserves are particularly prone to the effects of global warming because they cannot escape by migration. It is important to conserve genes that increase the ability of these species to cope with global changes. This will not be not an easy task. Although single genes can have major effects on stress responses, as in the case of genes controlling diapause in insects (Tauber et al., 1986), most traits associated with climatic responses are likely to be complex, involving a large number of genes (Langridge, 1963). For example, variation at several enzyme loci may be associated with temperature changes (Powers, in press). The identification of specific genes involved in responses to global warming will therefore be difficult, particularly in organisms in which genetic studies are not easy.

Several management steps could help to maximize the chances of conserving the "right" genes for adapting to climate changes. Widespread species will often have well-adapted genotypes in populations exposed to extreme conditions because populations often differ in climatic stress responses (Hoffmann and Parsons, 1991). As a result, large central populations from favorable conditions with high overall levels of genetic variation might be less important for conservation purposes than small populations from harsh environments with lower overall levels of genetic variation. Similarly, levels of asymmetry (Thoday, 1958; Zakharov, 1989) or metabolic rate may be useful phenotypic indicators of individuals and populations most suitable for conservation. Locating reserves in areas where the environment is heterogeneous may be useful for the preservation of species not only because this provides refuge areas during stressful periods but also because useful genetic variation may be conserved in heterogeneous environmental conditions. Although overall levels of genetic variation may not be higher in heterogeneous conditions (Hedrick, 1986), polygenic variation under selection in the different environments can be maintained, regardless of whether the environmental heterogeneity is spatial or temporal (Gillespie and Turelli, 1989).

If future research in evolutionary genetics is to contribute to these types of management decisions, the way we conduct our experiments needs to be modified. Laboratory estimates of heritabilities and genetic correlations may not reflect field values because the environment influences the expression of genetic variation. Consequently, such estimates predict the potential response of animals under a defined set of conditions.

A promising technique for estimating genetic parameters relevant to field conditions is to make measurements on animals from the field as well as on their progeny reared in the laboratory. In *Drosophila* this approach has been used to look at heritable variation in morphology (e.g., Prout and Barker, 1989). However, although the approach produces lower bounds for genetic variances, estimates are valid only if genes controlling variation in the field have a similar influence in laboratory conditions because of weak genotype–environment interactions, and if maternal effects are weak (Riska et al., 1989).

Laboratory experiments need to be made more relevant to the field. Consider the evolution of chemical resistance in insects. Although laboratory experiments suggest a polygenic basis for resistance, field studies generally indicate a monogenic basis. This disparity may reflect differences in population size and selection intensity between the field and laboratory environments (Roush and McKenzie, 1987). Recent quantitative genetic experiments suggest ways to simulate field conditions more accurately for factors such as population size (e.g., Weber, 1990).

Traits scored in genetic studies on stress resistance should be related more closely to ecologically important traits. Stress levels applied in many selection experiments may never be found in the field. In *Drosophila*, for example, heat resistance has been scored as survival after exposure to high temperatures that might never be experienced in the microclimates flies select in the field (Jones et al., 1987). Where temperatures correspond to those found in natural habitats, exposure times to achieve mortality may often be too long to be of much relevance to natural conditions. This partly reflects a preoccupation with mortality traits. Animals will suffer many deleterious effects before mortality levels are reached, such as a reduction in reproductive output, feeding rate, or development rate. These nonlethal fitness effects may be sufficient to cause large changes in the distribution and abundance of organisms, particularly when mediated via biotic interactions. It is not known whether genetic differences in mortality extrapolate to other fitness traits under less extreme conditions. Individual differences among lizards for running performance seem to be consistent between optimal and extreme temperatures (Huey and Hertz, 1984), but this has not been established at the genetic level.

CONCLUDING REMARKS

The above discussion outlines some of the evolutionary responses that may occur in animal populations exposed to climate change. At the moment relevant data are few, and we are a long way from predicting the extent to which a species can adapt rapidly and effectively. More research is needed in areas such as:

1. Field heritability estimates of traits relevant to global warming
2. Studies of the effects of population structure and interactions between traits on evolutionary responses
3. Comparative studies of evolutionary responses in different groups, such as endotherms vs. ectotherms and short- vs. long-generation species
4. Genetics of marginal populations and, in particular, estimates of genetic variance for those traits that may be responsible for present species distributions

These may seem like onerous tasks. However, they provide an opportunity for evolutionary genetics to be used in a predictive rather than a descriptive manner so that in the future, global warming will not simply provide more demonstrations of natural selection.

ACKNOWLEDGMENTS

We would like to thank Ray Huey, Joel Kingsolver, Peter Parsons, and Michael Turelli for many helpful comments and criticisms of earlier drafts. We are also grateful to Joe Graves and Dennis Powers for sending unpublished manuscripts.

CHAPTER 11

EVOLUTIONARY RESPONSES OF PLANTS TO GLOBAL CHANGE

Monica A. Geber and Todd E. Dawson

Research on plant responses to global change has focused almost exclusively on phenotypic responses (e.g., in physiology, allocation, growth) and on the scaling of phenotypic effects to ecosystem processes (Strain and Cure, 1985; Jarvis, 1989; Bazzaz, 1990; Ehleringer and Field, in press; Ayres, this volume). Little consideration has been given to the potential importance of adaptive evolutionary responses by plants to global change, except to air and soil pollutants (Bishop and Cook, 1981; Pitelka, 1988; Taylor et al., 1991).

Evolutionary biologists interested in global change have been concerned primarily with the problem of species extinction and with strategies for species conservation. They have largely ignored the potential for microevolutionary responses because of the widespread beliefs that climate change will proceed too rapidly for populations to adapt genetically and that populations are more likely to avoid extinction by migrating. Indeed, models by Lynch and Lande (this volume) suggest that, in the absence of migration, populations can only sustain adaptive evolution if the speed of environmental change is quite modest.

Although migration is likely to be the most common response to climate change and the best assurance for species persistence, it is certainly a fallacy to think that migration can entirely obviate the importance of adaptive evolutionary responses. Migrating populations will often encounter novel biotic and abiotic conditions, which may impose novel selection regimes. Furthermore, models show that the likelihood of extinction, with or without migration, is diminished when genetic variation for selectively important traits is present in populations (Pease et al., 1989; Lynch and Lande, this volume). And finally, where selection pressures are strong and genetic variation is present, plant and animal populations are capable of rapid evolution (LeBaron and Gressel, 1982; Bradshaw, 1984; Hoffmann and Blows, this volume).

In this chapter we focus on adaptive responses of plants to climate change. Holsinger (this volume) and McCauley (this volume) address non-adaptive evolutionary change by genetic drift, resulting from habitat fragmentation and changes in population size and structure.

Global change raises three questions of general interest concerning adaptive evolution. First and foremost, how important is adaptive evolution in preventing species extinction in the face of rapid change (Pease et al., 1989; Holt, 1990)? Second, what ecological and genetic factors affect the likelihood and magnitude of sustained adaptive evolutionary response? Finally, what traits are favored under climate change?

The most fundamental and least understood question is the importance of adaptive evolution in preventing extinction. It is a hotly debated issue in evolutionary biology (Gould, 1985). It is obviously pertinent to understanding the causes of species turnover in communities, both past and present. It is also relevant to the fate of species that are presently restricted to habitat islands and nature reserves, for whom migration is not an option and adaptive evolution is the only possible response to climate change (Lynch and Lande, this volume). Second, a better understanding of the ecological and genetic limits to adaptive evolution will help in the preservation of species with narrow ranges and in the regulation of invasions by weedy species. Finally, knowledge of what traits are favored under climate change will be important to the development of new crop varieties and to the success of restoration programs that must find suitable transplants to establish new populations.

We begin this chapter by reviewing fossil evidence about plant responses to past climate change. Because this evidence deals with time scales considerably longer than the current scale of change and lacks adequate taxonomic resolution, it is not very well suited for detecting microevolutionary changes. Nevertheless, it provides a useful view on the evolutionary history of plants. We also argue that evidence of microevolutionary responses to past climate change can be sought by comparing extant populations that have diverged since the last glaciation (Critchfield 1984). We then review the extent, implications, and sources of genetic variation in ecologically important traits in natural populations. Finally, we discuss the nature of selection on plant traits that have resulted from changes in abiotic and biotic factors and attempt to identify what traits will be favored. Throughout the chapter, we try to identify promising avenues of research and point out areas where information is lacking.

PLANT RESPONSES TO CLIMATE AND CLIMATE CHANGE

Climate as a selective agent

The geological record shows that tremendous climate changes have occurred during the 3- to 4-billion-year history of life. Even in the last 200 million years there have been large changes in temperature, precipitation, and atmospheric CO_2 concentration (Peng et al., 1983; Pearman et al., 1986; Jouzel et al., 1987; Clark, this volume). Atmospheric CO_2 levels have ranged from highs of 3000 ppm in the Cretaceous (120 mybp) to lows of approximately 200 ppm during the Holocene full-glacial (20 kybp) (Neftel et al., 1988; see Ehleringer et al., 1991). Furthermore, atmospheric CO_2 levels have been positively correlated with temperature (Kuo et al., 1990; Schneider, this volume). Climate change per se is not new, and plants have responded in a variety of ways to past climate change.

The strong association between climate and vegetation type, as well as the evolutionary convergence of morphological, physiological, and physiognomic traits among phylogenetically unrelated plant groups inhabiting regions with similar climate, are evidence that climate can act as a powerful selective agent (de Candolle, 1855; Raunkiaer, 1934; Box, 1981; Orians and Paine, 1983; Grace, 1987; Woodward, 1987). Evolutionary responses to climate are documented over a range of time scales. Over long time scales ($>10^4$ years), climate change is associated with changes in morphology and life history of plants during the Carboniferous (DiMichele and Phillips, 1992), with changes in leaf size, shape, and texture of the dominant flora during the Cretaceous and Tertiary (Upchurch and Wolfe, 1987), with the rise of wind pollination during the arid Eocene (Crepet, 1989), and with the increase in taxa possessing the C_4 photosynthetic pathway as atmospheric CO_2 declined in the Miocene (Ehleringer et al., 1991). At shorter time scales (10^4–10 years), plants have responded evolutionarily through the appearance of polyploid taxa in zones once covered by Pleistocene glaciers (Stebbins, 1984), through the formation of latitudinal or elevational ecotypes (Turresson, 1922; Clausen et al., 1940), and through ecotypic differentiation in response to anthropogenic changes in habitat (Bishop and Cook, 1981; Taylor et al., 1991).

The paleontological record: Evidence of extinction and migration

Extinction and migration are the best-documented responses of plants to past climate change. Plant species have gone extinct during gradual and abrupt changes in climate (Knoll 1986; Wolfe, 1991), both in the distant and more recent past (Davis, 1989; Betancourt et al., 1990; Delcourt and Delcourt, 1991). In addition, extinction has occurred following large (Niklas,

1986) as well as small climate changes (e.g., during the "little Ice Age;" Lamb, 1982).

Periodic mass extinctions appear to have been less pronounced in plants than in animals (Raup, 1986; DiMichele et al., 1987). Plants, and in particular seed plants, may be relatively resilient to catastrophic events associated with mass extinctions because of their seed banks or because of their extraordinary ability for vegetative propagation. The very old ages (10^3–10^4 years) of individuals from extant species, such as aspen, creosote bush, and bracken fern, are testimony to the fact that some individual plants can live through climate change (see Cook, 1985). It is unclear whether the current rapid climate change will have an impact on vegetation and floras similar to that of past "catastrophic events," such as those that might have followed meteor (bolide) impacts, or whether its effects will resemble those of a more gradual climate change (DiMichele and Phillips, 1992).

Plant species also migrate in response to climate change, at rates that are in part dependent on dispersal ability (Davis et al., 1986; Webb, 1988). Migration is a viable response, provided that dispersal routes and appropriate habitat are available. During the Pliocene, for example, forests of eastern North America and western Europe were similar at the generic level. With the expansion of ice sheets, North American species were able to migrate down the Appalachian chain and persist in refugia along the Gulf Coast and in Mexico (Davis et al., 1986). In contrast, European species were blocked from southward migration by the Alps, and extinctions were more common (but see Currie and Paquin, 1987; Adams and Woodward, 1989). The current fragmentation of habitats by humans will certainly restrict the migration of plant and animal species (Groom and Schumaker, this volume), and will place a premium on long-distance dispersal ability or on the ability to evolve genetically (Holsinger, this volume; McCauley, this volume).

Microevolution and past climate change

Evidence of microevolutionary response to past climate change can be found in extant populations that have diverged since the Pleistocene (Clausen et al., 1940; Mooney and Billings, 1961; Cwynar and MacDonald, 1987). Climate change may have selected for high dispersal ability. For example, the northward expansion of lodgepole pine in Canada during the postglacial period appears to have favored seeds with greater dispersal ability (Cwynar and MacDonald, 1987). Seed dispersal characteristics of populations can be related to the migrational history of populations, as judged from the fossil pollen record. Populations that have only recently been established at the northern end of the range have lighter seeds with large wings relative to their mass (small wing loading) compared with those of southern populations that have been established for 12,000 years and are far from the migrating front.

Migrating populations may have faced a variety of selection pressures, in addition to selection on dispersal ability. Because plant and animal species tended to migrate individually in response to climate change, community assemblages often changed (Davis, 1986; Davis et al., 1986; Graham and Grimm, 1990). Migrating plant populations would have been exposed to novel biotic conditions (e.g., competitors, herbivores, pathogens, pollinators, and dispersers) that could have altered patterns of selection (Turkington, 1989; Simms and Fritz, 1990; Galen 1989; Stanton et al., 1991). Similarly, migrating populations faced changes in abiotic factors (photoperiod, seasonality, precipitation, and soils). In plant species that now occupy regions once covered by glaciers, populations are extensively differentiated in photoperiodic responses, in phenology, in water relations, and in edaphic adaptation (Vaartaja, 1959; Rathcke and Lacey, 1985; Kruckeberg, 1986; Lacey, 1988; Dawson and Bliss, 1989). This supports our contention that migration does not preclude evolution. It is more difficult, however, to assess the importance of adaptive evolution to preventing extinction unless one can relate the magnitude or speed of range expansion to rates of differentiation in ecologically important traits.

Because the genealogical relationships among geographically differentiated populations are rarely known, and because the timing and extent of genetic isolation of populations are also generally unknown, the speed of population differentiation for selectively important traits has not been estimated. Molecular sequence variation (e.g., mtDNA or cpDNA) may enable us to infer the phylogenetic relationships and colonization history of populations (Slatkin and Maddison, 1989; Travis and Futuyma, this volume). The expression of ecologically important traits (e.g., photoperiodic response) could then be overlain on the phylogeny (Harvey and Pagel, 1991). Traits that vary little between populations may be phylogenetically conservative, selectively unimportant to the colonization of new habitats, or under similar stabilizing selection in all environments. Where variation is found in trait expression among populations, its functional significance may be inferred by comparing the pattern of trait variation to environmental differences in the population's sites of origin.

If the rate of molecular sequence divergence can be calibrated at a number of independent neutral DNA loci, then the time of isolation of populations as well as the rates of divergence in ecologically important traits can be estimated (Hudson, 1990). There are difficulties with such rate estimates. First, this approach may underestimate the actual divergence rate of ecological traits because it assumes that divergence occurred at a constant rate since isolation (an assumption that may be valid for neutral loci) rather than being concentrated during the period of initial colonization. In addition, if ancestral populations were polymorphic at DNA loci, with some of the variants being distributed among populations at each colonization, either the loss of DNA variants from descendant populations through drift or poor sampling of variants in populations will lead to overestimates of the time since isolation and to underestimates of divergence rates in ecological traits. Finally, even low levels of gene flow can prevent divergence at molecular loci and in ecological traits in the face of differential selection.

FACTORS AFFECTING THE LIKELIHOOD OF MICROEVOLUTION

Many factors affect whether a population will evolve, and how fast it can evolve in response to global change. These include existing levels of genetic variation for relevant traits as well as the rate at which new variation is generated by mutation, the genetic relationships among traits, the nature and intensity of selection, population size, breeding system, generation time, fecundity, and dispersal ability. Here we focus on the evidence for genetic variation in ecologically important traits in natural populations and what it tells us about the potential for microevolution. We also discuss sources of new variation through mutation and hybridization.

Genetic variation in ecologically important traits

Considerable amounts of genetic variation are found within wild plant populations for ecologically important traits, ranging from variation in growth, phenology, and life history (Weis et al., 1987; Cheplick and Quinn, 1988; Venable and Burquez, 1989; Geber, 1990), to gas exchange physiology (Geber and Dawson, 1990; Nevo et al., 1991), nutrient relations (Wanyancha and Morgenstern, 1987), and defense or resistance against disease and herbivores (Berenbaum et al., 1986; Maddox and Root, 1987; Alexander, 1989). Plant populations thus have the genetic potential to respond to selection unless prevented by unfavorable genetic correlations among traits. However, the nature of the genetic correlations among traits needs additional study (see Hoffmann and Blows, this volume). In addition, more information is needed on which traits are genetically variable and which ones are not, and on how genetic variation in ecological traits differs among species and environments.

Studies of phenotypic selection in wild populations show that selection intensities on plant traits are often of considerable magnitude, though they may vary from site to site and year to year (Kalisz, 1986; Stewart and Schoen, 1987; Farris, 1988; see also below). We found no study, however, that combined measurements of phenotypic selection on plant traits with measurements of a response to selection in the targeted traits (but see Bradshaw, 1991).

Genetic differentiation among populations in ecologically important traits is also common in plants. In a cursory survey of literature published since 1986, we found over one hundred studies of ecotypic or geographic differentiation for a wide variety of traits (Table 1). Character divergence was shown to be genetically based by means of common garden or reciprocal transplant experiments. In many cases, trait variation could be related to environmental differences in the populations' original habitats; and, where reciprocal transplant experiments were used, populations were often shown to perform best at their home site. These studies provide convincing evidence of adaptive evolution in the past that presumably extended the range of a species into novel habitats. In many of these examples and in others cited by Endler (1986), adaptive differentiation occurred in response to climate gradients or to anthropogenic changes in habitat.

Unfortunately, we cannot use this survey to tell whether population differentiation happens more readily in some types of plants (short versus long lived, herbaceous versus woody) or along some environmental gradients, because research has often focused on species with economic value (trees and grasses) or has been restricted to regions, such as northern Europe, where there is a long tradition of work on ecotypes. It is also difficult to gauge from this survey whether differentiation occurs more readily among populations at the centers than among those at the margins of a species' range in the face of similar environmental (and hence selection) gradients. If, as Hoffmann and Blows (this volume) suggest, the effects of global change will first be manifest at the edges of species ranges, the ability of marginal populations to adapt to novel circumstances will be important. Do marginal populations lack sufficient variation in important traits for adaptation to novel environments? With global change, selection may favor different kinds of traits at the higher and lower latitudinal limits of a species' range. For example, with global warming, populations at the higher latitude margins may actually experience an amelioration of climate, whereas those at the lower latitude margin will experience deteriorating environments.

As noted earlier, in most cases of geographic or ecotypic differentiation, little information is available on rates of divergence among populations. The exceptions involve examples of divergence in response to anthropogenic modifications of the environment, in which evolution has often been extremely rapid. The classic work on the evolution of heavy metal tolerance in plants near mines and smelters is by now familiar (see Macnair, 1981; Bradshaw, 1991). Similarly, mounting evidence documents rapid evolution in response to air pollutants (see Taylor et al., 1991). Inadvertent selection for resistance to pollutants in crops has apparently occurred (Horsman et al., 1979). For example, present-day cultivars of wheat are less sensitive to SO_2 and NO_2 than are cultivars from the early twentieth century (Gould and Mansfield, 1989).

Many examples of the evolution of resistance to air and soil pollutants

TABLE 1. Plant genera for which ecotypic differentiation in traits has been reported.

Plant trait	Tree	Grass/sedge	Herb/shrub
Phenology	*Picea, Populus, Pseudotsuga*	*Cyperus, Panicum, Poa, Sorghum*	*Abutilon, Capsella, Datura, Daucus, Dryas, Erigonum, Eriophorum, Galium, Gentianella, Lapsana, Plantago, Salix*
Morphology, growth form, life history	*Betula, Fraxinus, Picea, Populus, Robinia*	*Cyperus, Dactylis, Echinochloa, Hordeum, Oryza, Panicum, Poa, Sorghum, Themeda*	*Acacia, Abutilon, Amelanchier, Atriplex, Capsella, Cochlearia, Crepis, Daucus, Dryas, Dianthus, Eriogonum, Eriophorum, Galium, Gentianella, Heterosperma, Impatiens, Lapsana, Microseris, Plantago, Saxifraga, Stellaria, Symphoricarpos, Trifolium*
Temperature tolerance	*Pinus, Sequoidendron*	*Dactylis, Holcus*	*Artemisia*, ~80 legume spp, *Trifolium*
Drought tolerance	*Cercis, Juglans, Thuja, Sarcobatus*	*Dactylis, Pseudoroegneria*	
Gas exchange physiology	*Picea*	*Cyperus, Echinochloa, Holcus, Triticum*	*Dryas, Elodea, Fragaria, Nardus, Sedum*
Nutrient and edaphic relations	*Larix*	*Dactylis*	*Arenaria, Achillea, Fragaria, Gaillardia*
Tolerance to salts, metals, airborne pollutants	*Betula, Picea, Pinus, Prosopis*	*Agrostis, Andropogon, Arrhenatherum, Dactylis, Festuca, Lolium, Panicum*	*Geranium, Mimulus, Potamogeton*

(From a survey of studies published since 1986.)

come from herbaceous species, but some clear cases of rapid evolution of resistance to air pollution in long-lived trees are known (see Scholz et al., 1989). The evolutionary potential of long-lived species may be enhanced by the large amounts of genetic variation that are typically found within populations and because of their large seed crops. Long-lived plants may also have the potential to evolve by somatic mutation (Klekowski and Godfrey, 1989; see below). For many plants, and perhaps especially for long-lived ones, the ability to adjust to gradual environmental change through phenotypic plasticity may allow them to persist for considerable lengths of time (Bradshaw and Hardwick, 1989; Dickinson et al., 1991), and in effect increase the time available for advantageous mutations to arise (Travis and Futuyma, this volume).

Examples of pollution tolerance show that plants can evolve within a few decades to centuries when selection pressures are strong and variation for traits is present. However, the evolution of pollution tolerance may not be an appropriate model for the responses that can be expected in the face of global climate change. On the one hand, evolution of pollution tolerance may require changes in only one or a few traits, whereas adaptation to climate change is likely to involve a large number of traits (see below). On the other hand, because pollutants are both novel and toxic, genetic variation for tolerance may be rare in natural populations, if tolerance is costly (Wilson, 1988). Genetic variation for responses to climate variation may be much more common because natural environmental variation is ubiquitous.

A final point to be made about these examples of ecotypic or geographic differentiation is that even though they are impressive in number, they represent only the successes of evolution. Bradshaw (1991) argues quite strongly that we need to pay more attention to the failures of adaptation. He points out that for every species that has evolved tolerance to mine contamination, many more have not because they lack genetic variation for tolerance or for other traits that are required for colonizing mine tailings. Similarly, species vary in their ability to evolve resistance to pollutants, herbicides, and disease, or tolerance of high salinity, flooding, and liming of soils (Le Baron and Gressel, 1982; Bradshaw, 1984; Karnosky et al., 1989).

In order to evaluate the importance of adaptive evolution to the persistence and fate of populations in novel environments, the most direct approach will be the use of transplant experiments (Primack and Kang, 1989). In such experiments, the first generation of transplants will not yet have had an opportunity to evolve, and their performance can be used to judge the likelihood of persistence in the absence of evolution. Even if the transplants survive and reproduce in a foreign habitat, their performance may be so poor that persistence is unlikely. The persistence of a population depends on its rate of increase, which can be estimated from measurements of survivorship, growth, and fecundity of transplants (Caswell, 1989). The evolutionary potential of a transplant population can be assessed if measures of phenotypic selection on traits are coupled with an analysis of a response to selection across generations. This obviously limits study to plants with reasonably short generation times. If plants from one population are transplanted to a series of sites that are progressively less similar to their original habitat in climate or in community composition, one can begin to ask questions about the maximum amount of change that a population can endure and still persist and to which it might adapt (see Lynch and Lande, this volume). Plants could be moved both within and beyond their current range to determine what limits their geographic range. Finally, marginal and central populations could be used as sources of transplants to compare their evolutionary potential. Although transplant experiments have often been used in studies of local adaptation (see above) and of the limits to local distribution (Bell and Bliss, 1979), they have rarely been used over a broad geographic scale with the aim of studying the evolutionary potential of species (but see Schmidt and Levin, 1985; Rice and Mack, 1991).

Sources of new variation

We need to know more about mutation rates and mutational effects in ecologically important traits (Bradshaw, 1991; see also Travis and Futuyma, this volume). An intriguing area of investigation is the influence of environmental stress on mutation. Stress is known to cause elevated rates of somatic mutation in plants (Walbot and Cullis, 1985). Somatic mutations, if beneficial, can be propagated within a generation through enhanced growth of the shoots that contain them. In addition, somatic mutations can be inherited across generations if they occur in cell lineages that give rise to gametes. One class of stress-induced mutations involves changes in genomic size and organization (Cullis, 1991; Price, 1991). Similar changes in genome structure can result from hybridization between species or even between populations within species (Cullis, 1991).

Changes in genome size and organization often have profound effects on plant phenotypes. For example, nuclear DNA content affects cellular characteristics such as cell size and the rate or duration of cell development and replication. Variation in DNA content both between species and within species is strongly correlated with life history, phenology, and distribution across wet and dry environments (Price et al., 1981; Grime and Mowforth, 1982; Grime et al. 1985).

Plant species vary tremendously in the amount of nuclear DNA per cell (more so than can be accounted for by differences in ploidy; see Bennett, 1987), suggesting that changes in genome size have often occurred during evolution. In addition, examples of intraspecific variation in genome size are becoming common (Price, 1991). The frequency and phenotypic effects of changes in genomic structure should be evaluated more carefully in natural populations. This source of mutational variation may become more important because stress and hybridization will intensify when climate changes and formerly isolated populations come in contact through species migration.

PLANT TRAITS AND CLIMATE CHANGE

The importance of abiotic changes

Because the projected climatic changes are complex, and both the nature and magnitude of physical changes are quite uncertain at regional levels, meaningful predictions about biotic responses at local scales are difficult to make (Schneider, this volume). Nevertheless, we feel that three aspects of change in the abiotic environment will be universally important and will exert significant selective pressures on plants: (1) changes in precipitation and evapotranspiration; (2) changes in seasonality; and (3) increases in CO_2.

Global circulation models predict that while most regions will get warmer, some areas will become wetter and others drier (Schlesinger and Mitchell 1987; Sato et al. 1989; Schneider, this volume). We do not focus on the effects of rising temperature per se, because plants may acclimate to the projected increases of 2–3°C, and we feel that the correlated changes in evapotranspiration and in water and energy balance may be more important to plants (Nobel, 1991). Indeed, of all of the factors controlling plant performance and distribution, the availability of water (precipitation) for growth and metabolism and its loss from the land and leaf surfaces (evapotranspiration) are the most important (Box, 1981; Woodward, 1987; Stephenson, 1990). Seasonality will also be a key environmental characteristic to consider, because as climate changes the period that is favorable for biological activities will change. Lastly, because CO_2 is the primary substrate of plant productivity, the rise in CO_2 is certain to be important.

To identify the traits that are likely to be favored under climate change, one can examine current associations between climate and vegetation type. The strong association between vegetation type and climate implies that plant traits such as growth form, phenology, and leaf characteristics are likely targets of selection, and that different suites of traits confer high fitness under different regimes of temperature and precipitation (Box, 1981; Woodward, 1987). We present a simplified version of the association between plant traits and climate in Table 2. In addition to the traits used by Box (1981) to classify vegetation types, we have listed physiological and allocation characteristics of plants. We contrast the characteristics of plants in wet and dry climates at each latitude. In general, as climate changes, the traits that will be favored in a region are those currently prevalent at lower latitudes under the appropriate moisture regime. As a result, widespread immigration of species from lower latitudes will occur. In addition,

TABLE 2. Distribution of plant traits in regions differing in latitude and regional water balance.

Plant trait	Region	Wet[a]	Dry[a]
Morphology, growth form, life history	Arctic/boreal	Graminoids, forbs, shrubs	Woody shrubs, rhizomotous species
	Temperate	Trees, large shrubs	Annuals, graminoids, sclerophylous shrubs, succulents
	Tropical	Broad-leaved evergreen trees	Deciduous trees, graminoids
Phenology	Arctic/boreal	Seasonal, T & PHP dependent	Seasonal, T & PHP dependent
	Temperate	Seasonal, T & PHP dependent	Seasonal, precipitation & PHP dependent
	Tropical	Aseasonal	Seasonal, precipitation dependent
Leaf characteristics	Arctic/boreal	Evergreen and wintergreen	Mostly evergreen with many variants
	Temperate	Evergreen and seasonal/deciduous	Mostly evergreen and dry deciduous
	Tropical	Mostly evergreen	Dry deciduous
Physiology	Arctic/boreal	Higher A & g, lower WUE, poor drought tolerance	Lower A & g, higher WUE, drought tolerance
	Temperate	Very high A & g, low WUE, little drought tolerance	Low A & g, high drought tolerance/avoidance
	Tropical	Shade tolerant, poor drought tolerance	Shade intolerant, drought tolerant
Allocation	Arctic/boreal	Very high root:shoot	High root:shoot
	Temperate	Low root:shoot, large canopy	High root:shoot, moderate canopy
	Tropical	Low root:shoot, large canopy	High root:shoot, moderate canopy

[a]Abbreviations: A, rate of photosynthetic carbon assimilation; g, stomatal conductance to water vapor (water loss from leaves); WUE, water use efficiency (molar ratio of carbon fixed via photosynthesis to water lost via transpiration); T, temperature; PHP, photoperiod.

the abundance of local species that possess appropriate traits will also increase. Finally, evolutionary changes in resident and migrating species in the direction of favored traits can be expected.

Of the traits listed in Table 2, some are more likely to be modified by selection than others. For example, it is unlikely that woody species will evolve to become herbaceous in the next century even if the vegetation in a region changes from a woodland to a shrub-steppe. Some traits may also be more phylogenetically conservative than others. Comparative studies show that geographic ranges (a possible indicator of climatic tolerance) and flowering phenology are often conservative within genera or families (Ricklefs, 1989; Kochmer and Handel, 1986). Similarly, congeners are more likely to belong to the same rather than to different phenological guilds in forest understory herbs (Givnish, 1987). In this instance, guilds are distinguished in large measure by leaf size and type (wintergreen, evergreen, deciduous). The basis for conservatism in plant traits is not well understood: do conservative traits not respond to selection because they are genetically invariant or because they are genetically correlated with other traits that are under opposing selection? Alternatively, are traits conservative because they are unimportant to fitness? In any event, major changes in growth form, in life history, and in leaf type (evergreen/sclerophylly versus deciduous/broadleaf) are less likely than moderate shifts in allocation, in physiology, and in developmental phenology.

Traits favored under climate change can also be determined in field studies of phenotypic selection, for this approach identifies what plant traits are targets of selection (Kingsolver and Schemske, 1991), as well as the direction, magnitude, and form of selection on those traits (Lande and Arnold, 1983; Wade and Kalisz, 1990). Finally, the extensive agricultural literature provides information on the kinds of traits that enhance yield under a variety of abiotic and biotic stresses and on the responses of these traits to selection.

Selection in wet versus dry environments

Where climates become warmer and evapotranspiration increases, traits that enhance the tolerance or avoidance of water stress and excessive heat loads should be favored (see Table 1). Certain leaf characteristics (e.g., reduced size, dense leaf pubescence, reduced leaf area, steeper leaf angles, and the ability to change leaf angle) will reduce water loss and improve leaf water and energy balance and water-use efficiency (i.e., the molar ratio of carbon gained via photosynthesis to water lost via transpiration; Ehleringer and Clark, 1988; Nobel 1991). The efficiency of water use is thought to be under intense selection; thus, as habitats become more xeric, increased water-use efficiency should be favored in both annual and perennial plant species (Cohen, 1970; Smedley et al., 1991). Significant genetic variation for water-use efficiency exists in natural plant populations (Geber and Dawson, 1990; Comstock and Ehleringer, in press); and genetic variation in water-use efficiency has been successfully exploited in crop breeding programs, suggesting that water-use efficiency may be responsive to selection (Hall et al., 1990).

Other selectively advantageous traits would include the maintenance of leaf turgor and a favorable water balance through osmotic adjustment or high tissue elasticity, increased stomatal sensitivity to both soil and atmospheric water deficits, and lowered rates of gas exchange during drought (Morgan, 1984; Robichaux et al., 1986; Schulze, 1986). Finally, as the climate becomes drier and the cost of water acquisition increases, selection would also favor increased carbon allocation to roots (Caldwell and Richards, 1986).

In wetter regions, on the other hand, the importance of aboveground structures for light interception would increase. Selection would therefore favor long leaf lifetimes (e.g., evergreen or seasonally deciduous), large canopies, and low root to shoot ratios, as well as high rates of gas exchange at the expense of lowered water-use efficiency and drought tolerance (Chabot and Hicks, 1982; Cowan, 1982; Evans et al., 1988).

Several recent studies of short-lived species have measured phenotypic selection on phenology (e.g., the length of the prereproductive period), morphology, allocation, and/or physiology in wet and dry environments (Farris, 1988; Lechowicz and Blais, 1988; Dudley, 1991, M. A. Geber and T. E. Dawson, unpublished). In all cases, the magnitude and direction of selection on traits differed across environments. In addition, the direct and indirect effects of water-use efficiency on reproductive success varied across environments, with the effects being stronger in dry habitats (Dudley, 1991). Such changes in the pattern of phenotypic selection can arise because the expression of traits is actually modified by the environment, and because traits may have very different direct and indirect effects on fitness in different environments. All studies point to the importance of physiology, and of water-use efficiency and drought tolerance in particular, to performance in dry environments. In short-lived species, dry habitats may also favor accelerated phenology and development (M. A. Geber and T. E. Dawson, unpublished).

We need to understand better the causal mechanisms of selection and to see whether consistent patterns emerge in selection on traits across species. Do the same traits turn out to be important to performance in dry and wet environments?

Seasonality

With global warming, the length of time available for plant growth and reproduction will depend on the regional water balance. In addition, photoperiodic and temperature cues used by organisms to signal the onset of favorable periods may no longer be reliable. For example, a model by Hänninen (1991) suggests that climatic warming could increase the risk of frost damage in northern tree taxa as a direct result of ontogenetic shifts in bud burst toward midwinter. Finally, if plants move poleward, they will experience altered distributions of light throughout the day and year. In the summer, days will be longer but the total daily energy balance will be lower at higher latitudes. Therefore selection for altered phenology and for adjustments in the translation of seasonal cues is likely to be intense. Geographic variation in phenology and in response to photoperiod and temperature is widespread (Vaartaja, 1959; Bongarten and Hanover, 1986; Shaver et al., 1986; Lacey, 1988; Pelham et al., 1988; Erskine et al., 1990). Vaartaja (1959) suggested that photoperiodic ecotypes in species from eight genera of trees may reflect selection for altered phenology since the last glaciation.

Carbon acquisition

Climate change can be stressful because organisms are exposed to conditions outside their range of tolerance. Elevated CO_2 levels, however, are likely to benefit rather than to stress plants because the cost of carbon acquisition will decrease. Because increased CO_2 often suppresses photorespiration and increases quantum (light) use efficiency, water-use efficiency and nitrogen-use efficiency improve, especially in C_3 plants (Pearcy and Bjorkman, 1983; Sharkey, 1985; Cure and Acock, 1986; Sage and Pearcy, 1987). To the extent that plants can actually fix more carbon under elevated CO_2 levels and reallocate the carbon to other functions, they may be better able to tolerate stresses imposed by drought, flooding, low nutrients, or high temperatures (see McLaughlin and Norby, 1991). Exposure to elevated CO_2 can indeed ameliorate the effects of stresses such as high salinity, cold, air pollutants, and drought (Coyne and Bingham, 1977; Winner and Mooney, 1980; Carlson and Bazzaz, 1982; Bowman and Strain, 1987).

Selection should thus favor genotypes that are able to increase net carbon gain and to allocate the surplus carbon to the acquisition of more limiting resources. Although numerous studies document interspecific variation in response to changes in CO_2 level and water or nutrient availability (see reviews by Jarvis, 1989; Bazzaz, 1990; Ayres, this volume), many fewer studies have examined intraspecific variation in responses among populations or genotypes (Wright, 1974; Garbutt and Bazzaz, 1984; Wulff and Miller-Alexander, 1985).

In general, few phenotypic studies have been done in a way that permits an analysis of selection. In most cases, measurements are made of traits that are likely to affect plant fitness (e.g., physiology, morphology, growth, or phenology), but fitness itself is rarely measured adequately. In addition, studies report only the mean responses of traits to elevated CO_2, rather than the variation in response or the covariance between plant traits and fitness. Such information is required to evaluate and understand how novel environmental conditions might select on plant traits.

Another problem with phenotypic response studies is that plants are typically grown at CO_2 levels that are twice to three times ambient levels. Even though plants growing near the forest floor may routinely experience CO_2 concentrations between 600 and 1000 ppm, plants of open habitats are unlikely to experience CO_2 levels in this range; and their response to a sudden doubling or tripling of CO_2 may not be representative of responses to more realistic increases of 1 to 2 ppm CO_2 per year or per growing season. For example, the response of plants to air pollutants differs considerably depending on whether the exposure is acute (short exposure to high concentrations of pollutant) or chronic (prolonged exposure to lower concentrations), and the genetic basis of resistance to acute and chronic exposure may differ (Bell et al., 1991). Studies aimed at assessing the evolutionary response of plants to elevated CO_2 levels might be more relevant if plants were exposed to moderate increases in CO_2. Experiments on plants with very short generation times could also look at the effect of incremental increases in CO_2 between generations, and so serve as models of evolution in species with longer life spans.

The importance of biotic changes

Thus far we have focused on the direct effects of physical changes in climate on the selective regime of plants. In the next century, global change will also indirectly affect a host of biotic associations (Holt, 1990; Ayres, this volume). Habitat fragmentation and the differential migration, extinction, and success of plant and animal populations will alter community composition (Tilman, this volume; Clark, this volume), and change the identity and abundance of competitors, pollinators, dispersers, herbivores, and fungal associates. Even without changes in community composition, changes in seasonality and phenology may have a large effect on the outcome of biotic interactions with competitors, pollinators, dispersers, and herbivores (Rathcke and Lacey, 1985; Aide, 1988; Ayres, this volume). Finally, abiotic factors can affect plant quality, and so alter the susceptibility of plants to herbivores (Ayres, this volume) as well as the benefits and costs of mycorrhizal associations (Bazzaz, 1990). Although many of these biotic changes will be profound, how they will affect the nature of selection on plant traits is difficult to predict (Holt, 1990).

With respect to plant competition, selection is likely to act on genetic variation in resource use (Tilman 1988), and in particular on the ability to reallocate carbon to the acquisition of limiting nutrients or water (Ayres, this volume). With respect to plant–herbivore and plant–pathogen relationships, selection may favor increased allocation to carbon-based defenses, since the cost of these defenses will decline with elevated CO_2 (Ayres, this volume).

As suggested earlier, habitat fragmentation may favor high dispersal ability. In addition, the disruption of communities may well select for characteristics of opportunistic or weedy species, including rapid growth, high fecundity, self-fertilization or apomixis, and dispersal or dormancy (Davy and Smith, 1988; Gray, 1989). DiMichele and colleagues suggest that climate change and extinction in the paleobotanical record are followed by the appearance of taxa with "opportunistic" life histories, typical of early successional habitats (DiMichele et al., 1987; DiMichele and Phillips, 1992).

The extensive migration of species in response to climate change and the ensuing disruption of communities will result in some species and populations being invaders and others being subject to invasion. Therefore, a very important approach to learning about the evolutionary consequences of altered biotic environments on species evolution will come from the study of biological invasions (Baker and Stebbins, 1965; Parsons, 1983). Rapid evolution is well known in invading species (Travis and Futuyma, this volume; Hoffmann and Blows, this volume). How much of the evolutionary divergence among populations of invading species occurs in response to abiotic factors and how much is due to evolution vis-a-vis novel competitors, herbivores, or pathogens? What makes for a good invader? And within a species, which genotypes are more invasive than others? How do invading species affect the selection on residents?

SUMMARY

Little consideration has been given to the importance of adaptive evolutionary responses by plants to global change because of the widespread beliefs that climate change will proceed too rapidly for populations to adapt genetically and that migration is a more likely escape from extinction. Even though migration is likely to be the most common response and the best assurance for species persistence in the face of climate change, it is a fallacy to think that migration entirely obviates the importance of adaptive evolutionary responses. Migrating populations will often encounter novel biotic and abiotic conditions, which may impose novel selection regimes (see Cwynar and MacDonald, 1987). In addition, for species that are presently restricted to habitat islands and nature reserves, migration may not be an option and adaptive evolution may be the only possible response to climate change. Finally, many weedy species of special human interest have attributes that will enable rapid evolutionary response.

Evolutionary biologists can provide insights into three important questions that are relevant to global change. First and foremost, how important is adaptive evolution in preventing species extinction in the face of rapid change? Second, what ecological and genetic factors affect the likelihood and magnitude of sustained adaptive evolutionary response? Finally, what traits are favored under climate change?

The first question is the most difficult to answer. While there is ample evidence that plants have evolved in response to past climate change and have differentiated along environmental gradients, these examples represent the successes of adaptive evolution and do not inform us about the failures. Evidence that plants have evolved to past environmental change can be found in fossil data, in the convergence of plant form and physiology among unrelated groups inhabiting regions of similar climate, and in numerous cases of geographic and ecotypic differentiation along environmental gradients. Many of these examples involve time scales considerably longer than the current scale of change. In cases of ecotypic or geographic differentiation since the Pleistocene, estimates of rates of adaptive divergence are lacking because the timing and extent of genetic isolation of populations are generally unknown. Better estimates of rates of adaptive divergence might be obtained if the timing of population isolation can be inferred from divergence at neutral molecular loci. Instances of extremely rapid evolution are known in plants in response to anthropogenic modifications of habitat. Furthermore, extensive genetic variation for ecologically important traits is documented within and between wild plant populations, suggesting that plant populations have the potential to respond to selection unless prevented by unfavorable genetic correlations among traits. Additional study is needed on the nature of the genetic correlations among traits, on identifying which traits are genetically conservative, and on how genetic variation in ecological traits differs among species and environments.

In order to evaluate the importance of adaptive evolution to the persistence and fate of populations in novel environments, the most direct approach will be to use transplant experiments. The performance of first-generation transplants, which will not yet have had the opportunity to evolve, can be used to judge the likelihood of persistence in the absence of evolution. The performance of subsequent generations can be used to evaluate their evolutionary potential. Transplants can be moved within and beyond a species' current range and should also be drawn from central and marginal populations.

Because rapid adaptive evolution may be limited by lack of genetic variation, we need to know more about mutation rates and mutational effects in ecologically important traits (Bradshaw, 1991; see also Travis and Futuyma, this volume). An intriguing area of investigation is the influence of environmental stress on mutation, and, in particular, on changes in genome size. Genome size can have marked effects on plant developmental time, phenology, and life history, and is correlated with habitat association within and between species.

Three aspects of change in the abiotic environment will exert significant selective pressures on plants: (1) changes in precipitation and evapotranspiration, (2) changes in seasonality, and (3) increases in CO_2. We identify the traits that will be favored under climate change on the basis of current associations between climate and vegetation, results from phenotypic selection studies in contrasting environments, and the results of plant breeding programs in agriculture. Unfortunately, of the many studies on the phenotypic responses of plants to elevated CO_2 levels, few are performed in a way that permits an analysis of selection on traits, because plant fitness is rarely measured and the covariance between fitness and trait variation is not reported.

Because projected climate changes are complex, and both the nature and magnitude of physical changes are uncertain, meaningful predictions about biotic responses are difficult to make. Habitat fragmentation, species migration, and changes in seasonality will all disrupt biotic associations between plant competitors and between plants and pollinators, fruit dispersers, pathogens, and herbivores. Selection may well favor characteristics of opportunistic or weedy species, including rapid growth, high fecundity, self-fertilization or apomixis, and dispersal or dormancy. The disruption of communities will result in some species and populations being invaders and others being subject to invasion. Therefore, a very important approach to learning about the evolutionary consequences of altered biotic environments on species' evolution will come from the study of biological invasions.

ACKNOWLEDGMENTS

We thank M. B. Davis, P. A. Delcourt, W. A. DiMichele and K. J. Niklas for bringing to our attention relevant paleobotanical literature. We also thank C. Galen, R. B. Huey, J. G. Kingsolver, M. A. Peterson and two anonymous reviewers for helpful suggestions about the manuscript. This chapter arose from work supported in part by USDA Grant 90-37280-5460 to M.A.G and T.E.D., Hatch Funds NY(C)1837401 to M.A.G., and Mc-Intire-Stennis Funds NY(C)1837555 to T.E.D.

CHAPTER 12

THE EVOLUTIONARY DYNAMICS OF FRAGMENTED PLANT POPULATIONS

Kent E. Holsinger

Although we most often think of the greenhouse effect and global warming when the phrase *global change* is used, human activities are changing our world in another, equally profound way. Our activities dominate many of the world's ecosystems. Fewer and fewer ecosystems escape our influence. We manage them for agriculture or forest products, pave them for shopping malls and housing developments, or extract the resources in them for our own consumption and well-being. These activities not only produce the greenhouse gases that could lead to a 4–5°C increase in global temperature by the middle of the next century (Schneider, this volume), they are also creating islands of natural ecosystems in a sea of human-dominated ones. In some parts of the world, such as China, India, and western Europe, this process is far advanced. In others, it has barely begun. Since fragmentation of natural ecosystems may have a dramatic impact on both the ecological and the evolutionary dynamics of the species within them (Gilpin, 1991; Gilpin and Hanski, 1991; McCauley, 1991; Quinn and Hastings, 1987; Templeton et al., 1990), any effort to understand ecosystem responses to global change must consider the additional effect that fragmentation of those ecosystems may have.

In this chapter I review a little of what is known about the genetic structure of plant populations and patterns of gene flow between them, attempting to infer some of the possible evolutionary consequences of habitat fragmentation in plant species. Of course, there is a great deal that we do not know, and these areas of ignorance will become painfully obvious, but there are some conclusions to be drawn from what we already do know. I shall argue, for example, that the genetic structure of plant species that are distributed continuously across a broad geographic range is far more likely to be affected by fragmentation than that of species whose distribution is naturally patchy. Thus, the plant species most likely to suffer from additional habitat fragmentation are those that are abundant members of geographically widespread communities. Although the direct demographic effects of the genetic changes accompanying fragmentation are difficult to evaluate, the available evidence suggests they may be minor and limited to cases in which fragmentation reduces the effective population size in the remaining patches to 100 or less.

Far more problematic is determining the long-term evolutionary consequences of fragmentation. It will often reduce genetic diversity within populations, simply because it reduces their effective size, and it will retard evolutionary responses to environmental change to the extent that those evolutionary responses are limited by available genetic diversity (Hoffmann and Blows, this volume; Lynch and Lande, this volume). On the other hand, occasional migration between neighboring population fragments could release a flush of variability when migrants arrive if the populations inhabiting these fragments are genetically distinct. Similarly, fragmentation may either promote or retard genetic differentiation between populations, depending on the rate of between-population gene flow in fragmented populations, the size of population fragments remaining, and the frequency of population extinction and reestablishment (McCauley, this volume). It may result in either more or less variation in average individual viability among populations than there was among different parts of the continuous population from which the fragments were derived. Thus, it can either increase or decrease the opportunity for interdemic selection, even as it diminishes the opportunity for individual selection within populations. The balance among these effects is difficult to determine, but it seems likely that fragmentation will more often retard than promote adaptive responses to environmental change.

GEOGRAPHIC DISTRIBUTION, GENETIC VARIATION, AND GENE FLOW

The first task in any attempt to understand the consequences of habitat fragmentation must be to identify the species in which it is likely to have an effect. At its most basic, that means identifying species in which fragmentation will alter the distribution of populations, the migration rates between populations, and the sizes of local populations. More specifically, habitat fragmentation will have an effect on the genetic structure of a species, beyond its obvious effect of eliminating some populations entirely, only if it significantly alters patterns of gene flow within the species. To anyone who is used to thinking of habitat fragmentation in terms of its effect on large vertebrates like the northern spotted owl, it must sound like every species would qualify. The good news is that in plants there is reason to suspect that additional habitat fragmentation will have a significant genetic impact on only a minority of species. The bad news is that those species in which it may have a significant impact are often ecosystem dominants. To see why this is so it is useful to review some basic facts of plant geography and plant population genetics.

Geographic distribution

Deborah Rabinowitz suggested that there are three ways in which a species may be rare (Rabinowitz, 1981):

1. It may have a limited geographic range.
2. It may have highly specific habitat requirements.
3. It may have a low population density everywhere it occurs, even though it is widely distributed and lacks highly specific habitat requirements.

This typology has been widely used by plant conservation biologists as a framework for identifying alternative management strategies for rare species (e.g., Kruckeberg and Rabinowitz, 1985; Holsinger and Gottlieb, 1991), but we can also use it to help us identify those plant species whose genetic structure is likely to be affected by habitat fragmentation (Table 1).

Notice that rare plants in either of the first two senses Rabinowitz mentioned are unlikely to be greatly affected by habitat fragmentation. The serpentine endemics in genera like *Streptanthus* (Kruckeberg, 1957), *Layia* (Clausen et al., 1947; Ford and Gottlieb, 1989), *Clarkia* (Lewis and Raven, 1958), and *Linanthus* (Woodell et al., 1975) are classic examples of plants that are rare both because of highly specific habitat requirements and because of limited geographic distribution. Similarly, *Zizania texana* is restricted to a 2.5-km stretch of the San Marcos River in Texas where water temperatures vary only 5°C annually (Terrell et al., 1978). In each of these cases habitat destruction may threaten or eliminate populations, and this will have a dramatic impact on the species' chances for survival. But the threat comes as a direct result of those populations being destroyed, not through any indirect effect on the genetic structure of the species—that is, the threat is demographic, not genetic (Lande, 1988). The vast majority of plants with a small geographic range also have highly specific habitat requirements (Rabinowitz, 1981; Kruckeberg and Rabinowitz, 1985). Thus, fragmentation is likely to have an important *genetic* impact only on species with a large geographic range. Only in those species can the loss of some

TABLE 1. A typology of species distribution and abundance.

Geographic range		Large		Small	
Habitat specificity		Wide	Narrow	Wide	Narrow
Local population size	Large	Locally abundant over a large range in several habitats	Locally abundant over a large range in a specific habitat	Locally abundant in several habitats but restricted geographically	Locally abundant in a specific habitat but restricted geographically
	Small	Constantly sparse over a large range and in several habitats	Constantly sparse in a specific habitat but over a large range	Constantly sparse and geographically restricted in several habitats	Constantly sparse and geographically restricted in a specific habitat

(From Rabinowitz, 1981.)

populations affect the genetic structure of the species without directly threatening its survival.[1]

Of course, not all plants with a large geographic range will suffer equally from fragmentation. Many plants are patchily distributed, common where they occur and wide-ranging but never forming large, continuous populations. For example, the cardinal flower (*Lobelia cardinalis*) is distributed along watercourses from New Brunswick west to southern Ontario and south to Florida and Texas (Fernald, 1950). It occasionally forms populations of several hundred individuals, but nowhere does it occur in a continuous patch covering more than a few hundred square meters, and neighboring patches are commonly separated by ten kilometers or more. Similarly, *Platanthera psycodes*, one of our most attractive native orchids, occurs in widely scattered populations of less than half a dozen individuals each, but it is found from Newfoundland south to Georgia and Tennessee (Fernald, 1950). For plants like these, fragmentation may have little genetic effect, even if there is a large demographic effect, since their distribution is already fragmented.

Clearly, patterns of gene flow will be most severely disrupted when a

[1]Fragmentation may have a very important *demographic* impact on geographically restricted species. In fact, the demographic effects of fragmentation may often be more important than the genetic ones (Lande, 1988; Holsinger and Gottlieb, 1991; Menges, 1991a). I focus on the *genetic* consequences of habitat fragmentation here because they are the most relevant to an analysis of its evolutionary consequences.

continuous distribution is broken up into a fragmented one. Though a wide geographic distribution alone is not sufficient to make a species susceptible to habitat fragmentation, if we combine it with a more or less continuous geographic distribution, then fragmentation is almost certain to have some effect. Thus, in the Pacific Northwest, plants like Douglas fir (*Pseudotsuga menziesii*) are probably far more susceptible to genetic changes as a result of habitat fragmentation than are those like the farewell-to-spring (*Clarkia pulchella*), which has a highly patchy distribution (Hitchcock and Cronquist, 1973). In short, the plants most likely to undergo genetic change as a result of fragmentation can be succinctly described as dominant, or at least important, components of widespread communities.

Gene flow and genetic variation

The basic facts of plant geography suggest that the widespread, dominant members of a community are what we should focus our attention on to detect the genetic effects of fragmentation. But knowing which species are most susceptible to these effects is, of course, only part of the story. We can predict the magnitude of these effects only if we have some idea about how fragmentation will affect patterns of gene flow among the parts of the population that remain. Unfortunately, most of the effort on understanding patterns of gene flow in plant populations has been focused on herbaceous plants with patchy distributions (e.g., Ellstrand et al., 1989; Handel, 1982; Handel, 1983; Levin, 1984; Levin and Kerster, 1974). The little we do know about patterns of gene flow in widespread ecosystem dominants comes primarily from electrophoretic analyses of conifers (e.g., Adams, 1981; Friedman and Adams, 1985; Muller, 1977; Neale and Adams, 1985), though some information on other hardwood species is available. Our knowledge of the genetic structure of widespread species is founded on a broader base (Hamrick and Godt, 1990), but any conclusions we draw from this analysis must be tempered by the knowledge that our data base on patterns of gene flow is extremely weak. Let me add one more word of warning: This section is not intended as a comprehensive review of the literature in plants. It is an attempt to illustrate the patterns rather than to document them.[2]

Patterns of gene flow in widespread species and their effect on its genetic structure can be inferred in a variety of different ways. Most are based on analyses of allozyme variation. Although patterns of allozyme variation may or may not reflect patterns of genetic variation at ecologically and adaptively significant loci (Hamrick, 1989; Holsinger, 1991), they do provide the most

[2]For comprehensive reviews of the literature on gene flow in plants, see Levin (1984), Hamrick (1987, 1989), and Hamrick and Godt (1990). Levin's review reflects inferences based primarily on patterns of gene flow within continuously distributed populations. The reviews by Hamrick and his colleagues attempt to infer patterns of gene flow among populations from the patterns of genetic differentiation among them.

broadly available comparative tool for analysis of genetic structure and they can provide useful insight into patterns of gene flow. The most common method of analysis is to partition the genetic variation present into within- and between-population components using Nei's gene diversity statistics (Nei, 1973). Such analyses indicate that in a typical plant species about 78 percent of the genetic diversity in the entire species is found within any one of its populations (Hamrick and Godt, 1990). Only about 22 percent is a result of between-population differentiation. Of course, these average figures mask a tremendous diversity in genetic structures among species. In conifers, for example, the pattern is strikingly different. Only about 7 percent of the genetic diversity within species is a result of between-population differentiation, and 93 percent is found within populations. One obvious explanation for this difference might be that gene flow occurs over a broader geographic range in wind-pollinated conifers than in most other plants, and there is some support for this idea.

Govindjaru (1988) surveyed estimated levels of gene flow in a variety of plants, using the product of effective population size and migration rate (Nm) as his measure.[3] Restricting our comparison to the 40 forest tree species included in his sample, we find that the average Nm in animal-pollinated species is 1.296 (range 0.001–5.444, $n = 8$), while the average in wind-pollinated species is 3.159 (range 0.540–7.840, $n = 32$). The differences in estimated levels of gene flow are consistent with the hypothesis that higher rates of gene flow are responsible for reducing genetic differentiation among populations. Unfortunately, there is a problem with this analysis. The estimates of Nm used in this analysis are based on the pattern of allele frequency differences among populations. These differences are assumed to be the result of independent evolution in the populations, while similarity is assumed to be solely the result of interchange of individuals among populations. Unfortunately, similar allele frequencies may also indicate recent population divergence. Thus, these estimates may be biased upwards if a species has recently expanded its range (Slatkin and Barton, 1989). For example, red pine (*Pinus resinosa*) shows very little allozyme differentiation among populations, but this does not appear to be a result of extensive gene flow. Rather, it appears that most of the variation in the species was eliminated when populations were reduced to small refugia during the last glacial maximum and that range expansion in the last 10,000 years has occurred without the accumulation of new genetic variants (Fowler and Morris, 1977).

More reliable estimates of the extent of gene flow are possible by comparison of maternal and offspring genotypes. Direct approaches using single gene markers have been widely employed in clonal seed orchards (Friedman and Adams, 1985; Harju et al., 1987; Smith and Adams, 1983)

[3]If $Nm > 1$, populations will tend not to diverge at neutral loci. If $Nm < 1$, populations will tend to diverge at neutral loci (Wright, 1931).

and in some natural populations (Handel, 1982; Handel, 1983; Muller, 1977; Schaal, 1980; Smyth and Hamrick, 1987), but maximum-likelihood techniques using multilocus allozyme genotypes are much more promising (Adams and Birks, 1991; Devlin and Ellstrand, 1990). Fortunately, the conclusions from both types of analyses are quite comparable to one another. In seed orchards of Douglas fir with an area between 2 and 5 hectares, 40 to 50 percent of the pollen accomplishing fertilization comes from outside the orchard (Smith and Adams, 1983; Wheeler and Jech, 1986). Obviously, most of these pollinations occur near the margin of the orchard. A maximum-likelihood analysis of gene flow within a 20-year old seed orchard gives a more complete picture of the extent of gene flow within stands. It showed that 50 percent of the pollen accomplishing fertilization on any individual plant comes from plants more than 25 meters away and that the average distance between mates is about 50 meters, assuming a negative exponential distribution of pollen from the trees on which it is produced (Adams and Birks, 1991). Unfortunately, we have no way of knowing how typical these estimates are, even for other wind-pollinated species, much less for animal-pollinated ones. Because the techniques have only recently been developed, few analyses have yet been attempted. In addition, these methods provide only a lower bound to the extent of gene flow, since they consider only pollen dispersal, not seed dispersal.

In short, it is far too early to offer any broad generalization about the typical scale of gene flow in widespread ecosystem dominants. Many workers concluded from the small observed dispersal distances in ecological studies that gene flow must be highly localized (Ehrlich and Raven, 1969; Levin, 1984; Levin and Kerster, 1974; Schaal, 1980; Schmitt, 1980). As additional information has accumulated, especially information gleaned from analyses of population differentiation at electrophoretic loci, several workers have suggested that these initial guesses may have considerably underestimated the importance of gene flow in widespread species (Devlin and Ellstrand, 1990; Ellstrand et al., 1989; Hamrick, 1987; Hamrick, 1989; Hamrick and Godt, 1990). Many additional analyses of gene flow in wind-pollinated hardwoods, in animal-pollinated species, in grasses, in conifers, and in other ecosystem dominants must be completed before anything definite can be said, but rates of gene flow in plant species may be higher than previously thought. If so, then gene flow in widespread ecosystem dominants is likely to occur on an even broader scale, and fragmentation is sure to disrupt existing patterns of gene flow, leading to large changes in the genetic structure of widespread species.

CONSEQUENCES OF GENE FLOW

Determining levels of gene flow in widespread ecosystem dominants is of paramount importance if we are to have any hope of understanding the genetic and evolutionary consequences of habitat fragmentation. The size of fragments necessary to maintain populations that will continue to function as a natural unit depends critically on the average distance between mates in continuous stands. In addition, even determining something as basic as whether fragmentation will increase or decrease genetic differentiation between the patches that remain depends on the distances that separate the patches relative to the average distance between mates before fragmentation and the frequency with which pollen or seeds are exchanged between patches. Since the plant species we expect to undergo genetic changes as a result of fragmentation are those with more or less continuous distributions, it is useful first to understand the genetic structure of a continuously distributed population.

Genetic differentiation in a continuously distributed population can have either of two causes: (1) adaptive responses to different environmental conditions in different parts of the distribution (i.e., ecotypic differentiation), or (2) random genetic differentiation as a result of genetic drift in widely separated parts of a population subject to uniform selection (i.e., isolation by distance). Ecotypic differentiation has been known in plants since the classic studies of Turreson (1922). Adaptive differentiation can happen over a very small scale, as in *Agrostis tenuis*, where ecotypes tolerant to heavy metals grow within a few meters of ecotypes that are nontolerant (Antonovics and Bradshaw, 1970). It can also happen over a very broad geographic scale, as with Ponderosa pine (*Pinus ponderosa*) where only four major ecotypes are recognized across all of western North America (Read, 1980; Wells, 1964). The scale over which ecotypic differentiation occurs depends on the geographic scale of environmental variation, the extent of gene flow, and the strength of selection in each of the habitats. Thus, fragmentation may have an effect on the ecotypic structure of widespread plant species through its effect on the patterns of gene flow among the remaining fragments. Even in the absence of ecotypic differentiation, however, disruption of existing patterns of gene flow may lead to significant genetic changes in the remaining population fragments.

Isolation by distance

Wright (1943a,b, 1946) began the theoretical study of continuously distributed populations in a uniform environment. His analyses suggested that members of the same population can evolve independently of one another if they are separated by a sufficient distance. Later, Malécot (1948), Kimura and Weiss (1964; Weiss and Kimura, 1965), and Maruyama (1972) provided elegant theoretical demonstrations of both his conjectures and formulae, showing the rate at which genetic correlations decay as a function of distance for a neutral locus subject to recurrent mutation. Goldstein and Holsinger (1992) noticed that the combination of isolation by distance and uniform selection for a phenotype that could be encoded for by many different homozygous genotypes might lead to different parts of a continuous population becoming fixed for different multilocus combinations producing the same phenotype. In short, even when subjected to uniform selection, different parts of a population might become genetically distinct.

To investigate this possibility we studied a simple model of polygenic inheritance, with 16 loci acting additively within and across loci to determine the phenotype. Individual fitness was determined by a Gaussian fitness function with an intermediate optimum. Thus, nearly 13,000 different homozygous genotypes could produce the optimum phenotype. In a panmictic population selection will result in fixation on only one of those homozygous genotypes (Barton, 1986; Wright, 1935). In a population with limited dispersal, however, the results are quite different. Individuals that are close to one another will tend to share the same genotype, as a result of high rates of gene flow. Individuals that are distant from one another, on the other hand, evolve independently of one another. Since selection favors the same intermediate phenotype in all parts of the population, the phenotype of individuals will tend to be relatively homogeneous across the population. But distant parts of the population are evolving independently, so the genotypes underlying the same phenotype will almost certainly be different. To put it another way, the genetic correlation between individuals declines with distance, so that beyond a certain point the genotypes of individuals are uncorrelated (Figure 1). Just where that point is depends primarily on the mean dispersal distance and the mutation rate per locus, though the strength of selection may have a weak effect.

Because selection appears to have only a weak effect, the analytical results of Malécot, Kimura and Weiss, and Maruyama on neutral loci allow us to predict quite accurately the distance at which genetic correlations for polygenic traits subject to optimizing selection will approach zero. This is not as surprising as it might seem at first glance for two reasons. First, the strength of selection at any one locus is actually quite weak. With the most extreme selection conditions we examined the relative fitness of the most extreme phenotypes possible was about 2×10^{-9}, but even then the selection coefficient at any single locus was less than 10 percent for genotypes differing from the optimum by only a single allele. Second, although the patch structure as reflected in the spatial pattern of genetic correlations is very stable, the patches themselves are ill-defined and transitory. Even within a patch the genotypes are not identical to one another, only more similar than randomly chosen genotypes in the population. Thus, an allele that is disfavored for 10 or 50 or 100 generations may suddenly be favored if a different allele becomes fixed at another locus. In short, the combination of weak and fluctuating selection on each individual locus makes results assuming neutrality a reasonable approximation for quantitative traits subject to optimizing selection.

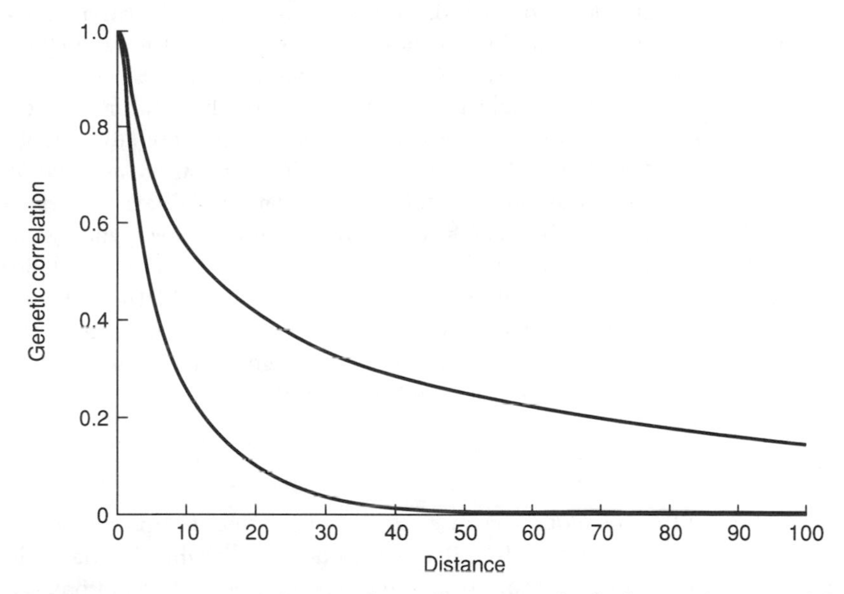

FIGURE 1. Average genetic correlation between individuals as a function of distance in a continuously distributed, two-dimensional population (Goldstein and Holsinger, 1992; Malécot, 1969). The lower curve is for a mutation rate of 10^{-3} per generation. The upper curve is for a mutation rate of 10^{-5} per generation.

Although the patches formed under such conditions are ill-defined, it is still useful to talk about a characteristic patch size. We define patch size in terms of linear dimensions as twice the distance at which the genetic correlation approaches zero. The analytical theory suggests that the correlation structure depends only on patterns of dispersal, not on population density. If we assume that the mutation rate per locus is 10^{-3}, which is consistent with estimates of the rate at which mutation introduces new polygenic variation (Lynch, 1988), the patch size is about 60 times the root mean squared dispersal distance. For negative exponential gene dispersal the root mean squared dispersal distance is equal to the mean dispersal distance. Thus, given that the mean distance between mates in Douglas fir orchards is on the order of 50 meters, we expect multilocus patches to be about 3 kilometers in diameter, covering an area close to 700 hectares. Our simulations suggested that populations can maintain such patch structure only when they are large enough to contain several, perhaps ten or more, patches. If these results can be applied to Douglas fir, it would suggest that natural stands of at least 3500–7000 hectares are required to maintain the patch structure.

In spatially structured populations only the additive genetic variance expressed when near neighbors mate will respond to selection. Since near neighbors are genetically correlated, the amount of genetic variance expressed in such matings may be substantially less than when distant individuals mate. Nevertheless, the additive genetic variance maintained among near neighbors by mutation-selection balance is actually greater than can be maintained by mutation-selection balance in an infinite population (Goldstein and Holsinger, 1992). More important, the population as a whole can store a tremendous amount of polygenic variation. Different parts of the population will often be near fixation on different, homozygous genotypes. As long as gene exchange between different parts of the population is possible, this store of variability can be released should the selective regime change. In short, a spatially structured population may be better able to respond adaptively to environmental change than a homogeneous one because of its greater store of genetic variability.

Fragmented populations

Understanding the consequences of gene flow in fragmented populations is more difficult than understanding it in continuously distributed ones. The problem is twofold. First, the spatial relationship among populations will have a dramatic effect on patterns of genetic differentiation. For practical reasons, however, theoretical analyses have concentrated on two extreme models of population structure: Wright's island model (Wright, 1951), in which all populations exchange migrants at the same rate, and stepping-stone models (Kimura, 1953), in which only adjacent populations exchange migrants. Second, gene flow can occur in two different ways in fragmented populations. It can occur either through the successful establishment of new individuals in existing populations, or through population extinction and recolonization.

The importance of the first problem is probably obvious. Wright's island model undoubtedly overstates the importance of gene exchange between populations, since it ignores any geographical structuring in populations and migration. Stepping-stone models probably underestimate its importance by allowing immigration only from geographically adjacent populations, populations that are likely to be genetically similar to one another. Unfortunately, there is little that can be done about this. If precise predictions are necessary for a particular species, there may be no alternative but to build and analyze specific, detailed models of the migration process.

The importance of the second difficulty, other than the obvious fact that establishment in an existing population and establishment of a new population are very different ecological processes, may be more difficult to see. Wright (1931) was the first to point out that allele frequencies tend to diverge in finite populations exchanging a small number of individuals. His prediction, however, depends on the assumption that migration results in the incorporation of individuals into an existing population. Its effect, of course, is to diminish allele frequency differences between populations, but if the rate of migration is low enough ($Nm < 1$), allele frequencies will diverge more rapidly as a result of drift than they can be homogenized by migration.

Suppose that gene flow occurs not by the incorporation of new individuals into an existing population, but rather as a result of the founding of new populations when previously existing ones have gone extinct. Maruyama and Kimura (1980) show that if the rate of local extinction and recolonization greatly exceeds that of migration between existing populations, the effective population size is much smaller than in a panmictic population with the same number of individuals. Thus, unlike the case considered by Wright, the whole set of populations may actually lose variation faster than a panmictic population with the same number of individuals. More recently Wade and McCauley (1988) and Whitlock and McCauley (1990) have shown that the particular mode of colony formation assumed for the recolonization process can also affect whether fragmentation increases or decreases differentiation. Thus, determining whether fragmentation will increase or decrease differentiation in parts of the population that remain will depend not only on determining the relative rates of migration, extinction, and recolonization, but on identifying the frequency with which different modes of colony formation occur (for more details see McCauley, this volume).

ECOLOGICAL AND EVOLUTIONARY CONSEQUENCES OF FRAGMENTATION

Predicting the effect of habitat fragmentation on the genetic structure of any species will obviously require detailed knowledge of its distribution and ecology. But such predictions are only a means to an end. What we are really interested in are the ecological and evolutionary consequences of the changes in genetic structure induced by habitat fragmentation. Fragmentation will almost always lead to a reduction in local effective population size, but knowing this is not enough. Much more important is that we know whether the magnitude of this effect is great enough to substantially increase the extinction probability for local populations, either as a result of inbreeding depression or through loss of adaptive genetic diversity. We also need to know whether the loss of rare alleles and genetic diversity within local populations will limit their ability to respond adaptively to environmental change. Although there is some reason to suspect that genetic effects will have only a minor direct effect on the persistence and success of local populations, we simply do not know how much fragmentation might retard adaptive evolutionary responses to environmental change.

Demographic consequences

A reduction in effective population size may have direct consequences for the probability of survival in one of two ways. Either the increased importance of drift may allow allele frequencies to depart significantly from their deterministic optimum, resulting in lower average individual viability, or the loss of genetic diversity associated with small population size may itself affect individual performance. The available evidence is equivocal on both points, but there is reason to suspect that for widespread ecosystem dominants the direct demographic consequences of fragmentation may be fairly minor, though for widespread species that are less common they may be important.

The long generation time of dominants in many ecosystems is, perhaps, the most important reason to suspect that any direct effect of reduced effective population size will be a long time in coming. The average life span in forested ecosystems is measured in hundreds of years, whether we are considering the old-growth Douglas fir forest of the Pacific Northwest or the mixed hardwood forest of the eastern United States. Similarly, in deserts and chaparral the average life span of ecosystem dominants is measured in centuries. Since changes in the genetic composition of a population require a generational turnover, drift effects may take a millennium or more to express themselves. Even in perennial grasslands much of the reproduction occurs by clonal propagation, and the frequency of sexual reproduction may be extremely low. This argument is, of course, only a proximate one. It simply asserts that the effects may be so long in coming that we need not worry about them now. There is, however, a more fundamental reason for suspecting that ecosystem dominants may not suffer greatly from direct demographic effects from reduced effective population size.

Analysis of a simple model with uniform selection against partially recessive deleterious alleles at 1000 loci suggested that average individual viability is reduced by less than 10 percent for populations with an effective size of 300 or greater (Holsinger and Gottlieb, 1991). Even with a population size as small as 100, average individual viability is reduced by less than 20 percent unless the selection coefficient against recessives is less than 0.01. Ecosystem dominants are likely to be reduced to such low numbers only in very small patches. Thus, direct effects on the average viability of individuals are likely to be even smaller than these. In addition, population viability analyses have suggested that typical plants must have a population size of at least 10^3 (Menges, 1991a). A population small enough that these influences are an important consideration is also small enough that it has probably already reached the point where it is demographically threatened. The genetic threat is entirely secondary (Holsinger and Gottlieb, 1991). Thus, fragmentation effects are most likely in widespread species that are common, but not dominant, members of their community. Only in those species are population numbers likely to be reduced enough to show such effects, and there is some evidence that the effects can be detected. Royal catchfly (*Silene regia*), for example, is a perennial prairie plant whose distribution has been drastically reduced by habitat fragmentation. In populations of more than 150 individuals seed germination rates were uniformly high, between 60 and 90 percent. In populations of fewer than about 150 individuals, however, seed germination rates varied between about 2 and 90 percent, with nearly a third of the populations having germination rates less than 50 percent (Menges, 1991b).

Evolutionary consequences

Although changes in the genetic structure of populations as a result of habitat fragmentation may have little direct demographic impact, they are likely to have a substantial effect on its evolutionary dynamics. Before discussing the possible effects, however, it is useful to comment on one effect that is unlikely to be important. Genetic diversity within a population can be measured either by the number of alleles present in the population or the average heterozygosity in the population. These two measures of genetic diversity are affected very differently by a reduction in population size. As Lewontin (1965) and Nei et al. (1975) pointed out, a population whose size is reduced is likely to lose its low-frequency alleles. Thus, its allelic diversity will be reduced, but its heterozygosity will be largely unaffected, unless the size reduction is severe and prolonged. Loss of the rare alleles is probably of little consequence for several reasons. First, most rare alleles are unconditionally deleterious and are maintained only as a result of recurrent mutation. Second, most adaptively significant variation is contained in alleles found in moderate to high frequency. Third, even if the low-frequency alleles might be valuable in adaptation to some future environment, they are likely to be lost before they are ever needed. Fourth, different populations are likely to lose different low-frequency alleles. In short, adaptive responses to environmental change are more likely to arise from new genetic variants or from existing genetic variants that are introduced from another population that did not lose them than from ones currently present in low frequency (Holsinger and Gottlieb, 1991; see also Hoffmann and Blows, this volume, and Lynch and Lande, this volume).

The most important evolutionary consequence of the changes in genetic structure induced by habitat fragmentation will result from the interplay of migration, extinction, recolonization, and the spatial scale over which these processes occur. Table 2 presents an initial attempt at a typology of possible consequences similar to the typology of rarity presented in Table 1 (see also McCauley, this volume). Notice that the categories of this typology are very vague and general, a necessity given how little we know. Still, it may provide a useful framework in which to discuss the possible consequences of habitat fragmentation. This typology is based on three simple observations.

1. If population fragments are separated by more than 30 times the root mean squared dispersal distance, little additional differentiation is likely,

TABLE 2. A typology of possible evolutionary consequences associated with habitat fragmentation.

Scale of fragmentation	**Large Separation > 30 × σ[a]**		**Small Separation < 30 × σ**	
Rate of migration	$Nm > 1$	$Nm < 1$	$Nm > 1$	$Nm < 1$
Size of population fragment: Small (single panmictic population)	Loss of diversity among populations; local diversity maintained, but lower mean fitness in all populations	Diversity among populations maintained; local diversity lost; lower mean fitness in all populations	Diversity among populations may remain similar; lower mean fitness in all populations	Diversity among populations increases; local diversity lost; possible operation of shifting balance
Size of population fragment: Large (several independent patches)	Loss of diversity among populations; substantial local diversity maintained	Diversity among populations maintained; local diversity maintained	Diversity among populations may remain similar; local diversity maintained	Diversity among populations increases; local diversity maintained, but possibly at lower level

[a] σ = Root mean squared dispersal distance

since the genetic correlation between individuals in these fragments will already be near zero (see above).

2. If the number of migrants between fragments is greater than one individual per generation, they will tend to remain genetically similar. If the number is less than one per generation, they will tend to diverge (Wright, 1931).[4]
3. If the remaining population fragments are too small to maintain several genetically independent patches, then a great deal of local genetic diversity will be lost (see above).

To make the discussion more concrete and to show how this typology might be used, I shall outline a scenario for the evolutionary consequences of fragmentation in a hypothetical old-growth Douglas fir stand. I can do so only in the vaguest terms, both because of our deep ignorance concerning the magnitudes of the relevant parameters and because the theoretical machinery necessary to make these predictions is still poorly developed. In developing this scenario I shall assume that the gene flow estimate of 50 meters per generation is a reasonable figure. I shall also assume that the original population was large enough to contain numerous genetically isolated patches within its boundaries, requiring that the population cover an area of perhaps 50,000–100,000 hectares. Of course, this far exceeds the size of any remaining old-growth stands. Thus, the scenario outlined below may actually reflect what is to befall old-growth stands in the coming centuries.

In 1981 the largest patch of old-growth forest in the Siuslaw National Forest was only 388 hectares (Harris, 1984). If these patterns are typical of old-growth forest elsewhere in the Pacific Northwest, each of the remaining fragments of old-growth Douglas fir forest is behaving as a single panmictic population, since the estimates derived above suggest that an area of between 3500 and 7000 hectares would be necessary to maintain evolutionarily independent patches given the observed pollen dispersal distances. Thus, the genetic diversity within the remaining patches is likely to decline substantially. Several important questions for determining the evolutionary dynamics of the system now become apparent. Will extinction and recolonization play an important role in determining the genetic structure of the remaining stands? What is the migration rate between existing old-growth stands? What is the genetic effect of managed stands on adjacent old-growth forest? What are the cumulative effects on the ability of these forests to respond adaptively to environmental change?

Although extinction of old-growth stands is likely to occur, either as a result of logging or natural catastrophe, establishment of new stands is far less certain. Thus, it seems likely that migration between stands will dom-

[4]This simple generalization obviously ignores the possible complications associated with different modes of migration.

inate the evolutionary dynamics. Whether the stands will tend to become more differentiated from one another depends on the distance between them relative to the scale of genetic differentiation before fragmentation. If genetic differentiation below the regional scale is primarily a result of isolation by distance, fragments separated by more than about 1.5 km (30 times the root mean squared dispersal distance of 50 m) already represent independent evolutionary entities. They will become more differentiated only if gene flow previously prevented locally adapted ecotypes from evolving. In addition, if gene flow between the remaining fragments is sufficiently reduced, much of the within-population genetic variation will be lost and, along with it, the ability to respond adaptively to environmental change.

The loss of local genetic variation could be exacerbated if managed stands surrounding the fragments of old-growth forest are drawn from genetically distinct stock. Gene flow into seed orchards has been studied precisely because such pollen contamination is expected to significantly reduce the performance of the seeds produced (Adams and Birks, 1991), but these effects are not limited to artificial seed orchards. Nursery-grown trees are commonly planted in native habitats, especially in intensive tree-breeding programs. These plantings can pose a significant threat to the genetic integrity of remaining old-growth fragments unless they are carefully managed (Millar and Libby, 1991). When the remaining fragments are very small, more localized introductions can also have an effect. In California, for example, Monterey pine (*Pinus radiata*) is planted extensively as an ornamental, and two of the five native populations receive a significant fraction of their pollinations from the ornamentals. The ornamentals are derived from New Zealand landraces that are themselves several generations removed from a third native population (Libby, 1990).

Since within-population genetic variation is reduced, the effectiveness of individual selection as an agent of adaptive evolutionary change is also reduced. Furthermore, genetic differentiation among the remaining fragments is unlikely to be significantly increased unless the fragments are separated by less than 1.5 km. If genetic differentiation among the remaining population fragments is increased, however, the opportunity for interdemic selection of the sort Wright envisioned in his shifting-balance theory may be increased (Wright, 1931, 1932, 1977). Although it requires a very special balance among the parameters, it is possible that the population structure produced by fragmenting the population might closely match the balance between drift, mutation, and selection that Wright identified as the optimal population structure for adaptive evolution. In short, conditions may exist in which habitat fragmentation actually *increases* the ability of a population to respond adaptively to changes in its environment. Of course, the remaining fragments of old-growth Douglas fir are so isolated from one another that fragmentation is much more likely to inhibit adaptive responses to environmental change.

CONCLUSIONS

It should be abundantly clear by now that there is almost nothing definite we can say about the effects of habitat fragmentation on the evolutionary dynamics of plant populations. Its precise effect depends critically on the scale over which gene flow occurs before fragmentation, the pattern of migration among patches after fragmentation, and the rate of local population extinction and recolonization. In most cases fragmentation will reduce a species' ability to respond adaptively to changes in its environment, though it may, in rare circumstances, enhance evolutionary responses by mimicking a population structure that promotes Wright's shifting balance process. It also seems safe to say that only a minority of plant species are likely to show dramatic effects on their evolutionary responses as a result of additional habitat fragmentation, since the distribution of many plant species is naturally fragmented. Unfortunately, many of those species on which a major impact is possible are the dominant members of their ecosystem. The time scale over which these effects will be noticed is likely to be centuries or millennia, but it appears likely that the process has already begun.

There are several areas where more work is clearly necessary if we are to improve our understanding. First, we need detailed analyses of gene flow in a variety of widespread species, both ecosystem dominants and those that are less common. Without a clear understanding of the probable spatial scale over which genetic differentiation may occur, we have no way to predict the consequences of fragmentation. Second, tests of the isolation-by-distance model for genetic differentiation in continuously distributed populations are required. Isolation-by-distance models appear to provide a solid theoretical foundation on which to build an understanding of the genetic structure of continuously distributed species, but this foundation is only as sound as the evidence that the scale of genetic differentiation is determined more by the scale of gene flow than by the scale of environmental differentiation. Third, we must obtain a better understanding of the ecological circumstances governing the dynamics of local population extinction and recolonization. Without it, we have no way to predict when populations will tend to diverge and hence no way to determine when fragmentation might actually enhance adaptive responses to environmental change. Finally, we need theoretical treatments of the process of population fragmentation that explicitly deal with the ecological dynamics of gene flow, local population extinction, and recolonization. Ideally, they will treat not only equilibrium properties of fragmented systems, but the transient dynamics of the change from a continuous population to a fragmented one. For management purposes, we must know not only what the end point of the process is—after all, that may not happen until several centuries after our demise—we must understand the transient dynamics that take us from here to there.

ACKNOWLEDGMENTS

I am indebted to Joel Kingsolver and an anonymous reviewer for comments on an earlier draft of this paper. This work was supported in part by grants from the National Science Foundation (BSR-9107330) and the University of Connecticut Research Foundation.

CHAPTER 13

GENETIC CONSEQUENCES OF EXTINCTION AND RECOLONIZATION IN FRAGMENTED HABITATS

David E. McCauley

Human disturbance has resulted in the continued fragmentation of numerous habitat types (see Croom and Schumaker, this volume). As habitats are fragmented, so are the plant and animal populations that inhabit them, with an increased likelihood of extinction too often a consequence. Recently, much effort has been directed toward identifying the conditions necessary for the persistence of those species occupying increasingly fragmented habitats (see, for example, references in Soulé, 1986, 1987). While over short time frames the crucial questions focus on the ability of a species to simply survive and reproduce in a fragmented habitat—i.e., on its demographic characteristics—longer-term questions must ask how the habitat fragmentation might affect the genetic characteristics of the species as well. The genetic consequences of habitat fragmentation merit special consideration as one considers the suite of environmental stresses likely to affect the biota as a result of other concurrent aspects of global change. For example, population responses to stresses imposed by global warming, ozone depletion, acid rain, and the introduction of exotic competitors, pathogens, and predators depend ultimately on the genetic characteristics of those populations.

From a population genetics perspective, habitat fragmentation can represent a radical and rapid change in population structure. Generally, habitat fragmentation reduces the size and increases the isolation of local populations, contributing to the erosion of genetic variability. The genetic consequences of this reduction of local effective population size and rate of gene flow could influence both the long-term ecological viability and the evolutionary potential of the species in question. Of most immediate concern from a conservation perspective is the prospect that a loss of heterozygosity will result in inbreeding depression and that inbreeding depression, resulting from either the phenotypic expression of deleterious recessive alleles or from the loss of heterozygosity per se, will threaten the viability of the population or species (Gilpin and Soulé, 1986). Vrijenhoek (1985, 1989), Allendorf and Leary (1986), and others review the evidence relating heterozygosity to fitness in natural populations, suggesting that a reduction of genetic variation could indeed threaten the viability of populations. For example, a recent study by Quattro and Vrijenhoek (1989) showed that variation in demographic characteristics related to fitness among populations of the endangered topminnow was correlated with levels of genetic variation. The population containing individuals with the lowest fitness displayed no electrophoretic enzyme variation.

It is well known that the potential for a population to evolve is governed by available genetic variation (Fisher, 1930). Thus, given sufficient loss of genetic variation owing to habitat fragmentation, the evolutionary response to selection pressures generated by future environmental challenge could also be limited. This could be a particularly important influence on the outcome of biotic interactions. For example, it is argued that genetic diversity mitigates the impact of predators, parasites, and diseases on the host population, especially in combination with frequency-dependent selection (Clarke, 1979; Hamilton, 1980).

Fragmentation can result in a variety of habitat-patch configurations, ranging from cases in which the habitat becomes so fragmented that individual patches are too small to support viable populations of the focal species, to cases in which patches are large enough to support quite stable populations. When the populations inhabiting individual habitat patches are stable and capable of long-term persistence, the effects of fragmentation on genetic structure are well described by standard population genetics theory (e.g., Hartl and Clark, 1989; see also Holsinger, this volume). However, one likely consequence of habitat fragmentation is the formation of an array of patches supporting relatively unstable populations subject to some extinction probability, but also capable of exporting dispersers destined to colonize unoccupied habitat patches. Under these circumstances the array of populations can persist for a much longer period than the lifetime of any one population. The conditions under which such metapopulations (or populations of populations) can persist has become the focus of much recent ecological research (Quinn and Hastings, 1987; Hanski, 1989; Hanski and

Gilpin, 1991), including the development of an emerging body of theory that traces its origins to the original models by Levins (1969, 1970). From a genetic perspective, the most important consequence of metapopulation dynamics is that each local population is only a few generations removed from its founding, and that the number of individuals founding that population is likely to be considerably smaller than the ultimate carrying capacity. Thus, even abundant species would be subjected to recurrent local population bottlenecks. The population genetics consequences of this cycle of extinction and colonization can be complex and have only recently been subject to much study. This chapter will discuss the genetic consequences of habitat fragmentation with an emphasis on the effects of local extinction/recolonization on the maintenance of genetic variation. It will first show that mathematical models predict that, in general, extinction/recolonization can contribute to the loss of genetic variation even in species that are otherwise abundant. It will then discuss the relevance of the models to the genetic structure and viability of natural populations.

MODELS OF POPULATION STRUCTURE

The field of population genetics enjoys a tradition of interplay between mathematical theory and empirical observation. A large body of theory has been developed to investigate the effects of population structure on genetic variation. The models are necessarily an abstraction and simplification of the complexity found in nature. One such simplification is to assume selective neutrality in order to clarify the joint effects of genetic drift and gene flow on genetic variation. These models can provide a framework for anticipating the effects of habitat fragmentation on genetic structure in nature, perhaps suggesting which conditions threaten population viability.

Equilibrium models

The changes in population structure resulting from habitat fragmentation, and the resulting genetic challenge to the persistence of affected animal species, have been well recognized by conservation biologists (e.g., Frankel and Soulé, 1981; Schonewald-Cox et al., 1983; Lande and Barrowclough, 1987; Lande, 1988; Simberloff, 1988) and will be summarized only briefly here. Subdivision of a large, widespread population into many smaller local populations, interconnected by migration, has several interdependent genetic effects. Genetic variance, as represented by either heterozygosity or the effective number of alleles, declines *within* populations, while *among*-population genetic variance, represented by variance in population-specific allele frequencies, increases. Globally, heterozygosity declines, though alleles are not necessarily lost from the global system. These results are obtained from various mathematical models, the best known of which is

Wright's island model (see Slatkin, 1985, for a recent review of population structure models). The net effect of the population structure on the partitioning of genetic variance is a result of the relative strengths of genetic drift, generated by small local effective population sizes N, and gene flow, generated by migration among populations at some rate m. Because plant and animal species can vary tremendously in the life-history characteristics that determine N and m, different species inhabiting the same fragmented landscape could have very different population structures. An additional effect of habitat fragmentation on the maintenance of global, or species-wide, levels of genetic variation derives from its reduction of the species-wide genetic effective population size. This depends on the number of local populations, n, in addition to local population size and migration rate.

The genetic consequences of population structure can be represented by several parameters in these models. The best known of these is F_{ST}, which is a measure of genetic differentiation, or the degree to which genetic variation is partitioned among populations. F_{ST} can range in value from zero to one. In a two-allele system, $F_{ST} = 1$ implies that a fraction p of all populations are fixed for one allele and a fraction $(1 - p)$ are fixed for the other. Generally, as F_{ST} increases, within-population heterozygosity decreases. Another measure of the loss of heterozygosity is the probability f that two alleles drawn at random from an array of populations will be identical by descent. This probability is itself a function of the probability f_0 that two alleles drawn from the same population are identical by descent, and the probability f_1 that alleles drawn from different populations are identical by descent. The probabilities are related by the equation

$$f = [f_0 + (n - 1)f_1]/n$$

where n is the number of populations. When the number of populations is large, f (and f_1) is necessarily near zero, even in cases where f_0 approaches one. In that case, f_0 is approximately equal to F_{ST}. This situation implies that heterozygosity can be lost from within populations even as allelic diversity is maintained by the larger system. When the number of populations is small, however, there can be genetic drift at the level of the entire system owing to a small global effective population size, as represented by $f > 0$. Slatkin (1985) discusses the relationship between f and F_{ST} when the number of populations is small.

One reason Wright's island model is often cited in discussions of population structure is that it predicts a simple relationship between genetic differentiation, local population size, and migration rate:

$$F_{ST} = 1/(4Nm + 1)$$

Obtaining this result requires several assumptions that need to be explored. First, it is assumed that migrants move among all populations with equal probability. Certainly there will be some isolation by distance among populations of all but the most vagile organisms. So-called stepping-stone models (Kimura and Weiss, 1964) assume most migration to be between adjacent populations. This tends to limit the effectiveness of gene flow as a force to counteract drift, though the results obtained from island and stepping-stone models are often qualitatively similar. More important is the assumption that the number of populations is large. A class of models known as *n*-island models allow the number of populations to be small (Maruyama, 1970; Latter, 1973). These *n*-island models can be used to predict the loss of genetic diversity, as well as heterozygosity, from the total system and from local populations. Finally, it is assumed that the local populations persist for many generations, long enough for the system to be in equilibrium with regard to the opposing forces of drift and migration. The simple relationship between F_{ST}, N, and m cited above is obtained at equilibrium. The approach to that equilibrium after a perturbation can require many generations (Varvio et al., 1986). The following discussion will focus on the consequences of relaxing the assumption of population persistence.

Metapopulation models

Given sufficient habitat fragmentation and the resulting reduction of population sizes, local populations could become threatened with extinction owing to demographic or environmental stochasticity. At the same time, new populations could be continually formed as colonists reinvade vacant habitat patches. This dynamic of extinction and recolonization forms the basis for metapopulation theory (Levins, 1969, 1970; Hanski and Gilpin, 1991), which seeks to describe the conditions under which an array of populations can persist even in the face of local population turnover. Increasingly, it is becoming apparent that metapopulation phenomenon must be incorporated into ecological theory, especially as applied to conservation issues such as habitat fragmentation (Hanski, 1989, 1991). If this is the case, then the population genetics of metapopulations must be considered as well (Ewens et al., 1987; Pimm et al., 1989; Gilpin, 1991; McCauley, 1991). Metapopulation dynamics could influence the partitioning of genetic variance both within and among populations, as well as the retention of genetic variation by the entire system.

Wright (1940) was perhaps the first to recognize that systems with frequent local extinction and recolonization could be relatively common in nature and that extinction/recolonization dynamics could be an important determinant of population genetic structure. He emphasized that colonization events could act as local population bottlenecks, reducing effective population size averaged over time and enhancing the genetic differentiation of populations. Slatkin (1977) incorporated extinction/recolonization dynamics into a more formal extension of Wright's island and *n*-island models of population structure. In these models, populations are founded by k individuals and grow to size N in the next generation. Once founded, the established populations exchange migrants at a rate m. All populations are subject to an extinction probability, e, with new populations founded at a rate matching those lost to extinction. The partitioning of genetic variation within and among populations can be cast as either F_{ST}, Wright's measure of differentiation, or f_0 and f_1, the probability that two gene copies are identical by descent as utilized in n-island models. Two general types of models were developed. In migrant-pool models, each gene copy in a colonizing group is drawn independently and at random from all possible extant populations. In propagule-pool models, all k individuals that make up a given colonizing group ($2k$ gene copies) are drawn together from one of the possible source populations. Slatkin pointed out that migrant-pool colonization includes an element of genetic mixing, as well as a population bottleneck, and that extinction/recolonization need not result in much genetic differentiation.

One way to characterize the effect of metapopulation dynamics on population genetic structure is to compare the genetic differentiation of an array of populations subject to extinction/recolonization to the genetic differentiation of an array of populations similar in every way except that the population occupying each available habitat patch persists for an infinitely long time. This is particularly appropriate when the number of local populations is fairly large. The comparison is possible using Slatkin's models because with $e = 0$ they collapse to cases described by Wright's island and the n-island models. Wade and McCauley (1988) conducted such an analysis. They showed that the net effect of extinction/recolonization on population differentiation depends on the relationship between Nm, the number of individuals moving between extant populations, and k, the number of individuals founding new populations. Specifically, in the migrant-pool model, extinction and recolonization will increase differentiation at equilibrium when

$$2k < (4Nm + 1)$$

Recall that the equilibrium referred to here includes the balance between extinction and recolonization and differs from the equilibrium between drift and migration in the so-called equilibrium models described in the previous section. Note that $2k$ represents the amount of binomial sampling variance expected among an array of newly founded populations, and determines F_{ST} if just that subset of the metapopulation is considered. The quantity $(4Nm + 1)$ is equivalent to the equilibrium F_{ST} expected without extinction. Thus with migrant-pool colonization, extinction/recolonization can either increase or diminish genetic differentiation, relative to the case with no extinction. This is illustrated in Figure 1, in which Slatkin's migrant-pool model has been solved numerically for F_{ST} with various representative combinations of values assigned to the ecological variables.

With propagule-pool colonization, extinction/recolonization increases

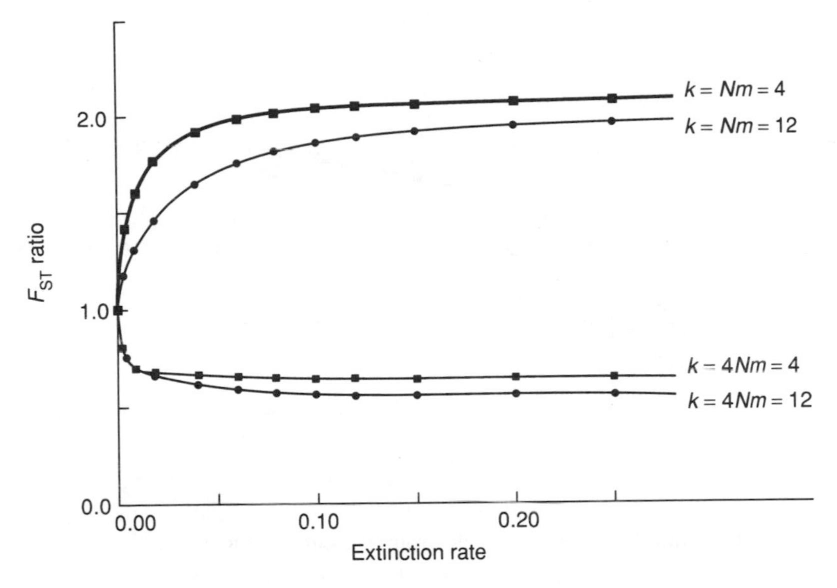

FIGURE 1. The ratio of the equilibrium F_{ST} derived from Slatkin's (1977) migrant-pool model at representative values of *Nm* to the equilibrium F_{ST} derived from Wright's island model, with identical values assigned to *Nm* but with no extinction. The F_{ST} ratio is presented as a function of the extinction rate. (From Wade and McCauley, 1988.)

differentiation under most conditions. The magnitude of differentiation is generally much greater with propagule-pool colonization than is permitted with migrant-pool colonization (Figure 2). In both cases, the effects of extinction/recolonization are amplified as the extinction rate increases, since the influence of founding events also increases. However, under most conditions increasing extinction probabilities above 0.2 per generation has little additional impact on genetic structure.

Whitlock and McCauley (1990) generalized Slatkin's models to include cases intermediate between the migrant- and propagule-pool modes of colony formation. Here the parameter ϕ defines the probability that two gene copies in a founding group are drawn from a common source population ($\phi = 0$ and $\phi = 1$ represent the migrant-pool and propagule-pool cases, respectively). This more general model shows that any tendency for gene copies to move in groups, such as in the movement of preinseminated females, increases the likelihood that extinction/recolonization leads to greater differentiation (Figure 3). At equilibrium, extinction/recolonization will increase differentiation if

$$k < [2Nm/(1 - \phi)] + 1/2$$

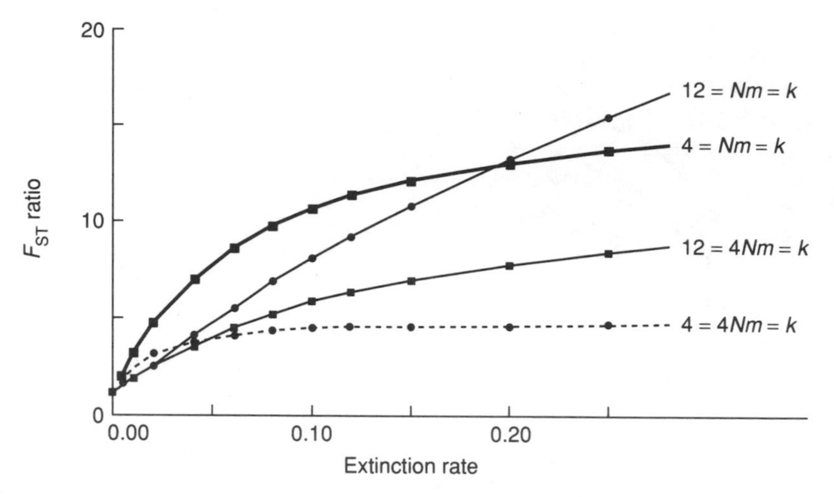

FIGURE 2. The ratio of the equilibrium F_{ST} derived from Slatkin's propagule-pool model to the equilibrium F_{ST} derived from Wright's island model when identical values are assigned to *Nm*. (From Wade and McCauley, 1988.)

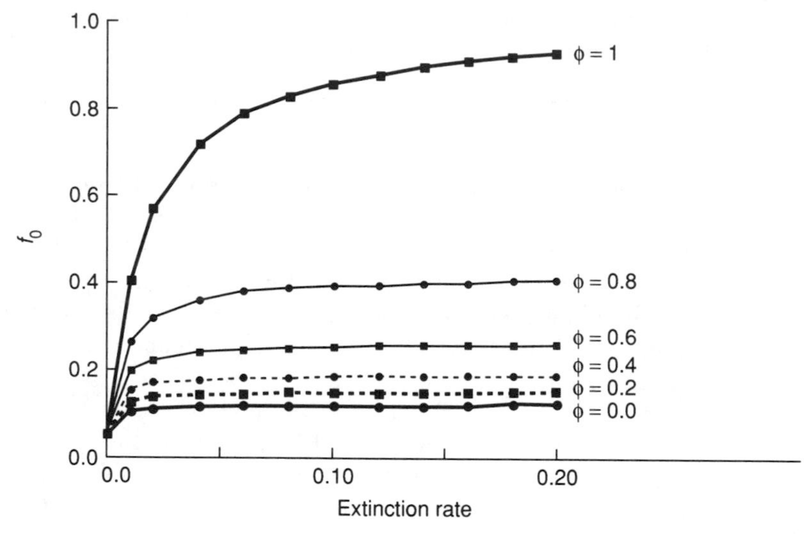

FIGURE 3. The probability of identity within populations (f_0) as a function of extinction rate and various probabilities of common origin (ϕ). In all cases $k = Nm = 4$. (From Whitlock and McCauley, 1990.)

Note that extreme differentiation is limited to cases where ϕ is nearly one (Figure 3).

From a conservation perspective, the rate of loss of genetic variation may be important as well. There is evidence that the expression of inbreeding depression is the most severe in cases in which the loss of a given amount of heterozygosity occurs rapidly (Lande, 1988). The models suggest that extinction/recolonization might increase the rate of loss early in the history of the metapopulation, even under conditions in which the equilibrium F_{ST} is equal to, and sometimes less than, the case without extinction (Figure 4).

As mentioned earlier, metapopulation dynamics can influence species-wide levels of genetic variation, as well as the partitioning of that variation within and among local populations. The repeated population bottlenecks associated with colonization are expected to reduce the species-wide genetic effective population size much below the actual number of individuals alive in the metapopulation at any one time. The effect of extinction/recolonization on global effective population size and maintenance of global genetic variation has been shown via modelling (Maruyama and Kimura, 1980; Ewens et al., 1987; McCauley, 1991) and simulation (Gilpin, 1991). The result is that species that are numerically abundant, but distributed into local populations subject to extinction/recolonization, are subject to a more rapid loss of genetic diversity from the system than would be found in the same population structure, but without population turnover. Unlike F_{ST}, f will approach one in nearly all cases, though the rate of approach varies widely, depending on conditions. This effect is most pronounced when the number of local populations is relatively small. McCauley (1991), using the Slatkin models, suggests that genetic variation is lost most rapidly when colonization is of the propagule-pool type (Figure 5). The loss of genetic

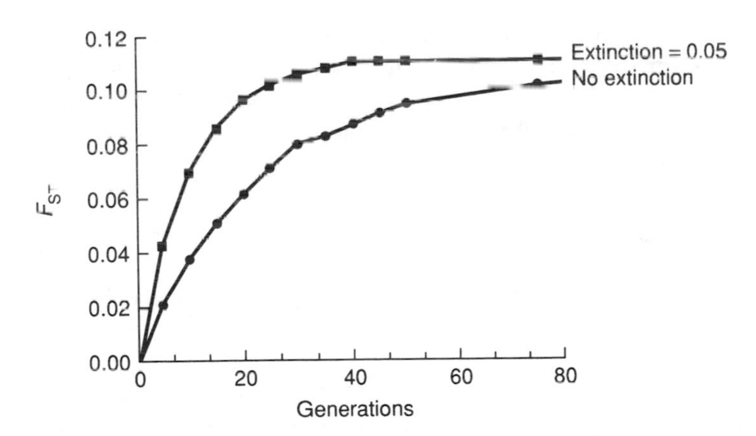

FIGURE 4. F_{ST} as a function of the number of generations since fragmentation when $e = 0.05$ and when $e = 0$. $N = 100$ and $m = 0.02$ in both cases. In the case with extinction, recolonization is of the migrant-pool type, with $k = 4$.

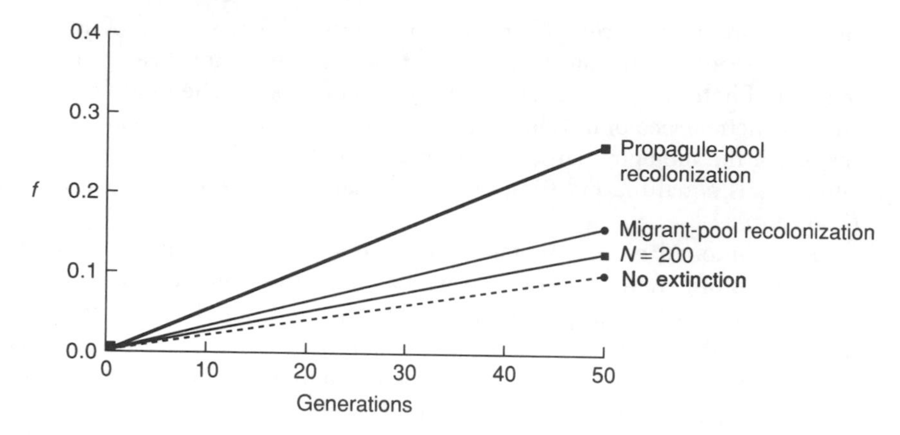

FIGURE 5. The probability f that two alleles, randomly chosen from among all populations, are identical by descent as a function of generations. Four cases are presented: one population of size 200, 10 populations of size 20 with no extinction, 10 populations of size 20 with extinction and migrant-pool recolonization, and 10 populations of size 20 with extinction and propagule-pool recolonization. In structured populations $m = 0.025$; $e = 0.05$ and $k = 2$ with extinction. (From McCauley, 1991.)

variation from the system at large continues to accelerate as e increases beyond 0.2 per generation (compare Figures 1 through 3 with Figure 6).

In sum, the models suggest that when habitat fragmentation results in a population structure with metapopulation dynamics, genetic variation can be lost from within local populations and from the metapopulation as a whole, even in species that are fairly abundant. This effect is greatest when extinction rates are high and the number of individuals in a founding group is small. Particularly important is the origin and composition of that group. The applicability of these models to the real world depends on the degree to which the dynamics of natural populations correspond to the "metapopulation" construct.

THE REAL WORLD

Clearly, habitat fragmentation has the potential to alter greatly the genetic structure of affected species. Theoretical models suggest that the nature of that effect can depend on the dynamics of local extinction and recolonization. However, there is a paucity of empirical work that demonstrates directly the effects of population turnover on genetic structure. A few studies do exist. For example, Vrijenhoek (1985) shows that the heterozygosity maintained by populations of the fish *Poeciliopsis monacha* from localities subject to frequent extinction and recolonization is less that than found in popula-

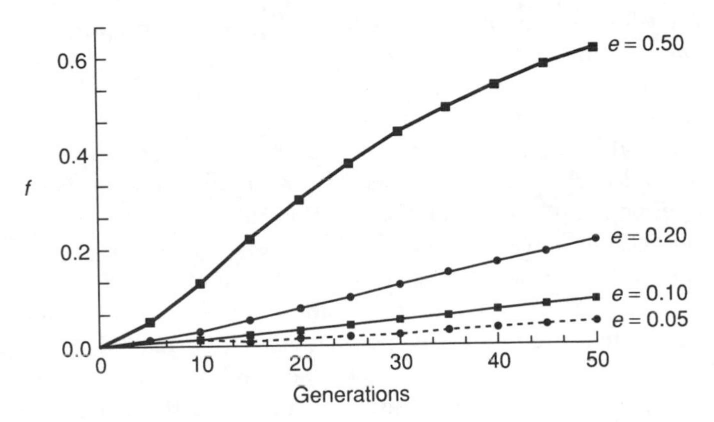

FIGURE 6. The probability of identity f as a function of time at various extinction rates. N, m, and k are identical to Figure 5 but there are 25 populations subject to propagule-pool recolonization.

tions from more stable habitats. Waller et al. (1987) attribute the lack of detectable genetic variation within or among four populations of Furbish's lousewort *Pedicularis furbishiae* to frequent local extinction and recolonization. Stiven and Bruce (1988) suggest that variation in heterozygosity among populations of the salamander *Desmognathus quadramaculatus* is a result of variation in the frequency of habitat disturbance resulting from logging. McCauley (1989) showed that turnover of local populations of a milkweed beetle, *Tetraopes tetraophthalmus*, was sufficiently frequent to influence genetic structure and that colonizing propagules were most likely small enough to allow for episodes of genetic drift in otherwise large populations. Finally, Whitlock (1992) showed in a study of the forked fungus beetle *Bolitotherus cornutus* that the frequent turnover of habitat patches owing to succession acted to increase local genetic differentiation. With the exception of the study by Stiven and Bruce, none of the cited examples were concerned directly with habitats fragmented by human disturbance.

What relevance do these models have to real populations occupying fragmented habitat? To answer that question one must first consider the degree to which natural populations conform to the domain of the metapopulation models, and which attributes of those populations would most contribute to the erosion of genetic variation. Second, one must consider whether the phenotypic consequences of genetic structure really threaten the short-term viability or long-term evolutionary potential of the species.

When does extinction/recolonization really matter?

The genetic consequences of extinction/recolonization are really only of interest in metapopulations that are otherwise capable of long-term persistence. The effect of population turnover on genetic structure increases as extinction rates increase, so the genetic models would appear to be most applicable to systems with high turnover. High extinction rates, however, might threaten the persistence of the entire metapopulation, obviating any concern over its genetic properties. A prediction from the Levins metapopulation model is that the persistence of the metapopulation itself requires that extinction rates not exceed colonization rates. It may be that the high extinction rates required for population turnover to have large effects on the erosion of genetic variation do not allow the persistence of the metapopulation. Further, the loss of genetic variation from the entire system is most pronounced when the number of populations, n, is small. Incorporation of a finite number of habitat patches into the Levins model further restricts the conditions that allow for persistence, owing to the possibility that all populations will go extinct simultaneously (Hanski, 1989, 1991).

One assumption of both Slatkin's and Levins's models is that all populations go extinct with equal and independent probability. Harrison (1991) discusses the empirical evidence supporting this assumption. She concludes that quite commonly there is a large variance in the expected persistence of populations. Particularly large or favorable patches often support quite stable populations that serve as the primary source of the colonists that reestablish populations on other patches subject to much higher population turnover. These have been described as mainland–island (Boorman and Levitt, 1973) and source–sink (Pulliam, 1988) dynamics. For example, Schoener and Spiller (1987) found that only the smaller of numerous spider populations were subject to frequent population turnover. Harrison et al. (1988) found a similar situation in populations of butterflies, in which a large, persistent "mainland" population appeared to be the source of most recolonizing individuals. From a genetic perspective this suggests that the genetic fate of a metapopulation sometimes rides on the fate of populations occupying just a subset of the habitat patches. It would seem that the fate of a subset of persistent and numerically dominant populations would limit the impact of colonization on population genetic structure and could be more simply modelled by the so-called equilibrium models.

Given that the population dynamics of a species occupying a fragmented habitat approximate those assumed in the models, which life-history and demographic characteristics would most contribute to a loss of genetic variation through the dynamics of extinction and colonization? Two critical determinants of the partitioning of genetic variance within and among elements of a metapopulation are the mode of colony formation and the relative likelihood that dispersal events will lead to colonization versus migration between established populations. The models suggest that extinction/recolonization dynamics influence genetic structure most strongly when colonization tends toward the propagule-pool mode (as ϕ approaches one) and when dispersers are more likely to move into established populations than to found new ones—that is, if k is small even with frequent migration (relatively high Nm). For example, colonization in which animals move in groups (such as social groups undergoing fission) will be more conducive to differentiation than colonization in which individuals move independently from one another. Included in this would be the stepping-stone type of movement, since it requires that individuals founding a given population be drawn only from adjacent extant populations, not from the metapopulation at large. Whitlock and McCauley (1990) also showed that propagules consisting of related individuals, such as might occur when population fission is along kin lines (or when many seeds disperse in single fruits), further increase the differentiating effects of propagule-pool type colonization.

Differentiation will also be enhanced to the degree that movement is more likely to lead to migration than to colonization, if this reduces the potential size of colonizing groups. An example might be a plant species in which genes can move as pollen or seeds. Both pollen and seed movement can lead to gene flow between extant populations, but colonization can only be accomplished by seeds. In animals, male-biased dispersal, such as is often seen in mammals (Greenwood, 1980), might result in more migration than colonization, since a single male needs to find a mate following movement to effect gene flow, whereas a pregnant female has the potential to found a population by herself. Smith and Peacock (1990) and Ray et al. (1991) suggest that conspecific attraction might direct a disproportionate fraction of dispersing individuals into occupied patches, limiting the number of available individuals that might serve as colonists. They cite several examples of animal species where this has been observed. Conspecific attraction could limit the size of colonizing propagules and enhance the bottleneck effect at population founding. Conversely, xenophobic social behaviors might limit the movement of dispersers into established populations and enlarge the pool of available colonists. In addition, colonists might be more likely to be successful than migrants when unoccupied habitat patches represent more abundant resources or when they are more likely to be free of predators and pathogens than patches already supporting conspecific populations. Recall that when movement is more likely to lead to colonization than migration, extinction/recolonization acts to diminish genetic differentiation.

Colonization can be quite a variable phenomenon. McCauley (1989) found a large amount of spatial and temporal variation in the number and sex ratio of adult milkweed beetles colonizing artificial patches of milkweed. Slatkin's original model assumes that the number of colonists is constant. Whitlock and McCauley (1990) show that when the number of colonists varies from patch to patch the effective number of colonists is equal to the harmonic mean among sites. This means that even occasional colonization events involving small numbers of individuals can contribute disproportionately to the bottleneck effects of colonization. Ebenhard (1991) points out, however, that when propagule size varies, smaller propagule sizes may be expected to have a lower probability of successfully founding populations than would larger propagules. This is because populations are expected to be particularly prone to demographic stochasticity in the early stages of colonization, the effects of which become more severe as initial propagule size declines. Ebenhard reviews both theoretical and empirical support for his contention. If there is indeed a critical threshold propagule size below which colonization cannot succeed, then the likelihood that extinction/recolonization will enhance the loss of genetic variation in nature becomes more limited because the most severe population bottlenecks do not contribute to the long-term characteristics of the metapopulation.

In summary, the models suggest that empirical studies must be conducted on real metapopulations, focusing in particular on how new populations are founded and on the conditions that promote stability. How often real metapopulations possess the characteristics that simultaneously promote stability yet threaten a severe loss of genetic variation remains an open question. Unfortunately, the answer may be elusive owing to the diversity of life-history characteristics found in nature.

Consequences for population viability

The relationship between genetic structure, genetic variation, and population/species viability must also be established more clearly. Understanding the population genetics of fragmented populations is only of value to the degree that we understand the phenotypic consequences of the genetic effects. The number of documented cases of populations whose viability is threatened by the consequences of reduced genetic variation is limited. Despite a generation of debate, the neutralist/selectionist controversy is not yet resolved, especially as it relates to the adaptive significance of heterozygosity. As evidence, witness the recent debate between Hughes (1991) and Vrijenhoek and Leberg (1991) over Hughes's contention that "at most loci loss of diversity should not be a cause for concern because the vast majority of genetic polymorphisms are selectively neutral" (p. 249). While this may be an extreme view, clearly more research into this most basic of issues in population genetics is required before we can fully evaluate the significance of any genetic consequences of habitat fragmentation.

For example, population geneticists still have not come to complete agreement as to the cause of inbreeding depression or as to which species might be susceptible to it (Shields, 1982; Charlesworth and Charlesworth, 1987). It is especially unclear how much erosion of genetic variation is required to threaten a population's viability. The best documented cases of inbreeding depression in both plants and animals involve close relative mating (or selfing), which results in losses of heterozygosity far more rapid than expected with all but the most extreme forms of genetic drift. Only with pure propagule-pool recolonization, or when a small number of populations is combined with a high extinction rate, do F_{ST} or f approach one rapidly. Is the loss of, say, 20 percent of the heterozygosity present before fragmentation sufficient to threaten viability? One outstanding question is the degree to which the expression of inbreeding depression is environment-dependent. There is evidence that the fitness differential displayed between inbred and outbred individuals is increased in stressful environments (Dudash, 1990; Schmitt and Ehrhardt, 1990; K. Helenurm, pers. comm.) Thus, any inbreeding effects resulting from habitat fragmentation must be evaluated in the context of the environmental stress likely to arise from other aspects of global change. More research into the nature of environment-dependent expression of inbreeding depression is called for. In addition, interspecific variation in susceptibility to inbreeding depression must be determined. Empirical and theoretical results suggest that inbreeding depression should be the most severe in those species that have not experienced inbreeding in their recent evolutionary history (Schemske and Lande, 1985; Lande, 1988). Plant and animal species found in large, widespread populations would not be expected to experience much inbreeding (except under some breeding systems) and therefore might be expected to be most prone to inbreeding depression as they experience the more structured populations that accompany human disturbance (Holsinger, this volume).

Additional negative effects of the loss of genetic variation arise if that loss precludes an adaptive evolutionary response to future environmental challenges that might accompany global change. It is not clear how the effects of selection, drift, and gene flow might interact if extinction/recolonization models were to incorporate selection as an added factor. One of the suites of selective pressures likely to accompany habitat fragmentation derives from changes in the biotic interactions in the altered communities. Habitat fragmentation is thought to result in a reduction of alpha species diversity (Wilcove et al., 1986; Simberloff, 1988; Groom and Schumaker, this volume), though different species may be lost from patch to patch. If so, this will certainly affect the nature and intensity of trophic interactions. In addition, edge effects can increase the intensity of predation, parasitism, and disease owing to contact with species previously rare or excluded from less fragmented habitat (Wilcove et al., 1986). Any selective response to alterations in the biotic environment depends on the genetic diversity underlying variation in the characters under selection and on the dynamics of the process. Particularly interesting are interspecific interactions that result in frequency-dependent selection. There has been much speculation that interactions with predators, parasites, and pathogens can result in rare advantage frequency-dependent selection (Clarke, 1979; Hamilton, 1980; Allen, 1988; Seger, 1988). If so, the negative impact of these interactions on prey populations can be mitigated by their reservoir of genetic diversity.

For example, Lively et al. (1990) showed a particularly high rate of parasitism by a helminth in a population of the fish *Poeciliopsis monaca* whose genetic diversity had been reduced by a recent founder event, relative to more genetically variable populations. It may be that the diminished capacity to respond to the stresses imposed by biotic interactions is a primary threat to viability imposed by a reduction in genetic variation.

One likely consequence of habitat fragmentation is an increase in spatial variation in the attributes of the environment contributing to natural selection. The aforementioned patch-to-patch variation in community structure is one example. In a species divided into a series of completely isolated populations, each population would experience an independent evolutionary response to selection determined by the selection pressures and genetic characteristics specific to that population. In the extreme, this should result in the evolution of locally endemic varieties, as seen as a consequence of fragmentation in South American rain forests during the Pleistocene (Turner, 1981). By definition, however, elements of metapopulations are connected by some amount of migration or gene flow. This would offset, to some degree, the response to local selection pressures. Furthermore, with frequent extinction and recolonization, the local adaptive mode would be reset with each instance of population turnover. It is likely that the capacity for localized responses to spatial variation in selection pressures is somewhat limited by the transient nature of the elements of metapopulations.

An additional evolutionary consequence of these metapopulation dynamics can be derived from theories proposed by Sewall Wright. Wright's interest in population structure stemmed primarily from his belief that adaptive evolutionary change was more likely in highly structured populations than in large, panmictic units (Wright, 1978; Provine, 1986; Crow, 1991). In species with a high degree of population structure, selection and drift can act jointly in local populations, even in species that are abundant. It is the joint effects of drift and selection that allow populations to undergo evolutionary transitions that are ultimately adaptive, but whose early stages are maladaptive. Put differently, drift allows a local population to deviate from the domain of one adaptive state to another. Novel adaptations are then exported to other localities by differential rates of migration or colonization. If Wright's theory holds, then one could speculate that upon occasion habitat fragmentation could actually enhance the rate of evolution (see also Vrijenhoek, 1985). Wright further stated (Wright, 1940, 1978) that it was under conditions of local extinction and recolonization that his theory was most likely to apply, owing to some of the effects on population structure discussed here. Exactly what combination of drift and gene flow would allow for shifting-balance evolution without inbreeding effects threatening the viability of populations is not yet clear. It has been argued that the necessary relationship between migration rates, population sizes, and selection intensities is so restrictive as to severely limit the operation of the shifting-balance process in nature (Hartl and Clark, 1989). Again, with high extinction rates, local populations might not persist long enough to respond to the selection needed to draw them into novel adaptive peaks. Recent theoretical (Crow et al., 1990) and empirical (Wade and Goodnight, 1991) studies suggest, however, that the required conditions may not be quite so restrictive. The mode of evolution in metapopulations remains largely unstudied.

SUMMARY

Habitat fragmentation generally results in a reduction in the size and an increase in the isolation of local populations. This change in population structure can lead to an increase in the genetic differentiation of local populations and a loss of genetic variation from within those populations. Reduced genetic variation could threaten population viability owing to the possibility of inbreeding depression and a reduced capacity to respond to natural selection. A fragmented habitat could further lead to an increase in the turnover of local populations or a cycle of local population extinction and recolonization. Recent mathematical models show that extinction and recolonization can allow for episodes of genetic drift in otherwise abundant species, enhancing the effects of the population structure on the partitioning and maintenance of genetic variation. The effect is particularly dependent on the size and composition of groups founding new populations.

Evaluation of the contribution of the models to understanding the genetic properties of species inhabiting fragmented habitats and predicting their viability requires answers to several outstanding empirical questions basic to population biology. It is critical that the properties of real metapopulations be better documented. Of particular importance are studies of the establishment of new populations in order to determine both the types of dispersal events that lead to colonization and the conditions that allow a colonizing event to translate into the successful founding of a persisting population. It is also important to better quantify the relationship between genetic variability and the short- and long-term viability of populations. Outstanding issues include environment-dependent expression of inbreeding depression and the role of genetic variation in mitigating interspecific biotic interactions.

ACKNOWLEDGMENTS

Much of my thinking about the genetic properties of metapopulations has been facilitated by conversations and collaborations with Michael Wade and Michael Whitlock, though any mistakes are my own. Financial support for some of the work described here was provided by Award BSR 86-20119 from the National Science Foundation.

CHAPTER 14

EVOLUTION AND EXTINCTION IN RESPONSE TO ENVIRONMENTAL CHANGE

Michael Lynch and Russell Lande

INTRODUCTION

Extensive geological and paleontological research indicates that life historically has experienced periods of major environmental change, resulting from astronomical and tectonic events, on time scales ranging up to hundreds of thousands of years (Hays et al., 1976; Ruddiman and McIntyre, 1981; Hide and Dickey, 1991). Species have often responded to these changes by migrating, adapting, or speciating (Bartlein and Prentice, 1989; Davis, 1981; Cronin and Schneider, 1990), but in many cases they have simply gone extinct (Nitecki, 1984; Elliot, 1986; Jablonski, 1991). Because the rates of environmental change now being induced by human activity are well beyond those normally experienced by most organisms (Malone and Roederer, 1985; Bolin et al., 1986; Abrahamson, 1989; Schneider, this volume; Groom and Schumaker, this volume), there is justifiable concern about the possibility of a global environmental catastrophe.

Due to the complex structure of ecological communities, it is conceivable that the inability of one or a few critical species to cope with long-term environmental change will have cascading effects on other interdependent species (Holt, 1990), possibly producing a major alteration of community structure. It is therefore of some importance that we begin to develop a general understanding of the constraints on the ability of populations to adapt to environmental change. Since almost all quantitative characters exhibit some genetic variation, prolonged directional change in the environment generally will result in adaptive evolution. The critical question is whether, in the face of unfavorable environmental change, a population can evolve rapidly enough to avoid extinction.

The salient issues are summarized in Figure 1, in which it is assumed that the source of environmental change is external to the ecosystem and uninfluenced by changes in local community composition. Populations can profit from environmental change (e.g., populations at their polar limits may benefit from global warming), but our focus will be on situations where the environment is shifting toward conditions that are clearly unsuitable for a locally adapted population. The initial *direct* impact of such environmental change will be a reduction in reproductive capacity and/or viability resulting from maladaptation to the new conditions. Provided the necessary genetic variation exists, natural selection will result in the gradual evolution of more suitable phenotypes, but any substantial reduction in population size provoked by the environmental change will reduce the opportunities for such adaptation. The amount of genetic variation that can be maintained in a population decreases as population size decreases, and when populations are reduced to very small sizes, the chance fixation of unconditionally deleterious alleles may lead to a further decline in fitness.

These circumstances may be modified in mobile species if the change in environment induces individuals to migrate in search of more suitable habitat. Any further reduction in local population size by emigration will reduce the local evolutionary potential even more, especially if the popu-

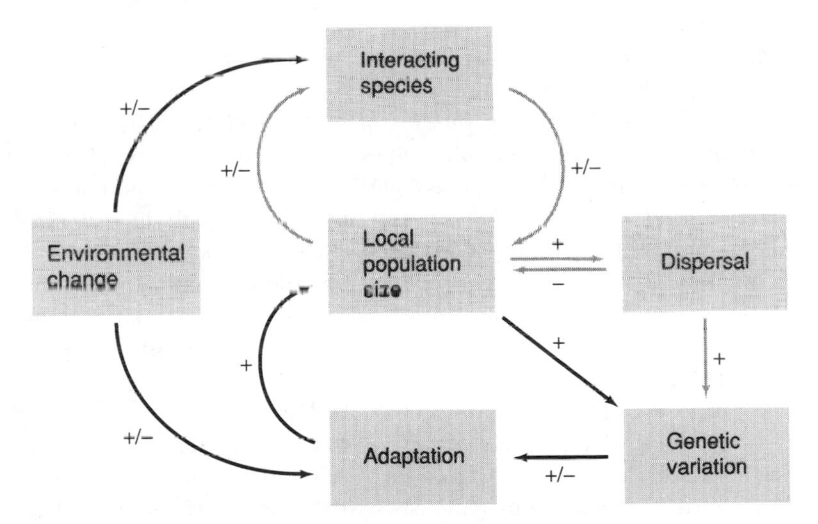

FIGURE 1. A simplified picture of the factors influencing the response of local population size to environmental change. Only the influences denoted by black lines are considered explicitly in this paper.

lation is geographically isolated (e.g., an island population or a preserve surrounded by unsuitable habitat). Provided the emigrants are successful, however, dispersal can replace evolution as a survival strategy, at least temporarily. But when environmental change operates over a broad enough geographic range or over a long enough time scale so that the current set of phenotypes is unsuitable in all habitats, evolution provides the only route to species survival.

In addition to the direct impact of an environmental change on a species' physiological, growth, or reproductive characters, it can exert substantial indirect influences through changes in the abundances of interacting species such as prey, predators, competitors, parasites, or hosts. In the long run, the net influence of such indirect effects can be detrimental or beneficial depending on the total change in community composition, and they may sometimes overwhelm the direct effects. Since the consequences of both dispersal and community structure change are likely to be highly species-specific, neither of those issues will be pursued any further here, although their potential significance should not be underestimated.

Ideally, an understanding of the evolutionary and demographic consequences of long-term environmental change ought to be provided by empirical research. But for many species this is impractical, if not impossible, due to long generation times, difficulties in rearing, ethical issues, etc. Because of the urgency of the problem, the development of mathematical models seems well warranted. Ultimately, such models will require the integration of population dynamics theory with evolutionary theory, an area that has received very little attention in the past. We present some initial results in this paper.

Consider a character under stabilizing selection in an environment changing such that the optimum phenotype increases (or decreases) with time. To a degree depending on the magnitude of the additive genetic variance for the trait, the mean phenotype in the population will lag behind the changing optimum. This lag in adaptation will cause a reduction in mean population fitness that, if substantial enough, will result in a decline in population size. If the rate of environmental change is sufficiently slow, the mean phenotype will remain close enough to the optimum that extinction can be avoided. However, there must be a critical rate of environmental change beyond which the selective load is so great that a positive population growth rate cannot be maintained.

The purpose of this paper is to evaluate how this critical rate of environmental change depends on certain aspects of the genetic system. For the sake of simplicity, we assume that individual fitness is determined predominantly by a single continuously distributed character subject to the standard laws of quantitative inheritance (Falconer, 1989). The character under consideration might be an attribute of size or shape, thermal or pH optimum, etc. We examine two extreme situations: sexual populations of arbitrary size, and asexual populations that are effectively infinite in size.

SELECTION IN A CHANGING ENVIRONMENT

Natural selection commonly favors an intermediate phenotype, with fitness declining approximately quadratically with the absolute deviation from the optimum (Johnson, 1976). The focus of this paper is on the situation in which the optimum phenotype θ exhibits temporal change. We will be considering continuously growing populations with overlapping generations, so the appropriate measure of fitness is the instantaneous rate of increase. Thus, at any point in time, the phenotypic fitness function (i.e., the expected fitness of an individual with phenotypic value z) can be represented by

$$r_z = r_m - \frac{(z - \theta)^2}{2\sigma_w^2} \tag{1}$$

where r_m is the rate of increase for the optimum phenotype, and σ_w is the width of the fitness function. Note that as σ_w increases, the curvature of the fitness function, and hence the intensity of selection, declines. A negative value of r_z implies that individuals of phenotype z do not produce enough offspring to replace themselves. From Equation (1), this is seen to occur whenever the absolute deviation of z from the optimum exceeds $(2r_m\sigma_w^2)^{1/2}$. The parameters r_m, θ, and σ_w, at least in principle, are all estimable properties of natural populations (Lande and Arnold, 1983; Mitchell-Olds and Shaw, 1987; Schluter, 1988; Lynch et al., 1991).

Quantitative characters are usually influenced by multiple genetic loci and environmental effects. Consequently, they generally exhibit a continuous distribution of variation that is approximately normal on some scale of measurement. The observed phenotypic value of an individual can be represented as the sum of two components: an additive genetic value (g), and a nonheritable residual deviation (e) from g caused by nonadditive gene action, environmental effects, and developmental noise. Assuming the distribution of additive genetic values has mean $\bar{g}$ and variance σ_g^2, and the distribution of residual deviations has mean 0 and variance σ_e^2, then the rate of population growth is

$$r = r_m - \frac{(\bar{g} - \theta)^2 + \sigma_z^2}{2\sigma_w^2} \tag{2}$$

(Lynch et al., 1991), where $\sigma_z^2 = \sigma_g^2 + \sigma_e^2$ is the phenotypic variance. Note that because the mean residual deviation ($\bar{e}$) is assumed to be equal to zero (i.e., we are modeling only genetic change), $\bar{g}$ is equivalent to the mean phenotype in the population.

We interpret r as the maximum rate of population growth that would be observed in the absence of density-dependent influences. Thus, if $r < 0$ for very long, the population is clearly doomed to extinction. Note that even if the mean phenotype is at the optimum ($\bar{g} = \theta$), the population rate of growth will generally be less than the maximum possible (r_m), since any level of phenotypic variance will result in the production of nonoptimal phenotypes. Thus, although genetic variance provides the fuel for adaptive evolution, it also imposes a load on the population.

We will assume that the phenotype is measured on a scale such that the initial optimal phenotype is equal to zero. Due to the environmental change, the optimum increases at an expected rate k per unit time. The actual optimum at time t is then

$$\theta(t) = kt + \epsilon_\theta \tag{3}$$

where ϵ_θ represents stochastic temporal variation in the optimum around its expected trajectory. We assume that ϵ_θ is normally distributed with no correlation in time, mean equal to 0, and variance equal to σ_θ^2. The width of the fitness function is assumed to be constant.

FINITE SEXUAL POPULATIONS

As a consequence of human disturbance, many species, especially vertebrates and vascular plants, have been reduced to isolated populations containing tens to hundreds of individuals. For purely demographic reasons, such species are highly vulnerable to extinction even in fairly constant environments (Lande, 1988), but when they are confronted with the additional problem of an environmental change, their vulnerability to extinction will be magnified even further.

The development of a theory for critical rates of environmental change for finite sexual populations introduces a number of technical difficulties. For infinite populations in a deterministic environment [i.e., $\theta(t) = kt$, $\sigma_\theta^2 = 0$], the dynamics of the mean phenotype can be evaluated explicitly (Lynch et al., 1991). But for the case under consideration, there is no single deterministic outcome. Random genetic drift of the mean phenotype, combined with stochasticity of the environment, results in a probability distribution of mean phenotypes realized among a hypothetical set of replicate populations. We have employed diffusion theory to evaluate this distribution.

We start with the stochastic differential equation

$$\frac{d\bar{g}}{dt} = \frac{\sigma_g^2(kt + \epsilon_\theta - \bar{g})}{\sigma_w^2} + \epsilon_{\bar{g}} \tag{4}$$

where $\epsilon_{\bar{g}}$ is the change in the mean phenotype caused by random genetic drift, and the nonstochastic portion of the equation follows from procedures for computing the change in the mean of a normally distributed character under quadratic selection (Lynch et al., 1991). The remaining results in this section (several steps to which we omit) are obtained by expressing this model in the form of a classical Ornstein-Uhlenbeck diffusion process (Karlin and Taylor, 1981), under the assumption that the genetic variance, σ_g^2, is essentially constant in time. Justification for this assumption is given in the appendix, where it is also argued that provided N_e is less than 10^3 or so, $\sigma_g^2 \simeq 2N_e\sigma_m^2$, with N_e being the effective population size and σ_m^2 the rate of input of genetic variance by mutation.

The expected dynamics of the mean phenotype conditional on a given value of $\bar{g}$ are

$$E\left(\frac{d\bar{g}}{dt}\,\middle|\,\bar{g}\right) = \frac{\sigma_g^2(kt - \bar{g})}{\sigma_w^2} \tag{5}$$

This shows that the expected mean phenotype always evolves towards the optimum, and it does so at a rate that is proportional to the additive genetic variance and to the inverse of the squared width of the fitness function.

Random genetic drift causes the variance of the distribution of mean phenotypes to increase by the amount σ_g^2/N_e per generation (Lande, 1976), and stochasticity in the environment further inflates the variance in the probability distribution of the mean phenotype by the amount $\sigma_g^4\sigma_\theta^2/\sigma_w^4$. Thus, the variance in the rate of change of the mean phenotype conditional on a given value of $\bar{g}$ is

$$V\left(\frac{d\bar{g}}{dt}\,\middle|\,\bar{g}\right) = \frac{\sigma_g^2}{N_e} + \frac{\sigma_g^4\sigma_\theta^2}{\sigma_w^4} \tag{6}$$

The asymptotic distribution of $\bar{g}$ for large t can be viewed as that expected after sufficient time has passed for a stochastic balance to be reached between the forces of selection, drift, and mutation. This distribution can be shown to be normal with mean

$$E(\bar{g}) = kt - k\frac{\sigma_w^2}{\sigma_g^2} \tag{7}$$

and variance

$$V(\bar{g}) = \frac{\sigma_w^2}{2N_e} + \frac{\sigma_g^2\sigma_\theta^2}{2\sigma_w^2} \tag{8}$$

Equation (7) shows that the expected mean phenotype evolves asymptotically at exactly the same rate as the change in the optimum (k) but lags behind the optimum by an amount $k\sigma_w^2/\sigma_g^2$. The variance of $\bar{g}$ consists of a random genetic drift term, $\sigma_w^2/(2N_e)$, identical to that found by Lande (1976), and an additional term due to the environmental stochasticity.

Taking the expectations of Equation (2) over the distributions of $\bar{g}$ and θ, the expected long-term fitness of a population in stochastic equilibrium with the changing environment is found to be

$$E(r) = r_m - \frac{\sigma_z^2}{2\sigma_w^2} - \frac{k^2\sigma_w^2}{2\sigma_g^4} - \frac{1}{4N_e} - \frac{\sigma_\theta^2}{2\sigma_w^2}\left(\frac{\sigma_g^2}{2\sigma_w^2} + 1\right) \tag{9}$$

Four types of loads contribute to the reduction in the expected population growth rate below its maximum possible value, r_m. In order of their appearance in Equation (9), first, as noted above, phenotypic variance results in a net loss of fitness due to the production of nonoptimal phenotypes. Second, a lag load results from the deviation of the expected mean phenotype from the moving optimum. With increasing σ_g^2/σ_w^2, the population mean tracks the moving optimum more closely, and the lag load is reduced. A third type of load results from the stochastic dispersion of the mean phenotype from its expectation due to random genetic drift. As shown by Lande (1976, 1980), the drift load has the interesting property of being independent of both the genetic variance and the width of the fitness function. While the amount of drift in the mean phenotype increases with the genetic variance, so does the response to selection, and the influences of those two factors on r cancel exactly. Similarly, although a flatter fitness function provides a greater opportunity for drift, the fitness consequences of a unit change in the mean phenotype are reduced, and again the two factors cancel. Fourth, environmental stochasticity causes a dispersion of the mean phenotype from the path of the expected optimum due to both the past selection response to the variation in θ and to the current deviation of the realized θ from its expectation. The first component of the environmental stochasticity load increases with the genetic variance, which causes the mean phenotype to carry a "memory" of past stochastic variation.

The critical rate of environmental change, beyond which the population crosses the extinction threshold, is the absolute value of k that satisfies $E(r) = 0$,

$$k_c = \frac{\sigma_g^2}{\sigma_w} \left[2\bar{r}_m - \frac{1}{2N_e} - \frac{\sigma_\theta^2}{\sigma_w^2} \left(\frac{\sigma_g^2}{2\sigma_w^2} + 1 \right) \right]^{1/2} \tag{10}$$

where $\bar{r}_m = r_m - (\sigma_z^2/2\sigma_w^2)$ is the maximum rate of population increase when the mean phenotype is at the optimum. For the special case of large population size ($N_e \to \infty$) and a deterministic environment ($\sigma_\theta^2 = 0$), Equation (10) simplifies to

$$k_c = \sigma_g^2 \sqrt{2\bar{r}_m/\sigma_w^2} \tag{11}$$

Noting that the genetic variance is necessarily less than the phenotypic variance, an upper bound to the critical rate of environmental change can be obtained from this expression. In units of phenotypic standard deviations,

$$k_c' = \frac{k_c}{\sigma_z} < \sigma_z \sqrt{2\bar{r}_m/\sigma_w^2} \tag{12}$$

Since it is much easier to estimate σ_z^2 than σ_g^2, and since $\bar{r}_m$ is also estimable, this expression may have some practical utility for identifying situations in which an extinction threshold will definitely be exceeded. For example, if the maximum possible rate of population increase is 0.5 per generation, and the width of the fitness function is equivalent to ten phenotypic standard deviations, then $k_c' = 0.1$, implying that the population could not possibly sustain a prolonged change in the optimum value of the selected trait in excess of 0.1 phenotypic standard deviations per generation.

Using a commonly observed value for mutational heritability, σ_m^2/σ_e^2, to compute the genetic variance, Figure 2 illustrates the relationship between k_c' (k_c in units of phenotypic standard deviations) and the effective population size for three values of σ_w^2 and σ_θ^2, for the case in which the potential rate of increase is $r_m = 0.67$ (equivalent to a potential population doubling per generation). For the parameter values used in this figure, there is a nearly log-linear increase of k_c' with population size, due largely to the dependence of σ_g^2 on N_e. Note also that, counterintuitively, the critical rate of environmental change actually declines with decreasing selection intensity (increasing σ_w^2). This is due to the increased lag load that develops when selection is weak.

LARGE ASEXUAL POPULATIONS

Since microbes such as bacteria, fungi, and algae play critical roles in all biogeochemical cycles and form the base of most food chains, their abundances influence most higher-level organisms either directly or indirectly. Therefore, it seems especially important to develop a general understanding of the constraints on the evolution of microorganisms. Random genetic drift is unlikely to play much of a role in the evolutionary dynamics of microbial populations, since their population sizes are usually effectively infinite, and the drift load can be assumed to be negligible. However, most microbial populations reproduce predominantly by asexual means, and the absence of segregation and recombination is expected to lead to reduced levels of genetic variation, and therefore to lower evolutionary potential relative to the expectation in otherwise similar sexual populations (Lynch and Gabriel, 1983).

As in the case of sexual populations, asexual populations asymptotically settle into an expected rate of evolution equal to the rate of change in the environment, lagging behind it by a constant amount (Lynch et al., 1991). The expected asymptotic dynamics of the mean phenotype are as described by Equation (7). Noting that $\sigma_g^2 = \sigma_m\sigma_w$ (Lynch et al., 1991; Appendix), the asymptotic distribution of the mean phenotype is obtainable from Equations (7) and (8). The extinction threshold is then simply

$$k_c = \sigma_m \left[2\bar{r}_m - \frac{\sigma_\theta^2}{\sigma_w^2}\left(\frac{\sigma_m}{2\sigma_w} + 1\right)\right]^{1/2} \tag{13}$$

Thus, the critical rate of environmental change increases with the rate of polygenic mutation, σ_m^2, which provides the genetic variance required for adaptive evolution. It also increases with the maximum population growth

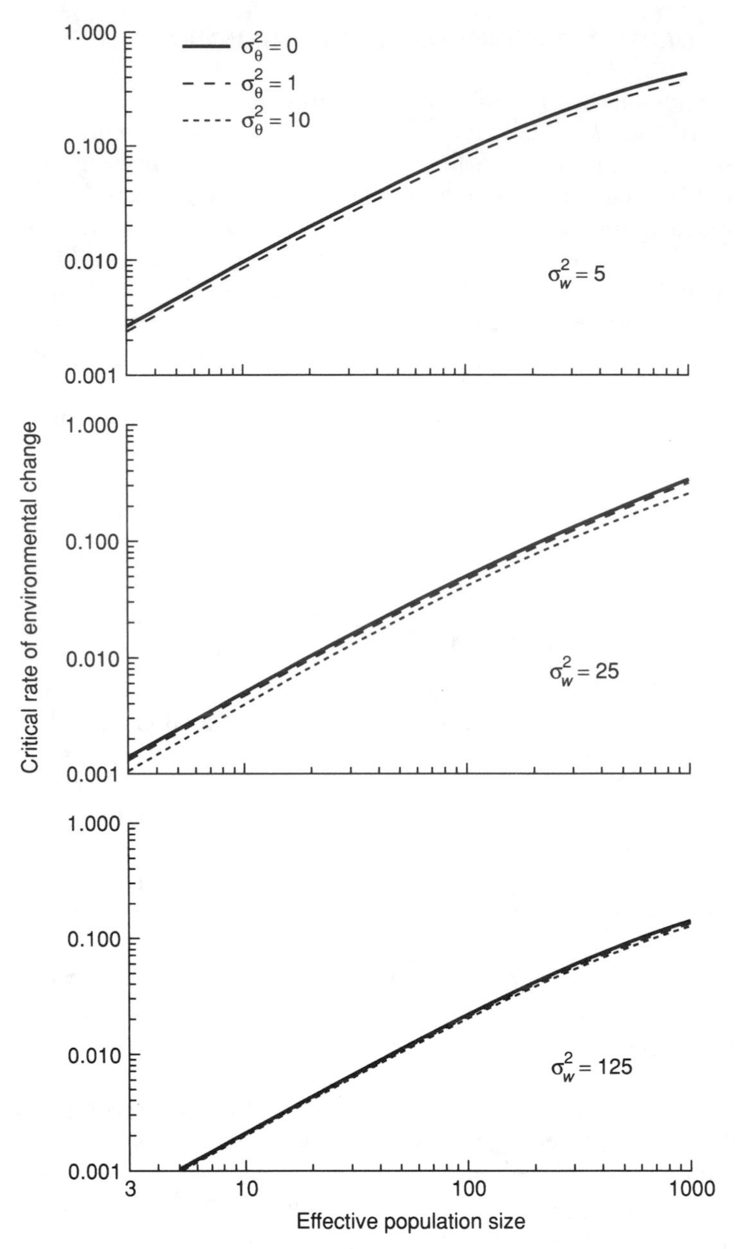

FIGURE 2. Critical rate of environmental change k'_c (in units of phenotypic standard deviations), beyond which the expected rate of population growth becomes negative, given as a function of the effective population size and the width of the fitness function. Results are given for $r_m = 0.67$ (equivalent to a potential population doubling per unit time), $\sigma_m^2 = 0.001$, $\sigma_e^2 = 1$ (a typical observed mutational heritability is $\sigma_m^2/\sigma_e^2 = 0.001$; Lynch, 1988), and letting $\sigma_g^2 = 2N_e\sigma_m^2$ (Appendix). In the upper panel, the population cannot maintain a positive growth rate at any N_e when $\sigma_\theta^2 = 10$.

rate, $\bar{r}_m$, and it decreases with environmental stochasticity, σ_θ^2. Note that when the temporal change in θ is deterministic ($\sigma_\theta^2 = 0$), k_c is independent of the width of the fitness function. Because the genetic variance is proportional to σ_w, and the lag load is proportional to σ_w^2/σ_g^4, the latter is independent of σ_w.

Using the results in the Appendix, the critical rate of environmental change per generation for an asexual species is found to be only about $(2n_e)^{-1/2}$, where n_e is the effective number of loci—as large as that for a sexual species. Thus, all other things being equal, recombination provides sexual species with an evolutionary advantage. However, with their shorter generation times, it is possible that asexual microorganisms can sustain larger values of k on an absolute time scale. For example, all other things being equal, on an absolute time scale, k_c for an asexual population would match that of a sexual population if the latter's generation time were $\sqrt{2n_e}$ longer.

TRANSIENT DYNAMICS

Up to now, the results that we have described apply to the steady-state situation in which the expected lag in the population mean phenotype has been attained. The general relevance of the theory depends on the time it takes to reach this state. The transient dynamics for an effectively infinite asexual population in a deterministic environment are worked out in Lynch et al. (1991). Here we present the more general case.

If it is assumed that the genetic variance is initially at its equilibrium value, the dynamics of the expected mean phenotype can be obtained from the solution to Equation (4),

$$E[\bar{g}(t)] = [\bar{g}(0) + (k/\alpha)]e^{-\alpha t} + kt - (k/\alpha) \tag{14}$$

where $\alpha = \sigma_g^2/\sigma_w^2$. This expression converges to Equation (7) for large t. The time scale for the approach to the equilibrium lag is proportional to $1/\alpha$. That is, a doubling of α speeds up the dynamics by a factor of two. Moreover, a doubling of σ_g^2 has the same influence on the expected evolutionary trajectory as a halving of σ_w^2.

Provided the effective population size and the environmental stochasticity have been constant in the past, the initial realized mean $\bar{g}(0)$ actually will be distributed about the expectation $\theta = 0$ with variance given by Equation (8). However, it will take some time for the evolutionary trajectory for any single population to develop the steady-state variance in time. Using diffusion theory, it is possible to solve for the transient dynamics in the variance of the population mean phenotype,

$$V[\bar{g}(t)] = \left(\frac{\sigma_w^2}{2N_e} + \frac{\sigma_g^2\sigma_\theta^2}{2\sigma_w^2}\right)[1 - e^{-2\alpha t}] \tag{15}$$

This converges to Equation (8) as $t \to \infty$, the time scale being half that noted above for the dynamics of the mean.

An example of the dynamics of the distribution of $\bar{g}$ is given in Figure 3. The expected lag of the mean phenotype behind the optimum, $\bar{g} - kt$, attains 50 percent of the equilibrium lag by $t = 0.7/\alpha$ generations, which for the example given ($\sigma_g^2 = 1.0$ and $\sigma_w^2 = 25$) occurs in only 17 generations. The variance of the distribution of $\bar{g}$ attains 50 percent of its equilibrium value in half that time. Note that due to the relatively high level of environmental stochasticity in this example, the expected equilibrium lag is approximately 1.8 phenotypic standard deviations, and there is always some possibility that the realized mean phenotype will equal or exceed the optimum.

DISCUSSION

Although the theory we have presented relies on a number of simplifying assumptions, it provides a useful starting point for investigating the potential for populations to adapt to long-term environmental change. Obviously, due to functional and developmental constraints, no character can be expected to evolve directionally for an indefinite period of time. Our results are

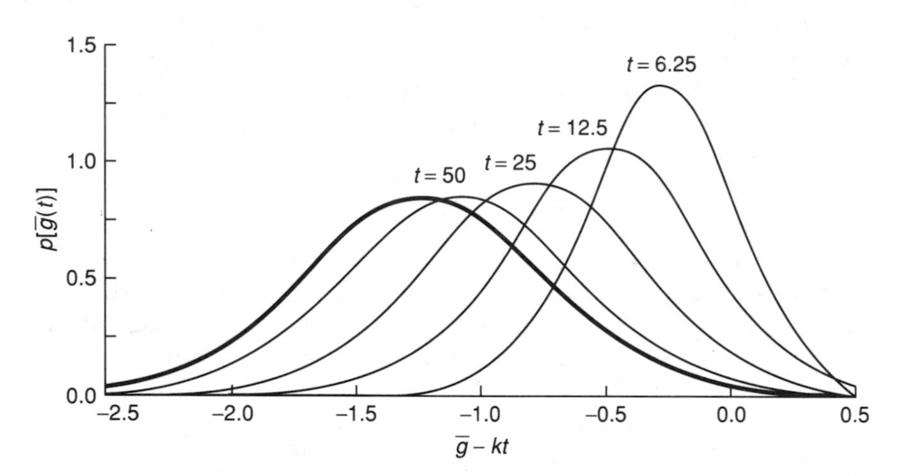

FIGURE 3. Probability distribution of the mean phenotype as a deviation from the expected optimum, $\theta = kt$, as a function of t generations. The initial mean is assumed to be $\bar{g} = 0$, the genetic variance to be at its equilibrium expectation $2N_e\sigma_m^2$, $\sigma_m^2 = 0.001$, $k = 0.05$, $\sigma_w^2 = 25$, $\sigma_\theta^2 = 10$, and $N_e = 500$. The heaviest (leftmost) curve is the asymptotic distribution approached as $t \to \infty$. If it is assumed that the environmental variance is $\sigma_e^2 = 1$, which is reasonable for the value of σ_m^2 used (Lynch, 1988), then the phenotypic variance is equal to 2, and each unit on the abscissa is equivalent to 0.7 phenotypic standard deviations.

meant to identify rates of environmental change that can be sustained when adaptive potential *is* present. For that reason and others discussed below, the critical rates that we have identified are likely to be upper bounds.

The models that we have employed are based on the premise, first suggested by Fisher (1918), that most complex traits are products of a large number of loci with individually small effects. This premise has stood up to a substantial body of empirical work (Falconer, 1989). Results from long-term selection experiments on numerous characters, in both plants and animals, show very clearly that directional changes on the order of 10 or more phenotypic standard deviations can often be accomplished in less than 100 generations (Jones et al., 1968; Kress, 1975; Dudley, 1977; Eisen, 1980; Weber and Diggins, 1990). Changes of this magnitude are well outside the range of variation seen in the base population at the start of an experiment, but they need not involve mutation. As selection jointly advances the frequencies of favorable alleles at multiple loci, very large changes in the mean phenotype can be accomplished with the genetic variance residing in the base population alone. But there is now compelling evidence that polygenic mutation makes significant additional contributions to long-term selection response (Lynch, 1988; Mackay, 1989; Keightley and Hill, 1990; Weber and Diggins, 1990; Caballero et al., 1991). Dykhuizen (1990) reviews these issues in microbial populations, which usually respond to selective challenges via multiple mutational steps.

These observations imply that the critical rates of environmental change that we have identified can be sustained evolutionarily for a large number of generations before functional or developmental constraints become important. For example, from Equation (13), k_c' for an asexual population is necessarily less than $(2\bar{r}_m\sigma_m^2/\sigma_z^2)^{1/2}$. Assuming $\bar{r}_m$ is on the order of 0.5 or less, and noting that σ_m^2/σ_z^2 is on the order of 10^{-3} (Lynch, 1988), then an upper limit to the rate of environmental change (and the steady-state rate of evolution associated with it) that can be sustained by an asexual population is roughly 0.03 phenotypic standard deviations per generation. For finite sexual populations, from Equation (11), k_c' is necessarily less than $2N_e\sigma_m^2$ $[2\bar{r}_m/(\sigma_w^2\sigma_z^2)]^{1/2}$, which is almost certainly less than $(2N_e\sigma_m^2/\sigma_z^2)(2\bar{r}_m)^{1/2}$ for a viable population. So for populations smaller than $N_e = 500$ the critical rate of environmental change will be less than a phenotypic standard deviation per generation, perhaps considerably so. Note that since the population mean phenotype is expected to evolve at exactly the rate k at equilibrium, these estimates of k_c' can also be viewed as maximally sustainable rates of evolution.

A number of additional issues need to be addressed in future research. For example, our work only deals explicitly with a single panmictic population in a spatially homogeneous environment, a situation particularly relevant to species that are fragmented into isolated demes. Another common type of population structure, one being increasingly imposed on organisms by human activity, involves semi-isolated demes interconnected by narrow corridors of suitable habitat. Little theoretical work has been done on stabilizing selection with a moving optimum in a subdivided population. It seems likely, though, that population subdivision would reduce the critical rate of environmental change due to an increase in the lag and drift loads in local populations. Pease et al. (1989) have examined the situation in which a spatial cline in the optimum moves across a continuous landscape inhabited by a population with restricted dispersal ability, pointing out that there is a critical rate of environmental movement beyond which the population's ability to migrate and/or locally adapt is overwhelmed by the selective load.

As noted in the Appendix, there is still considerable debate as to the appropriate way to model the genetic variance for characters under stabilizing selection. When there is a directional component to selection and population sizes are on the order of 10^3 or less, there are reasons to expect the predictions of the neutral model to approximate the equilibrium genetic variance closely, an assumption that we relied on in the preceding examples for sexual populations. But to verify this, the theory for Gaussian selection with a moving optimum needs to be further developed mathematically. In our derivations, we also ignored fluctuations in the genetic variance that might exist between generations. This should be roughly valid for effectively infinite asexual populations, but less so for small sexual populations. Further work needs to be done to evaluate how violations of these assumptions influence the quantitative predictions of our theory. Since the critical rates of environmental change identified above are nonlinear functions of the genetic variance, a high degree of variation in the latter would affect the predictions.

Throughout this chapter, we have treated the effective population size as a constant, thereby assuming that the population maintains a roughly constant density until the extinction threshold has been crossed. In the future, it will be useful to model the process more explicitly by allowing the population density to respond directly to the environmental change. One way of approaching this problem is to couple the equations for the dynamics of the mean phenotype and the genetic variance with a population dynamics equation, such as the logistic model. As noted above, when the population size is finite and the environmental change has a stochastic element, the realized population rate of increase varies around its expectation, sometimes becoming negative. This will cause the population to decline below its carrying capacity occasionally, which will temporarily increase the drift load, and if prolonged, will cause a reduction in the genetic variance. Recovery from the loss of genetic variance may then require a long period of replenishment by polygenic mutation. By leading to an even greater lag load in a changing environment, these two genetic consequences of a population bottleneck will further reduce the population's ability to maintain itself, and may cause it to become extinct. We anticipate that these kinds of synergistic effects will lead to critical rates of environmental change that are somewhat lower than those defined above.

These kinds of synergistic interactions between random genetic drift and the loss of population fitness have been explored recently in a somewhat different context—finite populations in a constant environment, with recurrent mutation to unconditionally deleterious alleles (Gabriel et al., 1991; Lynch and Gabriel, 1991). In small populations, there is always some possibility that a mildly deleterious mutation will increase in frequency, sometimes even going to fixation. This causes a further decline in population size, which makes it easier for the next round of deleterious mutations to drift to high frequency, which again reduces the population size, and so on, ultimately leading to population extinction. This process, referred to as a mutational meltdown, will obviously be exacerbated in a changing environment that precipitates a decline in population size, and should be explored in the future.

Because we have focused only on the long-term average maximum growth rate of a population under selection, rather than on the actual population dynamics, our results have an additional limitation. We have only identified the critical rates of prolonged environmental change beyond which a population is doomed to certain, rapid extinction. However, even if the long-term average growth rate is predicted to be positive, random genetic drift and/or environmental stochasticity can cause any population to temporarily deviate far enough from the optimum for a long enough time that extinction might occur. The likelihood of such an event will be magnified if the rate of environmental change is already close to the extinction threshold, but it will also depend on the effective population size, the rate of polygenic mutation, the intensity of selection, and the degree of environmental stochasticity. These issues can be addressed by extending future work to the expected distributions of extinction times.

Although our focus has been on Gaussian selection with a moving optimum, the approach can be extended readily to other forms of selection. For example, Hill (1982) showed that under long-term truncation selection (and under the assumptions of our genetic model), a population asymptotically settles into a steady-state evolutionary trajectory, with the mean phenotype lagging behind the advancing truncation point by a constant amount. The asymptotic rate of evolution is simply $\sigma_g^2 i/\sigma_z$, where i is the standardized selection intensity, and as noted above, $\sigma_g^2 = 2N_e\sigma_m^2$. Provided the fraction of the population eliminated by selection per generation (p) is in the range of 0.25 to 0.995, then $i \simeq 0.8 + 0.4 \ln[p/(1 - p)]$ (Smith, 1969). In units of phenotypic standard deviations, the critical rate of increase (or decrease) of the truncation point is simply $k_c' = h^2 i_c$, where h^2 is the equilibrium heritability of the trait ($2N_e\sigma_m^2/\sigma_z^2$), and i_c is the critical selection intensity beyond which the population cannot maintain itself. As a simple example, consider a low-fecundity population that requires at least 50 percent survival (from selection) to maintain itself. Then, $p = 0.5$, $i_c = 0.8$, and $k'_c = 0.8h^2$—the population would be doomed to certain extinction if the rate of movement of the truncation point exceeded $0.8h^2$ phenotypic standard deviations per generation. Because σ_m^2/σ_e^2 is on the order of 10^{-3}, when N_e is small h^2 is on the order of $2N_e \times 10^{-3}$. For the above example, this implies a critical rate of environmental change of approximately $1.6N_e \times 10^{-3}$ phenotypic standard deviations per generation.

In closing, we note that our heuristic models suggest some simple experiments that might be done to further clarify our understanding of the consequences of environmental change for evolution and extinction. For practical reasons, such experiments will need to be performed with species that have short generation times and that are grown easily in controlled laboratory or field environments. By exposing a large set of replicate populations to a stepwise or continual change in an important environmental parameter, such as temperature, it should be possible to acquire some information on the mean and variance of evolutionary rates and of extinction times. It will be especially useful to perform such experiments at several different effective population sizes and at several levels of stochastic variation of the environmental parameter. Obviously, the results of such "natural selection" experiments are likely to be rather species-specific, but given the nearly complete absence of critical work on the problem (with the exception of that on bacteria) and the significant contribution that such studies would make to the further development of the theory, a well-designed series of empirical studies is essential to further progress in this area.

SUMMARY

Due to the ubiquity of genetic variation for quantitative characters, populations that are exposed to environmental change are expected to adapt gradually via evolutionary changes in one or more traits. However, natural selection will result in a reduction in average viability and/or fecundity. If the rate of environmental change is sufficiently fast, the reduction in fitness due to selection will overwhelm the population's ability to maintain itself, even though adaptation may be occurring, and extinction will ensue. We present some simple models that lead to predictions for the critical rate of environmental change beyond which extinction is inevitable. Results are given that suggest the relationship of this rate to the intensity of selection, the magnitude of polygenic mutation, the effective population size, and the reproductive system.

APPENDIX: ADDITIVE GENETIC VARIANCE UNDER GAUSSIAN SELECTION

Following the procedures of Kimura (1965) and Latter (1970), an explicit expression can be obtained for the dynamics of the genetic variance expected under the joint action of random genetic drift, mutation, and Gaussian selection. Under the assumptions of normality, the dynamics of the expected genetic variance are given by

$$\frac{d\sigma_g^2}{dt} = \sigma_m^2 - \frac{\sigma_g^4}{2n_e\sigma_w^2} - \frac{\sigma_g^2}{2N_e} \quad \text{(A1)}$$

where the three terms respectively describe the input of new variance via polygenic mutation, and the losses of variance by selection and by random genetic drift. $\sigma_m^2 = 2\mu E(a^2)$ is the rate of input of new genetic variance via mutation, and n_e is the effective number of segregating factors, which is roughly the chromosome number plus the mean number of crossovers per meiotic event. The above expression is a Riccati equation, the solution to which is

$$\sigma_g^2(t) = \hat{\sigma}_g^2 + \left[\left(\frac{1}{\sigma_g^2(0) - \hat{\sigma}_g^2} + \frac{1}{2\phi n_e\sigma_w^2}\right) e^{\phi t} - \frac{1}{2\phi n_e\sigma_w^2}\right]^{-1} \quad \text{(A2)}$$

where $\phi = 1/(2N_e) + \hat{\sigma}_g^2/(n_e\sigma_w^2)$. The expected genetic variance for sexual populations under drift–mutation–selection equilibrium, $\hat{\sigma}_g^2$, is obtained by solving the preceding equation as $t \rightarrow \infty$,

$$\hat{\sigma}_g^2 = \sqrt{(n_e\sigma_w^2/2N_e)^2 + 2n_e\sigma_w^2\sigma_m^2} - (n_e\sigma_w^2/2N_e) \quad \text{(A3)}$$

When drift prevails over selection (N_e small, σ_w^2 large), $\hat{\sigma}_g^2 \simeq 2N_e\sigma_m^2$, the neutral result (Lynch and Hill, 1986). The same result is approached asymptotically as n_e becomes very large, since the intensity of selection operating on individual loci becomes infinitesimally small.

There has been considerable debate as to the validity of Kimura's "infinite-allele" approach versus an alternative non-Gaussian "house-of-cards" approach to modelling the genetic variance of quantitative traits in sexual populations (Lande, 1975; Turelli, 1984; Barton and Turelli, 1987; Slatkin, 1987). However, most of the debate has concentrated on populations of effectively infinite size under stabilizing selection with the mean phenotype coinciding with the optimum. For finite populations under directional (truncation) selection, the equilibrium genetic variance is remarkably close to the neutral expectation, provided the distribution of mutational effects is symmetrical about zero (Hill, 1982; Keightley and Hill, 1987; Hill and Keightley, 1988). Favorable mutations make a higher average contribution to the heterozygosity while they are segregating than do neutral mutations, but this is essentially balanced by the shorter time to fixation for the former. These results, combined with other recent observations that the "house-of-cards" model also yields solutions close to the neutral expectation for $N_e < 10^3$ or so (Keightley and Hill, 1988; Bürger et al., 1989; Houle, 1989), provide fairly strong justification for the use of $\sigma_g^2 \simeq 2N_e\sigma_m^2$ for small sexual populations. This is a particularly useful result because both N_e and σ_m^2 are estimable parameters (Lande and Barrowclough, 1987; Lynch, 1988). More generally, for nonsymmetrically distributed mutational effects, $\sigma_g^2 \simeq 4PN_e\sigma_m^2$, where P is the fraction of mutations with effects in the direction of selection (Keightley and Hill, 1987).

The assumption that the genetic variance is constant in time cannot be strictly true for finite populations, but some justification for such an approximation can be given for sexual populations of at least moderate size. Imagine a set of replicate populations all exposed to the same stochastic processes, and assayed for their additive genetic variance at the same point in time. For neutral quantitative characters, the squared coefficient of variation of the additive genetic variance among replicates is approximately $\{8 + [E(a^4)/4\mu E^2(a^2)]\}/(3N_e)$, where μ is the total gametic mutation rate for the trait, and $E(a^2)$ and $E(a^4)$ are the second and fourth moments of mutational effects (Keightley and Hill, 1989; Zeng and Cockerham, 1991). If the distribution of mutational effects is highly leptokurtic [i.e., $E(a^4) >> E^2(a^2)$] and/or the gametic mutation rate is low (i.e., $\mu << 1$), the second term can dominate. When mutational effects are normally distributed with mean zero, i.e., $E(a^4) = 3E^2(a^2)$, the above expression reduces to $[8 + (3/4\mu)]/(3N_e)$. Computer simulations of Keightley and Hill (1989) showed that these expressions work quite well for populations under stabilizing selection when the mean is at the optimum, but the model that we are considering (with a moving optimum) has not been investigated. The most we can say at this point is that the coefficient of variation of the additive genetic variance is unlikely to be less than $(3/N_e)^{1/2}$, and it may be several times higher.

For effectively infinite asexual populations, the explicit definition of the equilibrium genetic variance needs to be modified. Letting $N_e \rightarrow \infty$ in the previous equation, the equilibrium level of genetic variance for an effectively infinite sexual population is found to be is $\sigma_g^2 = \sigma_m\sigma_w(2n_e)^{1/2}$. In the absence of sex, the genome functions as a single nonsegregating unit, so $2n_e = 1$, and $\sigma_g^2 = \sigma_m\sigma_w$ (Lynch et al., 1991).

ACKNOWLEDGMENTS

We thank W. Bradshaw, T. Hansen, and E. Martins for helpful comments. This work has been supported by NSF grants BSR 8911038 and BSR 9024977, and PHS grant GM36827 to ML, and by PHS grant GM27120 to RL.

CHAPTER 15

GLOBAL CHANGE: Lessons From and For Evolutionary Biology

Joseph Travis and Douglas J. Futuyma

Evolutionary biologists bring a unique perspective to the examination of how global change will affect the earth's biota. They carry the burden of knowing that rapid environmental changes have been associated with extinctions throughout life's history. However, they also know that some organisms have adapted to a bewildering array of ecological challenges and have often done so at astonishing rates. Whether a species adapts or becomes extinct depends on whether it is able to mobilize sufficient genetic variation for key phenotypic traits before being overtaken by the pace of environmental alteration. The factors that determine this capability can provide not only the lessons that evolutionary biology has to offer to the study of global change, but also the focus of the research that evolutionary biologists must undertake to understand what global change will do to the earth's biota.

The perspective of evolutionary biology must also influence how the managers of the earth's biota respond to global change. From that perspective, there is no such entity as a "functionally redundant" species. Each species is a unique collection of genetic and phenotypic variation and so no species can be replaced exactly in a community or ecosystem. In some cases a replacement would have only a small effect on productivity and nutrient cycling, but in many cases the replacement of a species by a congener, or even by a different genetic stock of the same species, would have profound effects (Carpenter et al., this volume; Schimel, this volume). Consequently, the conservation of as much biotic diversity as possible is the best insurance for the metaphoric "health" of communities and ecosystems.

These considerations set the research agenda for evolutionary biologists in the face of global change. To understand when global change will bring extinction and when it will bring adaptation, we must understand what determines the potential rate of sustained adaptive change in a species as a whole. This question is not easily answered. A species consists of an ensemble of populations that are already differentiated genetically and ecologically but are connected by gene flow. As a result, research directed at this question must be broad in its conceptual and geographic scope. To understand how to restore those species that we have already driven to the brink of extinction, we must understand how to manipulate extant genetic variation in the ensemble of a species' populations to minimize further loss and to maximize the chance for future adaptation.

EVOLUTIONARY LESSONS FOR ECOLOGICAL SKEPTICS

We confront two paradoxical sets of lessons, which testify on one hand to the great adaptability of species and on the other to strong limits on their adaptability, and therefore offer paradoxical arguments for a role for evolutionary biologists in the study of global change.

Lessons that reinforce skepticism about a role for evolutionary biology

1. *Paleontology and systematics reveal that extinction has been the fate of most species.* The most conspicuous effect of climatic changes during the Pleistocene was not adaptation in situ but shifts in geographic ranges. Species tracked the movements of their habitats by extinction of some populations and establishment of others in new locations, and were frequently reassembled into different communities through differential responsiveness of species (Davis, 1981). At least in morphological features, many species (e.g., of beetles; see Coope, 1979) did not evolve in response to the drastic environmental changes of the Pleistocene. This is not to deny that many populations of some species that were spatially segregated during the Pleistocene diverged in some features.

2. *The existence of stable species ranges indicates that most species cannot mobilize sufficient genetic variation to adapt to environmental conditions immediately beyond their borders.* Many agriculturally important insect pests in the United States move northward each summer by the millions and then die, not having formed resident populations capable of surviving winter at high latitudes (Fitt, 1989). Every salt marsh and alpine tundra receives a constant rain of seeds from plant species at its border, species

that have not adapted to those habitats and that therefore do not invade (Bradshaw, 1991). Even bacterial "species," despite huge population sizes and incessant mutation, are limited to characteristic habitats. Whether the range of a species is limited by biotic interactions, abiotic factors, or their interactions (Dunson and Travis, 1991), the point remains.

This lesson appears at odds with the data from six decades of studies in population and quantitative genetics that show that most populations harbor a great deal of genetic variation, however it is measured (as quantitative variation in phenotypic characters, allozymes, or DNA sequences). Yet we know surprisingly little about the levels of genetic variation for ecologically critical characters under natural selection (Hoffmann and Blows, this volume), and theory and data from laboratory studies suggest that we might expect populations to harbor very low levels of genetic variation for some such characters (Roff and Mousseau, 1987). Some characters do not respond to artificial selection (Maynard Smith and Sondhi, 1960), and response to selection is often countered by factors such as antagonistic pleiotropy and linkage disequilibrium (Lerner, 1954; Loeschcke, 1987; Falconer, 1989). Moreover, there are clear connections between the level of genetic variation and a population's sustained viability. For example, certain genetically impoverished asexual or inbred natural populations suffer from high parasite loads, low survival and fecundity rates, and developmental instability (Quattro and Vrijenhoek, 1989; Lively et al., 1990). Thus the failure of a species to move beyond its borders can be traced ultimately to a lack of relevant genetic variation, which in turn inspires pessimism about a species' ability to respond to rapid environmental change.

Lessons that mandate a role for evolutionary biology

1. *Rapid evolutionary change is a well-documented response to rapid environmental changes.* There is an enormous roster of case histories in plants and animals that document rapid adaptation. Such cases include adaptation in response to novel localized selective pressures, sustained wide-scale alterations in native environments, and physical and biotic features of geographic regions that species have recently invaded (Bishop and Cook, 1981; Geber and Dawson, this volume; Hoffmann and Blows, this volume). Species have evolved rapidly in highly diverse ways in response to both biotic and abiotic factors. Indeed, different species have often generated idiosyncratic adaptations to a common novel challenge (e.g., pesticide resistance through different mechanisms; see Georghiou, 1972; Plapp, 1976; Roush and McKenzie, 1987).

2. *A species is not a genetically or ecologically uniform entity.* This lesson has three crucial consequences. First, the genetic variation among popula-

tions within a species often results from local adaptation, which illustrates again the ability of selection to find solutions to ecological problems (Hoffmann and Blows, this volume). The subtleties of such differentiation can be striking; populations of many species are known to differ in their responsiveness to many kinds of environmental variation, and in a number of cases such differentiation appears to be a product of adaptive evolution (Travis, in press).

Second, because of this variation, it will not be easy to predict biotic responses to various types of global change. Populations of grasses with a history of grazing alter their patterns of allocation to shoots and roots in response to grazing in a way very different from that of populations without a history of grazing (Schimel, this volume). This difference between genetic stocks produces a dramatic difference in predicted levels of CO_2 assimilation and carbon cycling, which in turn generates an enormous difference in the predicted feedback from vegetation to climate (Schimel, this volume).

Even though we argued above that species often seem unable to mobilize enough genetic variation to expand their ranges, population differences provide some counterexamples. In some cases, experiments have shown that individuals are capable of thriving outside the present species range but that profound, genetically based differences in ability to do so exist between individuals from different parts of the species' range (Levin and Clay, 1984; Schmidt and Levin, 1985). The distinctive ecological difference between central and marginal populations of a single species can produce striking differences in the ability of one genetic group to thrive in the environment of the other, indicating that constituent populations are not interchangeable even within a species' range (Grant and Antonovics, 1978; Geber and Dawson, this volume). Predictions about altered species ranges and similar effects of global change will have to take into account the variable properties of populations within a species.

The most profound consequence of this lesson is that the genetic differentiation of populations can increase the likelihood of rapid evolutionary change in the species as a whole. Work on this problem was pioneered by Wright (1931), and recent theoretical (Crow et al., 1990) and experimental (Wade and Goodnight, 1991) investigations have refocused attention on it. The quantitative genetic and demographic parameters that govern the rate of sustainable change in a single continuous trait in a rapidly changing environment have been brought into recent focus by Taylor (1990), Pease et al. (1989), and Lynch and Lande (this volume). All of these studies reveal that an adaptive response to rapid environmental change is possible but that it requires that very specific conditions be met. Unfortunately, the likelihood that they will be met is as yet a poorly known empirical issue. This ignorance sets the research agenda for evolutionary biology in the face of global change.

A RESEARCH AGENDA FOR EVOLUTIONARY BIOLOGY

In this section we outline the specific questions and new programs that we envision as most likely to contribute to our understanding of whether extinction or adaptation will result from global change. These questions and programs are focused on the factors that determine the maximum sustainable rate of adaptive change in the species as a whole.

Problems in systematics

A comprehensive understanding of the sustainable rate of evolutionary change must begin with a firmer grasp on the lessons of history. Research in systematics should focus on identifying labile and conservative traits in a variety of taxa and should attempt to delimit the maximal rates of trait change. One way to do this is to place ecologically important traits atop a well-defined and well-supported phylogenetic hypothesis (Futuyma and McCafferty, 1990; Brooks and McLennan, 1991; Donoghue, 1989; Harvey and Pagel, 1991). For example, the position of habitat type (salt water or fresh water) within a phylogenetic hypothesis for the fundulid and cyprinodontid fishes (Parenti, 1981), along with studies of comparative physiology, reveals that osmotic physiology is extremely labile within these groups and that the potential for future adaptation to altered water chemistry is probably significant for some species (Dunson and Travis, 1991). In contrast, specific host-plant associations show striking phylogenetic conservatism in many taxa of herbivorous insects (Mitter et al., 1991), some of which appear to have low (undetectable) levels of the genetic variation that would be required to shift to certain other, even rather closely related, plants (D. J. Futuyma, in preparation).

Problems in genetics

Models that describe the sustainable rate of change of a character have helped to isolate several factors that can limit that rate (Taylor, 1990; Pease et al., 1989; Lynch and Lande, this volume). The genetic factors that come into focus are those that surround the genesis of variation and its organization and expression as phenotypic variation.

The most obvious limit to the rate of sustainable change is the rate at which new variation is introduced into the several populations of a species. Although mutational input is often viewed as unlikely to limit the rate of change in a directionally selected phenotypic trait (e.g., Ewens, 1979), the projected rapid rates of environmental change may impose both higher selection intensities and higher rates of change in the optimum value of a trait than those found in classic conceptualizations of the process of directional evolution. Consequently, mutational input could well be important. In at least some cases de novo mutation has contributed to the responses of experimental populations to strong artificial selection (Lynch and Lande, this volume).

The distribution of mutational effects on continuous phenotypic traits, which are the traits most likely to be of ecological significance, is largely unknown outside of a few well-studied species. The magnitude of mutational effects is as important as the rate at which they are introduced. However, altered environmental milieux are likely to introduce physiologically stressful conditions to many populations. Because some evidence suggests that mutational input increases in environments that are stressful (Hoffmann and Parsons, 1991), attention must be focused on the distribution of mutational effects in stressful and nonstressful conditions, both for traits with a history of lability and of conservatism, and for separate populations throughout the range of a species. Environment-specific mutation rates, especially if they are higher in stressful environments, will increase the likelihood of sustainable, rapid evolutionary change. What constitutes a "stressful" environment will differ between populations within a single species because of the heterogeneity of small-scale climatic or other anthropogenic changes (Schneider, this volume) or because of the genetic variation already present among populations in their norms of reaction (e.g., physiological races; see Garland and Adolph, 1991). Such a comprehensive examination of mutational effects may not only alter our view of whether adaptive evolution is a likely response to global change but may also contribute to resolving some fundamental gaps in our basic understanding of the structure of quantitative genetic variation (Lynch and Lande, this volume).

The expression of existing genetic variation in a species' populations is not independent of environmental conditions during ontogeny (Sultan, 1987; Hoffmann and Parsons, 1991). The levels of additive variation in a trait that are exhibited in the population's current condition may under- or overestimate the levels that will be expressed in altered conditions. As a result, and because of environment-specific patterns of nonadditive variation in the phenotype (e.g., nonadditive genetic variation, random environmental variation), present heritabilities may be poor predictors of future levels (Hoffmann and Blows, this volume). This is not a call for random investigations of environment-specific patterns of gene expression, but rather a call for studies that focus on populations throughout a species' range, that examine traits likely to be the ecologically critical ones under the altered conditions that species will face, and that contrast patterns of expression in current conditions with those in the most plausible future conditions. For some species, future conditions might be warmer climates throughout the range; for others, the most relevant change might be moisture levels. In species with broad geographic ranges, some populations may experience significant changes while others do not. Hence the contrast between central and marginal populations will be critical for this area of inquiry. This topic could also be characterized as a call for understanding the distribution of norms of reaction among populations of a species in response to very specific environmental conditions.

Finally, the number of loci that govern the variation in ecologically critical traits should be estimated. Polygenic control provides more latitude for character evolution than does oligogenic control. However, the genetic constraints on adaptation that arise from the number of effectively segregating loci cannot be specified independently of the sizes of the populations in which selection occurs. For example, resistance to insecticides usually has a polygenic basis in small, artificially selected laboratory populations, but in natural populations of arthropods it is usually attributable to one or two alleles of large effect, the initial frequencies of which are so low that they are available to selection only in large populations (Roush and McKenzie, 1989). Responses to artificial selection occur more rapidly in large than in small laboratory populations (Weber and Diggins, 1990; see also the review by Simberloff, 1988), and the interplay between population size, selection intensity, and the effective number of loci determines whether a response to directional selection can be sustained for many generations without intermittent plateaus (Gill, 1965; Latter, 1965).

The simple presence of genetic variation is not enough to guarantee successful adaptation: it must also be responsive to selection. Although theory and laboratory studies have taught us much about the factors that impede the response to selection by a genetically variable character (factors such as the nonadditive component of genetic variance, antagonistic pleiotropy, and linkage disequilibrium; cf. Lerner, 1954; Loeschcke, 1987), we have an inadequate knowledge of their importance in natural populations. In addition, because it is likely that novel ecological conditions will create novel selection pressures on many characters, there is a need for more detailed study of the responses to simultaneous selection on several characters. Genetic load theory (Haldane, 1957) provides only one of several reasons to think that the evolution of individual characters will be slower if directional selection is acting on many independent characters. In the context of global climate change, a population is likely to experience alteration not only of temperature and rainfall (which may impose selection on several physiological and phenological traits) but also of the array of competitors, predators, parasites, and resource species with which it must contend. Although it may be possible for a single trait to evolve rapidly, a realistic assessment of the complexity of adaptive change in the entire phenotype may lead to greater pessimism.

Despite the real constraints on rapid evolution, its repeated occurrence should alert us to yet another threat to species: the evolution of their parasites and pathogens. From examples of the evolution of virulence in the myxoma virus of rabbits (Ross, 1982); the rapid adaptation of fungal pathogens (Rayner, 1991), nematodes (Trudgill, 1991), and insect pests (Maxwell and Jennings, 1980) to previously resistant crop varieties; and the adaptation of human pathogens to antibiotics (Hardy, 1986), we know that rates of evolution of microparasites can be great enough to threaten their hosts. As parasites shift their geographic ranges with climatic change, and as host species become fragmented into small, perhaps genetically impoverished populations, the little-understood processes of coevolution may be critical for the long-term survival of many species.

The horizons of future research are not strictly empirical. Although several extant models offer a solid and promising start on the problem of the maximal sustainable rate of response (see Lynch and Lande, this volume), much remains to be explored. First, a load on the maximal sustainable rate will be produced by the pleiotropic effects of the segregating elements that determine the phenotypic value of a trait under selection. Second, a lag in response may result through gene flow from populations within a species' range that are not experiencing the new selective regime. The empirical importance of the gene-flow lag is unclear. The effect may be most likely in a species with extensive preexisting differentiation among its constituent populations. Yet if such a species occupies a limited geographic range, like many tropical species, new environmental conditions may affect all populations equally. In this case, gene flow may enhance an evolutionary response because it will mobilize the full range of available genetic variation to respond to the new conditions.

Problems in ecological genetics

Several key issues lie at the interface of population genetics and population ecology. The first is the need to understand the distinctions, for a variety of taxa, among the rates of routine dispersal of individuals, the rates of colonization of new habitats, and the rates of gene flow among extant populations. These differences can have profound effects on population structure and on the ability of a species to mobilize its total extant levels of genetic variation to respond to a novel environmental challenge (Holsinger, this volume; McCauley, this volume). This issue can be especially important if different populations accumulate distinct rare alleles that may be advantageous in the novel environments experienced by some of a species' populations (Hoffmann and Blows, this volume). Knowledge of the distinctions among these processes can be used to build models for individual species that estimate the critical rate of habitat fragmentation above which the individual populations of a species are no longer connected and can thus no longer contribute genetic variation to one another. Fragmentation rates above this critical threshold will condemn the individual populations of a species to their own fates unless humans actively intervene and manage the genetic resources.

Global change will induce dramatic changes in environmental conditions at species boundaries as well as in the centers of species ranges, although this process will differ from group to group. The ecological determinants of a species' range are largely unknown; this is a striking gap in both ecological and evolutionary knowledge. A species' range may be a conservative property of the larger clades in which species are embedded (Ricklefs, 1989). Moreover, range sizes differ between tropical and temperate taxa and are often correlated with many other species attributes (Brown and Maurer, 1987). Even less is known about the genetic reasons for species' range limits, even though the importance of this topic was pointed out (in a quite different context) almost 30 years ago (Mayr, 1963). We do not know whether species are unable to adapt to conditions a few kilometers beyond their borders because of lack of genetic variation in one or a few critical traits, because of a "cost" of further adaptation, because of a multitude of selection pressures imposed by competing species (or other biotic agents), or because of a breakdown of coadapted gene complexes by gene flow from interior populations. This issue is critical to the assessment of the likelihood that populations at present range limits will adapt to latitudinal shifts in climate or will become extinct (resulting in contraction or shift in the species' range).

These considerations point to the need for carefully guarded transplantation experiments in an appropriate ecological context. Such experiments can reveal the extent of genetic variation among populations of a species for the characters that play the key roles in preventing the expansion of the species' range. Transplant experiments can help us to understand how a species extends its range, that is, whether the adaptations of ecologically marginal populations (e.g., ecotypes) are necessary prerequisites for invading marginal habitats, or whether they evolve only after population establishment has occurred.

Ecological genetics can also offer methods for assessing rates of gene flow among populations, which is not only important in itself for several reasons noted above but also can provide information on the capacity of species to colonize new areas. We may wish to know whether a species with an insular distribution, such as a mountaintop salamander, is at all capable of colonizing other areas, for this ability may prove critical to its survival in a changing world. Several methods have been developed for estimating gene-flow rates from gene frequencies, such as the distribution of rare alleles (Slatkin, 1985), but these may be confounded by recent range expansion (Holsinger, this volume). However, DNA sequence variants may provide more useful information. Especially for nonrecombining sequences (such as mitochondrial DNA), phylogenies of intraspecific sequences are readily constructed. Clades of genealogically related sequences are sometimes geographically segregated, implying low rates of exchange between regions (e.g., Avise, 1989). The number of colonizations between regions can be approximately estimated from the discordance between geography and sequence phylogeny (Slatkin and Maddison, 1989). A time dimension can be provided in some cases (yielding a rate) by coalescence theory (Hudson, 1990), which, together with an external time calibration, can be used to estimate the age of divergence of the clades of DNA sequences. Sequence variation among isolated populations could thus be used to estimate the time of isolation and the number of exchanges (colonizations) among them.

As a final note in ecological genetics, if a gradual transition allows sufficient physiological acclimatization, species are often capable of maintaining populations in environments in which they do not often occur (Hoffmann and Blows, this volume; Dunson and Travis, 1991). This process is, in effect, another characterization of phenotypic plasticity, of whether populations are sufficiently plastic in physiological processes that sufficient adjustments can be made even after ontogeny is completed. The role of acclimatization, especially an acclimatization that may be transgenerational (Kinne, 1962; Forster-Blouin, 1989), should be given more attention as one mechanism that may not only permit individuals to adjust to a low rate of environmental change but also increase the likelihood that advantageous genetic variants can arise on a useful physiological background.

Issues in management

For many species, the pace of global change will be insurmountable in an evolutionary sense, either because the limiting factors cannot be overcome or because prior anthropogenic effects have reduced them to one or a few isolated populations of small numerical size. Some opportunities may arise to restore populations of a species to a habitat or region from which it has been extirpated but which is newly suitable for it. These opportunities may occur through the creation and restoration of appropriate habitat (e.g., created wetlands) or the acquisition of appropriate habitat as a new preserve.

The appropriate synthesis of a new population in these contexts raises several questions for which evolutionary biology has yet to provide clear answers. If a species is to be introduced into a new area, should colonists be drawn from several equally adjacent populations or from a single population? If the genetic material from several populations is to be used, should the colonists be the F_1 progeny of a bout of random mating among representatives from each population or should the colonists be the representatives themselves? These questions arise from considerations of the importance of locally coadapted complexes of genes, the extent of partial reproductive isolation among distinct populations of a species, and the best way to organize extant genetic variation to maximize the likely response to the novel selective pressures of the new environment. The resolution of these issues will only come from experimental studies of different synthetic methods. The active programs of habitat restoration and land acquisition that have been inspired by the growing appreciation for the impacts of global change in the broad sense will bring this management problem into sharper focus in the years ahead.

RESEARCH LINKS TO OTHER SUBDISCIPLINES

If evolutionary biologists are to understand and anticipate the impact of global change on the earth's biota, they must interact with colleagues in other disciplines. Several linkages come readily to mind. Although all these areas fall within the subject of ecology *sensu lato*, it is convenient to distinguish among them.

Population biologists who are largely concerned with genetic variation and selection should find it useful to integrate genetic information with information from paleontology, including paleoecology. Fine-grained stratigraphic records (e.g., of some Foraminifera) may provide useful information on rates of evolution. Biotic responses to Pleistocene events, as noted by several contributors to this volume, offer perhaps the best model of the effects of relatively rapid environmental change, and many contemporary patterns of species distribution and genetic variation cannot be understood without reference to Pleistocene history. Population biologists will want to know about the relative incidence of extinction, range shift, and adaptation during the Pleistocene, as well as rates of genetic differentiation of populations and the kinds of characters that have become differentiated.

Physiological characters are important both directly (e.g., temperature tolerance) and indirectly. For example, activity budgets may be constrained by thermal regimes, as has been shown in butterflies (Kingsolver and Watt, 1983; Kingsolver, 1985), lizards (Grant and Dunham, 1988), and other animals; these in turn can affect the persistence of populations. The outcome of competition among species is determined, ultimately, by physiological capacities (see, e.g., Tilman, 1982), the mechanistic bases of which are generally poorly understood. Even at a phenomenological level, we know little about genetic variation in physiological features (Geber and Dawson, this volume), still less about the genetic architecture of physiological traits, and very little about the precise role of physiological determinants of species range limits (Hoffmann and Blows, this volume). Enhanced collaboration between evolutionary biologists and physiologists is long overdue (Feder et al., 1987).

Despite an ongoing liaison between evolutionary biology and population ecology, a more meaningful relationship is necessary in two areas. First, the ecology and evolutionary biology of microorganisms have been greatly neglected. Microorganisms may not have much esthetic appeal to many people, but because of their roles as decomposers, gut symbionts, root symbionts, and pathogens, they exert a great influence both on ecosystem processes and on the dynamics of plant and animal populations. Most of the questions we have posed about plants and animals are equally germane to microorganisms, and to these we may add questions about how changes in either the genetic properties of populations or the species composition of microfloras affect animals and plants.

Second, the population genetic models of finite populations that examine the limits to the rate of adaptive change are notable in that virtually every key parameter is a function of population size. Most of our extant models deal with steady populations of arbitrary size, but real populations fluctuate continuously as a function of density-dependent and density-independent effects. Some processes may be unaffected by small fluctuations in the effective size, other processes will change markedly as the effective size fluctuates. Moreover, changes in the actual size may induce behavioral and ecological changes that change the ratio of the genetically effective size to the actual size (Travis and Mueller, 1989; Travis, 1990). Dynamic models of fluctuating effective size, based on ecologically observable patterns of fluctuating census numbers and dissections of the ecological relationship of the effective size to the census number, will be necessary if we are to make realistic and robust predictions of the likely evolutionary responses to global change.

SUMMARY

Evolutionary biology offers several lessons for those who would predict the effects of global change on natural populations. The first lessons are the paradoxical ones of history. Rapid environmental changes have always been associated with heightened rates of extinction; yet some organisms have adapted to a bewildering array of environmental challenges and have often done so at astonishing rates. The remaining lessons are the sobering ones of population biology; they are derived from the observation that a species is not a genetically or ecologically uniform entity. A species' range will be limited by different ecological factors at different locations, and climatic conditions will determine only a small part of a range terminus. Genetic differentiation among conspecific populations will produce significant heterogeneity in their responses to common alterations in environmental conditions.

Whether a species adapts or becomes extinct depends on whether it can mobilize sufficient genetic variation for key phenotypic traits before it is overtaken by the pace of environmental change. This principle sets the research agenda for the evolutionary biologists who will refine our understanding of the biotic effects of global change and develop the management principles necessary to preserve as much biological diversity as possible. Systematists must identify traits with evolutionary histories of lability and clades with histories of adaptive modifications. Geneticists must probe the origin and expression of genetic variation for ecologically critical traits, particularly under conditions of environmental stress. Ecological geneticists must understand which factors most influence the rate of gene flow among populations and the degree to which gene flow will herd the individual populations of a species toward a common fate. Mathematical population geneticists must examine the roles of pleiotropy and gene flow in producing a lag in a species' response to selection and develop dynamic models for the genetics of finite populations. Physiological characters deserve special attention in these contexts; these characters not only affect viability directly but also act indirectly to mediate the outcome of biotic interactions.

Without the lessons of evolutionary biology, ecological predictions of the effects of global change are naive. But without an effort to understand the evolutionary consequences of global change, future lessons will disappear with the organisms that might have taught them to us.

PART FOUR

COMMUNITY RESPONSES TO ENVIRONMENTAL CHANGE

CHAPTER 16

SPECIES DYNAMICS AND GLOBAL ENVIRONMENTAL CHANGE:

A Perspective from Ecosystem Experiments

Stephen R. Carpenter, Thomas M. Frost, James F. Kitchell and Timothy K. Kratz

Global environmental change acts on ecosystems through its effects on organisms. It is almost inevitable that changes in the environment will alter metabolic and demographic rates and the distributions of populations. The new configurations of species that arise from global change will have no evolutionary history of association and will not have been studied previously by ecologists. Their interactions, stability, and resistance to future changes are difficult to predict. Yet predicting these community properties is fundamental to anticipating and mitigating the effects of global change on ecosystems.

In this chapter, global environmental change is defined as anthropogenic stress that acts over large regions of the earth and may affect biospheric fluxes of radiation, gases, water, and solutes. These stressors include global climate change, global land use change, many pollutants, species invasions and extirpations, and others. The ecological consequences of global climate shifts must be gauged against the background of ongoing change driven by land use, pollution, species replacement, etc. In a statistical sense, these factors are confounded: independent variables are changing simultaneously, and their separate effects on the dependent variables are difficult to discern. However, some information on the effects of perturbations is available from case studies. The lessons learned have important implications for our efforts to anticipate consequences of global climate and land use change.

The capacities of ecosystems to withstand stress lie on a continuum between robust and fragile. In some cases, ecosystem processes like productivity and nutrient cycling may resist change because intolerant species are replaced by more tolerant ones that are functionally equivalent. At the other extreme, relatively minor stresses may lead to dramatic, discontinuous changes in ecosystems. Surprising changes in ecosystems often revolve around species interactions (Walters, 1986; Holling, 1988). The societal significance of changes in ecosystem structure is most obvious when species linked directly to humans, such as trees, crop plants, fishes, pests, or pathogens, are involved. However, more complex ecosystem functions of great importance, such as feedbacks to air, water, or soil quality, can also be affected by population and community processes.

This chapter shows that the dynamics of individual species can have substantial effects on ecosystem processes and can serve as optimal indicators of impending changes in ecosystems. Given this reality, the key question then becomes, how can we identify a priori the species that will be good indicators or perform pivotal roles in ecosystem processes? To address this problem, we draw our evidence mainly from whole-lake manipulations.

In the sections that follow, we first outline the scope of large-scale ecological experimentation. We then summarize the evidence that population dynamics influence ecosystem processes, emphasizing our experience in lakes. This review leads to a comparison of stress indicators ranging in degree of aggregation from species to trophic level. We then offer a conceptual framework to explain intersystem differences in the feedbacks between population processes and ecosystem processes.

LARGE-SCALE EXPERIMENTS

Large-scale experiments are a powerful way to study ecosystem responses. In order to study the interactions of population and ecosystem processes, experimental data spanning a considerable range of spatial and temporal scales are needed. Most ecological experiments occur in relatively small spaces for relatively short periods of time (Tilman, 1989; Kareiva and Andersen, 1988), and are too fine-grained for credible inferences about ecosystem dynamics. Nonetheless, ecologists have gradually gained considerable experience with manipulations at the scale of whole ecosystems (e.g., lakes, watersheds) over time spans of years to decades. Manipulated variables have included nutrients (e.g., nitrogen, phosphorus), contaminants (e.g., acid, heavy metals), and system structure (e.g., clear-cutting, predator additions or removals). Response variables have included ecophysiological and demographic rates, population and community dynamics, and ecosystem processes. Information on the history and scope of large-scale, long-term experimentation is given by Likens (1985), Tilman (1989), and Schindler (1988). Discussions of the design, analysis, and interpretation of large-scale experiments are found in Matson and Carpenter (1990).

In large-scale experiments, perturbations can act through multiple pathways of indirect effects that may not be easily distinguished from one another (Tilman, 1989). Fortunately, both long-term observation (Tilman, 1989) and smaller scale experiments embedded in large-scale studies (Frost et al., 1988) can help to resolve indirect effects.

A more formidable challenge in large-scale experiments is the fact that communities and ecosystems are extremely variable experimental units. Consequently, moderate perturbations may produce responses that cannot be clearly separated from the routine variability of undisturbed ecosystems. Thus, there is considerable risk that important ecological effects will not be detected when manipulations are mild (Carpenter, 1989). However, this does not imply that the experimentalist should adopt perturbations that exceed a realistic range of conditions (Carpenter et al., 1991). Ideally, ecosystem manipulations should be near the boundaries of known variability, and should be sustained for multiple generations of the longest-lived organisms involved in ecosystem response.

Large-scale experiments are most powerful when important arguments revolve around contrasts in system state or inputs to systems. For example, erosion was recognized as an important problem on grazing lands of the southwestern United States early in the twentieth century, but managers disagreed bitterly about the cause of the problem and its solution (Flader and Callicott, 1991). The prevailing view was that fire caused the problem, and that the solution required fire control and reforestation. Cattle, by thinning grasses, could be used to suppress fire and favor growth of woody vegetation. Leopold (1924) took the contrary position. He argued that overgrazing caused erosion, and that dense grass growth was required to stabilize the soils. Grasses could be enhanced by reducing cattle densities and allowing wildfires to burn. Though global change was a distant issue in the 1920s, the transformation of the Southwest had clear connections to both climate and land use (Leopold, 1979). The management controversy could have been resolved by experimental grazing and burning at the watershed scale. Such experiments would also have resolved the basic ecological disagreement about the interactive effects of fire, cattle, grasses, and woody vegetation on soils (Leopold, 1924). We believe that many debates about the functioning and management of large ecological systems can be resolved experimentally. Often, basic researchers and management agencies will benefit from collaboration on large-scale experiments of mutual interest

(Kitchell, 1992). Restoration ecology offers remarkable opportunities for large-scale experimentation (NRC, 1992).

The major advantage of ecosystem experiments is that they are scaled appropriately, in time and space, to allow direct inferences about effects of specific stresses or management actions. Unlike smaller scale experiments in plots or enclosures, no assumptions are needed to extrapolate results to the ecosystem scale. Analyses of ecosystem perturbations show that species change can critically affect ecosystem processes, and that certain species are sensitive indicators of impending ecosystem change.

THE EFFECTS OF SPECIES ON ECOSYSTEM PROCESSES

Increasing evidence shows that ecosystem processes depend significantly on plant and animal population dynamics under certain circumstances (Naiman, 1988; Vitousek, 1990). Indeed, fluctuations in certain populations may reverberate throughout all trophic levels, causing changes in productivity, nutrient cycling, and fluxes of contaminants and pollutants. For example, beaver populations determine water levels and hydrologic residence times over vast watersheds (Naiman et al., 1988). Changes in beaver density may even have significant implications for river flows at continental scales. Since river output is coupled to ocean dynamics and climate (Idso and Brazel, 1984; Broecker et al., 1990), one might argue that beavers influence global climate change. Other case histories documenting massive effects of single species may be found by studying species invasions (Lodge, this volume). For example, plant species invasions in tropical ecosystems have fundamentally altered nitrogen cycling, fire frequency, successional dynamics, and the spatial mosaic of seral stages on the landscape (Vitousek, 1990).

In lakes, the trophic cascade hypothesis couples food chain dynamics with ecosystem processes (Carpenter et al., 1985). The trophic cascade hypothesis, which owes its name to Paine (1980), was devised to explain the considerable amount of variability in lake productivity that cannot be attributed to nutrients. In any given lake, temporal fluctuations in fish predation are essentially independent of fluctuations in nutrient input. We proposed that variance in fish predation was transmitted through the food web, causing variance in primary production by phytoplankton independently of, and additive to, the variance explainable by nutrient input (Carpenter et al., 1985). Basic research on the trophic cascade has implications for biomanipulation, a water management idea due to Shapiro et al. (1975). Biomanipulation attempts to control nuisance algae in lakes by suppressing planktivorous fishes, allowing herbivorous zooplankton to proliferate and crop the algae. Major research issues at this time include the roles of refuges, behavior, morphology, and life-history in key predator–prey interactions; alternative stable states of primary producer communities; and the interactions of nutrient input and food web effects (Gulati et al., 1990;

McQueen, 1990; Kitchell, 1992; Carpenter and Kitchell, in press). These issues are familiar ones for population and community ecologists, and we will not attempt to review them here.

Surprises are common in whole-lake experiments on biomanipulation or trophic cascades. Often such surprises have involved behavioral or morphological plasticity that gives prey unexpected resistance to predation; irreversible losses of key species; and explosive growth of previously rare or unknown species. We will examine two examples, each with significant implications for ecosystem processes.

When introduced rainbow trout replaced largemouth bass as the top predators in Peter Lake, the large effective grazer *Daphnia pulex* and the invertebrate zooplanktivore *Chaoborus* spp. were greatly reduced in density (Figure 1). The reduction of these species can be traced to the fact that the trout invaded a refuge that had previously provided a safe haven from largemouth bass. In particular, *Chaoborus* avoids significant predation from bass by migrating to deep waters or profundal sediments during the day. At night, *Chaoborus* migrate to surface waters where they feed voraciously on small copepods and rotifers. *Daphnia pulex* is large enough to escape significant predation by *Chaoborus*, and avoids bass predation by migrating to deep, unlighted waters during the day. Because of its broad diet, *D. pulex* is able to suppress many of the large algae that occur in the lake, mainly gelatinous colonies of green and blue-green algae. Unlike bass, rainbow trout feed effectively at the cool temperatures found deep in the lake, and consequently were able to dramatically reduce populations of both *Chaoborus* and *D. pulex*. The chain of effects did not, however, stop there.

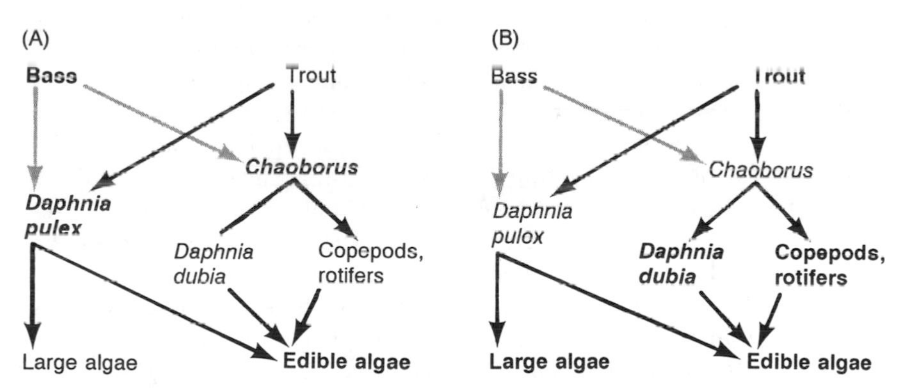

FIGURE 1. Major food web interactions in Peter Lake when the top predator is largemouth bass (A) or rainbow trout (B). Taxa present in high densities are shown in boldface; taxa that are absent or present at low densities are shown in regular type. Black arrows show strong interactions (Paine 1980); gray arrows show weak interactions.

Soon after *Chaoborus* and *D. pulex* declined, we observed a remarkable irruption of *Daphnia dubia*, a species never before seen in the lake. Whereas *D. dubia* is normally held in check by *Chaoborus*, it can evade trout predation and upon the introduction of the trout began to flourish. Finally, since *D. dubia* cannot feed effectively on large algae, this chain of events led ultimately to algal blooms. In summary, then, trout introduction caused increases in herbivore biomass, algal biomass, and primary production (Carpenter and Kitchell, in press). These significant changes in the ecosystem stemmed from the unexpected population explosion of a species not previously known to occur in the lake.

Peter Lake lies on private land, and both the lake and its catchment were protected from unwanted disturbances during the course of our experiment. However, many large-scale experiments will lack the luxury of protected, pristine conditions. When ecologists exploit experimental opportunities offered by large-scale management or restoration projects, the public will usually have access to the study sites. It may therefore be essential to regard humans as a component of the system under study. The next example shows that humans can be the agent of surprise in large-scale experiments.

Lake Mendota, Wisconsin, has been the focus of an ecosystem experiment designed to test the efficacy of biomanipulation in a large, eutrophic lake subject to intensive recreational use (Kitchell, 1992). The manipulations involved the stocking of an extraordinary number of piscivorous fishes (walleye pike and northern pike) in combination with the most restrictive fishing regulations in the state of Wisconsin. The regulations, like those of most sport fisheries, restricted the numbers of fish anglers could keep (bag limits) and required that those fish exceed a specified size (size limits). The goal of the experiment was to increase piscivore stocks and trigger a trophic cascade that would reduce densities of small panfish, increase densities of large zooplankton, and reduce algal concentrations to improve water clarity in the lake. The publicity that attended the stocking, regulations, and research activity brought a more potent predator, anglers, into play. The numbers of anglers and the time they spent fishing increased fourfold in three years. The regulatory mechanisms control only the numbers and sizes of fishes harvested by each angler each day; there is no control over the number of anglers or the number of days they fish. Angling mortality of both stocked predators and the panfish increased sharply. Management has responded to increased angling pressure through even more stringent size and bag limits, and by publicizing the fact that the lake is overfished. At this time, the outcome of the experiment is uncertain, largely because of the human response. The Lake Mendota experience emphasizes that experimentation at large scales will require more effective integration of humans into ecological thinking. To develop a perspective on humans as ecosystem components, ecologists will need to collaborate with economists and sociologists.

These case studies show that population dynamics and species change can substantially alter nutrient cycling and productivity of ecosystems. The interaction between global environmental change and species-level processes is not one-way; significant feedbacks are possible. When species are affected by global environmental change, their reactions can affect the abiotic environment and cause further change. In addition to participating in critical feedbacks, species can also serve as indicators of change. The discussion below summarizes information about ecological indicators that has emerged from ecosystem experiments in lakes.

SPECIES AS OPTIMAL INDICATORS OF ECOSYSTEM CHANGE

While synthesizing the results of more than 20 years of ecosystem experiments on the effects of chemical stressors on lakes, Schindler (1990) drew an important conclusion. The most sensitive harbingers of ecosystem stress were certain population and community variables, such as species diversity, food chain length, proportion of *r* strategists, and sizes and life spans of organisms. The traditional variables of ecosystem ecology, such as production and respiration, were among the least sensitive to chemical contaminants. In fact, ecosystem variables sometimes did not change until severe, even irreversible damage had occurred. To some ecologists, it is paradoxical that key responses to large-scale ecosystem perturbations should be found at the level of species populations.

Schindler's analysis suggests that we should seek indicators of ecosystem stress at the population level. Although populations can be quite sensitive to change, their densities or biomasses also tend to be quite variable. Detection of change may be hampered by high variability that is independent of any ecosystem stress. On the other hand, while ecosystem variables may be less responsive to stress, they also tend to be less variable and changes may therefore be easier to detect.

To critically examine the relative merits of indicators at different levels of aggregation, we calculated the sensitivity of zooplankton to a whole-lake acidification and a whole-lake fish manipulation. Each experiment was conducted in two lakes for six years. In the acidification experiment, Little Rock Lake was divided in half with a plastic curtain. One basin was maintained as a reference while the second, after a one-year baseline period, was acidified during three bienniums from an initial pH of 6.1 to levels of 5.6, 5.2, and 4.7 (Brezonik et al., 1986; Watras and Frost, 1989). For our analyses of the acidification experiment, we calculated natural variability in the reference basin during the baseline, pH 5.6, and pH 5.2 manipulation periods, and differences between the treatment and reference systems during the pH 5.2 manipulation. In the fish manipulation, small planktivorous fishes were removed and replaced by large piscivorous ones in the experimental lake (Tuesday Lake) after a one-year baseline period, while the reference lake (Paul Lake) was not manipulated (Carpenter and Kitchell, in press). Analyses reported here are based on one year of premanipulation data and five years of postmanipulation data for both reference and experimental lakes.

Zooplankton biomass data from the lakes were analyzed at four levels of aggregation: species, genus, functional group (rotifers, copepods, cladocerans), and total biomass. Our choice of functional groups is arbitrary; other criteria, such as size, are possible. Our functional groups are consistent both taxonomically and with the practices of most limnologists. Life histories are similar within each functional group. All species in our data sets are planktonic. Almost all are herbivores, but other zooplankton occur in the diets of one rotifer species and several copepod species.

In each experiment, time series for each zooplankton variable were converted to interlake differences (experimental lake minus reference lake) (Carpenter et al., 1989). Sensitivity was calculated as the manipulation effect divided by the baseline variability. The manipulation effect was the absolute value of the mean premanipulation interlake difference minus the postmanipulation interlake difference (Carpenter et al., 1989). The baseline variability was the standard deviation of residuals from a seasonal autoregressive model fit to the series of interlake differences, centered about the pre- and postmanipulation means (Wei, 1990; Frost et al., in preparation).

In both experiments, the strongest responses to manipulation occurred at the species level (Figure 2). The greatest range in sensitivity also occurred at the species level. Although some species were relatively insensitive to the manipulations, one could always find a few species that were the optimal indicators of ecosystem perturbation. Levels of aggregation that are more traditional for ecosystem analyses, functional groups and total zooplankton biomass, did not respond as strongly as the most sensitive species. Losses of sensitive species were compensated by increased growth of ecologically similar species, which damped responses at higher levels of aggregation (Frost et al., in preparation).

Indicators are essential to anticipating effects of global environmental change. Effective mitigation of ecosystem stress requires indicators that respond before ecosystem processes are impaired. Our results agree with those of Schindler (1990) in showing that certain species are the most sensitive indicators of ecosystem perturbation. What are the prospects for a priori identification of good indicators? The next section suggests an approach to this question.

HOW CAN WE ANTICIPATE WHICH SPECIES ARE IMPORTANT?

The preceding sections show two ways that species may have key roles in global environmental change: through pivotal effects on ecosystem proc-

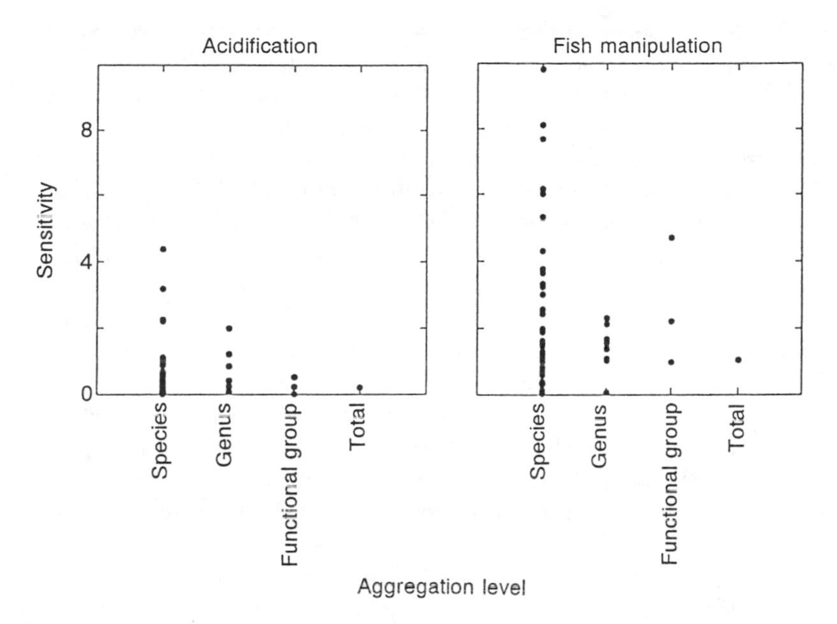

FIGURE 2. Sensitivity vs. aggregation level for zooplankton in whole-lake acidification (Little Rock Lake) and fish manipulation (Tuesday Lake) experiments. Sensitivity is the manipulation effect divided by the variability of each time series. Each dot is the sensitivity calculated for a time series of a separate species, genus, or functional group, or for the total zooplankton biomass.

esses, and as indicators of environmental change. However, only a few species in a given community are likely to have pivotal effects or be useful indicators, while most species play neither role. Ecologists therefore face the challenge of predicting which species are likely to have pivotal effects or to be useful indicators. These two roles of species may be mutually exclusive. If a species has pivotal effects, then by definition ecosystem processes change substantially when that species is added or removed. Thus a change in such a species indicates that shifts in ecosystem processes have already occurred. In contrast, a species that does not have pivotal effects, but changes before ecosystem processes change, is a useful leading indicator of future change. We suggest that indicators and pivotal species are linked to material cycles of ecosystems in fundamentally different ways.

From an ecosystem perspective, species can be viewed in terms of their linkages in cycles of materials such as limiting nutrients, carbon, or contaminants (DeAngelis et al., 1989). A fundamental distinction exists between cyclic and parallel linkage patterns (O'Neill and Reichle, 1979) (Figure 3). Components in parallel might represent populations at the same trophic level that interact mainly through a shared, limiting resource and are substitutable with respect to their effects on ecosystem processes. The cyclic configuration might represent populations at different trophic levels (and with different turnover times) that interact through cycling of a nutrient. Most ecosystems are a mixture of parallel and cyclic linkages.

Perturbations of ecosystems lie on a continuum between input perturbations (e.g., drastic change in the limiting resource) and state perturbations (e.g., drastic reduction in the biomass of one component). In parallel systems, a state perturbation has little effect on system output, because other components compensate for the perturbed one (Figure 3). Input perturbations to parallel systems, on the other hand, tend to have substantial effects on system output. The expected effects of these two different categories of perturbation are reversed, however, when we examine cyclic systems. In particular, for cyclic systems, a state perturbation can create a bottleneck that leads to substantial change in system output. Effects of an input perturbation, on the other hand, are constrained by the turnover time of the slowest compartment and may therefore be damped or delayed (DeAngelis et al., 1989).

Nutrient additions to ecosystems are input perturbations. In lakes, nutrient addition increases primary production and the rates of nutrient cycling. Food web structure does not change appreciably despite changes in plankton species composition (Schindler, 1990). Detritus and the major functional groups of the food web are linked cyclically, while parallel structures exist within each major node of the cycle. This assumption—parallel species structures embedded in the nodes of nutrient cycles—is a common one in ecosystem ecology. It leads logically to the view that species are largely substitutable, and that the substitutions have limited effects on ecosystem processes. Results of input perturbations often support this inference. However, the view that species are substitutable begins to break down when ecosystems are subjected to state perturbations.

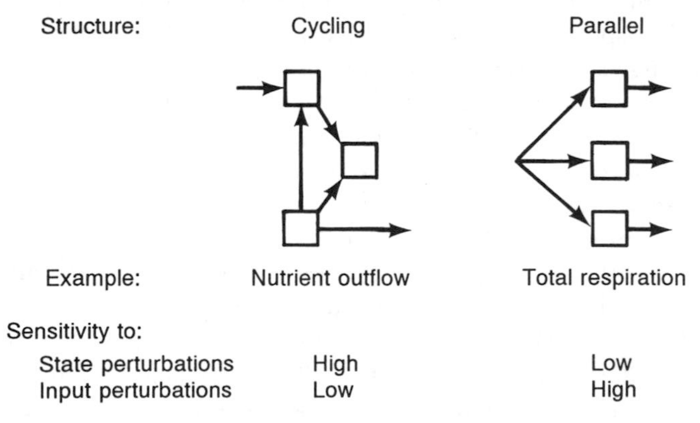

FIGURE 3. Cyclic and parallel system structures. (After O'Neill and Reichle, 1979.)

Food web manipulations, species invasions and extirpations are state perturbations of ecosystems. Contaminant additions can also act as state perturbations by deleting sensitive species (Howarth, 1991). When species act in parallel, extirpations can occur with little change in ecosystem processes. With increased stress, enough species may be suppressed to change ecosystem processes. Thus, the species that are affected at early, low levels of stress forecast future change. When a species acts alone as a node of a cycle, then addition or removal of that species causes immediate ecosystem changes. Examples of all these situations are known from chemical contamination of lake and marine ecosystems (Howarth, 1991). Schindler (1990) and Vitousek (1990) note severe effects on ecosystem processes when the sole occupant of a key niche is added or removed.

Species change is significant to ecosystem analysts for another reason distinct from those discussed so far. Ecosystem process rates, and parameters of ecosystem models, aggregate the collective activity of many species (Schimel et al., 1991). For example, the aggregate functional response of the fish component of an ecosystem model may be derived from a stochastic model of a large population of individual fishes (Madenjian and Carpenter, 1991). The best estimates of the aggregate parameters can change as the underlying species composition changes. Walters (1986) refers to this problem as "parameters that aren't." Parameter drift is a major challenge in modeling fisheries, best resolved by occasional perturbations that enable the modelers to update parameter estimates (Walters, 1986). Manipulations of ecosystems offer analogous opportunities to improve estimates of parameters for process models.

How can we learn to forecast which species will play pivotal roles in ecosystem processes? Such species are especially critical because they can transmit global environmental changes to changes in ecosystem processes. We hypothesize that pivotal species will occupy unique positions in cyclic ecosystem processes. The ecosystem will contain no other species capable of substituting for the pivotal species in one or more critical processes. The most direct way to learn whether a species is pivotal is to alter its abundance substantially in an ecosystem experiment.

How can we learn to forecast which species will be useful indicators of impending environmental change? We hypothesize that indicator species will act in parallel with other species capable of substituting for the indicator's role in ecosystem processes. Good indicators must also be sensitive to environmental change. Sensitivity depends on a species' autecology as well as its interactions with other species (Ives and Gilchrist, this volume). Ecologists must also consider whether a given indicator responds to a wide variety of stressors or a few specific ones. Finally, the costs of sampling must be considered in choosing indicators. Ecosystem experiments are a powerful means of comparing the merits of indicators (Schindler, 1990; Howarth, 1991; Frost et al., in preparation).

Since most species will be neither pivotal nor indicators, we need research toward principles for identifying these classes of species. Ecologists need to resolve fundamental questions about the linkages between species and ecosystem processes. Inevitably, the answers will require species-level studies in an ecosystem context. We believe that progress will follow most rapidly from carefully designed manipulations of ecosystems that include studies of responses at physiological, population, community, and ecosystem levels. Few of the ongoing experimental programs in ecology consider this breadth of response variables. Nevertheless, the tools and expertise for the necessary research already exist in the subdisciplines of ecology. In our view, ecologists are fully capable of carrying through an ambitious program of large-scale experimentation if funding were made available. This is an area of science that is poised for major advances, where a moderate investment in funding will offer remarkable yield.

CONCLUSIONS

Global environmental change is a biospheric syndrome that encompasses climate change, land use change, pollution, and invasions and extirpations of species. These changes often profoundly alter physical, chemical, and biotic fluxes from ecosystems, thereby affecting adjacent landscapes, waters, and the atmosphere. Usually the resulting perturbations to ecosystems involve multiple causes. Thus, global environmental change is truly a feedback process, in which multiple causes are often confounded and the distinction between independent and dependent variables is not always apparent.

Species dynamics are often the nexus between global environmental change and ecosystem processes. The case studies reviewed here suggest that species are significant in two ways: as transmitters of effects and as indicators of effects. Global environmental change is often transmitted to ecosystem processes through effects of the environment on certain organisms. In some cases, a single species governs a key flux of energy, gas, water, and/or solute. Such species have pivotal effects on ecosystem processes. The resulting changes in biogeochemical rates can feed back at larger scales. Other species are often sensitive harbingers of ecosystem stress. Since early detection of change may facilitate mitigation, we should seek some way of identifying ideal indicator species.

Significant uncertainties surround the roles of species as both transmitters and indicators of global environmental change. Our state of knowledge resembles a collection of case studies more than a coherent theory backed by clear experimentation. Effective theory must distinguish between state and input perturbations and must cope with the mixture of cyclic and parallel configurations found in ecosystems. Effective experimentation must be scaled appropriately for global environmental change and must embrace species dynamics and ecosystem processes.

Large-scale experiments are an essential tool for global environmental change research. Such experiments can be scaled to encompass both population and ecosystem dynamics. Case studies suggest that these experiments are likely to reveal surprising links between species and ecosystem processes as well as insights about indicators of environmental change. We emphasize that successful large-scale experiments require substantial manipulations (creating strong contrasts in the manipulated state variable or input) that are sustained for a long period (relative to the return time of the ecological processes under study).

Given the inherent variability and compensatory capacity of most communities, we see little hope for detecting subtle effects in large-scale experiments. However, the ecological effects of climate change, land use change, resource management, contaminants, and species invasions or extirpations are not subtle. In fact, changes in natural systems due to anthropogenic effects are often both substantial and sustained. The changes forecast by the current global climate models portend even greater effects, which are in addition to those stressors already operating. A reductionist approach to subtle effects of global environmental change may demand more time than we have. Natural systems may undergo rapid changes, with unknown consequences for both their intangible values and the resources they yield. These are the problems that rightly concern the society that pays for the research we do. We must offer responses commensurate with the challenges.

ACKNOWLEDGMENTS

We are grateful to the Andrew W. Mellon Foundation for supporting our comparative analyses of ecosystem experiments, and to the agencies that support the individual manipulations: the National Science Foundation, the Environmental Protection Agency, the Wisconsin Department of Natural Resources, and the U.S. Department of Interior through the Federal Aid to Sport Fish Restoration Act. Peter Kareiva and three anonymous reviewers provided helpful comments on the manuscript. We thank Alejandro Muñoz del Río for performing statistical computations.

CHAPTER 17

EFFECTS OF GLOBAL CLIMATE CHANGE ON NORTH AMERICAN BIRDS AND THEIR COMMUNITIES

Terry L. Root

No disagreement exists in the scientific community about the existence of the greenhouse effect. The debate arises over the precise ways in which increased amounts of greenhouse gases will influence the climate (see Schneider, this volume). Without knowing exactly what will happen, scientists have tried to put bounds on the maximum and minimum possible effects. For example, two extreme scenarios have been constructed by an international assessment team (Jaeger, 1988). On the low end, the overall change in global temperature is projected to rise only 0.6°C by the end of the next century, whereas the high estimate is a dramatic 5°C change before the end of the century, or roughly an 8°C increase by 2100. On at least a local or regional scale, the biological consequences of even the low-end scenario will be significant enough to precipitate measurable range changes and perhaps some extinctions (e.g., MacDonald and Brown, in press). Forecasts of the possible biological consequences of the projected rise in global temperature could provide the guidance needed to avert, or at least to help plan for, incipient disruptions of communities and ecosystems.

The possible biological changes would not only be due to increased temperature per se, but, probably more important, to the rate of the increase. The anticipated changes in global climate are expected to occur at a rate most biologists acknowledge as too rapid to be tracked by evolutionary processes such as natural selection (but see Lynch and Lande, this volume). Therefore, species' abilities to adapt will be significantly inhibited, which in turn could enhance the probability of extinction for some species. Indeed, the importance in the difference between evolutionary and ecological time scales was noted by Davis (1990):

> The fossil record shows that most forest trees were able to disperse rapidly enough to keep up with most of the climatic changes that took place in recent millennia. These changes were much more gradual than the climatic changes projected for the future. Even so, there were occasional periods of disequilibrium between plant distributions or abundances, soils, and climate that lasted a century or more. The most rapid dispersal rates known from the fossil record, however, are an order of magnitude too slow to keep up with the temperature rise expected in the coming century.

Quite simply, many tree species take hundreds, even thousands, of years for substantial dispersal. Indeed, the tree with one of the fastest known dispersal rate is spruce, which expanded its range an average of only 20 kilometers every 10 years about 9000 years ago (Roberts, 1988). Additionally, land use by humans has fragmented the natural environment in such a way that dispersal of individuals and expansion of ranges will be limited without substantial human intervention (Main, 1988; Arnold, 1988). Hence, the biological consequences of the impending warming event could easily be significantly more severe than those in prehistoric times. Many more species could go extinct, and the contractions and expansions of ranges could be more extreme. Such biological disruption could greatly influence the functioning of ecosystems, which in turn could feed back and help escalate the disruption of communities.

Even though we do not know what the exact increase in the global temperature will be or how regional climates (e.g., precipitation patterns) will change, various models exist that can help us predict possible changes (e.g., IPCC, 1990). Using these models, we can make projections of the potential biological consequences of the changes based on our knowledge of species' ecology, behavior, evolution, physiology, and the like. Not only is there species-level variation in ecological and physiological needs (Strain, 1987), but individual variation within species can be quite extreme. Both of these factors will cause differential disruption within and among communities (Cohn, 1989). Consequently, a tearing apart of communities will probably occur because the ranges of some species will move farther and faster than others.

The exact disturbances that will occur cannot be determined, but the magnitude of plausible trends, and perhaps the extreme possibilities, can be forecast with some credibility. Such forecasting is necessary given the many decades it would take to do the studies required to learn directly about the consequences of global change. Obviously, utilizing only direct studies would be risky because the amount of destruction that could easily occur during this time could be significant and much of it could be irreversible. The purpose of this chapter is not to provide a complete evaluation of all the possible effects of global climate change or to do a comprehensive review of all research done thus far on the biological effects of global warming. Instead, it is intended to provide examples of the type of work that has been done, highlighting work on birds wintering in North America, and to suggest avenues for future work.

ECOSYSTEM RESPONSE TO RAPID CHANGE IS PROBLEMATIC

Vegetation

Understanding the consequences of global warming for birds requires at least a rudimentary understanding of the influence climate change will have on vegetation, because the essential characteristics of ecosystems are shaped by vegetation (Graetz et al., 1988). Climate strongly affects the distribution of plants; many studies have shown that vegetation changes as the climate does (e.g., Woodward, 1987 and references therein; COHMAP, 1988; Gates, 1980). Even small changes in ambient temperature can cause dramatic effects owing to changes in the physiological demands on plants. Lower energy demands due to warmer temperatures could easily change the delicate dominance structure that is assumed to be maintained by competition (Strain, 1987). Moreover, species disperse at different rates (Cohn, 1989), which will result in dramatic alterations of the species composition of basically all biological communities.

Estimates of transient forest response have been made using so-called gap models, first developed by Botkin et al. (1972) and then modified, expanded, and applied by many others (e.g., Pastor and Post, 1988; Bonan et al., 1990; Botkin and Nisbet, 1992). Figure 1 shows the results of one such later-generation model (Martin, 1990) in which the model-generated succession of aspen, maple, birch and so forth can be seen for "normal" climate for the "landscape" of Minnesota over 400 years. Then, after 400 years a 3°C per century warming trend scenario is imposed, and the forest response is quite significant in that it suggests the "disappearance of the forest" (Martin, 1990). The simulated disappearance is quite dependent on internal model assumptions, such as air humidity and biophysical parameterization of leaf temperatures. Although such internal dynamics or sensitivity to internal model formulation advises us not to take any one scenario or response projection like Figure 1 literally, the results should be taken seriously inasmuch as they show how major shifts in forest species abun-

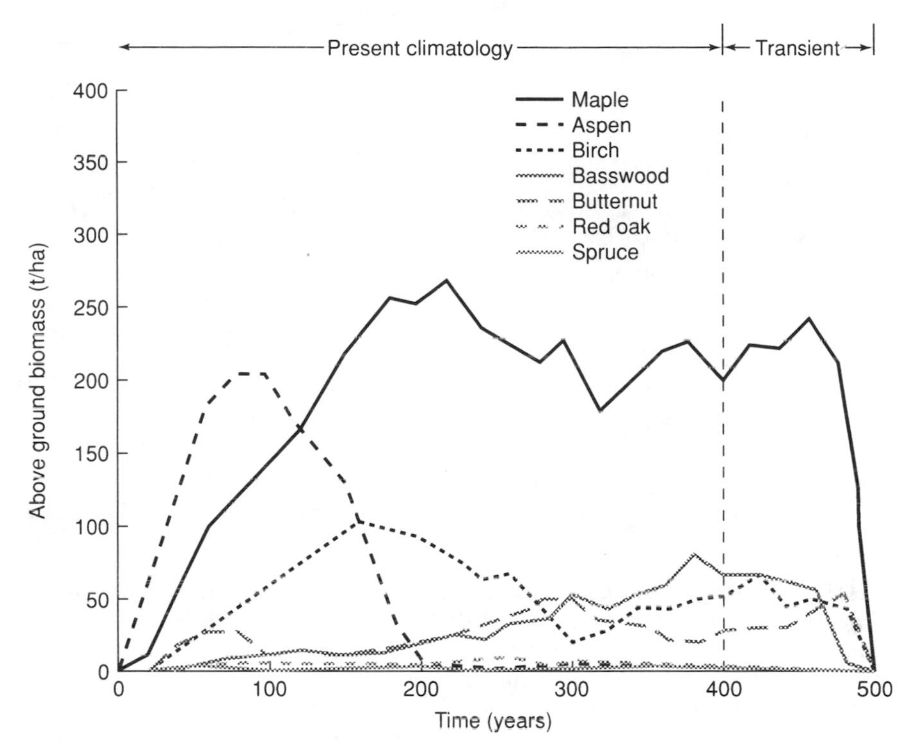

FIGURE 1. A forest model was used to simulate the succession of trees which might occur in the climate and soils of St. Paul, Minnesota, over a 400-year period. A transient climate change scenario with a linear change in temperature of +3°C and in precipitation of −10 percent over 100 years (from year 400 to year 500) was used to simulate the effects of a CO_2-induced climate change. Incoming shortwave and longwave radiation, as well as specific humidity, were kept at their values for present climatic conditions. (From Martin, 1990.)

dances could occur with a few degrees of warming and that these shifts would take decades or centuries to unfold—or undo.

The projected shifts in the range of the sugar maple for two CO_2-doubling scenarios are presented in Figure 2, along with the present range. Sugar maple not only expands farther north under these conditions, but massive local die-outs could occur in the south (Davis and Zabinski, 1992). The extent of the predicted distributional shift is less extreme when the Goddard Institute for Space Studies climate model scenario is used (Figure 2A) than when the climate is predicted by the Geophysical Fluid Dynamics Laboratory model (Figure 2B). However, even though the details are not the same, both models indicate significant range changes, which in turn will probably cause the structure of both northern and southern communities to change dramatically.

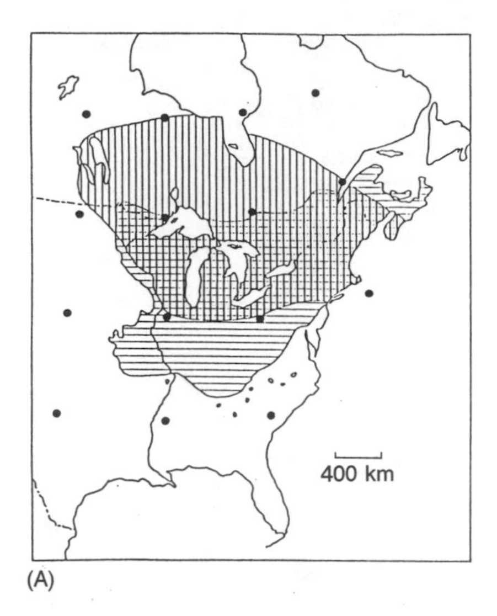

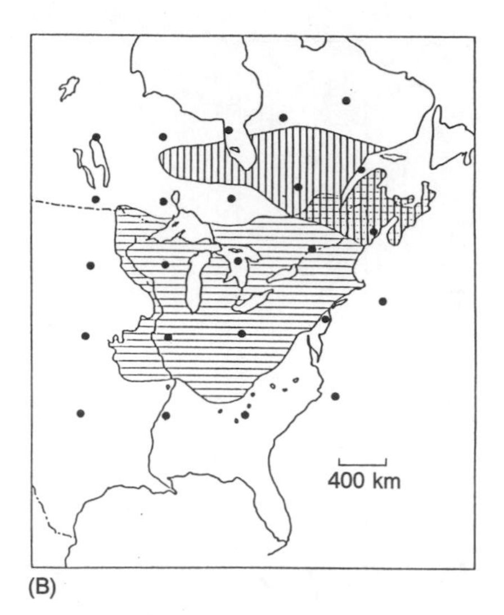

FIGURE 2. The present geographic range of sugar maple (horizontal lines) and its potentially suitable range under doubled CO_2 (vertical lines). Cross-hatching indicates the region of overlap. (A) Predictions using a climate scenario derived from the Goddard Institute for Space Studies general circulation model. (B) Predictions using a climate scenario derived from the Goddard Fluid Dynamics Laboratory model. Gridpoints are sites of climatic data output for each model. (From Davis and Zabinski, 1992.)

The anticipated changes in the ranges of plants will likely have dramatic effects on animals, both on the large, biogeographic scale and on the local, community scale. The ranges of many animals have been found to be strongly linked to vegetation. For example, red-cockaded woodpeckers (*Picoides borealis*) are endemic to mature pine and pine–oak forests (Mengel and Jackson, 1977), and the winter range of Sprague's pipit (*Anthus spragueii*) is coincident with *Andropogon* grass (Root, 1988c). Consequently, the ranges of various animals that rely on vegetation will change as the plants' ranges shift, assuming, of course, that some other factor is not limiting the animals. If, however, the climate changes more rapidly than the dispersal rate of the plants—resulting in individuals dying in the south before individuals can be established in the north—then the ranges of animals relying on these plants will be compressed. Indeed, in some cases extinction could occur. For instance, the red-cockaded woodpecker needs mature living trees for nesting sites (Jackson, 1974). If the rising temperature causes a large majority of mature trees to die before the newly established trees reach maturity, then this rare woodpecker would go extinct.

Many species of animals have ranges that are not directly limited by vegetation, but instead are restricted by temperature. This is true for most ectotherms (e.g., insects and amphibians) as well as for some endotherms, such as the eastern phoebe (*Sayornis phoebe*), which winters in areas warmer than 4°C (Figure 3; Root, 1988b). As the earth warms, those species directly limited by temperature will be able to expand northward as rapidly as their dispersal mechanisms will allow, again assuming that other factors are not limiting. Animals limited by vegetation will be able to expand their ranges only as rapidly as the vegetation changes. Consequently, the potential for significant disruption among communities is quite high. For instance, some animals may no longer be able to coexist because an invading species disrupts the "balance" between competing species or between predator and prey species. Therefore, to understand the ecological consequences of global warming on animals in general and birds in specific, the link between animals and plants needs to be understood, and some notice of how particular plants will respond to a warming climate is needed.

Birds

Birds are potential "victims" of rapid climate change. One representative case, which has been described by Botkin et al. (1991), is that of the Kirtland's warbler in northern Michigan. Its range is restricted to a narrow area of Jack pine trees that grow in sandy soils in the region. Ecological models of growth and decline of Jack pine forests suggest that even a small climatic change would be enough to devastate this habitat. The Jack pines will move north, but the warbler will not be able to survive in the more northerly areas. This bird nests on the ground under relatively young pines.

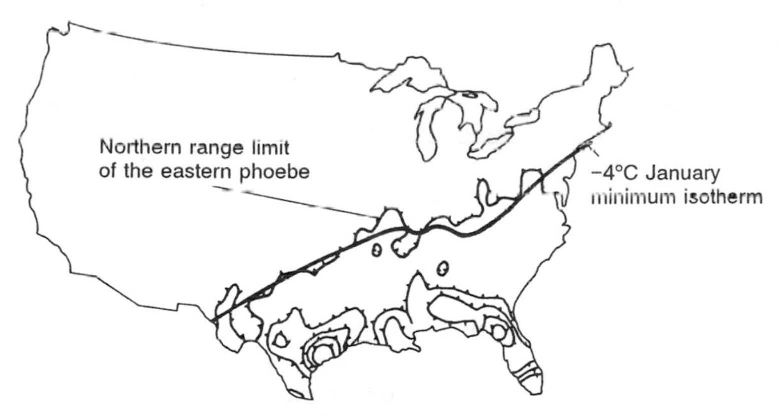

FIGURE 3. Distribution and abundance map of the winter range of the eastern phoebe. The northern boundary lies very close to the −4°C isotherm of average minimum January temperature (heavy solid line). (From Root, 1988a.)

The soil to the north of the present range is not sandy enough to allow sufficient drainage for successful fledging of young (Cohn, 1989). This scenario almost certainly dooms the warbler to extinction within 30–60 years. This potential extinction is indicative of how the already high rate of extinctions around the world will be substantially exacerbated by climate changes occurring more rapidly than species can adapt (e.g., Wilson, 1989; Peters and Lovejoy, 1992).

Although it has long been thought that climatic factors govern broad biogeographic patterns (Andrewartha and Birch, 1954), rarely has this hypothesis been rigorously quantified (Brown and Gibson, 1983). To redress this gap I undertook a massive study of the biogeographic patterns of all wintering North American birds. I found a high degree of statistical correlation between the distribution and abundance of many bird species and large-scale environmental factors such as temperature and precipitation (Root, 1988a). Figure 3, for example, depicts the strong association between the average minimum January temperature and the northern range limit of the eastern phoebe. I compared the distribution and abundance maps for 148 land birds wintering in North America with six environmental factors: average minimum January temperature, mean length of frost-free period, potential vegetation, mean annual precipitation, average humidity, and elevation. The species maps were generated from 10 years of the National Audubon Society's Christmas Bird Count data, from the winter of 1962–1963 through the winter of 1971–1972. Associations were quantified by finding the area between the range boundary and an environmental isopleth (e.g., Figure 3) and then dividing by the length of the range boundary. Less than 1 percent of the possible associations between the species' ranges and environmental factors are expected to occur by chance. Other methods can be used to quantify the association between birds' northern boundaries and environmental isopleths, such as examining the range of temperatures spanned across a given boundary or the standard deviation of temperatures along the boundary (Repasky, 1991). Such one-dimensional, heuristic measures may be easier to calculate than my two-dimensional average area deviation method (Root, 1988b), but they cannot provide the same level of detail and focus necessary for certain questions. Because the methods used can lead to differing inferences, a method must be chosen that is the most appropriate to specific scientific questions.

I did not examine species' southern boundaries because over 80 percent of the species have range limits abutting the southern edge of the study area (Root, 1988b). Nevertheless comparisons along the northern, eastern, and western boundaries of species' ranges reveal that environmental factors show frequent associations with range boundaries (Figure 4). Indeed, more than one factor often associates with a given range limit, which is expected because the factors are not independent.

Various climatic and vegetation variables are often associated with winter

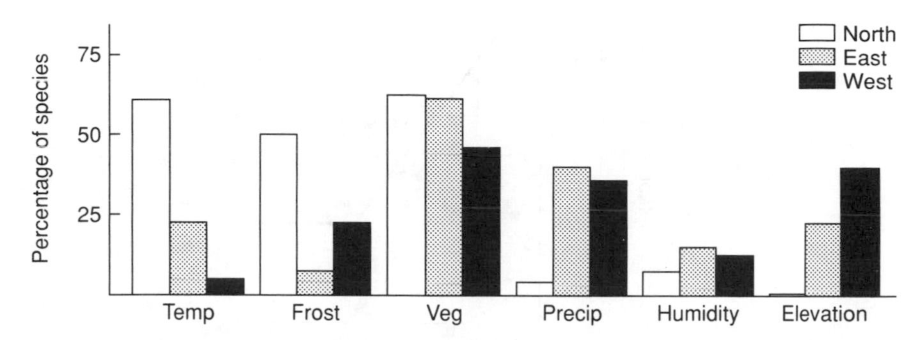

FIGURE 4. Plot of the percentages of species' northern, eastern, and western range boundaries associated with six environmental factors. See text for explanation of terms. (After Root, 1988a.)

range boundaries of birds, implying that changes in global climate could easily reshape and relocate those ranges. For instance, I found that the northern boundaries of the birds' ranges were most often associated with two temperature variables (average minimum January temperature was associated with 60.2 percent of all the species examined, and mean length of frost-free period with 50.4 percent) and with the vegetation variable (63.7 percent). Vegetation also exhibited frequent associations with eastern and western range boundaries (62.8 percent and 46.0 percent, respectively), as did mean annual precipitation (39.7 percent and 36.0 percent, respectively). Because changes in global climate are expected to shift the geographic position of temperature and rainfall regimes (IPCC, 1990), the aforementioned patterns suggest that the boundaries of bird distributions are likely to be reconfigured in concert with climate changes.

I investigated whether or not physiological constraints offered a likely explanation for the observed associations between average January minimum temperature and northern range limits in North American birds (Root, 1988b). Of the 51 species of songbirds (Passeriformes) whose northern range limits were strongly associated with ambient temperature (Root, 1988a), winter physiology studies had been reported in the literature on 14. From these studies I obtained lower critical temperature (TCRIT, the ambient temperature below which an individual must increase its metabolic rate to maintain thermal homeostasis), the basal metabolic rate (BMR, the metabolic rate of a night-resting individual at an ambient temperature a few degrees above its TCRIT), and conductance (COND, the heat loss of an individual as the ambient temperature drops below TCRIT).

Figure 5 depicts the relationship between the measured physiological values, temperature at the northern boundary of the distribution (TDIST), and the value for the metabolic rate at the northern boundary of the distribution (NBMR). The equation used to derive NBMR is

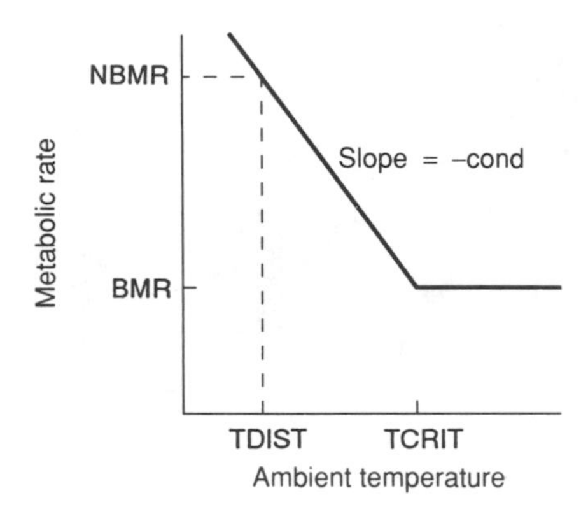

FIGURE 5. A schematic representation of the relationship between metabolic rate and ambient temperature. BMR, basal metabolic rate; COND, conductance; TCRIT, lower critical temperature; NBMR, northern boundary metabolic rate; TDIST, average minimum January temperature at the northern boundary of distribution. (From Root, 1988b.)

$$\text{NBMR} = [(\text{TDIST} - \text{TCRIT}) \cdot \text{COND}] + \text{BMR} \tag{1}$$

with TDIST and TCRIT measured in degrees Celsius, BMR and NBMR in kilojoules per day, and COND in kilojoules per day per bird per °C. For all the birds examined, I found that the metabolic rate at their northern boundaries was nearly a constant multiple of their basal rate. The relationship between these two metabolic rates is

$$\text{NBMR} \approx 2.5 \text{ BMR} \tag{2}$$

The ratio of NBMR to BMR shows little variation among species; the mean of this ratio is 2.49 with a standard error of ± 0.07. The 95 percent confidence limit around the regression line defining the NBMR and BMR relationship provides values that range from 1.95 to 2.93 (Root, 1989).

Basal metabolic rate is strongly related to body mass (Aschoff and Pohl, 1970; Root, 1988b). Consequently, the relationship stated in Equation 2 suggests, as Repasky (1991) correctly points out, that larger species should occur farther north than smaller species. This is exactly what is found for those species that have northern range boundaries associated with temperature isotherms (T. L. Root and J. Price, in preparation). Small-bodied birds that do not have a northern range boundary limited by temperature do, of course, occur in the north. Contrary to Repasky's suggestion (1991), by being exceptions that are explainable by additional factors, these small-bodied birds help "prove" rather than negate the rule. For example, many of the more northerly small-bodied birds have developed special energy-conserving physiological mechanisms, such as hypothermia in the black-capped chickadee (*Parus atricapillus*) (Reinertsen, 1983), or behavioral mechanisms, such as cavity roosting as in the brown creeper (*Certhia americana*) (Ehrlich et al., 1988).

The calculated metabolic rates at edges of species' distributions (Root, 1988b) are based on resting metabolic rates, and they do not account for extra heat generated, for example, during digestion or activity. One of the assumptions of these calculations is that ambient temperature is an adequate index of an individual's thermal environment. Certainly microhabitats and wind conditions have great influence on the energy expended to keep warm (Buttemer, 1985). Nevertheless, average minimum January temperature appears to describe, to the first order, the wide-scale thermal environment of many species. Further investigations will help determine if a connection exists between the seemingly ubiquitous value of 2.5 BMR for the metabolic rate at the northern boundaries and the daily energy expenditures found by other workers (e.g., Walsberg, 1980; Drent and Daan, 1980; Peterson et al., 1990) or if the similar values are just an unusual coincidence.

This type of analysis implies that as climate changes, the physiological tolerance of some birds could cause them—habitat permitting—to change their ranges as rapidly as the climate changes. The range changes in those species for which the habitat is not permissive will probably not be as extreme, and those species with ranges not associated with ambient temperature will probably not exhibit an immediate change in range. Such differential movements of species will certainly cause a tearing apart of communities, thereby forcing potentially dramatic restructurings and reorganizations.

FORECASTING POTENTIAL CHANGES IN SPECIES COMMUNITIES

Three major research activities are needed before we will be able to forecast reliably the effects of global climate change on species communities throughout North America. First, much can be learned about the effects of global warming by studying prehistoric species communities that underwent a similar warming event from a glacial to an interglacial period. However, important differences between prehistoric climate changes and those expected today, such as the presence of glaciers and the much slower rate of temperature change in ancient times, certainly prevent direct comparisons. Consequently, generalities rather than specifics about such periods will be the most helpful. For instance, Graham and Mead (1987) note that environmental changes associated with the last deglaciation (called "Termination 1" in the marine record) had profound effects on the restructuring of biotic communities in North America. Vertebrate species, especially mammals, are particularly useful indices for these changes, so they provide excellent documentation of the climatic fluctuations of the late Quaternary.

In general, mammalian fauna responded to the last deglaciation in North America by shifting their ranges relatively quickly (Graham and Mead, 1987). Throughout most of North America during the full-glacial era, boreomontane and arctic species extended their ranges to much lower latitudes and altitudes than they occupy today. Communities near the ice front contained species like arctic shrews, lemmings, voles, and ground squirrels, which today inhabit arctic tundra. Several full-glacial faunas from Iowa and Wisconsin were the most tundralike of any community south of the ice sheet. These small mammal faunas, however, are not directly analogous to those of the modern tundra, because the prehistoric communities included species that do not inhabit the tundra or northern boreal forest today (e.g., prairie vole, sagebrush vole, thirteen-lined ground squirrel, eastern chipmunk). The presence of these species in ice-marginal habitats suggests that winter temperature extremes were not as severe as those in the modern Arctic; in fact, they may not have been any more severe than those currently found in the northern United States. Furthermore, these findings imply that seasonal variation during full glaciation was not excessive; summer temperatures were cold enough to support arctic species, but the winters were not cold enough to eliminate all amphibian and reptile species.

Graham and Grimm (1990) caution against extensive reliance on past conditions to forecast future patterns. They argue that predicting community response to greenhouse warming becomes particularly uncertain because the rate of the forecasted global warming exceeds that of any period of the last 120,000 years. Future climates may lie outside not only the existing climatic domain, but also outside our paleoclimate database and outside the "climate space" to which existing species are evolutionarily adapted. Therefore, inferences from past changes can only be taken as a heuristic guide to possible future changes. However, if such past changes are used to calibrate and validate models of climate–mammal interactions, then such models may provide credible projections of the effects of climate change on mammalian distributions and abundances.

A second kind of research needed is on the ecology, behavior, and physiology of individual species, with an emphasis on increasing our understanding of how biogeographic patterns may shift as the climate changes. Certainly large-scale studies need to be encouraged; such studies can then be used to indicate the types of small-scale studies that are needed. For example, large-scale studies can indicate which species have patterns that suggest their ranges may be temperature limited. In-depth studies on the behavior and physiology of these sensitive species are needed in order to understand the mechanisms that are acting to shape broader patterns. Multiple-species studies of communities probably will not be as beneficial as single-species studies because species will show unique, differential responses. Vertebrate communities are not tightly bound aggregates of species; rather, they are collections of species occurring within a given habitat

(Whittaker, 1970). Consequently, environmental change will not elicit a uniform response from the community as a whole, but instead each species will respond according to its own tolerances. Climatic change may cause the species composition of a community to be constantly in flux, and significant environmental fluctuations will cause major biotic reorganizations. Of course, to fully understand the effects of global warming on species communities, biogeographical, ecological, behavioral, and physiological studies need to be done not only on birds but on all organisms occurring within various communities. For example, the prey base of some birds (e.g., invertebrates) in a community may be strongly affected by changing climate. Indeed, Coope (1977) has shown that beetle assemblages are highly sensitive to climatic change.

On a third research front, the scientific community needs to foster interdisciplinary work, or, at a bare minimum, multidisciplinary work, that will combine information from a broad spectrum of research: climatology, soil science, entomology, mammalogy, ornithology, botany, and other fields. By integrating information from all these fields, a reliable model can be built that will forecast the spectrum of consequences that continued release of greenhouse gases will have for ecosystems. This will require actively opening communications among scientists in different fields. This can be a time-consuming activity, not only because the jargon varies significantly among fields but also because the groups within the scientific community are structured primarily around disciplines rather than around problems. Additionally, the reward system within the scientific community is such that interdisciplinary work is not valued as highly as disciplinary work (e.g., single-authored papers "count" more toward promotion than multiauthored ones, and granting agencies are set up to fund disciplinary work more easily than interdisciplinary work). Such barriers are common at the cutting edge of problem solving, but they must be breached before the complex problems caused by global climate change can be adequately addressed.

OUTREACH

The possible biological consequences of global warming are catastrophic. The best guess of knowledgeable scientists is a 60 percent chance that the world will warm 1.5 to 4.5°C in the next century (IPCC, 1990, as explained by Schneider, this volume). Some people may say that the odds of this happening are too low to justify acting now. This argument focuses only on the probability rather than on the consequences of catastrophe due to global warming. A good analogy is that of a person having a 1 percent chance of getting a cold if a certain activity is performed. The cost of that action (i.e., a cold) is often low enough that a probability as low as 1 percent does not modify the person's behavior. However, if the cost of the activity is contracting a fatal disease, then a 1 percent probability should be of major concern, and dramatic steps might be taken to avoid the activity.

The scientific community has the ability to determine a range of probable consequences of global warming. If these consequences (whether they are mild or catastrophic) are clearly identified, policymakers will be able to use this information to structure the type of actions that are needed. If we do not provide outreach to policymakers and the general public, we as a scientific community abdicate our responsibility, leaving the field to people who often have specific agendas or selfish interests. Hence, the scientific community as a whole needs to acknowledge the importance of providing the research results needed by decision makers, and of participating in the dissemination of that information. Finally, we must support (and reward) those in our scientific community who competently fill this role.

SUMMARY

Within the next 100 years the global temperature is expected to increase between 0.6°C and 8.0°C (Jaeger, 1988; IPCC, 1990). An increase in temperature on the same order of magnitude occurred between the last glacial and present interglacial period. During that time period, communities exhibited major disruptions due to differential expansion and contraction of ranges (Strain, 1987). In addition, numerous extinctions occurred (Graham and Mead, 1987). The rate of this temperature change was at least an order of magnitude slower than that expected for the impending temperature change. Consequently, dramatic disruption of communities can be expected to occur in the next century. Single-species studies are needed to investigate the effects that climate may have on large-scale biogeographic patterns and on the ecology, behavior, and physiology of all species, including plants and animals. These results must be linked with information from climatologists, geologists, and others, in order to design and validate models that will allow us to forecast more reliably the possible biological consequences of the looming global warming. These forecasts can then be used by policymakers and the general public to determine what types of actions are needed to impede the rapid increase in the global temperature, or to ameliorate its effects on natural systems.

ACKNOWLEDGMENTS

I would like to thank Ray Huey, Peter Kareiva, and Joel Kingsolver for having the insight to convene this workshop and for helpful criticisms on earlier drafts of this paper, Stephen Schneider for encouraging our interdisciplinary collaboration, and Ted LaRoe for facilitating this work. This work was partially funded by grants from the Winslow Foundation, the USFWS Cooperative Research Center, and the National Science Foundation (BSR-9058031).

CHAPTER 18

IMPLICATIONS OF CLIMATE CHANGE FOR STREAM COMMUNITIES

Nancy B. Grimm

Running waters are transport systems that receive and move materials exported from adjacent terrestrial landscapes. However, streams and rivers do much more than simply transport organic matter and nutrients—through biotic processes and interactions they also transform these substances and integrate outputs from a wide variety of ecosystems. Indeed, through their activities, streams can profoundly influence downstream characteristics, and affect even estuaries and oceanic systems. Since water fluxes such as precipitation and evapotranspiration determine the hydrologic regimes that set the physical template for stream processes, it is obvious that climate should influence the character of streams. In this chapter I explore linkages between global climate change and its likely effects on streams. I also examine stream systems more generally, as models for developing conceptual approaches to investigating the biotic consequences of global change. I focus primarily on small to mid-sized lotic ecosystems. Although research in stream ecology has historically centered on small streams in temperate forests, I draw examples primarily from the work of my laboratory and colleagues on streams in deserts. In a sense, desert streams are ideal for investigating connections between climate and stream function because the influence of the physical world can be so much more transparent in deserts than in other biomes.

PHYSICAL RESPONSE TO CLIMATE CHANGE: SETTING THE TEMPLATE FOR STREAM COMMUNITIES

Stream ecosystems are characterized by several unique features: a linear shape; unidirectional flow; disturbances that are primarily hydrologic (at both the dry and wet ends of the spectrum); lateral, longitudinal, and vertical connections with adjacent systems; organisms with short life spans; and diverse communities. Responses of stream communities to climatic change depend upon a host of interactions, controlling or "forcing" functions, and feedbacks. The simple diagram in Figure 1 identifies the most likely physical changes that will affect stream communities. Organisms must cope with both the thermal and the hydrologic regimes. Changes in the components of the hydrologic regime (increase or decrease in total and seasonal runoff as well as increased or decreased streamflow variability; Williams, 1989) inevitably influence stream communities. Less obviously, the axiom that changes in extremes will be more influential than changes in means is likely to be true of streamflow patterns, simply because extremes of hydrology define disturbance in lotic ecosystems. When a stream dries up or floods, the consequences dwarf the effects of moderate daily or seasonal variation (i.e., ±1 S.E.) in flow rates. Unfortunately, regional hydrologic change is most difficult to predict because of the uncertain influences of warming on precipitation, vegetation feedbacks (such as evapotranspiration), and soils (see Karl and Riebsame, 1989). Poff (1992) provides a thorough discussion of regional hydrologic responses to climate change for the continental United States.

ECOLOGICAL RESPONSES TO CHANGE IN THERMAL REGIME

Increased water temperature may directly affect biogeographic distributions of aquatic organisms. Sweeney et al. (1992) have suggested that both the magnitude and rate of temperature change *and* the genetic structure of populations will determine the capacity of aquatic insects to respond to changes in thermal regime through migration or adaptation. They quantified population genetic variability (as measured by allozyme electrophoresis) for six species of mayflies occurring along a latitudinal transect in eastern North America, and found that genetic variability was lowest in high latitudes, where temperature increases are expected to be greatest (Schneider, this volume). They also pointed out that phenology and life history changes are other means of coping with higher temperature and associated changes in food quality.

In addition to direct temperature effects, oxygen solubility declines with increasing temperature, which means that species with high oxygen requirements may be especially vulnerable to increases in maximum water tem-

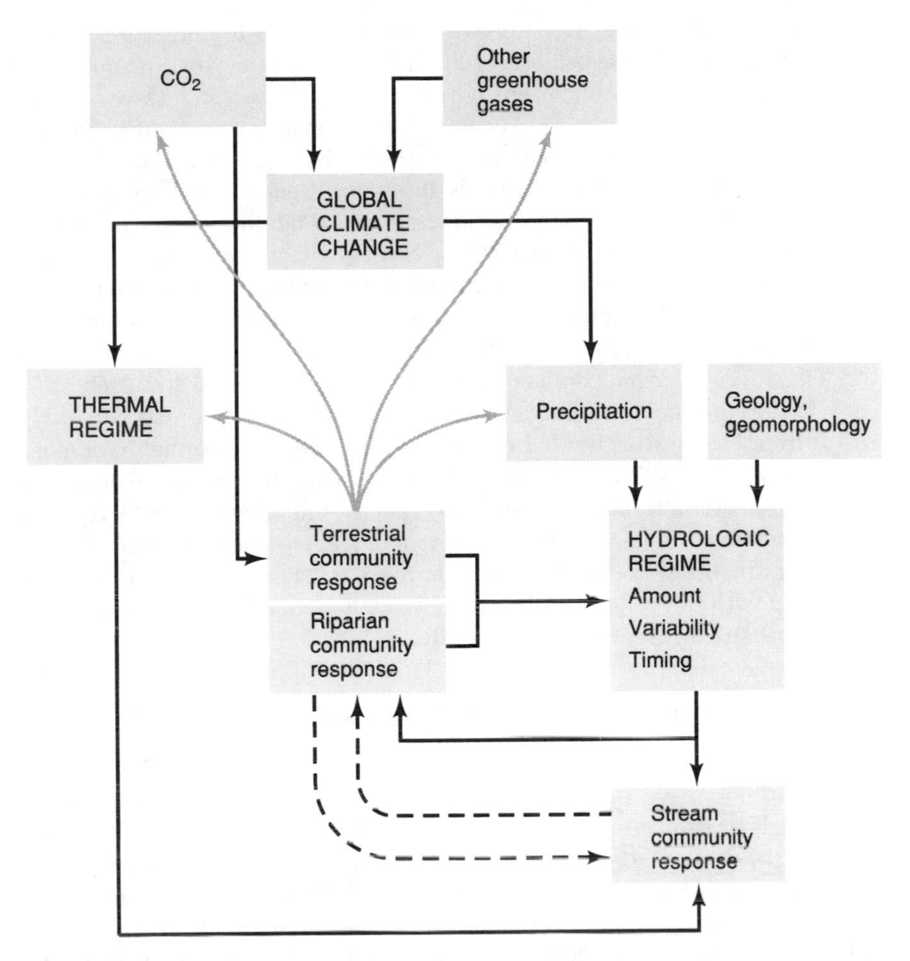

FIGURE 1. Factors likely to influence the impact of global climate change on stream communities. Two major regimes (thermal and hydrologic) plus ecological exchanges with the surrounding terrestrial environment define the physical template of streams. Direct relationships (solid arrows)—for example, those linking global climate change, precipitation, the hydrologic regime, and stream community response—are controlling factors that will directly affect this template. Direct, controlling relationships may be modified to an as yet unknown degree by feedback (gray arrows) and altered ecological exchanges (dashed arrows).

perature. Impacts of temperature on particular organisms can have a cascade of associated effects; for example, a shift in the range of a trout species may change invertebrate community structure where the animal is an important predator (e.g., Hemphill and Cooper, 1984) as well as affecting the fishery (Carpenter et al., in press).

In the riparian zone, tree species may respond to temperature change individualistically, as the paleorecord indicates they have over past millennia (Foster et al., 1990; Graham and Grimm, 1990; Geber and Dawson, this volume). Subtle shifts in species composition of riparian forests, though they may not affect community biomass or productivity, can significantly alter the quality of organic matter that is input to streams. Leaf species vary widely in their rates of decomposition and their suitability as food for aquatic macroinvertebrates (e.g., Kaushik and Hynes, 1968; Ward and Cummins, 1979). Of course, fragmentation or complete loss of the riparian forest would have far-reaching consequences for stream communities, which may be dependent upon trees for food, shade, and in some cases, habitat structure (reviewed in Meyer and Pulliam, 1992, and Carpenter et al., in press).

Changes in species assemblages in streams and riparian forests will likely result from changes in thermal regime because species' thermal tolerances differ. However, any changes due to thermal perturbations may be obscured by the dramatic alterations in streams already under way due to riparian canopy removal, channelization, watershed logging, flow regulation, and pollution. In order to predict the relative importance of climate change compared with direct human disturbance, we will need a clear understanding of how streams function dynamically.

ECOLOGICAL RESPONSES TO CHANGES IN HYDROLOGIC REGIME

Several authors have suggested that changes in extremes or increased variability will disrupt ecosystems more than will changes in mean conditions (Walker, 1991; Poff, 1992). This hypothesis is a central assumption of the following discussion, but further analyses and tests of the hypothesis are essential (Ausubel, 1991). Because the three components of hydrology (amount, timing, and variability; see Figure 1) make up the "disturbance regime," I contend that in streams, *the major impact of global climate change will be a shift in the degree to which disturbance or biotic interactions control ecosystem functioning*. This shift may arise from increased or decreased spate frequency and magnitude, changes in timing of high- and low-discharge events, changes in frequency of drying events, or prolongation of drying.

The question of whether biotic or abiotic factors determine aquatic invertebrate community structure has dominated research regarding streams, as it has for other communities. Peckarsky (1983) suggested that streams can be placed on a continuum from harsh to benign, which dictates whether biotic or abiotic factors are most important in determining community structure. Although useful heuristically, this concept is simplistic in not allowing for temporal and spatial variability in abiotic and biotic controls within the same system. It also distracts researchers from the more general question of *when* (under what conditions) abiotic or biotic factors are operating in a given system, or whether they might be operating simultaneously and interactively (Power et al., 1988). More important, the distinction between biotic and abiotic is difficult, particularly in streams. Traditional "abiotic" factors may in fact be altered substantially by biotic activities, and in turn may significantly affect other biota. Should these factors then be classed as biotic or abiotic?

A classification scheme developed by Poff and Ward (1989) for streams of the continental United States represents an expansion of the "harsh–benign" hypothesis in that it assigns measurable hydrologic characteristics to the terms *harsh* and *benign* (Figure 2). Their analysis is especially useful in providing a suite of measurements that describe flow variability, predictability, and disturbance (flood) frequency, and in emphasizing the im-

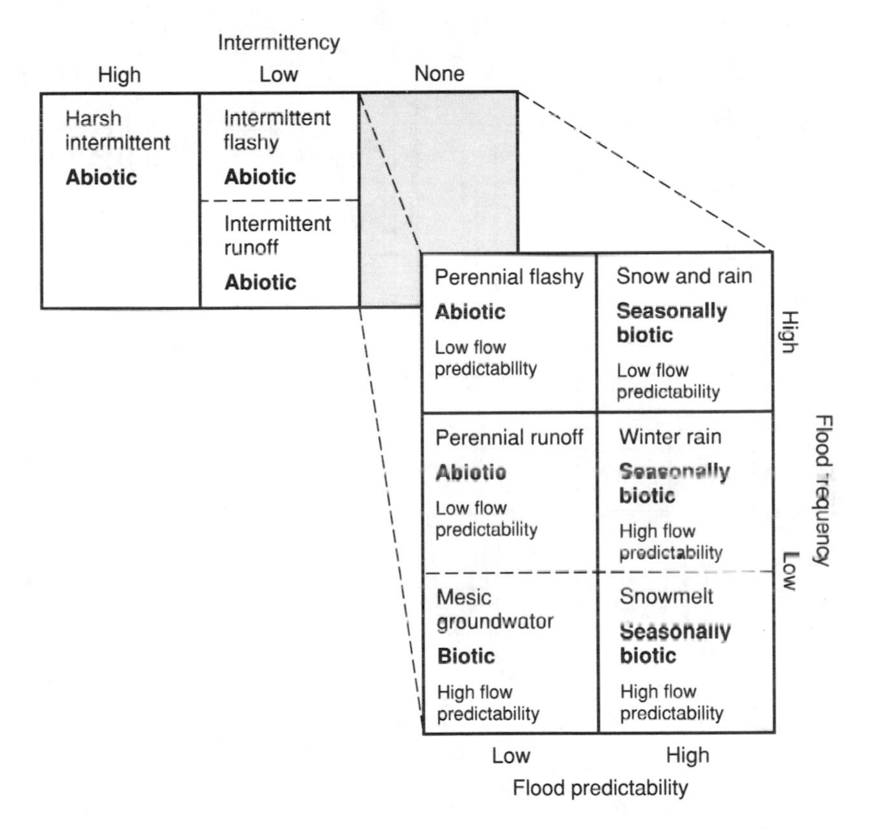

FIGURE 2. Classification of streams of the United States along axes of intermittency, flood frequency, and flood predictability. Predictions of whether communities are predominantly controlled by abiotic or biotic factors are based on position along these axes and on flow predictability (inside box). (From Poff and Ward, 1989.)

portance of drying as a stress or disturbance. Poff and Ward rely on flow records from U.S. Geological Survey gauging stations at single points in streams, and do not consider spatial variability or deal explicitly with disturbance magnitude. As with the Peckarsky model, Poff and Ward's conceptualization retains an abiotic–biotic distinction that may have the problems noted above.

In order to illustrate the challenge of describing streams along a simple biotic–abiotic dimension, it is useful to look at a few streams in detail (Figure 3). Sycamore Creek (Maricopa County, Arizona), a spatially intermittent desert stream, is "flashy" compared with Fort River (Massachusetts), a

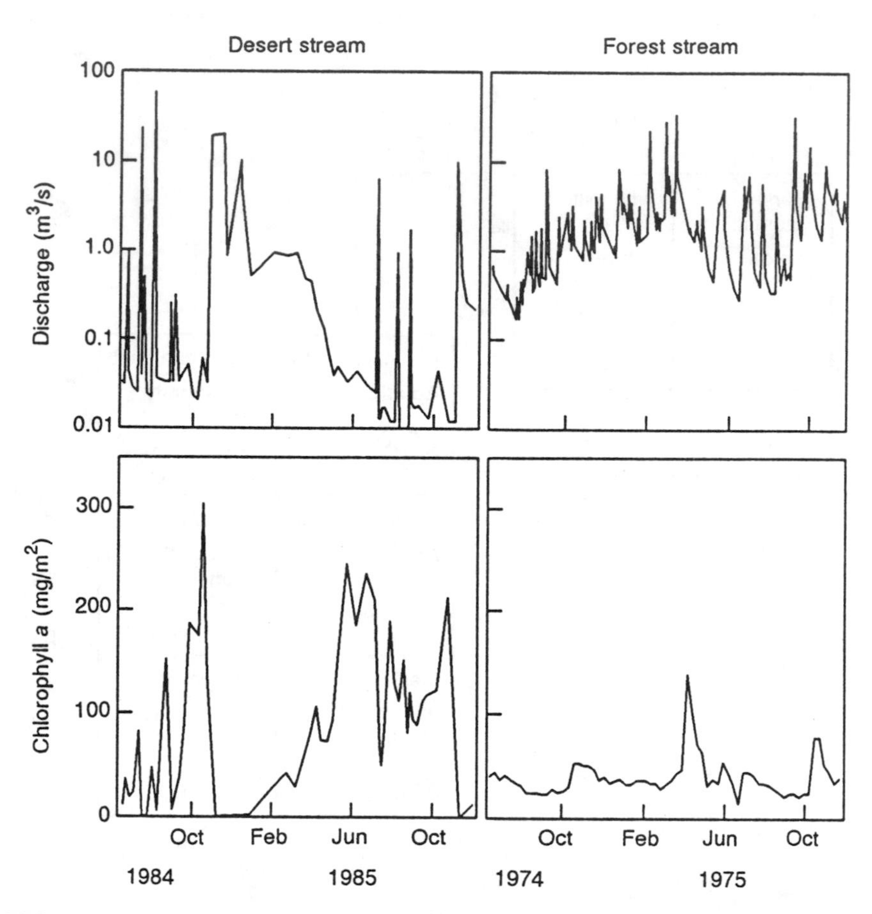

FIGURE 3. Comparison of 1.5-year hydrographs (discharge) and trajectories of algal biomass (chlorophyll *a*) for Sycamore Creek, an "intermittent flashy" desert stream, and Fort River, a "perennial runoff" forest stream. Note log scale for discharge. (After Fisher and Grimm, 1988.)

perennial runoff forest stream. The hydrographs of these two streams over an annual period illustrate this difference: peak flows in Sycamore Creek may exceed base flows by several orders of magnitude, and occur infrequently relative to the common, small-magnitude discharge peaks of Fort River. Annual variation in algal biomass is clearly related to hydrology in the former system, where disturbance-related variables (time since spate, peak spate discharge, and current discharge) explain nearly 70 percent of the variance in chlorophyll *a*, a measure of algal biomass. In contrast, for Fort River, none of the variability in chlorophyll *a* is explained by these hydrologic variables (Fisher and Grimm, 1988). At this annual scale, abiotic factors (in this case, spates) appear to be more important in the flashy Sycamore Creek system than in the more moderated Fort River system. We do know that disturbance and subsequent succession are not evident in Fort River's algal biomass patterns. Clearly we need a more inclusive model than one that simply predicts biotic or abiotic control if we are to understand differences between ecosystems (Fisher and Grimm, 1991).

Disturbance has recently captured the attention of stream ecologists (Resh et al., 1988), and questions of abiotic vs. biotic controls have yielded to questions of whether disturbances or biotic interactions structure stream communities (e.g., Townsend, 1989). Although the distinction between disturbance and biotic interaction is more realistic, like the abiotic–biotic dichotomy it represents an "either–or" approach (Power et al., 1988). The disturbance regime is of course a subset of the more general suite of abiotic factors that determines the limits of organism distributions. Our studies of desert streams have suggested that there are temporal shifts between control of communities by biotic interactions and control by disturbance within a single stream that exhibits high year-to-year variability (Figure 4A; Grimm and Fisher, 1992). Flash floods decimate biota and are followed by a period of "recovery" of about 30 days. If interflood periods are long, drying (particularly in summer) becomes an important disturbance. Between the times when the system is in postflood succession or undergoing accelerated drying, we suspect biotic interactions are important. The spatial variability of most streams implies a parallel situation in space, with discrete disturbance-controlled and biotically controlled patches. Since periods of disturbance and biotic control appear to be broadly overlapping in the desert streams we have examined, it is clear that the two "controls" should not be viewed as mutually exclusive.

How will global climate change affect the temporal and spatial distribution of biotic controls and disturbances? Hydrologic changes in response to increased or decreased total precipitation, extreme events, or seasonality will affect the amount, timing, and variability of flow. Changes in total amount or timing of runoff may result in increased or decreased incidence of intermittency, as well as altering the frequency or severity of flash flooding. Using data on the annual distribution of spates with peak discharges

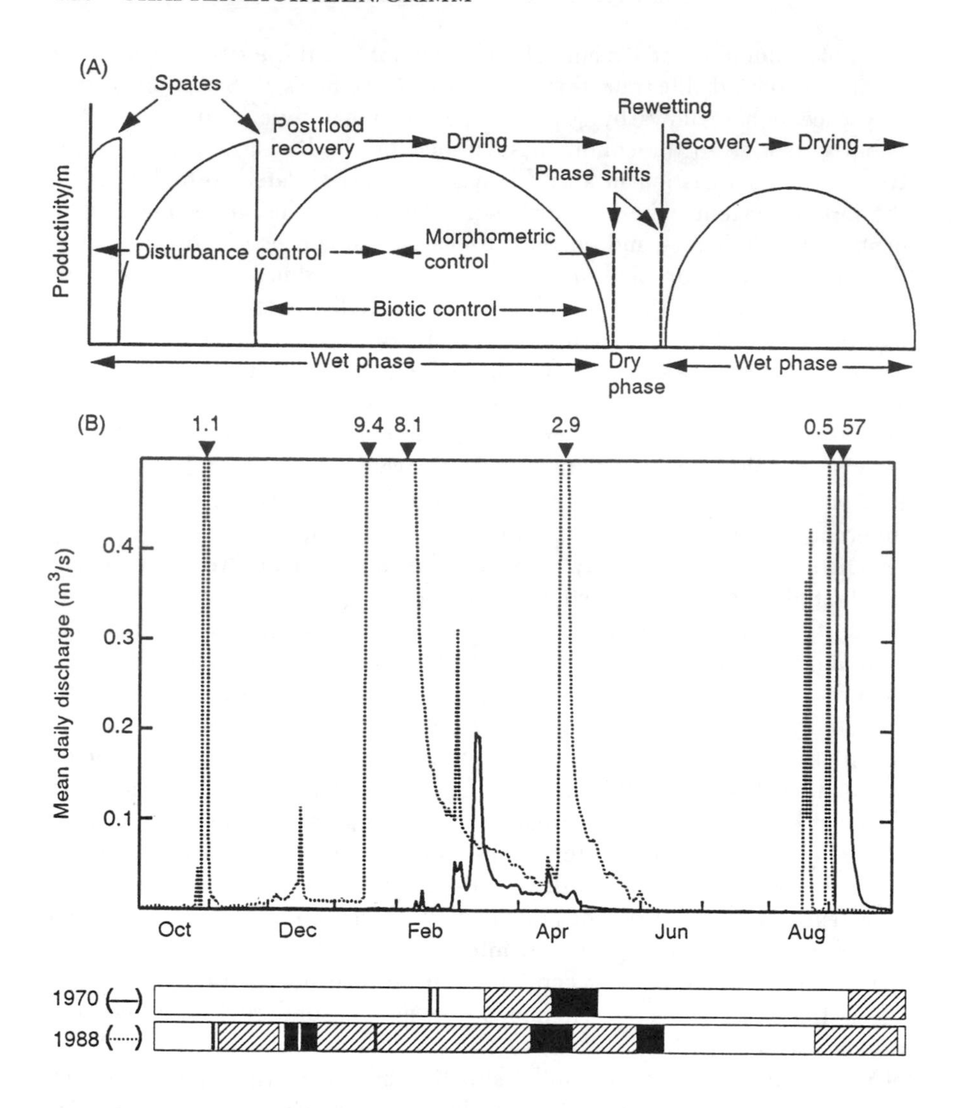

≥11 m^3/s (floods with an annual recurrence interval), I calculated the percentage of time Sycamore Creek was in an early to middle successional stage (≤30 days after flooding), in a state of drying (>200 days elapsed since last flood), or in neither state for the 5 driest and the 5 wettest years of a 29-year record. In wet years the stream had significantly more spates, significantly shorter maximum and average interflood intervals, and was in the early to middle postflood successional state for an average of 128 days (35 percent of the year), significantly more often than in dry years (Table 1). During dry years, on average 167 days (46 percent of the year) were in late successional stages (>200 days since flood) and likely experienced accelerated drying conditions. An interesting aspect of this analysis is that

◀ **FIGURE 4.** (A) Conceptual model of temporal shifts in control for ecosystem functioning in a stream influenced by flash flooding (spates) and drying. Dependent variable is productivity per linear meter; thus the decline during drying periods is due in part to system shrinkage. During the dry phase, surface water is absent and stream productivity is zero. A controlling agent is defined as the best predictor of system functioning; note the broad overlap in disturbance, morphometric (drying), and biotic control. (From Grimm and Fisher, 1992.) (B) The effects of different seasonal distributions of streamflow on potential for biotic or disturbance control in a desert stream (Sycamore Creek, Arizona). A single flash flood in September of 1970 (solid line) accounted for 97 percent of the total annual runoff (7.0×10^6 m^3), whereas in 1988, 88 percent of the total annual runoff (5.8×10^6 m^3) occurred during January, February, and April spates. Time lines at bottom denote shifts in control by flash floods (hatched bar), drying (open bar), or biotic factors (shaded bar). Times were assigned as follows: Any day in which mean discharge increased rapidly to >0.1 m^3/s plus the following 30 days were considered "successional," and thus under flash flood disturbance control; days on which mean daily discharge fell below 0.01 m^3/s were considered to be influenced primarily by drying disturbance; and days on which discharge ≥ 0.01 m^3/s but time since flood (>0.1 m^3/s spike) exceeded 30 days were considered under biotic control. Maximum instantaneous discharges for each year were: Water year 1970, 684 m^3/s (5 September, 1970); water year 1988, 40 m^3/s (18 January, 1988). Numbers at top of figure are peak daily mean discharges exceeding the scale of the graph.

the percentage of time during which neither postflood succession nor drying were occurring is relatively constant (overall mean = 177 days, or 48 percent of time; Table 1) across these hydrologically very different years. Thus, if we equate the absence of drying or postflood succession with a predominance of biotic interactions, it appears the portion of a year dominated by biotic factors remains similar over a broad range of climatic conditions. In mesic forest streams, increased frequency of drying combined with reduced frequency of flooding may similarly conserve the prevalence of biotic controls.

This state-frequency analysis compares only extremes of total annual runoff; variations in seasonality and temporal dispersion of flash floods clearly can alter the conclusions. Water output is only loosely correlated with percentage of time in drying or successional states because a single high-magnitude flood can produce the same annual water yield as several evenly spaced events. For example, 1970 and 1988, two years with similar annual runoff, had markedly different numbers of spates (2 vs. 9), annual spate distributions (fall vs. winter–spring) and thus percentages of time in postflood succession or drying states (Figure 4B). Similarly, two sites can differ during the same year according to their susceptibility to disturbance. The point of this discussion is that even within a single stream, there is potential for pronounced year-to-year and spatial variation in the temporal alternation of disturbance and biotic controls of community structure and ecosystem functioning.

TABLE 1. Hydrologic characteristics of extreme wet and dry years in Sycamore Creek, Arizona.

Year[a]	Annual runoff ($m^3 \times 10^6$)	Number of floods	Days in succession[b]	Days in drying[c]	Days biotic[d]
		DRY YEARS			
1971	0.56	2	47	148	170
1967	0.96	1	30	119	216
1990	1.14	1	30	317	18
1964	1.19	3	41	151	173
1972	1.44	2	55	100	210
Mean[e]	1.06 (0.33)	1.8 (0.8)	41 (11)	167 (86)	157 (81)
		WET YEARS			
1983	73.9	11	149	48	168
1980	91.8	10	89	92	184
1973	91.8	10	191	0	174
1978	94.5	5	79	10	276
1979	100.0	6	133	52	180
Mean[e]	90.4 (9.83)	8.4 (2.7)	128 (46)	40 (37)	196 (45)
P^f	0.0079	0.0079	0.0079	0.0079	0.6905

[a]USGS gauge #09510200, Sycamore Creek near Fort McDowell. Period of record: 1962–1990 water years.
[b]Number of days in water year ≤30 days since spate of ≥11 m^3/s peak discharge
[c]Number of days in water year ≥200 days since spate of ≥11 m^3/s peak discharge
[d]Number of days under "biotic control" = 365 − (days in succession + days in dyring).
[e]Mean hydrologic characteristics for 5 driest (dry) and wettest (wet) years of record. Rank based on annual runoff. Standard deviation given in parentheses.
[f]Probability that wet and dry years do not differ; Mann–Whitney test, exact *P*.

Hydrologic disturbance and biotic control: Flash flooding

What is the evidence that temporal and/or spatial variation in system state (e.g., postflood succession or drying) produces corresponding variation in communities and in ecosystem functioning? Power et al. (1985; Power and Matthews, 1983) have demonstrated strong biotic interactions across three trophic levels in a small Oklahoma stream susceptible to flash floods. The predator, largemouth bass, is restricted during low flow periods to deep pools but is stochastically redistributed among pools during high flows so that there is pronounced spatial variation in trophic interactions: in the pools lacking bass, algae are sparse because of effects of the grazer *Campostoma anomalum*, whereas pools containing bass are "green" because algivorous fish are scarce (Figure 5). An added dimension is that algae are often nutrient limited and unable to "outrun" effects of grazing (Stewart, 1987). These studies have clearly demonstrated that disturbance and biotic interactions can shift temporally and spatially in controlling an ecosystem function as fundamental as primary productivity. These studies are particularly noteworthy because they address variation at several scales. Too often studies of biotic interaction are conducted at small spatial or temporal scales in ignorance of important constraints that dominate at larger scales (see also Menge and Olson, 1990). A multiscale approach will be essential for deciphering any signal of a changing importance of disturbance in stream communities due to climate change.

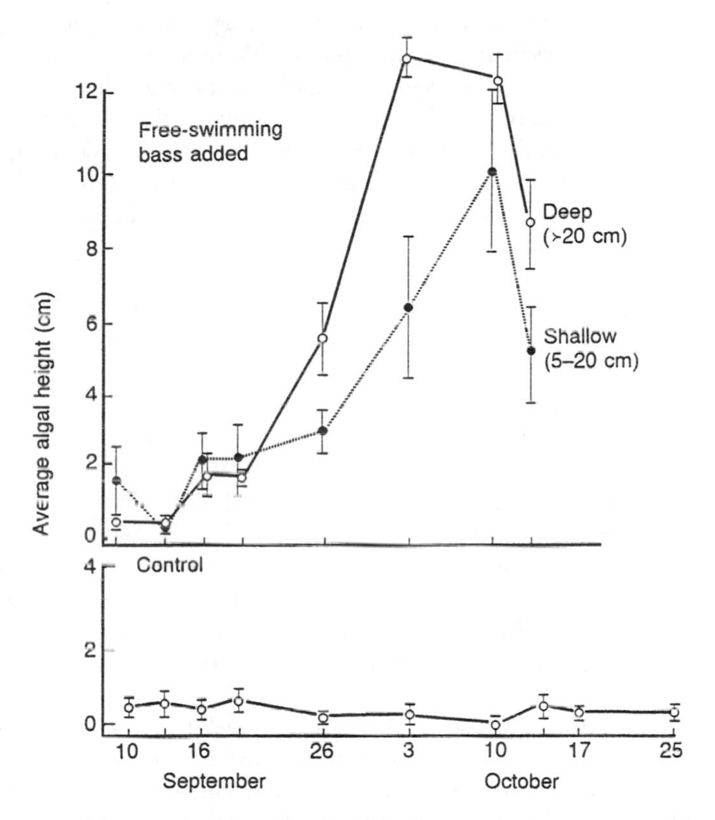

FIGURE 5. Effect on algal height of adding bass to pools containing the grazing fish *Campostoma anomalum*, in Briar Creek, Oklahoma. The addition of the predator reduced grazer densities, which permitted algal growth in the experimental pool (top panel) to exceed that in the control pool (no bass), in which grazing kept algal abundance low (bottom panel). (From Power et al., 1985.)

Another aspect of stream functioning that warrants examination is the connection between nitrogen and algal production. Desert stream algal communities are frequently limited by availability of nitrogen (Grimm and Fisher, 1986; Peterson and Grimm, 1992). Nitrogen availability is spatially variable, with distinct zones of nitrate enrichment where subsurface water upwells from sediments to become surface flow. At these sites, algae recover faster after spates than at downstream sites (Valett, 1991). As water flows downstream, algae take up most of the nitrogen; thus their activity directly influences the availability of nitrogen to downstream communities, which are dominated by cyanobacteria. This important biotic interaction is interrupted by spates, which reduce algal standing crops by more than 80 percent (Grimm and Fisher, 1989).

Similar biotic control of algal assemblages occurs in time. Early successional communities (dominated by diatoms and filamentous green algae) reduce nitrogen levels through uptake and incorporation in accruing biomass, and late-successional communities are dominated by cyanobacteria (Fisher et al., 1982), many species of which are nitrogen fixers (Peterson and Grimm, 1992). Repeated flooding may prevent the establishment of communities dominated by nitrogen fixers for two reasons: flooding opens new space and improves individual access to nutrients, and floods often carry high concentrations of inorganic nitrogen (Grimm, in press).

For invertebrates, successional patterns in community biomass and numbers of individuals also may be related to nitrogen availability. For example, we examined twelve successional sequences and found that numbers and biomass increased throughout the sequence in six cases, but "crashed" in six others. The biomass declines appeared to be related to nitrogen limitation, since late-successional (>30 days after flood) invertebrate numbers were correlated with nitrogen flux (a measure of availability). We hypothesized that the biomass crash was caused by declining food quality due to N limitation of microbial processing in the detrital food (Grimm and Fisher, 1989). It is clear that detritus must be continually reingested by invertebrates, since their consumption rate exceeds primary production (Fisher and Gray, 1983); and egested material is lower in nitrogen than is food (Grimm, 1988). Thus, by decreasing reduced-nitrogen availability, biotic processes (i.e., algal uptake) influence both algae and consumers during late succession after flash floods.

In streams, perhaps unlike terrestrial systems, it is hard to recognize continuous, long-term successional trends. Instead, a better conceptual model is to imagine alternating periods of disturbance and biotic control. A vivid illustration of this point can be found in a series of experiments in Rattlesnake Creek, a southern California stream that experiences high flows during the winter wet season and prolonged periods of low flow. High winter flows reduce algal and invertebrate standing crops (Dudley et al., 1986). Early spring establishment of blackflies occurs on substrata exposed by scour, but these animals are competitively displaced by hydropsychid caddisflies and maintain high densities only with repeated substratum disturbance (Hemphill and Cooper, 1983). During low-flow periods, fish and invertebrate predators strongly regulate the composition and abundance of lower trophic levels (Cooper, 1984; Hemphill and Cooper, 1984), and grazers can influence macroalgae, especially during establishment on newly opened substrata (Dudley and D'Antonio, 1991).

Hydrologic disturbance and biotic control: Drying

Although floods clearly can influence stream communities, initiating successional sequences in algae and invertebrates and permitting competitively inferior or preferred prey species to persist, hydrologic disturbance in the form of drying also strongly influences streams (Boulton and Suter, 1986; Stanley and Fisher, 1992). In fact, drying is potentially more severe in its effects because it results in fragmented or lost habitat. Increased incidence of stream drying may arise from changes in both amount and seasonality of runoff. Arid regions would be particularly sensitive in this regard, not only because many arid-land streams are already intermittent, but because slight changes in rainfall can produce large changes in streamflow (Figure 6; Karl and Riebsame, 1989). Hydrologic data from Sycamore Creek provides another example of the complex linkages between variation in weather patterns and the character of streams.

In most years, Sycamore Creek is continuous during the winter months (December–February) along a 25-kilometer reach between 500 and 1100 meters elevation. In summer, however, permanent reaches are restricted to six sites where bedrock is near the surface of the alluvium. The stream becomes discontinuous when discharge falls below approximately 0.05 m^3/s. Mean daily discharge (by month) varies between extreme wet and dry years and "average" years (Figure 7). Winter rainfall is more variable than summer precipitation, and pronounced differences in runoff among years are due mainly to the occurrence and duration of winter rains (Figure 7; Grimm and Fisher, 1992). Further, prolonged drying reduces available habitat to a few short, permanent reaches in years when summer floods account for most of the water output, because in those years the fate of most rainfall is evapotranspiration. Monthly means of daily discharge were $\leq 0.05\ m^3/s$ (indicating discontinuous flow) for an average of 10 months in dry years, 5 months in average years, and 2 months in wet years. We are just beginning to understand the consequences of this habitat reduction for biotic communities and ecosystem functioning (Stanley and Valett, 1992; Boulton et al., in press).

We examined invertebrate community structure changes associated with season, floods, and drying over a three-year period using detrended correspondence analysis (Boulton et al., in press). We have previously shown

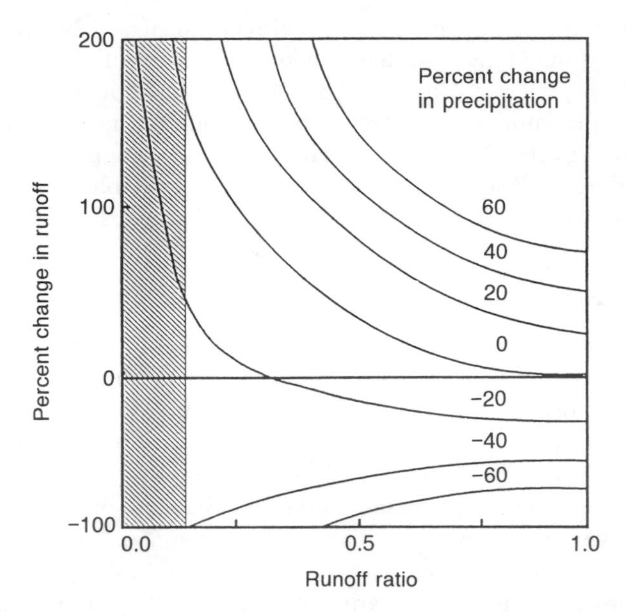

FIGURE 6. Predicted change in surface runoff for various precipitation change scenarios at runoff ratios (runoff/precipitation) ranging from 0 to 1.0. The stippled area encompasses typical runoff ratios for arid-land rivers; thus in deserts extreme changes in runoff are predicted for any given change in precipitation. (After Wigley and Jones, 1985.)

that whole-system variables are not resistant to floods and drying but are very resilient. We wanted to determine whether a community-level property such as community structure (assessed as position of the assemblage in multivariate species-space) showed similar stability characteristics. We found distinct seasonality in the invertebrate community that was little altered by spates. Collective (i.e., system-level) attributes of the invertebrate community (biomass and total density) change dramatically with flooding and exhibit distinct patterns of change during postflood succession (Gray and Fisher, 1981; Fisher et al., 1982; Grimm and Fisher, 1989), yet diversity is seldom reduced by spates (Gray and Fisher, 1981) and community structure is displaced only slightly (Boulton et al., in press). Over the three-year period of the study, however, a shift in fall communities provided an exception to the conclusion that community structure was stable in the face of disturbance (Figure 8). This shift, essentially a decline in mayflies and an increase in the snail *Physella*, corresponded to increased temperature and algal biomass and decreased discharge and nitrogen flux, which were likely related to prolonged drying periods in the summers of 1985 and 1986 (Figure 8).

Surprisingly, then, community structure seems to be more robust in the

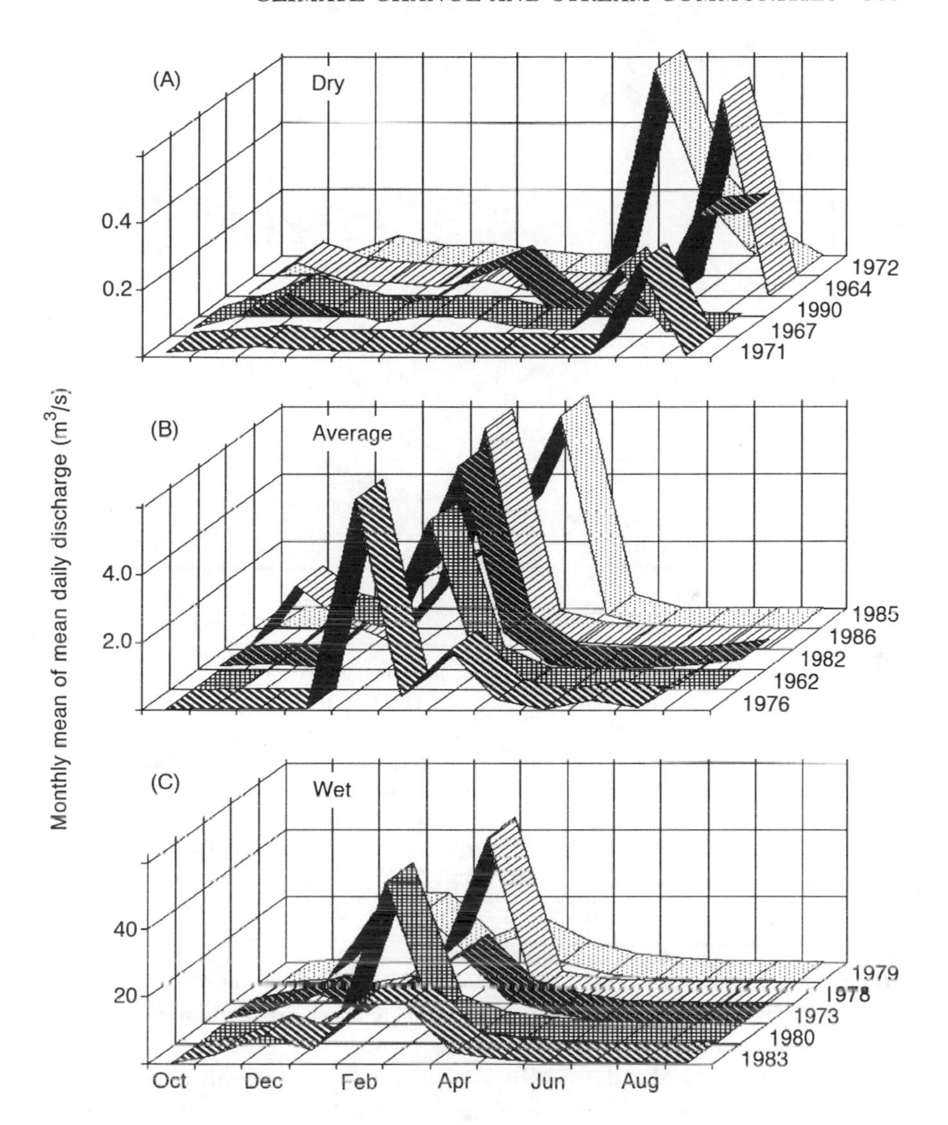

FIGURE 7. Seasonal distribution of daily discharge (monthly means) in Sycamore Creek, Arizona. (A) Dry years. (B) "Average" years. (C) Wet years. The years are ranked along a third axis from lowest (front) to highest (back) total annual runoff. Dry and wet years are those listed in Table 1; normal years are within 2 standard errors of mean annual runoff ($11.9–36.9 \times 10^6$ m^3) for the 29-year record. Note that scales of ordinate increase an order of magnitude each from dry to average to wet years.

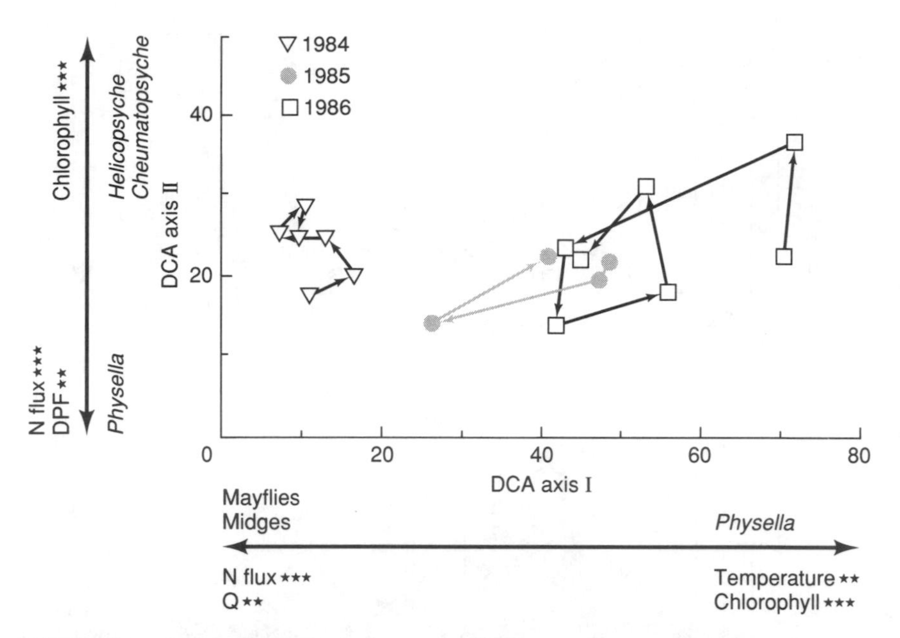

FIGURE 8. Ordination plot of changes in fall invertebrate communities between 1984 and 1986 in Sycamore Creek, Arizona. Points are centroids of five replicate samples. Arrows linking points reflect temporal change (approximately 1 week) in community structure during fall postflood sequences. DCA axes I and II are also labeled with significantly correlated environmental variables; negatively correlated variables are listed at the low end of the axes (**, $P < 0.01$; ***, $P < 0.001$). Indicator taxa (e.g., mayflies, *Physella*) for low and high ends of DCA axes I and II are also listed. (After Boulton et al., in press.)

face of flash flooding than are collective attributes such as numbers, biomass, and productivity. Drying, however, is a more severe disturbance in that it results in community change. Periods of very low flow may introduce stresses including wider daily swings in dissolved oxygen and temperature, higher concentrations of toxic substances, and reduced permeability of substrates to water due to clogging by algae-derived detritus. We do not know whether the observed community change was caused by direct physiological stress or by increasing strength of biotic interactions, but we suspect both were important. As streamflow is reduced during drying, ratios of biota and their waste products to water volume increase. During early stages of drying, this relationship may lead to intensified biotic interactions; for example, mobile grazers move as waters recede and thereby increase in density, whereas algae are sessile and unable to adjust by migrating. There is likely a point at which these stresses exceed the physiological tolerances of some organisms, however, and ultimately, the system may dry completely. Unlike flooding, if dry conditions persist, the aerial source of colonists (adult insects; Gray and Fisher, 1981) is reduced or eliminated.

Biotic control in highly seasonal streams

A special case is presented by streams with a strongly seasonal hydrologic regime. For example, many West Coast streams have winter rainy seasons and summer low flows. Streams fed by snowmelt, such as those in the montane West, and glacial-fed streams (Oswood et al., 1992) have seasonal hydrologic regimes. The dry-season period over which biotic interactions may become significant in these systems is long and its occurrence is predictable (Poff and Ward, 1989). For example, in the South Fork Eel River of northern California, there is a regular seasonal phenology of algae and their consumers (M. E. Power, 1990a and unpublished). In late spring, *Cladophora glomerata* forms extensive turfs of long filaments. As summer progresses, these turfs become heavily infested with tuft-weaving midge larvae if fish that reduce the densities of the midge's predators are abundant. Late summer senescence of the alga may be hastened by these invertebrate colonizers, but regrowth occurs following winter rains and associated high flows that reduce invertebrate populations (M. E. Power, unpublished). Experimental studies of competition in stream insects suggest that this biotic interaction is only seasonally important in two other California streams (Feminella and Resh, 1990; Hemphill, 1991). In one, wet-season flow varied over a five-year period, and competitive interactions were hypothesized to persist through the winter in a dry year (Feminella and Resh, 1990). For these "seasonally biotic" (*sensu* Poff and Ward, 1989) systems, the effects of change in precipitation/flow seasonality, such as those evident in analyses of snowmelt-fed streams (Gleick, 1987; Schaake, 1990), would be profound.

SPATIAL AND TEMPORAL HETEROGENEITY IN STREAM ECOSYSTEMS

The foregoing examples of temporally and spatially alternating and/or overlapping "controls" of stream functioning and community structure illustrate well that streams are heterogeneous systems. This heterogeneity represents a key feature that must be incorporated into any predictive theory about how streams respond to external perturbations. In particular, a beginning point for any study of stream ecosystems ought to be the identification of patches or subsystems. I use these terms to refer to a spatial unit consisting of organisms and abiotic elements, usually bounded in some reasonable way (for example, as a function of relative interaction strength within a patch vs. across its boundaries [White and Pickett, 1985]). Once we have identified the component patches, we can then scale detailed mechanistic studies up to predictions about how organismal interactions with the physical environment will influence ecosystem functioning. A patch can be a small or very large piece of the earth, depending upon the focus of the investigation. For example, landscape ecologists may view whole ecosystems (e.g., lakes) as patches, while in a river such as the Eel, benthic turfs and floating mats of *Cladophora* are discrete patches (Power, 1990b). Whatever the scale of investigation, a patch perspective places small-scale species interactions into a context whereby their effects on larger properties can be readily assessed. Furthermore, interaction between different patches may be critical to large-scale processes. By contrast, the more traditional community approach, in which species interactions per se are the focus, is more relevant to questions of how global environmental change will alter biodiversity.

Patchiness has captured the attention of both community and ecosystem ecologists. Many stream ecologists advocate a patch dynamics approach to studying streams because it provides a link between small-scale, mechanistic research (i.e., "microecology," *sensu* Orians, 1980) and large-scale process measurements (Pringle et al., 1988). In a recent paper, Menge and Olson (1990) called for a predictive framework of community structure that is a hierarchical arrangement of local-scale models (frequently species interactions) nested within more complex, large-scale models (incorporating variability in abiotic factors at larger spatial and temporal scales). Such an approach recognizes that communities of organisms are inextricably linked with the physical environment, and that species interactions are a subset of system dynamics that are particularly evident at certain spatial and temporal scales.

Using a patch dynamics approach for streams, we can begin to understand how adjacent patches interact and thereby predict changes in response to environmental change. Naiman and colleagues (Naiman et al., 1988; Johnston and Naiman, 1990) have done this very successfully for beaver-influenced stream and riparian landscapes by showing how increases in beaver populations have altered the distribution and abundance of various habitat types (e.g., beaver pond, riffle, riparian forest). Historical reconstruction of fluvial landscapes has also revealed how human use has drastically altered distributions of habitat types in stream and riparian systems (Sedell and Froggatt, 1984; Bravard et al., 1986). It has become clear that the "natural" effects of beavers or the "unnatural" effects of humans on fluvial landscapes profoundly alter ecosystem functioning and fish productivity (Pringle et al., 1988; Sedell et al., 1990).

The patch perspective also draws our attention to linkages between different habitats. For example, desert riparian zones are renowned for their bird diversity and are clearly productive systems (relative to the desert matrix) because of their proximity to streams. One of the most profound effects of global climate change and habitat fragmentation might well be alteration of linkages between adjacent ecosystems. In mesic watersheds, stream biota rely on organic matter supplied by surrounding forests for their food base (Fisher and Likens, 1973; Hynes, 1975), but in deserts, there is abundant evidence that terrestrial consumers heavily use stream productivity (e.g., emergent insects; Jackson and Fisher, 1986). Is this because the relative productivity of the adjacent systems determines the linkage direction between them (Grimm and Fisher, 1992)? Perhaps a measure that describes how adjacent systems interact can be developed and used to evaluate future changes. Pulliam (1988) has suggested that population birth/death ratios might indicate connections between source and sink populations. Similar measures for ecosystems and communities (within patches) should be explored with a view to predicting biotic response to environmental change.

DO SPECIES AND BIOTIC INTERACTIONS MATTER?

Do species matter in stream ecosystems? There is considerable evidence from other systems for the importance of individual species to ecosystem functioning (e.g., Vitousek, 1990; Vitousek and Walker, 1989; Carpenter et al., 1985). What characteristics do species that matter share? The collective properties of a spatial unit (ecosystem, landscape element, subsystem or patch) determine its link with surroundings (i.e., what it "does"). Within each unit, species identities may be irrelevant if there is substantial functional redundancy, that is, if several species perform the same process (e.g., inorganic nitrogen uptake) at approximately the same rate. Species matter when they possess unique abilities, when they play a keystone role and/or are the only species of their kind in the species pool, or when they confer physical structure on a system.

Examples of these characteristics of species that matter can be found for stream ecosystems; indeed, many have already been mentioned. Cyanobacteria that are able to fix atmospheric nitrogen enjoy a competitive advantage over nonfixers when nitrogen availability is low. In Sycamore Creek, biomass of primary producers (and presumably, rates of primary production) continues to increase despite limiting nitrogen levels because of increased representation of nitrogen fixers in the community. Under severe nitrogen limitation, when inorganic nitrogen concentration is low—even immediately after flooding—a species of diatom containing cyanobacterial endosymbionts (*Epithemia sorex*, Epithemiaceae) dominates early successional communities (Peterson and Grimm, 1992); this species and three other species of the Epithemiaceae are unique among early successional colonists in their presumed nitrogen fixing ability.

Perhaps the best examples of keystone species in lotic systems are grazing fishes. They deserve that title because they can severely reduce algal biomass. In North America, this group is depauperate in species, especially relative to the South American fauna. Some streams contain a single grazing fish species, while many others support no grazers. In the

Oklahoma stream studied by Power and colleagues (1985; Figure 5) *Campostoma anomalum* is an example of a "strong interactor," for which there is no substitutable species.

Finally, processes or organisms that provide physical structure can be important in ecosystems. Coral reefs are a well-known example of organisms giving a system its structure. In streams, long-term accumulation of wood and its redistribution into "debris dams" defines the structure and retention capacity of small streams in forested regions, particularly in old-growth forests (Bilby and Likens, 1980; Swanson et al., 1976). Beaver dams represent another example of a structure with profound consequences; indeed, beaver activity alters stream hydrology, geomorphology (White, 1990) and thereby nutrient dynamics and productivity (Naiman et al., 1988; Coleman and Dahm, 1990). Algal mats and macrophyte beds are yet another example of ecologically important structures—they alter surface flow patterns and hydrologic exchange between surface and subsurface waters in many streams (White, 1990).

Do biotic interactions affect stream ecosystem functioning? Biotic interactions clearly are important in many streams, but their importance varies in space and time. A lesson from streams is that a patch perspective may be more appropriate for spatially heterogeneous systems than a community perspective. Indeed, the best way to uncover whether species identity matters a great deal is to first distinguish the patches within an ecosystem, and then look at how species function within patches or link together different patches. This perspective allows processes occurring within a patch (biotic or abiotic) and across its boundaries to be placed in a larger context. It also recognizes the potential simultaneous existence of, for example, disturbance-controlled, nutrient-controlled, predator-controlled, and oxygen-controlled spatial aggregations of organisms plus the abiotic environment. The patch perspective is analogous (in space) to the shifting controls of ecosystem functioning (in time) discussed earlier (see Figure 4). Finally, in streams, but perhaps in most ecosystems, biotic interactions often are mediated through or ameliorated by the physical environment (e.g., effects of upstream nutrient uptake on downstream algal communities, fish feeding on drifting aquatic insects, the prevalence of detritivory). In these instances, biotic factors are inseparable from abiotic factors.

CONCLUSIONS

My objectives in this chapter were to outline a general hypothesis for stream community responses to global climate change, and to highlight insights from stream research that are relevant to the consequences of global change for community ecology in general. Here I will briefly summarize some elements of a research agenda that I believe will improve understanding and conservation of species and communities not only for their own inherent worth, but for their part in the functioning of whole ecosystems. Three overall recommendations seem imperative: (1) expand scale; (2) conduct comparative studies; and (3) merge abiotic and biotic realms in conceptual models.

Expanding the scales of our investigations will require that we use whole-system experiments (see Carpenter, 1989; Carpenter et al., this volume) or creative experiments within systems that allow evaluation of factors at a variety of scales without misleading artifacts. This has been a problem for stream ecologists—many common experimental methods such as caging are inappropriate for running waters (Cooper et al., 1990). We should also consider how patches are linked within a system at one scale, and across scales. O'Neill et al. (1986) suggested seeking an explanation for an observation at higher (for context) and lower (for mechanism) adjacent scales or hierarchical levels. We need to expand scale in time, and this requires support of, and attention to, long-term descriptive studies. Coupling experiments with descriptive studies strengthens the informational value of each.

One powerful tool in many fields of science, the comparative method, has been infrequently used in ecology (see Cole et al., 1991). Comparative studies might provide a reasonable means of evaluating future change, if problems of comparing diverse systems can be surmounted. For example, species distributional patterns (Pickett, 1988; Grimm and Fisher, 1992) along spatiotemporal gradients of aridity might reveal trends expected for scenarios of future drought. Comparisons of extant communities with those of other epochs, and of streams today with streams described in the historical record, have given us a fairly clear picture of how communities respond to environmental change (including habitat destruction by humans). Finally, if we are to predict how changes brought about by human activity will affect communities, we should study ecosystems that have already been affected by human activity. This approach can also take advantage of knowledge—in advance of the event—of major anthropogenic disturbances.

As community ecologists, our thinking should include the abiotic realm as a matter of course. Abiotic controls (if they exist, or can be defined) are one of a set of alternative hypotheses. In many ecosystems (streams are a good example), the distinction between biotic and abiotic factors is blurred, and changes in abiotic factors frequently are attributable to biotic processes. Abiotic and biotic realms can be merged by visualizing ecosystems as a spatial aggregation of patches, and focusing on what each patch does in context (i.e., is it a source or sink of phosphorus? of organic matter? of species X? of predators?). The coupling of local-scale and large-scale models suggested by Menge and Olson (1990) is another means of accomplishing this goal.

These recommendations are not comprehensive, nor do they address all objectives of global environmental change research. However, I believe they are essential to successfully linking population and community studies to ecosystem and larger-scale questions. This link is vital because it can establish the importance of species dynamics and biodiversity for essential ecosystem processes and services.

ACKNOWLEDGMENTS

Thanks to the organizers and the National Science Foundation for putting together this enlightening workshop. I thank Steve Carpenter, Emily Stanley, and Peter Kareiva for comments on the manuscript. Stuart Fisher and Steve Carpenter suggested the idea of the "state frequency analysis" and provided other helpful advice.

CHAPTER 19

PALEOECOLOGICAL PERSPECTIVES ON MODELING BROAD-SCALE RESPONSES TO GLOBAL CHANGE

James S. Clark

Models that describe the dynamics of plant communities can help us anticipate long-term effects of environmental change on vegetation. Indeed, because we cannot do large-scale manipulative experiments on climate change, models are probably necessary tools for investigating the ecological consequences of CO_2 enrichment and increasing temperature. Of course, modeling in itself is insufficient to produce a general understanding; any model must be grounded on a firm experimental and observational base. For global climate change, the paleobotanical record offers an especially valuable data base with which to test or formulate models. In particular, the record of recent vegetation and environmental change provides an indication of the vast spectrum of possible responses to changing environments. Thus this chapter begins with a sketch of key insights suggested by the paleobotanical record. Then I turn to a selection of models I think offer promise for exploring how plant communities are likely to respond to future climate change.

MODERN ENVIRONMENTS AND PLANT ASSEMBLAGES ARE UNUSUAL

Existing environmental conditions are strikingly unusual in the context of earth history. Our present-day combination of climates (MacCracken et al.,

1990) and levels of atmospheric carbon dioxide (Berner, 1990; Jasper and Hayes, 1990), methane (Lorius et al., 1990), oxygen (Robinson, 1989), and other constituents differs greatly from climates of the preceding ice ages (the Pleistocene) and of the more distant past. This observation underscores the fact that vegetation patterns have been in a continual state of flux, at a variety of temporal scales. For example, the Holocene represents only the most recent of many short-lived (10^4 yr) interglacial intervals that have punctuated the relatively long glacial stages (10^5 yr) of the late Cenozoic. Because the climate has been in a constant state of change, we can hope to learn about the dimensions of future biological responses by looking at past responses.

Recent vegetation responses to climate change in eastern North America

Biogeographic patterns since the Cretaceous (Crane and Lidgard, 1989), especially in the late Quaternary (Davis, 1978; Jacobson et al., 1987), include vegetation types that no longer exist. The full-glacial spruce woodland of interior North America is an example of a widespread vegetation type that lacks an analogue in our modern vegetation. More open than the boreal forest of the last 6000 years, this biome was also unusual in its co-occurrence of temperate hardwoods and spruce during the early Holocene. Many broad-leaved deciduous tree species probably have northern range limits defined by minimum winter temperatures (Sakai and Weiser, 1973; Larcher and Bauer, 1981; Larcher, 1982; Arris and Eagleson, 1989). Conifers, in contrast, have lower maximum photosynthetic rates at higher temperatures (Sprugel, 1989), and so may not be competitive at lower latitudes. These unique woodlands may be explained by a combination of cold and dry climatic conditions quite unlike anything we know for contemporary interior North America. Changing seasonality during Glacial and early Holocene times (Kutzbach and Guetter, 1986), produced by a different latitudinal distribution of solar radiation and by ice-sheet influences on circulation patterns, could have generated this unusual interior climate.

Full-glacial tundra vegetation also differed from the modern wet tundra. Along the ice-sheet margin were taxa characteristic of xeric steppe environments, e.g., *Artemisia* (Lamb and Edwards, 1988; Webb, 1988a), perhaps reflecting dry easterlies produced by high pressure over the ice sheet (Kutzbach and Wright, 1985). The wet tundra common in modern North American Arctic did not develop until 8000 years B.P. Here again, a different seasonal pattern of temperatures and precipitation could explain different combinations of species than are widespread today.

An explanation for species combinations in the western United States since the Pleistocene requires consideration of topography, a generally more negative water balance, and the climatic consequences of a large continental ice sheet as it influenced the position of the jet stream. The dominating effects of moisture deficit at low elevations (e.g., Stephenson, 1990), combined with topographic complexity, would have produced diverse local responses at small geographic scales superimposed on broader shifts in storm tracks and monsoonal influences. A thick ice sheet directly in the path of the jet stream could have blocked its eastward flow during the full-glacial, sending one branch north along the Arctic coast and another branch to the Southwest. The southern branch of this split jet stream may have supplied moisture to the southwestern United States. In addition, anticyclonic circulation over the ice that brought dry easterlies to the Pacific Northwest would explain xeric tundra along the ice margin (COHMAP, 1988).

Transient responses of vegetation to global change are also evident in the fossil record (Davis, 1978; Webb, 1988a). In the eastern United States, the dominant tree species came to occupy their current ranges by different routes and at different rates. For example, the modern Great Lakes forest contains a mixture of taxa having very different late Quaternary histories. *Ulmus*, *Fraxinus*, and *Ostrya/Carpinus* forest were dominant around the Lake States at the beginning of the Holocene (Figure 1). Other species migrated into the Great Lakes region, albeit via different pathways. For example, while *Quercus* and *Carya* ranges moved northward, partly west of the Appalachians, *Pinus strobus*, *Tsuga*, and *Fagus* migrated up the Atlantic seaboard and then spread westward into Michigan and Wisconsin. *Castanea* arrived in New England only recently, but soon became one of the dominant tree taxa. The different combinations of tree species, together with their different responses to climate change, suggest that new climate patterns could result in new kinds of forest communities having no counterparts on existing landscapes.

Indirect consequences of climate change

Although aspects of these past vegetation changes make some sense in terms of our current understanding of prevailing climates (Webb et al., 1987), that understanding does not translate directly into predictive capability for new climate combinations, partly because climate affects other aspects of the environment as well. For example, when drier conditions prevailed in western Washington during the early Holocene, one would expect *Alnus* to be rare; in contrast, pollen evidence indicates the opposite—an abundance of *Alnus* (Cwynar, 1987). The reason for this "surprise" is perhaps that water-balance does not tell the whole story for *Alnus*. *Alnus* was probably abundant at that time because fire, and therefore early successional environments, favored the species. Fire and/or humus accumulation may also explain the rather sudden decrease in *Abies* pollen and rise in *Picea* in eastern Labrador after 6000 years B.P. (Lamb, 1980; Engstrom and Hansen, 1985). Contemporaneous expansion of the Southeastern pine forest over

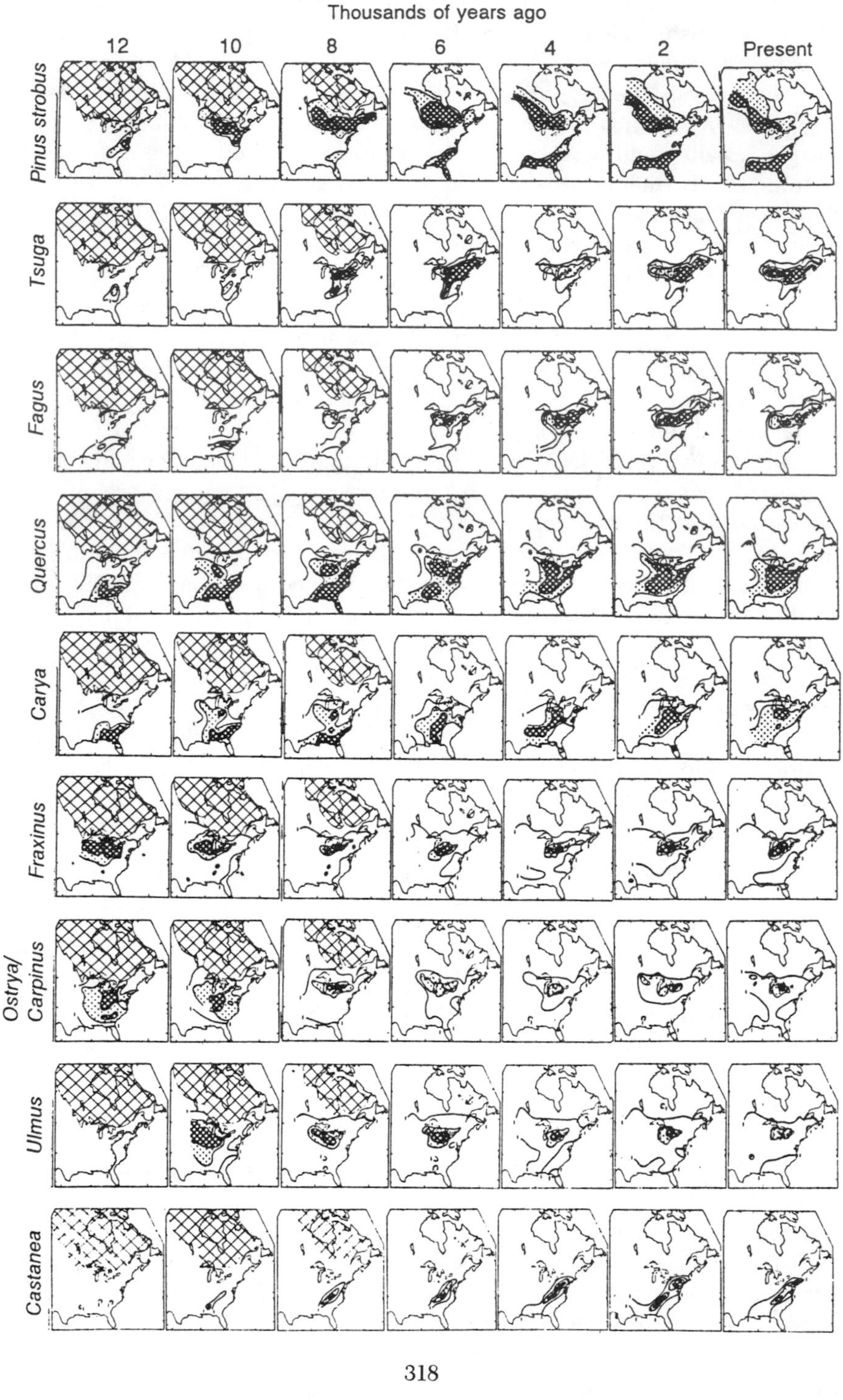

Thousands of years ago

12 10 8 6 4 2 Present

Pinus strobus

Tsuga

Fagus

Quercus

Carya

Fraxinus

Ostrya/ Carpinus

Ulmus

Castanea

oak-dominated assemblages could reflect increasing winter temperatures and perhaps fire (Watts, 1980; Webb, 1988a). In the absence of more precise knowledge of past climates and of their effects on fire and soil organic matter, it is difficult to use paleobotanical observations to predict future vegetation responses to climate change.

Fire becomes an increasingly important indirect consequence of climate along a gradient from mesic closed forest to open woodland to savanna environments. Fire and moisture stress together appear to explain modern grassland/woodland ecotones (Medina and Silva, 1990; Menaut et al., 1990). Dramatic geographic shifts occurred in these ecotones throughout the upper Midwest during the mid-Holocene (McAndrews, 1966; Dean et al., 1984), when the climate was more continental than it is today. Changes in vegetation and fire regime occurred again at 300–400 years B.P. in northwestern (Clark, 1990) and southeastern (Grimm, 1983) Minnesota. More modest changes in fire importance in northwestern Wisconsin (Gajewski et al., 1985) and an absence of changes in northeastern Wisconsin (Swain, 1978) suggest an eastwardly diminishing response of vegetation to climate change. These patterns are consistent with model predictions of increasing sensitivity of ecosystem processes to moisture availability when moisture availability is low (Parton et al., 1987; Pastor and Post, 1988). Observation (Daubenmire, 1937), experimental evidence (Scholes, 1990), and model predictions (Pastor and Post, 1988; Schimel et al., 1990) suggest a complex interaction of water balance, seasonality, fire, soil texture, and topography, due partly to feedbacks between climate and litter quality or nutrient return.

Not only does fire regime influence vegetation, but the plant community can determine local fire frequency. In addition, once established, some vegetation types can alter fire regimes in ways that resist transition to other vegetation types (Grimm, 1983; Kalisz and Stone, 1984). These varied influences make prediction of vegetation response to climate fluctuations difficult, despite the fact that plant communities are highly sensitive to climatic variables.

Atmospheric chemistry may also have influenced the evolutionary history of plants and the emergence of particular plant strategies. For example, C_4 photosynthesis, which allows for higher water use efficiency (WUE) in xeric environments, may have provided no real benefit until atmospheric CO_2 levels began to decline near the end of the Cretaceous (Ehleringer et al., 1991). Indeed, grassland expansions at the end of the Cretaceous may reflect a combined influence of changing climate and CO_2 levels on the

◀ **FIGURE 1.** Maps of selected pollen percentages at 2000-year intervals since 12 thousand years ago. Dense shading indicates higher pollen percentages. The cross-hatched area in the northern section of each map prior to 8 kya represents the Laurentide ice sheet. (From Webb, 1988a.)

fitness of now-dominant C_4 grassland species. However, predicting responses of grasslands and savannas as CO_2 increases even further in the future will not be easy, because any prediction is confounded by the contrasting effects of raised temperatures and CO_2 enrichment on C_3 and C_4 species. In temperate grasslands, WUEs of early-season C_3 grasses are more sensitive to CO_2 than are those of C_4 species; elevated CO_2 might allow C_3 grasses to extend growth later into the season, when C_4 species are now most active. Alternatively, higher temperatures might benefit C_4 species, allowing earlier onset of growth. The combined effects of higher temperatures and CO_2 concentrations might include increased competition between C_3 and C_4 species. Tropical savannas, where trees and shrubs are C_3 and grasses C_4, pose a similar situation (Walker, 1991). The range of CO_2 and climate combinations that have prevailed since the angiosperm radiation of the late Mesozoic suggests that the level of niche overlap between C_3 and C_4 species could have changed many times in the past.

Human activity represents another indirect influence that further frustrates prediction of vegetation responses to global change. A paleobotanical record of responses to climate change in the face of contemporaneous climate effects on human cultures is available from most continents. The expanded role of *Eucalyptus* at the expense of the more fire-sensitive *Casuarina* during the last two interglacials in southeastern Australia might reflect aboriginal burning (Singh et al., 1981; Kershaw et al., 1991) as opposed to more direct climate effects (Clark, 1983). Broad changes in vegetation and fire regime with changes in land use are clearly evident in the fossil record at the onset of the Neolithic in western Europe (e.g., Iversen, 1941; Behre, 1981) and Japan (Tsukada et al., 1986) and following European colonization of North America (Brugam, 1978). Less obvious examples include the responses to aboriginal use of fire that attended human immigration to grasslands and woodlands of Madagascar (Burney, 1987a,b), the potential maintenance of *Corylus* in western Europe by Mesolithic burning (Smith, 1970; Clark et al., 1989; Tallis and Switzer, 1990), and Indian agriculture in North and Central America (McAndrews, 1988). The decimation and subsequent regrowth of North American forests since European colonization has changed communities in ways far more severe than any climate changes that have taken place since the end of the Pleistocene. The current demise of tropical rainforests can be viewed as a continuation of the logging and agricultural practice that saw the near extirpation of forests from western Europe by the Middle Ages, and that rapidly stripped the eastern United States of its natural vegetation within a span of two centuries.

Given the complexities illustrated by the preceding examples, I believe it is unlikely that simple and predictable vegetation changes will attend the environmental changes forecast for the next century. It is nonetheless plausible that models might make adequate predictions for some aspects of population- and community-level responses. Paleoecological efforts can already claim some success at interpreting past climate–vegetation relationships on the basis of modern associations, sometimes even when those past environments do not possess analogues in the modern data. In addition, the increasing predictive capabilities of models for production, decomposition, and water use at broad geographic scales (Running and Coughlan, 1988; Schimel et al., 1990) indicate a growing understanding of ecosystem-level processes. However, we need much more experimental work in order to quantify growth (Jarvis, 1989; Bazzaz, 1990) and nutrient cycling (Norby et al., 1986) responses to rising CO_2. Also, substantial model development is required to better link processes at different scales, such as leaf gas exchange and microbial decomposition to global atmospheric chemistry (Andreae and Schimel, 1989). In the next two sections I discuss directions in which modeling efforts might evolve to better reflect the effects of climate change on plant populations and communities over broad spatial and temporal scales.

MODELS AS A TOOL FOR UNDERSTANDING LONG-TERM EFFECTS OF CLIMATE ON VEGETATION

If models of long-term vegetation responses to environmental change are to be validated, they must make predictions compatible in scale and detail with fossil data. Fossil pollen and charcoal extracted from lake sediments or mires are used to reconstruct vegetation changes at time scales of decades to millennia and at spatial scales of tens of meters to subcontinents. The assumption is that the resulting time series evident in pollen cores from an individual lake represents vegetation changes occurring on the surrounding landscape (e.g., Tauber, 1977; Prentice, 1985). If we accept this assumption, then time series for a location and maps of patterns over large regions (see Figure 1) provide a basis for interpreting changing abundances of tree taxa (e.g., Davis, 1978), for reconstructing climate changes at the same spatial and temporal scales (Bartlein et al., 1984; Webb et al., 1987; Gajewski, 1988), and for assessing models of vegetation dynamics (Dexter et al., 1987; Delcourt and Delcourt, 1987). The interpretation of long-term vegetation change in the foregoing section is an example of one kind of use of these mapped patterns.

A second use of paleoecological data involves tests of general circulation model (GCM) predictions of climate change. Given inputs for past solar radiation, ice-sheet locations and thicknesses, sea surface temperatures, and atmospheric CO_2 concentrations, these models can be used to predict circulation patterns and other aspects of past climate systems (Kutzbach and Wright, 1985; Kutzbach and Guetter, 1986). At the same time, by relating current climates to modern pollen distributions, one can build climate–vegetation correlations that can in turn be applied to fossil pollen data to infer past climates. We can then ask whether GCMs aptly predict these inferred past climates (Bartlein et al., 1984; Webb et al., 1987). Independent evidence from geomorphology and other kinds of fossil evidence (fossil plants, vertebrates, and insects) serves to further refine and/or challenge the results of this approach. Although not specifically aimed at population- and community-level questions, these efforts facilitate understanding of vegetation and climate relationships at broad spatial and temporal scales.

Finally, pollen data can be used to assess the predictions of forest population models. However, predictions at this level of detail may surpass the quality of pollen data. In particular, pollen production and dispersal biases make it difficult to quantify the relationship between pollen abundance in lake sediment and the abundance and distribution of a *population* on the surrounding landscape. What population variable should be used for comparison with pollen abundance? Density, biomass, and basal area (which is proportional to crown area) are all potential candidates. Crown area within codominant height classes probably best reflects pollen production. The inherent size structure of plant populations together with flowering patterns that depend on canopy status make density an unlikely predictor of pollen abundance. Also, the spatial resolution of pollen data does not adequately match the mosaic character of most plant communities. Although simple models of pollen dispersal (e.g., Prentice, 1985) appear promising for interpreting forest composition on homogeneous landscapes, assumptions are violated where topography and shifting spatial patterns confer heterogeneity. Lastly, because few pollen grains can be identified to the species level, the pollen signal may reflect the confounded dynamics of a whole genus. For example, the pollen record of *Quercus* may combine up to a dozen separate species at some locations in eastern North America.

Despite these obstacles, there are some overwhelming patterns in the fossil pollen record that can be used to test semiquantitative predictions of vegetation models. Fossil pollen represents perhaps the only spatially and temporally extensive documentation of vegetation change that is sufficiently abundant to permit statistical analysis of spatial and temporal patterns. Pollen data have demonstrated vegetation responses to ice ages, shorter-term fluctuations in seasonality, cultural impacts, and the invasion of new species, and they hold promise for evaluation of more detailed vegetation models.

The expansion phase of tree populations as they have changed their range limits since the Pleistocene has been the subject of several simple population models. Most commonly, paleoecologists have attempted to describe the phase when pollen first begins to increase with a simple growth equation that ignores spatial dynamics, i.e.,

$$dp/dt = \alpha p(t) \qquad (1)$$

where p is a measure of pollen abundance, and t is time (e.g., years). Rate α may be taken as a constant or as a function of pollen density (e.g., logistic).

A constant α has often been interpreted as a Malthusian parameter or intrinsic rate of increase. This term is not a density-independent population growth rate, because the focal population is changing its spatial distribution, expanding into an area that includes many plant and animal species with which it interacts. The observation that α represents complex interactions does not diminish the potential usefulness of fitting and comparing parameter values (Bennett, 1983; Delcourt and Delcourt, 1987); it simply points out that the interpretation is not straightforward. On the other hand, such models provide no aid for the interpretation of spatial pattern (i.e., migration) vs. local population growth. In particular, an increase in sediment pollen over time could be a consequence either of migration into a region or of population growth sustained entirely by a resident population. Both of these factors may be important, yet the nonspatial model provides no hint as to their respective contributions.

Geographic (spatial) patterns are more readily interpreted in terms of climatic controls and dispersal patterns, although there has been little attempt to relate these patterns to specific population models. The few exceptions include Skellam's (1951) application of diffusion models to examine arguments concerning the Holocene expansion of oaks into Great Britain. This model represented population growth rate as the sum of local growth rate and diffusion in space,

$$\partial p/\partial t = \alpha p(x,y,t) + D[(\partial^2 p/\partial x^2) + (\partial^2 p/\partial y^2)] \tag{2}$$

where p and α are as in Equation 1, (x,y) represents a spatial coordinate, and D is a dispersal rate, or more technically, the diffusion coefficient. One interesting prediction of this model is that the asymptotic rate of expansion for the range of any taxon obeying Equation (2) (assuming some threshold density required for detection) is:

$$\text{Velocity of range expansion (distance/time)} = (4D\alpha)^{1/2} \tag{3}$$

Skellam used this prediction to suggest that oak expansion into Great Britain must have been facilitated by animal vectors, since the observed velocity implied a dispersal rate too great to result from the average distance of dispersal by wind. This theme arose again, independently, in the paleoecological literature some years later (Davis, 1981; Webb, 1986; Birks, 1989).

This model is worth reexamination considering the synthesis of fossil pollen data that has occurred since Skellam's early work. Consider, for instance, Davis' (1981) rough estimates of Holocene tree range expansions at rates from 100 to 400 m/yr^{-1}. Average dispersal distances, $(4D)^{1/2}$, depend on seed characteristics and on the nature of the dispersal vector, such as drag and height of release (for wind dispersal) and foraging behavior (for animal dispersal). Wind dispersal in closed stands produces average values on the order of 10^2 m for seeds of many forest trees (e.g., Okubo and Levin, 1989). The observed rates of pollen increase in lakes as pollen taxa first appeared are of the order 10^{-2} to 10^{-3} yr^{-1} (Bennett, 1983; Delcourt and

Delcourt, 1987). If a diffusive process alone were responsible for tree migrations during the Holocene, the observed rates of 100–400 m/yr^{-1} imply (1) higher population growth rates than are suggested by fossil pollen data, (2) higher dispersal rates than are typically observed for wind-dispersed seed in closed stands, or (3) both (Figure 2). Thus, the model places bounds on parameter values that can be compared with modern observation to suggest that average dispersal distances cannot explain the rapid spread.

A second example is Dexter et al.'s (1987) use of an advection–diffusion model to fit parameters for the expansion of *Fagus* pollen across eastern North America in the early to mid Holocene. Here α is taken as density independent, but is permitted to vary in space and time. A biological definition of the advection term is not altogether clear, unless perhaps dispersal is to be viewed as being driven along by prevailing winds. This diffusion–advection analysis represents an important first step toward quantitative application of simple spatial models to fossil pollen data. In addition, the geographic data base used by Dexter et al. (1987), together with growing appreciation for seed dispersal patterns, suggests there is room for further development of the approach. For example, future efforts might consider leptokurtic seed dispersal, which differs from diffusion by allowing rare long-distance dispersal of propagules. The model given by Equation 2 could

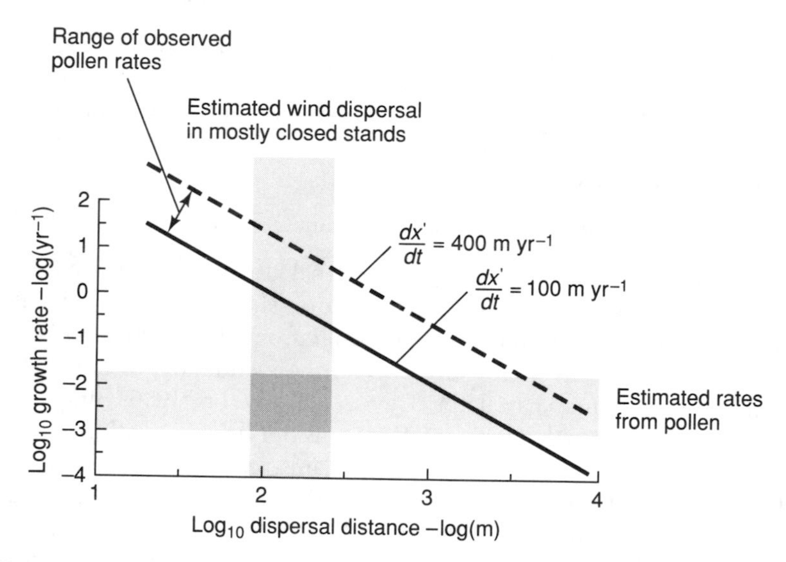

FIGURE 2. Rates of tree range extensions dx'/dt during the Holocene (ranging from 100 to 400 m^2 as reported in Davis, 1981) compared with predictions of Equation 3 given a selection of parameter estimates from pollen data and modern dispersal distances. The region where shaded areas overlap falls below observed rates.

thus be made more versatile by introducing the function $s(u)$, which is a probability density describing the likelihood that seed moves a distance u away from its parent tree. We can now represent population growth and dispersal as follows:

$$\partial p/\partial t = \alpha p(z,t) + \int_{-\infty}^{+\infty} s(u)p(z - u,t)du \quad (4)$$

(Murray, 1989). In this model, $s(u)$ describes how pollen density at location $z - u$ affects density (through dispersal) at location z. Upon expansion of $p(z - u)$ about z we arrive at a series of terms containing the moments of s. If s is symmetric, odd moments equal zero, and we can approximate Equation 4 by truncating our series expansion after the fourth moment,

$$\partial p/\partial t = \alpha p(z,t) + s_0 p + s_2(\partial^2 p/\partial z^2) + s_4(\partial^4 p/\partial z^4) \quad (5)$$

where s_j is the jth moment of s, s_4 being kurtosis. Although this model has not been used for modeling population expansions since the Pleistocene, consideration of the leptokurtic nature of seed dispersal might sharpen debate concerning interpretation of population expansions (Watts, 1973; Davis, 1981; Bennett, 1988a,b; Webb, 1986, 1988b). For instance, leptokurtic dispersal can allow a population to be almost immediately present as climate ameliorates far from its initial range, despite restricted average dispersal distances.

These considerations lead to several recommendations for further development of spatially structured models for the study of Quaternary vegetation patterns.

1. Nonspatial models (e.g., Equation 1) are unlikely to provide much further insight into the processes of shifting ranges. On the other hand, the utility of nonspatial models for addressing other kinds of problems may expand as parameterizations become available for climate and atmospheric effects on population processes (see next section).
2. Density dependence must ultimately be incorporated in some fashion in the growth term α, unless the model is applied only to the advancing "front" of the population. (Interestingly, density dependence does not alter the prediction about the speed of range expansion or "front movement.")
3. The leptokurtic aspect of dispersal may be critical for modeling population expansions and is consistent with observations of spread more rapid than can be explained by mean dispersal distances (e.g., Skellam, 1951; Davis, 1981; Webb, 1986).
4. Efforts to identify climate changes by methods independent of fossil pollen data could help us distinguish between dispersal processes and climate effects as the primary determinants of spatiotemporal patterns in pollen abundance.

Some of these research directions will be difficult to pursue because the biotic or abiotic environment is poorly understood, or because the pollen–vegetation relationship is difficult to quantify. Density dependence, for example, encompasses a complex set of interactions among many species that is better addressed using more mechanistic models (next section). Although point 4 appears obvious, it is crucial to realize that shifts in pollen abundance can be due to either migration or local population growth, and our models must allow us to examine both of these possibilities. Thus, by assuming that geographic patterns are driven largely by spatial gradients in abundance, rather than by climate dependencies in growth rate α, models like Equations 2 and 4 may fail to capture the true climate effects on local population dynamics.

Multispecies models are also needed in order to understand biotic interactions at broad spatial and temporal scales. Although difficult to quantify, the decrease of some pollen taxa with the arrival of others may well reflect competitive interactions that could be related to specific model predictions. Changing species abundances reflect direct climatic effects on individuals together with indirect biotic factors (e.g., competition, herbivory, host–parasite interactions). In order to isolate the biotic component, models must contain both aspects, which would demand parameterization of growth responses to climate for many species. Thus far, simple (e.g., Lotka–Volterra competition) models have been applied to pollen data, but not with consideration of climate effects on growth; instead, community-level responses have been addressed using models containing more mechanistic detail and at smaller spatial scales, as discussed in the next section.

POPULATION AND COMMUNITY MODELS AT SMALLER SCALES

Many of the more mechanistic models of gas exchange and plant growth emphasize the effects of CO_2 and climate change on physiology and ecosystem-level processes, including photosynthesis and respiration (Jarvis, 1989; Eamus and Jarvis, 1989; Ryan, 1991; Long and Hutchin, 1991), nutrient and hydrologic cycles (Running and Coughlan, 1988; Schimel et al., 1990; Ågren et al., 1991; Walker, 1991), and spatial patterns that determine how local processes scale to landscapes (e.g., Norman, 1989; Hicks, 1989). Much less attention has been devoted to population- and community-level processes at broad regional scales (Pastor and Post, 1988; Shugart, 1990). Here I consider population- and ecosystem-level processes that are potentially highly sensitive to global change. I focus on relationships that require better understanding if we are to parameterize models that can serve as tools for understanding biotic consequences of global change.

The modeling tradition in forest ecology began with simulation approaches that strive for realism (Botkin et al., 1972; Shugart, 1984), and this tradition continues to make a strong contribution to problems of forest dynamics (Shugart, 1984; Huston and Smith, 1987; Smith and Huston, 1989), nutrient cycling (Pastor and Post, 1986), and the effects of global change (Solomon, 1986; Pastor and Post, 1988; Shugart, 1990). These "gap" models contain large numbers of functional dependencies and parameters for simulating the growth, mortality, and recruitment dynamics of patches the size of a mature individual tree. It is not possible to adequately measure most of the parameter values, a problem that is handled in different ways depending on the particular process. For example, seed production and dispersal is highly variable and difficult to quantify in closed stands. The gap models therefore assume a globally available seed pool. Climate effects on tree growth are not known to an extent that could be used to model growth responses throughout the life of a tree. Thus, some simple dependencies drawn from correlations between a species' geographic range and corresponding degree-days are used to parameterize growth of individual trees.

An advantage of these models is the promise they hold for exploring a portion of the rich spectrum of indirect effects of climate as channeled through biotic interactions and subtle effects on nutrient cycling (e.g., differential limitation and quality of litter produced). Pastor and Post (1988), for example, demonstrated that the positive feedbacks of nutrient cycling on species composition might exaggerate the responses of forests to climate change, depending on soil moisture availability. Indirect effects of growth and resource competition on successional patterns have also been postulated from such models by Huston and Smith (1987) and Smith and Huston (1989). Increasing application of this class of models to predicting effects of global change suggests several areas for further development, some of which are already under consideration.

Recruitment

Although seed and seedling sensitivity has been recognized as having a potentially important role during climate change, the lack of representation of seed production and dispersal in the models used to assess forest response reflects a deficiency that must be addressed, particularly in light of the highly fragmented landscape occupied by much of the world's forest vegetation (Groom and Schumaker, this volume). The profound importance of seed production and dispersal for plant population dynamics has been demonstrated repeatedly, both theoretically (e.g., Shmida and Ellner, 1984; Comins and Noble, 1985) and in the real world (Harper, 1977; Grubb, 1977). Numerous studies have suggested that climate-sensitive regeneration success can alter the entire physiognomy, composition, and population structure of plant associations at regional scales for periods lasting decades to centuries (Wardle, 1963; Piggott and Huntley, 1981; Payette and Gagnon, 1985; Crow, 1988).

Seed production varies temporally, spatially, and among species. Because of the intermittent seed production (masting) of many tree species, extreme interannual variability in seed set is commonly observed. In addition, seed production depends on climate, nutritional status, and developmental history of the parent plant (e.g., Harper and White, 1974; Piggott and Huntley, 1981). Climate change can result in poor recruitment success because of negative effects on seed development (Piggott and Huntley, 1981) or on seedling establishment (Wardle, 1963). Given the episodic nature of regeneration opportunities in forests and of good years for seed production, probabilities that high seed production and seed establishment will coincide can be extremely unlikely.

The interactive effects of climate variability, disturbance, and seed production on recruitment may be strikingly obvious at range boundaries. For instance, the regeneration success of *Picea* at its northern range limit (Payette and Gagnon, 1985) and near alpine tree lines (Agee and Smith, 1984) appears closely tied to climate and fire. Another example involves *Tilia*'s range limit in northwest England, which may be set by low temperatures that cause failure of seed production in *Tilia* (Piggott and Huntley, 1981). Although these effects are especially apparent at population frontiers, seed production and dispersal probably have strong effects on species composition and heterogeneity in all forests.

Tree growth and competition

The competitive abilities of trees depend on growth rates, resource use efficiencies, stored energy reserves, phenology, root distribution, canopy architecture, and an allocation schedule that directly and indirectly determines changing rates of growth, fecundity, and mortality (i.e. "life history"). The advantage of a particular schedule of energy allocation to somatic growth, defenses, storage, and reproduction depends on physical and biotic factors that are expected to change in coming decades as a result of altered climate and habitat destruction. The effects of climate changes on growth and competition depend on the time of year in which those changes occur (Cook and Cole, 1991). Lack of data regarding these influences at the appropriate spatial and temporal scales (Jarvis, 1989; Long and Hutchin, 1991) presents one of the principal challenges to experimentalists and modelers of population-level responses to global change.

The transient dynamics of growth and competition are also poorly understood. The effects of variability (Jarvis and Sanford, 1986; Grulke et al., 1990; Long and Hutchin, 1991) such as changing cloud cover, droughts and extreme winter temperatures depend on response times at the individual and population levels. For example, the effects of cloudiness on gas exchange and therefore on soil moisture depend on how clouds are distributed throughout the day and with the changing seasons. Low-temperature extremes and droughts can have long-lasting impacts that are difficult to parameterize from existing evidence. The effects of climate on disturbance regimes are also poorly understood (see next section). Because responses to environmental fluctuations can be protracted, extreme events may be among the most important influences on future community structure.

Although primarily discussed by physiological and ecosystem ecologists, the role of vegetation feedback on climate and resource availability should not be ignored by population biologists. Heat balance depends on albedo and surface roughness, controlled by canopy reflectance and architecture. Transpiration has a feedback effect on itself through its contribution to CO_2 concentration and vapor-pressure deficit within the boundary layer. Litter amount and quality affects microbially mediated decomposition processes, which in turn affect the availability of nitrogen. For example, CO_2-enhanced productivity on poor soils could lead to diminished litter quality, thereby favoring better nitrogen competitors (Walker, 1991). These "ecosystem effects" could easily influence competitive relationships—a realization that should foster close coordination of population- and ecosystem-level models.

Disturbance

Disturbance is a common and important factor in tree recruitment and mortality; thus, some treatment of disturbance in population models is necessary. Because climate and atmospheric chemistry affect the intensity and time and space distributions of most kinds of disturbances, global change is expected to alter forest composition through its influence on disturbance regimes. Some of the ways in which climate influences disturbance regimes are rather direct. Gusty winds and saturated soils, for example, are among the more direct climate effects contributing to blowdown of trees. Fire regimes also depend on rather direct climate influences, including lightning ignition and moisture availability.

More complex and unpredictable are the indirect climate effects on disturbance regimes represented by the ways in which climate influences the biota. Tree vigor, size, and canopy structure are important biotic components of blowdown regimes. Fire depends on indirect climate controls over fuel production and decomposition. Fuel moisture content depends on particle-size distributions of woody debris, litter type and amount, and humus depth. Plant tissue mineral content, which determines flammability, partly reflects overriding effects of regional climate on vegetation composition. These biotic components of disturbance regimes integrate climate and atmospheric CO_2 influences on growth and allocation.

The distribution of disturbance events in space and time depends on an interaction between these direct and indirect effects. If climate were the sole cause of disturbance, and the climate system contained no internal time scales of its own (e.g., the El Niño southern oscillation), we might expect a constant probability of disturbance from year to year. If lightning-caused fires were independent of fuel conditions, the probability of fire from year to year might also remain rather constant (disregarding for the moment trends in climate conditions). Such a constant probability of disturbance would result in a distribution of waiting times between events that is exponential. Alternatively, changes in the biota following disturbance would cause the probability of another disturbance to vary with the amount of time that has elapsed since the last disturbance event. For example, the probability of blowdowns may remain low within a young stand until trees reach a certain size (Foster, 1988). Blowdown probability can also rise because of decreased growth rate and vigor that predisposes a tree to other mortality agents (e.g., Loehle, 1988). The probability of two fires in close succession may be low due to inadequate time for the development of suitable fuel conditions (e.g., Heinselman, 1973; Romme, 1982; Agee and Huff, 1987; Clark, 1989).

Together, the direct and indirect effects of climate on disturbance determine the severity and distribution of disturbance events in space and time. If the frequency of convection events increases, as is forecast by GCMs for some regions, the average frequency of blowdowns and fire might increase (Overpeck et al., 1990; Gates, 1990). Any change in the disturbance regime is likely to affect population and community dynamics (Johnson, 1979; Sousa, 1984; Clark, 1991). For example, the higher moments of fire regimes changed with climate during the last 750 years in northwest Minnesota (Figure 3). The warm and dry fifteenth and sixteenth centuries were characterized by frequent fires at regular intervals. In contrast, with cooler and moister conditions beginning about 1600 A.D., fires became clustered at longer intervals, so that the regime included times when fires were frequent and other times when they were not (i.e., the variance in waiting times between fires became high).

These effects of climate and biota on disturbance combine with human influences to affect the size distribution of disturbances. There are many obvious ways in which this can occur, both through intentional management of fuels and through manipulations unrelated to disturbance hazard. For example, in regions where fire is suppressed in southern California, burns are much larger and more catastrophic than in neighboring Baja California, where fires are permitted to burn (Minnich, 1983). Indeed, several studies suggest that burns tended to be smaller prior to the fire suppression efforts of the twentieth century (Agee, 1981; Clark, 1990).

Only if we can quantitatively represent the direct and indirect effects of climate and atmospheric chemistry on disturbance regimes will we be able to effectively predict biotic responses to climate change. For example, a strongly nonlinear response of vegetation to climate at the prairie–forest

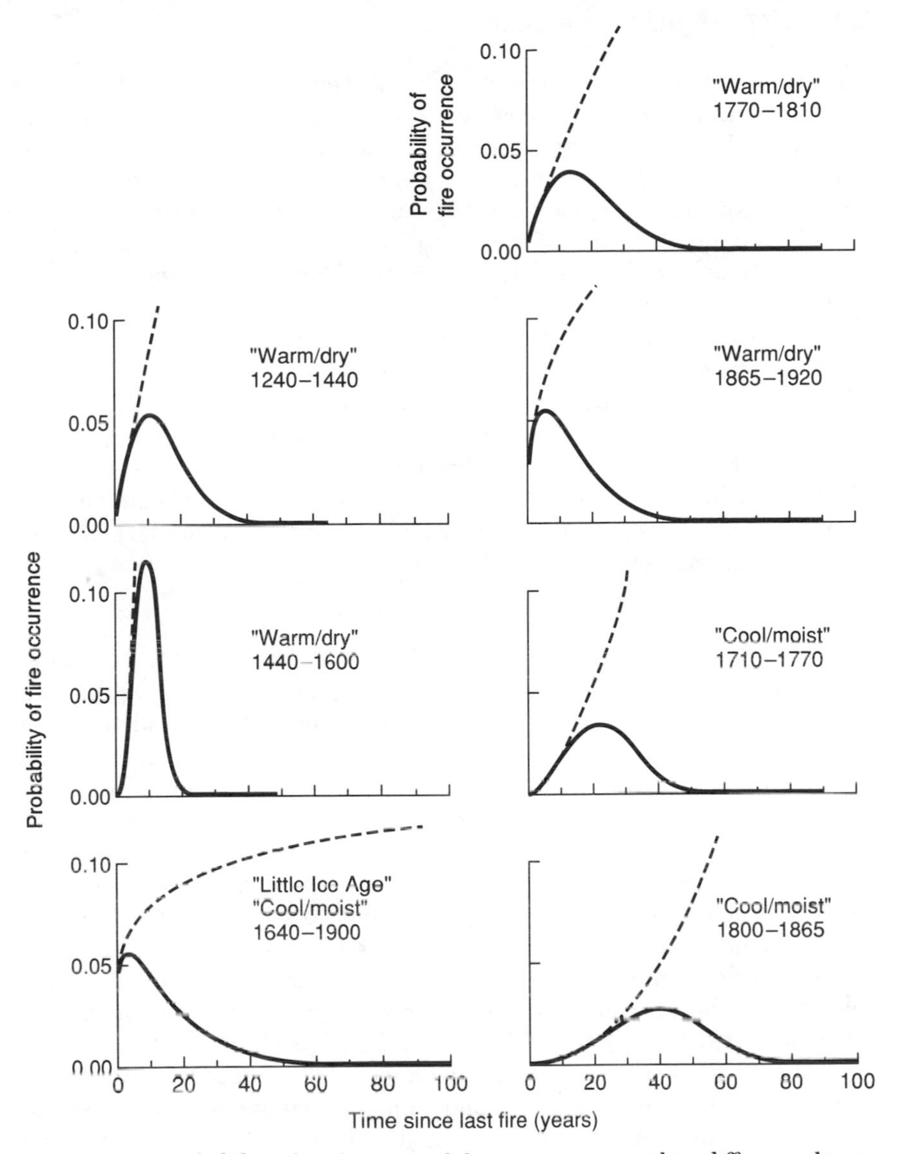

FIGURE 3. Probability distributions of fire occurrence within different climate settings over the last 750 years in northwestern Minnesota. At longer time scales (left), the warm, dry period of A.D. 1440–1640 is characterized by frequent fire and low variance for return intervals. Since the onset of cooler, moister conditions beginning about 1640, a larger variance for return times reflects periods when fires were frequent and other times when intervals between fires were long. Within this most recent period, shorter-term climate changes occurred that also influenced distributions of fire (right). The warm, dry times of 1770–1810 and 1865–1920 had generally shorter intervals between fires (top right) compared with the intervening periods when the climate was cooler and moister (bottom right). (After Clark, 1989.)

border is probably controlled by fire. The fire effect is suggested by abrupt vegetation transitions that follow fire breaks (Grimm, 1984) and striking vegetation changes that attend changes in fire regime over time (Grimm, 1983; Clark, 1990). Trees readily invade former tallgrass prairie once fire is excluded (Grimm, 1984; Crow, 1988), further suggesting that fire, rather than climate directly, explains the prairie–forest border. On the other hand, gap models also predict a strong sensitivity of forest vegetation to climate at the prairie–forest border, but do so without any consideration of fire. The sensitivity of gap models at the prairie–forest transition reflects instead the direct effects of climate on tree growth. By assuming that the climate effect is on tree growth, gap models cannot adjust their predictions to different patterns of fire suppression, and in fact assume that fire suppression is irrelevant to the community dynamics at prairie–forest boundaries. Clearly, in order to predict the ultimate consequences of climate change, we will need to sort out these direct and indirect ways in which climate influences disturbance regimes.

SUMMARY

Models must play a key role in research aimed at understanding the consequences of global change for vegetation at broad spatial and temporal scales. Despite many limitations, fossil evidence for vegetation responses to climate change during the late Quaternary represents one of the better opportunities for hypothesis generation and model testing. Local changes in pollen abundances and shifting geographic patterns provide opportunities to explore the complexity of responses to environmental change. Spatially structured models might be further developed to better accommodate aspects of climate and dispersal processes at the appropriate spatial and temporal scales. An understanding of indirect climate effects on vegetation will require the refinement of more mechanistic models that operate at more local spatial scales. The feedback effects of vegetation on the environment are perhaps critical, even for population-level models, because of their importance for nutrient cycling, and hence, competition. The need for quantitative estimates of growth rates, disturbance regimes and so forth places a continuing burden on data collection and the need for model parameterization. Model development must proceed in step with experimental work, both as a guide for continuing research and as a tool for translating local phenomena into broad, long-term vegetation patterns.

ACKNOWLEDGMENTS

For their helpful discussions I thank S. Pacala, P. Kareiva, C. Reid, and the conference participants. Preparation of this chapter was supported by NSF grant BSR-9107272.

CHAPTER 20

CARBON DIOXIDE LIMITATION AND POTENTIAL DIRECT EFFECTS OF ITS ACCUMULATION ON PLANT COMMUNITIES

David Tilman

The burning of fossil fuels has caused the atmospheric concentration of CO_2 to increase, from 280 ppmv to 350 ppmv during the Industrial Revolution, with a current rate of increase of about 4 percent per decade (Siegenthaler et al., 1987). This increase may have a profound effect on global climate (e.g., Schneider, 1989), and thus, indirectly, on plant communities (e.g., Huntley, 1991; Webb and Wigley, 1985; Schneider, this volume). In addition, CO_2 accumulation may have a direct impact on plant communities because CO_2 is an essential nutrient for plant growth.

The potential effects of elevated atmospheric CO_2 might be inferred from other cases of nutrient enrichment. Phosphate is a major limiting nutrient in freshwater lakes, rivers, and streams. Increased rates of phosphate loading have led to dramatic changes in aquatic habitats (e.g., Edmondson, 1961, 1969; Edmondson and Lehman, 1981). As a result of phosphate inputs, many previously unproductive lakes have become highly productive, with diatoms and green algae being replaced by cyanobacteria (blue-green algae), which often form dense surface scums. In another example, many heathlands in the Netherlands have been replaced by grasslands or forest, apparently as a result of increased rates of atmospheric nitrogen deposition (Berendse and Aerts, 1984; Aerts and Berendse, 1988). Nutrient additions have also been shown to change grassland species composition and decrease grassland diversity (Tilman, 1982). In the Park Grass Experiments of Rothamsted, England, for instance, many plots that initially contained over 40 species have become virtual monocultures following long-term, heavy fertilization.

In general, the addition of a limiting nutrient is thought to cause a shift in the composition and diversity of a plant community because each plant species responds to the addition in a different manner. All plants require carbon, nitrogen, phosphorus, potassium, calcium, magnesium, sulfur, and other elements for growth, but different species require these nutrients in different ratios (Tilman, 1982). A species is predicted to become competitively dominant if the ratio of supply rates of its limiting nutrients corresponds to the ratio in which that species uses them. An increased supply of any one limiting nutrient changes the ratios of nutrient supply and thus can have profound effects on competitive interactions and community composition (Tilman, 1982). If CO_2 limits growth in terrestrial or aquatic habitats, then the increased atmospheric CO_2 concentrations produced by anthropogenic global change will cause species to be limited by a resource other than CO_2. The best competitor for that newly limiting resource would displace other species limited by it.

Thus, if CO_2 is a growth-limiting resource, both experimental evidence and theory suggest that CO_2 accumulation might have a direct effect on the composition and diversity of plant communities. However, existing theory regarding resource competition includes several critical assumptions that may not apply to CO_2. This chapter proposes some alternative theories that may provide insights into the possible effects of CO_2 accumulation on the composition of plant communities. Although such resource-based theory may eventually be able to predict the outcome of CO_2-dependent interactions, its most immediate contribution is to lay bare critical assumptions and processes, and thus to guide field experiments.

IS CO_2 LIMITING?

Before developing this theory, it is important to ask whether CO_2 is a limiting resource in terrestrial or aquatic habitats. If CO_2 were not limiting, changes in the supply of CO_2 would not have any direct effect on resource-dependent interactions among plant species. An essential plant nutrient is considered limiting if increases (or decreases) in the abundance of the resource cause changes in the specific growth rate of a species. For plants, the specific growth rate is usually best measured as $dB/dt \cdot 1/B$, where B is biomass per unit of habitat area. A resource is considered limiting in a habitat if one or more of the species in the habitat are limited by it.

Literature reviews by Bazzaz (1990) and by Strain and Cure (1985) show that many wild terrestrial plants can be limited by CO_2, at least when grown in greenhouses or environmental chambers. Crop plants also can be CO_2-limited under controlled conditions (Acock and Allen, 1985; Cure and Acock, 1986). The underlying physiological causes of such limitation are complex because CO_2 influences numerous processes, including photosynthesis, dark respiration, stomatal conductance, transpiration, allocation, nutrient conservation, phenology, and reproduction (Bazzaz, 1990). Because plant species may differ in their growth responses to CO_2, CO_2 enrichment may influence the composition of plant communities. Zangerl and Bazzaz (1984) showed that different CO_2 concentrations changed species composition in laboratory communities of annual plants, especially on fertile soils or at high light intensity. In a greenhouse experiment, Wray and Strain (1987) showed that CO_2 levels increased the growth rate of a C_3 relative to a C_4 species, but did not alter their competitive interactions. Hunt et al. (1991) compared the responses of 25 native British plants to CO_2 in a greenhouse study in which all resources other than CO_2 were amply provided. They found major interspecific differences in the degree to which plant growth was influenced by CO_2 concentrations. Plants that they classified as "competitive" (generally taller pasture species that do well in short-term competition experiments in fertile soils) had the greatest growth response to CO_2.

Under controlled laboratory conditions, CO_2 limitation is more prevalent when water, other nutrients and light are provided at high levels (Bazzaz, 1990), as would be expected for any essential resource (Tilman, 1982; Bloom et al., 1986). These results suggest that CO_2 limitation might be more common in productive than in unproductive ecosystems. This prediction is in accord with the small growth responses to CO_2 enrichment observed in unproductive arctic ecosystems (Oberbauer et al., 1986; Billings et al., 1983; Peterson et al., 1984), and with the greater growth increases observed in *Scirpus* (but not in *Spartina*) in a more productive estuarine marsh ecosystem in Chesapeake Bay (Curtis et al., 1989a,b). Shapiro (1973) found that injection of CO_2 into a lake caused more pronounced changes in the abundances of the dominant species when the treatments also included other nutrients.

However, there have been few field experiments in which CO_2 has been added to natural plant communities and the responses of individual plant species observed (see Bazzaz, 1990; Woodward et al., 1991). Thus, at present, it is impossible to determine whether CO_2 limitation is more prevalent in particular types of habitats, or even whether many natural plant communities have a significant number of CO_2-limited species. Although the available evidence suggests the plausibility of CO_2 limitation for many plant species, it is imperative to test this experimentally in the field. By definition, essential plant nutrients affect plant growth; therefore it is not at all surprising that CO_2, or any plant nutrient, can influence interspecific competitive interactions under appropriate laboratory conditions. However, for an essential nutrient to influence the dynamics or structure of natural plant communities, that nutrient must be limiting under field conditions (Tilman, 1982, 1988). Future research must go beyond laboratory demonstrations by directly observing the effects of CO_2 enrichment in natural communities.

SHOULD CO_2 BE LIMITING?

My emphasis on demonstrating limitation in nature as a prerequisite for further work on CO_2 is also based on the empirical observation that sometimes the addition of essential resources does not produce any change in plant growth rates or community composition. In particular, when a resource is not limiting (and is not brought to toxic levels), supplementing that resource will have little, if any, direct impact on a plant community. For instance, for the past decade I have added phosphorus, potassium, calcium, magnesium, and sulfur, singly and in combination, to replicated plots in three successional grasslands and a stand of native prairie. There has never been a significant effect of any of these nutrients, singly or in combination, on the aboveground biomass either of individual species or of the total community (Tilman, 1990). In contrast, other plots that received nitrogen showed significant increases in total community biomass and significant shifts in species abundances and diversity (Tilman, 1988). In numerous aquatic and terrestrial habitats, nutrient addition experiments have produced similar results: Only one or a few nutrients are limiting, and only these nutrients influence community structure (e.g., Milton, 1934; Thurston, 1969; Spect, 1963; Shaver and Chapin, 1980; Tilman, 1982; Chapin and Shaver, 1985; Tamm, 1985).

This finding seems inconsistent with a growing body of theory on plant optimal foraging (Rapport, 1971; Tilman, 1982, 1988; Bloom et al., 1986; Chapin et al., 1987). One prediction from this theory can be called the "equal limitation hypothesis," which states that a plant should forage for essential resources so as to become equally limited by all of them. The premise underlying "equal limitation" is the presence of allocation plasticity in plants, which can be based on physiological and/or genetic variation. If a plant is limited by one essential resource and not limited by some other one, it could gain fitness by allocating more to the structures or functions involved in the acquisition or efficient use of its limiting resource and allocating less to acquiring or using nonlimiting resources. The optimal outcome of such shifts in allocation is the morphology and physiology that leads a plant to be limited simultaneously by all essential resources.

There is an apparent marked contrast between this prediction and the empirical observation that only one or a few resources are demonstrably limiting to plant species in most habitats. However, Gleeson and Tilman (1992) have shown that optimal foraging by plants does not necessarily result in equal growth responses to the addition of all essential resources. In particular, modifications that add biological realism to plant optimal foraging models lead to the prediction that some essential resources may not be limiting in a habitat. There are two major reasons for this. First, a plant's growth response to the addition of a resource should be proportional to the allocation that can be saved because of the higher abundance of that resource. If there is plasticity for patterns of allocation within the plant, then this savings could be reallocated to the acquisition of other resources. If a resource is relatively inexpensive for a plant to obtain, the savings caused by resource enrichment should be small, and therefore the growth response to its addition would be small, perhaps immeasurable. Second, many essential plant resources are acquired simultaneously by a single foraging structure. Roots, for instance, forage for nitrogen, phosphorus, potassium, calcium, magnesium, water, and other soil resources. Leaves forage for light and CO_2. Such simultaneous foraging can cause growth to be limited by a single resource (Gleeson and Tilman, 1992). The root mass of a plant, for instance, might be determined mainly by its requirement for nitrogen. Given enough root mass to equalize nitrogen versus light limitation, a plant might have coincidentally produced sufficient root mass to preclude limitation by phosphorus, potassim, calcium, or other soil resources. In this case, adding PO_4 would not decrease the amount of root mass needed to forage for nitrogen, so there would be no growth response to the added phosphate. The only potential savings to the plant from the addition of phosphate might be from the plant decreasing allocation to the actual enzyme systems (or mycorrhizal fungi) involved in active uptake of phosphorus by root hairs. This, however, probably would be a minor savings that would convert into a small growth response.

What, then, is the potential for limitation by CO_2? Because CO_2 is required in large amounts (carbon being the major constituent of dry plant mass), it seems plausible that CO_2 could limit the growth of many plant species. However, it is not obvious whether wild plants would experience major allocation shifts in response to CO_2 addition. CO_2 is obtained through stomata on leaves. If the major function of leaf area is light capture, plants may have more leaf area than they need for CO_2 capture. In this case, the addition of CO_2 would not result in smaller leaf area, but might provide some minor savings to a plant from decreased production of stomata. The energy saved, however, would probably have an immeasurably small impact on growth rates. Second, there may be a trade-off between leaf area and biochemical photosynthetic capacity. In productive habitats, CO_2 uptake by leaves may be limited by the rate of its diffusion through the boundary layer, and plants with greater total leaf area or different leaf shape or placement may have an advantage. In this case, increased CO_2 concentrations could increase growth and favor a shift in allocation to roots. However, predicting field responses to CO_2 elevation is complicated by the water cost associated with CO_2 uptake. Water diffuses out of open stomata as CO_2 diffuses in; therefore plants living in a high-CO_2 environment need less water. If the root mass of a plant were determined by its demand for water, a plant living in elevated CO_2 could allocate less to roots, and thereby increase its growth rate. The relative importance of the root and leaf costs associated with CO_2 uptake should depend on environmental conditions and plant traits. Perhaps CO_2 will be most limiting in dry habitats with rich soils (which are rare habitats in nature, at least before global warming) and in moist, fertile habitats. Alternatively, the confounding of both above- and below-ground costs, and the aspects of simultaneous uptake by both below- and aboveground structures, may mean that CO_2 is rarely limiting in nature. Answers to these questions await appropriate field experiments.

THEORY OF CO_2-DEPENDENT INTERSPECIFIC INTERACTIONS

For any individual plant, CO_2 may function in one of two ways: as a depletable resource or as a nondepletable physical factor. Different theories are needed for each case.

There is evidence of CO_2 depletion by plants on a variety of temporal and spatial scales. The uptake of CO_2 by leaves causes a CO_2 concentration gradient around each leaf. CO_2 is lowest near the stomata and in the boundary layer around a leaf, and higher further from the leaf surface. The steepness of this gradient depends on the rate of CO_2 uptake and the thickness of the boundary layer, which in turn depends on wind speed and leaf shape. Thus, CO_2 uptake by a leaf can decrease the amount of CO_2 locally available for further uptake by that leaf. On a larger scale, the CO_2 concentration inside the canopy of a stand of plants is lowered during photosynthesis. On a global scale, the atmospheric CO_2 concentration at Mauna Loa Observatory has an annual depression corresponding with peak terrestrial photosynthesis (Keeling, 1986). Such annual depressions suggest that plants can affect global CO_2 concentrations.

Models of competition for resources assume that species interact only through their effects on the concentrations of shared, depletable resources (e.g., Leon and Tumpson, 1975; Petersen, 1975; Taylor and Williams, 1975; Tilman, 1977, 1982). Such theory seems to apply to many cases of interspecific plant competition (Tilman, 1982, 1990; Tilman and Wedin, 1991a,b), and may apply to competition for CO_2, if it functions as a depletable resource.

However, it is not clear whether such CO_2 depletion by any one plant is sufficient to allow interspecific competition for CO_2. Because the atmosphere is relatively well mixed, it seems likely that CO_2 consumption by one individual plant might have little, if any, effect on neighboring plants (Begon et al., 1986: 85). In this case there would be no direct competition for CO_2. However, even if plants did not compete directly for CO_2, ambient CO_2 levels could influence plant growth rates, allocation patterns, nutrient conservation, and other plant traits. These traits could, in turn, influence the outcome of competition for a different resource, or the outcome of plant–herbivore interactions (Ayres, this volume), and thus influence the abundances of species in natural communities. In this case, the concentration of CO_2 would influence growth rates and other traits, much as might temperature, pH, or some other physical factor. I refer to this possibility as the case of "CO_2 as a physical factor."

Let us now consider two alternative theories. In one it is assumed that CO_2 is not depletable but acts as a physical factor; in the other, it is assumed that CO_2 is a limiting, depletable essential plant nutrient. First, though, we need to review the concept of competition for a single limiting resource (Tilman, 1982). Consider a plant species that is limited by a single nutrient resource, such as soil nitrogen (i.e., nitrate and ammonium). The growth rate of such a population will depend on the concentration of this resource. The population will also experience losses from herbivory, senescence, mortality, and other sources. As the population grows, its resource consumption will reduce the concentration of its limiting resource. At some point the population will attain a size at which it has reduced the concentration of the resource sufficiently that the resource-dependent growth rate of the population balances its loss rate, and the rate of resource supply balances consumption. At this point, a monoculture of this species will have reached equilibrium, and will have reduced the resource concentration down to a level R^* (Tilman, 1982). R^* represents the lowest concentration of available resource at which the species can survive. Now consider two plant species competing only for this same limiting resource. The plant with the lower R^* for the resource will be able to reduce the resource concentration to a point below that required for the survival of the other species, and should thereby competitively displace it from the habitat (Tilman, 1977, 1982). This theory has been tested for various combinations of five species of grasses competing for soil nitrogen (Tilman and Wedin, 1991a,b). Monocultures of the grasses differed in their R^* values for dissolved soil ammonium plus nitrate. When grown together in various pairwise combinations, the species with significantly lower R^* values displaced all others from unproductive, low-nitrogen field plots within 3 to 5 years. Such displacement occurred independent of the initial relative densities or initial life history stages of the competing plant species. Thus, experimentally observed R^* values can predict the outcome of interspecific resource competition among terrestrial plants.

CO_2 as a physical factor

Consider a simple case in which there is a single limiting, depletable resource. Let us assume that this resource is available soil nitrogen (i.e., nitrate and ammonium). How might CO_2, acting as a physical factor, influence the outcome of nitrogen competition? Several studies have shown that higher concentrations of CO_2 can lead to increased biomass allocated to roots, to higher C:N ratios in leaves, to higher photosynthetic rates, and to lower dark respiration rates (Bazzaz, 1990). All of these traits are predicted to influence the R^* of a plant for nitrogen (Tilman, 1990). The R^* of a species is a summary variable that incorporates the effects of physiology, allocation pattern, and nutrient conservation ability on competition. Indeed, five models that differed in their physiological and morphological complexity all predicted that a single number, R^*, was the only appropriate way to summarize the total effect of plant traits on nutrient competitive ability (Tilman, 1990). The most complex of these models predicted that the greater biomass a plant allocated to roots (as long as it still had enough leaf to meet root and leaf respiration needs), the greater its rate of photosynthesis and the greater its nutrient conservation ability (i.e., higher C:N ratios), the lower should be its R^* for a limiting nutrient (Tilman, 1990). Consistent with this, Tilman and Wedin (1991b) found that the observed nitrogen R^* values of five grass species were inversely correlated with root mass. Thus, R^* for nitrogen is likely to be a decreasing function of the atmospheric CO_2 concentration, if increased CO_2 leads to increased root allocation and high leaf C:N ratios.

Now consider two species whose nutrient-dependent growth is influenced by CO_2 concentration (Figure 1A). This is not a purely hypothetical case, since Bazzaz (1990) has found that different species do respond differently to CO_2 enrichment. Carrying through our specific example, it can be seen in Figure 1 that species A will be competitively dominant (will have lower R^* for nitrogen) if CO_2 concentration is less than 350 ppmv, and species B will be competitively dominant for higher CO_2 concentrations. Thus, if CO_2 acts as a physical factor, and if two species respond differently to CO_2, it is possible for changes in atmospheric CO_2 to tip the competitive balance from one species to another. This occurs even though the plants do not compete for CO_2. Different magnitudes of physiological responses to CO_2 are the norm (Bazzaz, 1991), and thus this case seems plausible. Similarly, if numerous species differ in the dependence of their R^* values on CO_2, increasing CO_2 can lead to a successional sequence of species dominance (e.g., dominance first by species A, then by B, C, and D, as in Figure 1B).

The scenarios in Figure 1 assume an interspecific trade-off between competing well for nutrients at one CO_2 level versus competing well at some other level. If no trade-off exists, atmospheric CO_2 levels would have

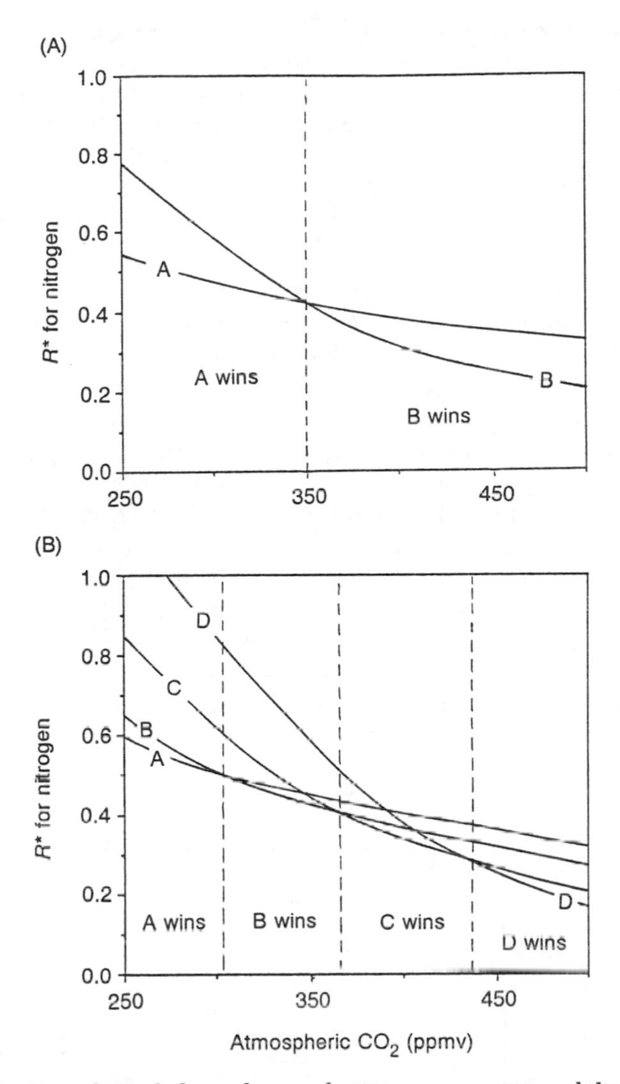

FIGURE 1. Hypothetical dependence of nitrogen-competitive ability on atmospheric CO_2 concentration, assuming that CO_2 acts as a nondepletable physical factor. The Curves show the R^* for dissolved soil nitrogen (ammonium plus nitrate) for species A, B, C, and D. At any given CO_2 concentration, the species with the lowest R^* for nitrogen is predicted to competitively displace all other species.

little direct effect on plant competition. Thus, it is important to determine whether such trade-offs exist, or whether morphological or physiological allocation patterns might make such trade-offs unavoidable. If trade-offs are unavoidable, then CO_2 enrichment could have profound direct effects on plant community composition and diversity. CO_2, acting as a physical factor, could also influence interspecific interactions in habitats in which there are two or more limiting resources. This would occur if one or a few of the species in a community benefitted more than the others from a CO_2 increase.

CO_2 as a depletable resource

Alternatively, plants may compete directly for CO_2 through either local or global depression of atmospheric CO_2 concentrations. If such competition were to occur, it would likely be a slow, weak process. However, even slow and weak processes can have major long-term implications for the dynamics and structure of communities. Consider first the highly hypothetical case in which it is assumed that CO_2 is the only resource limiting plant growth worldwide. If this were so, plants would eventually reduce atmospheric CO_2 down to a quasi-equilibrium level, R^*. At R^*, CO_2-dependent growth would just balance biomass loss (when viewed on an annual or longer time scale). Increased rates of CO_2 release, such as from the burning of fossil fuels, would lead to higher equilibrial plant biomass on the earth, but, once a quasi-equilibrium was reattained, would not change equilibrial CO_2 concentrations, assuming that loss rates remained constant. In this case, global plant biomass would directly regulate CO_2 (Figure 2), albeit with an as yet

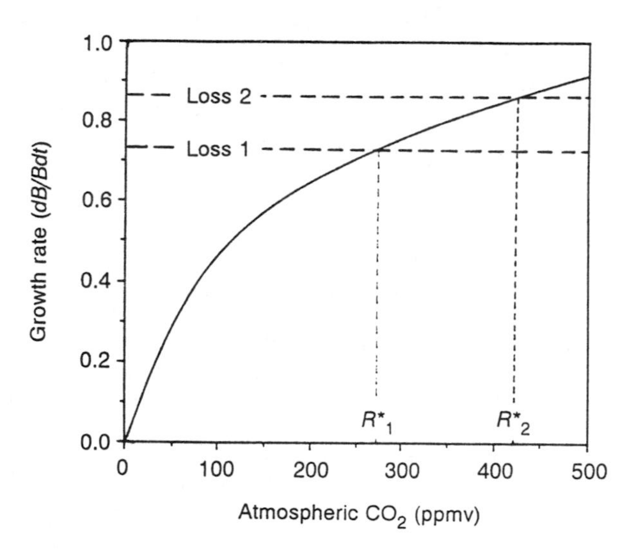

FIGURE 2. Here it is assumed that CO_2 is a depletable, essential plant resource. The solid curve is the CO_2 dependence of plant growth. The two dashed lines show different plant biomass loss rates, and the two dotted lines show the equilibrial CO_2 concentration associated with each loss rate.

unknown time lag following atmospheric enrichment. Habitat destruction, biomass harvesting, and other actions that increased loss or mortality rates for plants would increase the quasi-equilibrial atmospheric CO_2 concentration (Figure 2).

Now consider a more realistic case in which CO_2 is a limiting, depletable resource and there are other limiting resources. If soil nitrogen were also limiting, and if species were different in their abilities to compete for CO_2 versus nitrogen, changes in the supply of either resource would change the composition of the community (Figure 3). For instance, if the equilibrium concentrations of soil nitrogen and CO_2 in a habitat before the Industrial Revolution were at point 1, species A and B would be competitively dominant. An increase in the CO_2 supply rate could shift concentration to point 2, where species B and C are dominant. Further increases could lead to dominance by other combinations of species.

In this case, as CO_2 increases there is a progression of dominance from species that are better CO_2 but poorer nitrogen competitors toward species that are better nitrogen but poorer CO_2 competitors (Figure 3). As this

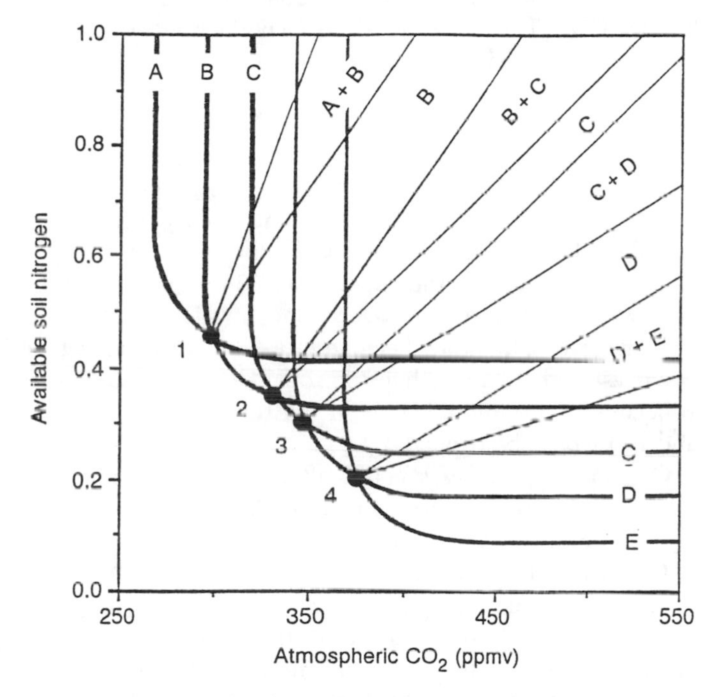

FIGURE 3. CO_2 is assumed to be a depletable, essential plant nutrient. The bold curves are resource-dependent growth isoclines for species A through E. These isoclines, along with the consumption vectors (straight lines), define regions in which a species or species pair is dominant. See Tilman (1982) for a detailed explanation.

occurs, the average environmental concentration of CO_2 increases and average soil nitrogen concentrations (dissolved ammonium plus nitrate) decrease. Each of these concentrations represents a potential quasi-equilibrium between the supply rate of CO_2 and its rate of consumption. A similar argument could be made in which the two limiting resources were CO_2 and light. If the best CO_2 competitors were the poorest light competitors, they would be replaced, as CO_2 accumulated, by taller, leafier species that were better light competitors, but poorer CO_2 competitors. These processes could occur on any spatial scale over which there was significant CO_2 depletion.

Biotic regulation of global CO_2 would occur because of shifts in the relative abundances of species that were different in their abilities to compete for CO_2 versus other limiting resources. At equilibrium, the species that were better competitors for CO_2 would be held in check by their high requirement for some other limiting resource. This would prevent them from further depleting atmospheric CO_2 levels. The species that were better competitors for the other resource would be held in check by their high requirement for CO_2. This would prevent them from reducing the availability of the other resource to a level below that required by the better competitors for CO_2. Thus, the processes hypothesized in Figures 2 and 3 could act to regulate global CO_2 concentration.

FEEDBACK EFFECTS

There are several additional ways that CO_2 enrichment could influence plant community structure. Optimal foraging theory predicts that CO_2 enrichment should favor traits that increase the ability of a plant to deal with limitation by resources other than CO_2. Many plant species increase root biomass and decrease leaf nitrogen concentration in response to CO_2 enrichment, suggesting that they become more limited by nitrogen (Bazzaz, 1990). In contrast, in habitats with fertile soils, CO_2 enrichment could cause a plant to become more limited by light. In this case, increased allocation to stem and leaf would be expected.

These CO_2-induced changes could affect nutrient cycling within the ecosystem, since the rate of nitrogen mineralization is influenced by the C:N ratio of litter. Litter with a higher C:N ratio, such as that produced by some CO_2-enriched species, will have a lower decay rate (Woodward et al., 1991). Low decay rates can lead to decreased availability of nitrogen. Wedin and Tilman (1990) observed that initially identical soils diverged as much as tenfold in their rate of nitrogen mineralization after 3 years, depending on the C:N ratio of the resident grass species. CO_2-induced changes in litter C:N ratios could potentially cause major shifts in nitrogen dynamics. The rate of nitrogen supply, in turn, could have a long-term impact on community structure. The impact would be the greatest if CO_2

were a depletable resource and if there were interspecific trade-offs, as illustrated in Figure 3. In this case, CO_2 enrichment would favor better nutrient competitors, and this would in turn favor nutrient conservation, which would lead to lower rates of nutrient supply, and thus further favor better nutrient competitors. Such a positive feedback loop could magnify the effect of CO_2 enrichment on changes in community composition.

There could be similar feedback effects mediated by herbivory (see Fajer et al., 1989; Bazzaz, 1990; Ayres, this volume). For instance, if CO_2 enrichment were to favor lower tissue nitrogen content in some species, herbivory on these species might decline in the long term, because herbivores might shift their food preferences or suffer population reduction due to the poorer food source. The decreased rates of herbivory experienced by these species would provide an advantage in their interactions with other plant species. Such herbivory shifts could further complicate the effects of CO_2 enrichment on competitive interactions and community structure.

CONCLUDING QUESTIONS

Both the known effects of CO_2 on plant physiology and a variety of theories regarding the effects of CO_2 on plant competition raise many more questions than they answer. The most important of these questions is the extent to which CO_2 is a limiting resource. Field studies are needed to investigate this question in a wide variety of natural terrestrial and aquatic habitats. Such field studies should simultaneously observe the responses of herbivore populations and soil nutrient concentrations and mineralization rates.

If it is found that CO_2 is frequently limiting in nature, it will then be important to determine whether there are observable interspecific trade-offs between responses to CO_2 concentration and other environmental constraints. For instance, are there unavoidable physiological or morphological costs associated with increased efficiency of CO_2 utilization? What are the traits of plants favored by elevated CO_2? If there are broad, underlying trade-offs that define the possible ranges of all plants, then the description of these trade-offs could provide a framework (as outlined in Tilman, 1988) that would eliminate the necessity of studying the response of each species separately. Rather, a species could be considered as a point or a region on a trade-off curve or trade-off surface. The general effects of an environmental change could then be predicted by knowing how it mapped onto the trade-off response surface. To focus on unavoidable trade-offs, if they exist, is to lift ecology from the study of the peculiarities of millions of species into a simpler, more general, predictive science.

Another question raised by the foregoing discussion is which plant traits should be studied to understand the effects of CO_2 on plant communities. The answer depends on the level of mechanistic detail desired. Currently there are no simple, logically consistent ways to extrapolate from a few measures of physiological parameters to the effects of CO_2 on competition or community structure. An approach that seeks an abstracted variable, such as R^*, and develops and tests theory that relates R^* to underlying physiological processes (see Tilman, 1990), is thus needed to provide a logical link between physiological mechanisms and their impact on community-level patterns.

Conservation and resource management in the face of global change require simple models that can give robust qualitative predictions of the effects of CO_2 enrichment, as well as those of other environmental perturbations, on the composition and diversity of plant communities. Such models will likely be resource-based, and thus will share some qualitative features with the models presented here. The level of complexity needed in such models is unclear at this time, as is whether they should assume that CO_2 is a depletable resource or that CO_2 is a physical factor.

CHAPTER 21

A SALTY AND SALUTARY PERSPECTIVE ON GLOBAL CHANGE

Robert T. Paine

Changes in global climate of unprecedented rapidity are anticipated within the coming century. A capacity to understand and predict the biotic consequences of such change is a necessity if humanity and its associated biota are to persist in an altered yet organically rich and appealing state. Marine organisms, especially those inhabiting the ocean's margins, may be an especially appropriate focus for studies seeking predictions about biotic responses to global change. First, it is difficult to imagine another environment as persistent throughout geological time as the ocean. There has probably always been a land–sea margin; by contrast, there may not have been lakes in recently glaciated regions or montane forests. Second, the history of such marine environments is rich in temperature changes of documented biological consequences occurring over time intervals of less than a decade to only a few thousand years. Last, since many marine species are linked by strong biological interactions, the ratio of signal-to-noise is increased and hence we are better able to detect both direct changes and secondarily derived or cascading, indirect effects.

Although this volume generally excludes marine ecosystems, my chapter is intended to identify some key observations and lessons regarding global change that can be gleaned from examining marine communities. I highlight two primary themes. First, the biological consequences of marine temperature increase are relatively well documented. I draw conclusions from prehistoric temperature shifts, recurring El Niño events, periodic warm water incursions, and those chronic but local environmental influences associated with the heated effluent of nuclear power plants. Second, I describe in selected detail the consequences of human intervention in, especially the exploitation of, nearshore marine resources. My ultimate intention is to confront the question, "Will physically driven climatic influences on the globe's biota be unambiguously recognizable against the existing background of increasing human devastation?"

BIOLOGICAL CONSEQUENCES OF ELEVATED TEMPERATURES IN MARINE ENVIRONMENTS: FOUR SCENARIOS

In marine ecosystems, elevated surface water temperatures are often associated with decreases in both oxygen tension and dissolved nutrient concentrations. All three factors, in concert or independently, are biologically important. In addition, the spatial magnitude of the shifts in these interacting physical factors is important because marine ecosystems are potentially coupled by long-distance transport of propagules (e.g., Scheltema, 1986). Such dispersal can serve to buffer assemblages and thereby reduce the probability of local extinctions. However it is not my intention to suggest that the world's oceans are a well-mixed soup: many species lack the capacity to disperse great (or even short) distances, and for those that do disperse, successful settlement from the plankton to the benthos is risky. I simply want to point out that an important variable in any case study of temperature change must be the spatial extent of that change. In this section I provide a variety of scenarios for the biological consequences of ambient temperature shifts. Because of the uncertainty associated with current predictions of global climate change, I have selected examples to bracket a range of rates, spatial extents, causes, and consequences.

Temporal scale, 3000 years; spatial scale, geographic; temperature change, 6°C

Kennett and Stott (1991) describe an abrupt warming trend that occurred about 57.33 mya, based on temperature profiles in deep-sea cores off the Antarctic. Plausible explanations involve different global geography and climate shifts associated with elevated temperatures and increased precipitation at high latitudes. These factors seem sufficient to have altered the character of the deep ocean water. The triggering mechanism remains unknown.

This global event caused substantial changes in the species composition of an assemblage of benthic foraminifera (protozoa that secrete calcareous shells, which become microfossils). Species richness decreased abruptly, especially during the early phases of the temperature increase, from about 60 to about 17 species. Partial recovery occurred later, with reappearance of some species. However, 35 percent of the assemblage appears to have gone extinct. A parallel pattern of decline characterized benthic ostracods, adding an element of taxonomic generality to the event.

Temporal scale, 50 years; spatial scale, geographic; temperature change, 0.5°C

Southward (1980) has reviewed the apparently cyclic changes in the chemistry and biology of the western English Channel ecosystem. Subtle climate shifts appear to drive this marine change, especially variations in wind and temperature. Water mass characteristics show only slight variations: 0.5°C in temperature, no obvious shifts in salinity, and less than a 25 percent alteration in the concentration of inorganic phosphate. These nearly imperceptible environmental fluctuations are of major economic consequence since they are associated with reduced phytoplankton production, changes in the composition of the zooplankton assemblage, and alterations in fish stocks. Abundance often shifts by orders of magnitude and species replacements are commonly observed. However, in spite of these dramatic numerical shifts, no extinctions have been recorded.

Temporal scale, about 1 year; spatial scale, geographic; temperature change, 2–10°C

The 1982–1983 El Niño (ENSO) was a major event that influenced oceanographic conditions and weather throughout the Pacific and altered food chains in both hemispheres. Perhaps the most surprising feature of this event was that, despite the very strong physical signal, generalizations about its biological impact were hard to develop. Sea surface temperatures off equatorial South America were more than 10°C above the long-term mean (Arntz, 1980); in western North America, depending on latitude, sea temperatures were elevated, nutrient concentrations diminished, and anomalies in ocean current flows prevailed (Wooster and Fluharty, 1985). Biological consequences were also highly variable and unanticipated. In South America, scallop and octopus populations increased, which created momentarily lucrative fisheries, while local finfish income plummeted. In California, giant kelp productivity diminished significantly, in large part due to increased nutrient limitation associated with elevated water temperatures. At higher latitudes, biotic changes due to El Niño were comparatively minor, and were virtually undetectable in the intertidal zone (Paine, 1986).

El Niño temperature shifts, though both periodic and variable in intensity, represent a natural short-duration warming pulse. They vary greatly between events. The severe 1982–1983 ENSO appears to have produced one extinction and some range reductions of reef-building hydrocorals along the west coast of Central America (Glynn and deWeerdt, 1991). Elsewhere, temporary range extensions northward were commonplace. If the temperature rise were chronic rather than periodic, one might expect an increasing frequency of temperature-dependent range extensions and extinctions. In addition, associated with—but not precisely attributed to—the warm water incursions were coral bleachings (Roberts, 1987) and massive echinoid mortality (Lessios et al., 1984; Carpenter, 1988), both of which generated obvious changes in community structure at geographic scales for nearshore marine assemblages.

Temporal scale, 10–20 years; spatial scale, local; temperature change, 4–6°C

Power generating stations can increase ambient temperatures substantially, with the effect driven by the volume of the heated effluent and the openness of the site. Nuclear power plants provide especially instructive examples of how resident assemblages respond to temperature increases. As long as the power plant is operating, the impact is chronic, though local, and thus provides a useful comparison with geographically broader but temporary natural increases, for instance those associated with El Niño events.

One immediate effect of power plant discharge is range extension of tropical or subtropical species into more temperate habitats. This opportunity is provided to species with long-distance dispersal capability, or to those surviving transportation on boat hulls or in ballast waters. For instance, Naylor (1965) recorded substantial range extensions of at least ten warm-water species to Wales (United Kingdom). Hoagland and Turner (1980) have similarly identified the establishment of potentially destructive shipworms along the eastern seaboard of the United States. These studies imply an increasing tempo of successful invasions by warm-water species associated with the projected increases in sea surface temperature. Replacements should eventually occur.

Lubchenco and her co-workers (Tissot et al., 1991) have begun to explore the extensive "before" and "after" data base associated with coastal nuclear power stations for insights about the direction and magnitude of change associated with chronic environmental warming. At the Diablo Canyon site (California), characterized by a local 4–6°C warming, biotic changes were considerable but could not be accurately predicted on the basis of either physiological studies or knowledge of a species' biogeographic origin. Instead, changes brought about by pathogens, herbivory, and cascades of indirect effects complicated and even dominated interpretations of the observed community response.

Conclusions from the "warming" scenarios

These four scenarios embody a broad range of environments and taxa. The temperature shifts vary in duration, periodicity, intensity and geographic extent. Range expansions, a relatively bland biotic response, seem commonplace, and in these nearshore assemblages usually lead to local enrichment. Extinctions appear to be unusual, occurring only after long exposure to the new temperature regime or when the stress is intense and the affected population of limited geographic distribution. The anticipated global warming trend by itself, at least at the median rate and predicted intensity, should have a barely perceptible effect within the next century on the structure and organization of nearshore communities, although eventually assemblages may be displaced geographically, species boundaries altered, and abundance patterns changed.

DIRECT HUMAN INTERVENTION IN MARINE COMMUNITIES: FOUR SCENARIOS

The human population continues to expand, having doubled to 5 billion between 1950 and 1987, and the majority of those humans live within 100 kilometers of a seashore. One result is a growing demand for marine products. The four scenarios sketched below were chosen with purposeful bias, two to suggest pervasive and negative changes directly attributable to human greed or necessity, one to describe a devastating natural phenomenon whose increasing frequency is probably anthropogenically driven, and the last to hint at possible salvation.

Sea otter decline and recovery

Historically, sea otters ranged from Japan to Baja California; however, by 1911, they had become restricted to 13 scattered localities due to hunting pressure driven by a demand for otter pelts. Comparisons of sites with and without otters, studies before and after reinvasions and, eventually, experimental studies of the sea otter–kelp association (Van Blaricom and Estes, 1988) provide a compelling example of community change attributable to a single species of high trophic status. Otters regulate the populations of herbivorous invertebrates, especially sea urchins, and these in turn control the productivity and species composition of the benthic algal community (Estes and Palmisano, 1974; Duggins, 1980). In addition otters influence the mosaic nature of kelp stands, which in the presence of otters support a greater abundance and diversity of fishes (Ebeling and Laur, 1988). Simenstad et al. (1978) even find support for this scenario in archaeological excavation of prehistoric Aleut middens.

Because many sea urchin species interact strongly with benthic algal populations, factors that determine urchin distribution and abundance are certain to cascade to their algal prey. Such is apparently the case throughout much of the 6000 km of habitat potentially occupied by otters. Although dramatic changes in community organization are associated with otter removals, these can be and have been reversed. No extinctions have been associated with either otter absence or high densities of these important consumers.

Oceanic shark removal

Some large sharks snack on people. To eliminate this hazard to swimmers, a system of gill nets was installed around popular swimming beaches adjacent to Durban, South Africa. Netting began in 1952; records kept since 1964 indicate the system's effectiveness and a significant decline in both sharks caught per net and inshore populations of large sharks. Another indication of net efficacy is "the once famous '1000 club' to which big-shark fishermen were affiliated, has been disbanded owing to a paucity of large sharks" (van der Elst, 1979, p. 358).

Large sharks prey on, and apparently control, the distribution and abundance of a variety of smaller, inshore elasmobranchs, which in turn feed on bony fishes, the backbone of shore-based sport and food fisheries. During the interval 1956–1976, catch per unit effort of teleosts declined from 90 to 51 percent of the total by number, and from 51 to 16 percent by mass. Over the same period, small sharks, initially accounting for only 5 percent (numerically) or 16 percent (by mass) of the catch, have increased to 37 and 46 percent, respectively.

Although the netting of large sharks was modest in geographic extent (32 km), the numbers caught (13,700 in 11 years) and the subsequent direct reverberations throughout the food web provide important insights about the susceptibility of such assemblages to changes in densities of selected species. Community-wide responses were substantial and have had significant economic consequences. Although the cascade was not tracked to the benthic assemblage, there surely must have been some effect. No extinctions were observed. Human intervention in the dynamics of an open, large-spatial-scale system produced immediate direct influences and numerous subsidiary effects.

Coral-eating starfish

The crown-of-thorns starfish has been associated with Indo-Pacific coral reefs for at least 1–3 million years (Birkeland and Lucas, 1990). It has attained notoriety in the last two decades because of sudden population outbreaks characterized by 4–6 orders of magnitude increases within a single year. These irruptions commonly devastate entire coral reefs. Indeed, the crown-of-thorns has been called ". . . *the* major management problem in a number of coral-reef areas" (Birkeland and Lucas, 1990, p. 202). For instance, during a major outbreak at Guam, about 90 percent of the corals on 1 km of fringing reef were killed per month. Secondary effects include increases in cover of algae and nonprey items, increases in herbivores (except corallivorous fishes), and decreases in topographic complexity.

The crown-of-thorns is a native predator with a naturally dominant role in coral communities. However, the frequency of starfish outbreaks appears to be increasing in correlation with increases in adjacent human populations. The expansion of the human population has accelerated island deforestation and increased the runoff of nutrient-rich waters into lagoons, which consequently may become plankton-rich and thereby enhance the success of larval starfish. Although no local extinctions have been recorded, the effects due to this single species are dramatic, mechanistically understood, and bear substantial economic implications for the affected regions.

Human exclusion manipulations

The exposed rocky shores of central Chile have been inhabited and exploited by humans for thousands of years. Humans act as consumers at several trophic levels in these communities, and their accumulated impact has been quantitatively documented by Moreno et al. (1984) and especially Castilla and his co-workers (e.g., Duran and Castilla, 1989). The quantities of algae, marine invertebrates, and fishes harvested annually per 1.5 km of shore are awesome (Duran et al., 1987): over 11,000 kg of the "loco" or southern abalone, approximately 10,000 kg of limpets, 9100 kg of sea urchin, and 330 kg of an endemic clingfish. Casual observation of exploited shores fails to suggest how important a role humans play in these communities. However, the exclusion of humans triggers a chain of dynamic responses influencing both animal and plant community components, expressed as quantitative changes at relatively small spatial scales (500 linear meters of shore) and within relatively brief time frames (the major exclusion has been in force only since 1983). In particular, exclusion of human influence initiates a series of dynamic changes, many of which are unpredictable: mussels decrease; benthic algae and their consumers increase; previously uncommon, undersized, or rare species increase; spatial patterns change; and so forth. Human exploitation of the entire region has generated no apparent extinctions. On the other hand, the nature of the assemblage at all trophic levels has shifted dramatically. These studies provide yet another example of the human capacity to dominate the ecological character of nearshore assemblages. The controlled exclusion of humans reveals that important interactions within the assemblages were often masked by reduced abundances, and highlight the potential for recovery and sensible management.

Conclusions from "human intervention" scenarios

These four scenarios imply a "top-down" view of what controls marine community structure, although when and if global warming begins to alter oceanic circulation and upwelling patterns, nutrient provision (a bottom-up perspective) will become an important influence. However, because animals of high trophic status tend to be large, commercially harvested, or feared, and hence subject to focused human intervention in their ecologies, such a top-down bias is appropriate. None of the studies identifies extinction as a consequence of radical change in community structure. The prevalence of cascades subsequent to single-species perturbations suggests that strong interactions are relatively common. Economic considerations permeate all four scenarios. The otter and starfish examples indicate that local cause–effect observations or experimental studies can be generalized to much greater geographic scales. Additional studies by Witman and Sebens (1992) on regional community effects due to codfish depletions in the Gulf of Maine and Springer's (1992) analysis of the walleye pollock's critical role in North Pacific food webs support all the above conclusions. Human involvement, as exploiter, is directly implicated in three of the studies, and indirectly in the fourth. All the studies suggest rapid responses over geographic spatial scales involving a wide variety of taxa at all trophic levels. Strong responses to human involvement in marine food webs will continue to generate unpredictable yet high temporal and spatial variation in the impacted assemblages.

One further consideration

Global change implies worldwide alteration of natural communities. The study of these is not apt to be coordinated and therefore will be piecemeal, idiosyncratic, generated by an investigator's personal passions or funding, and geographically local. All studies must account for the effects of biogeographical mixing enhanced by human activities because such mixing is certain to involve novel interspecific combinations and thus unanticipatable, surprising alterations of community structure. Two general examples are given.

In the extreme case, an opportunity for invasion is offered simultaneously to many taxa, and that opportunity is continuous. The best-studied marine example (Vermeij, 1991) is the stream of invasions associated with the completion of the Suez Canal. Completion of the canal in 1869 permitted hundreds of species to enter the Mediterranean from the Red Sea; exchange in the opposite direction has been unusual. Major invading taxa include crabs, molluscs, and fishes. One robust conclusion is that such faunal (and floral) exchanges tend to be highly asymmetric, with identifiable donor and recipient assemblages. The successful invaders include both opportunists characterized by high reproductive rate, and species with superior competitive or defensive abilities. An intriguing and important observation is that extinctions are uncommon, at least during the initial phases. Rather, the local biotas appear to be well below species saturation levels, and the additional species simply dilute their new assemblages.

The second example, with more globally general implications, involves the passive dispersal of marine species. The primary mechanisms are the attachment of fouling organisms to ship hulls, the survival of planktonic species in ballast water, or the transport of species associated with commercial ventures, especially molluscs and fishes. The effects are massive and ubiquitous. For instance, over 100 species have been successfully introduced to San Francisco Bay in the last 130 years (Carlton, 1979). Economic impact there has been substantial, expressed by increased shoreline erosion and destruction of wooden docks. Carlton (1989, p. 265) considered such invasions "unabated on a large scale throughout the world." No model of community ecology known to me is capable of predicting the impact of such an array of exotic species. Their increasing presence generates both an environmental threat and an important challenge to ecological theory.

A PERSPECTIVE

The tempo of marine community change, driven by human need and exacerbated by an increasing world population, will accelerate. Although immediate marine extinctions are apt to be rare, the condition of shoreline communities is certain to become increasingly muddled by exploitation-caused population shifts, by successful introductions of novel species, and by unpredictable ecological surprises. The units of change in many instances will be in orders of magnitude, with variation being great. These events are occurring now. Against this background of direct human disruption, will the effects of global climate change be striking? I think not. Indeed, scientific harangues about the ecological crises predicted as a result of global change miss the point. They are misleading because they direct our attention from the far more devastating effects directly attributable to overexploitation, coastal pollution, and continuing parades of successful introductions of exotic species. Climate change, probably, will ultimately become important. Until then, however, the immediate crisis is the retention of the globe's biotic resources in some semblance of its natural state and a concurrent development of insights and tools sufficient to predict the added complication of physical alteration of the biosphere.

ACKNOWLEDGMENTS

I thank the gang of three for allowing me to crash, at the last minute, their exclusive gathering, Peter Kareiva for both his translations from the original Italian and perceptive reviews, and the NSF Program in Biological Oceanography for financial support.

CHAPTER 22

FORECASTING ECOLOGICAL RESPONSES TO GLOBAL CHANGE: The Need for Large-Scale Comparative Studies

Michael L. Pace

It is obvious to most ecologists that the unprecedented changes now affecting global ecological systems demand a new level of research. A crucial question is what level of disruption populations, communities, and ecosystems can withstand before these systems are either destroyed or changed irrevocably. I will argue that research seeking to answer this important question requires large-scale comparative studies as one of its elements. I will also argue that in the near future, mechanistic studies of populations and communities will *not* provide the predictive knowledge necessary to forecast the effects of environmental change at large scales.

WHAT DO WE NEED TO KNOW?

Research addressing global ecological change must begin with an appreciation of pragmatic issues. In particular, we must recognize at the outset that research designed to assess the ecological impacts of global change is research with a purpose, as opposed to pure research, in which the goal is simply new knowledge. What we need to know in the most general sense is how to forecast the ecological consequences of global changes such as climate warming and landscape fragmentation. The key word here is *forecast*: we must be able to make predictions given future scenarios of environmental change.

At a specific level, ecologists will be asked how to conserve species and communities. Here detailed knowledge regarding the ecology of specific organisms is needed for population models. Such models can be used to project habitat requirements and to design strategies of habitat management (e.g., McKelvey et al., this volume). We must recognize, however, that it will not be possible to develop detailed strategies for every threatened species. In addition, if climate change or other global disturbances are sufficiently rapid, even the best species conservation plans may be void.

Beyond specific efforts then, ecologists must try to build large-scale views of how changes in the environment will affect the distribution and abundance of species and how such changes will, in turn, feed back on ecosystem processes. These are forecasts. This is a tremendous challenge for a field that has traditionally conducted descriptive and explanatory research on small plots over short time periods (Tilman, 1989). To confront global change issues, ecological research must consider drastically broader scales than it has considered in the past.

Contributions to this volume emphasize the need to develop and calibrate models that forecast the effects of global change (see Pacala and Hurtt, this volume; Murdoch, this volume; Quinn and Karr, this volume). But while there is wide agreement on the need for models, there seems to be a split among ecologists regarding the best strategy for developing models. One approach is to base models on a detailed, mechanistic understanding of ecological dynamics that can be scaled up to forecast processes over broad regions or long time scales (Pacala and Hurtt, this volume). Another approach is the development of models from comparative studies at larger scales that do not contain mechanistic detail. Both approaches will be necessary (Root, this volume). But I worry that because ecologists have such a strong tradition of working on mechanistically focused studies at small scales, they will fail to appreciate the limitations of mechanistic models used by themselves. Conversely, ecologists may also inadequately exploit opportunities to develop models via comparative studies. In the remainder of this chapter, I discuss the shortcomings of mechanistic models and turn to some of the benefits gained from the comparative approach.

WHY MECHANISTIC MODELS ALONE WILL NOT WORK

The term *mechanistic model* is somewhat deceiving, since by definition a model is an abstraction of some part of reality into a mathematical framework. What ecologists usually mean by the word *mechanistic* is that models should be built from important aspects of individual physiology or, alternatively, detailed population demography. For example, we can imagine a model based on individual trees describing how vegetation will shift across the landscape as a function of changing climatic parameters. It may then be possible to aggregate some key features of these individual-based models to develop large-scale forecasts of vegetation dynamics at regional or continental scales. Pursuing the development of such models is a worthy and important effort and should be a priority for future research.

We also need to recognize that such models are likely to fail. Ecological systems are not simple nested hierarchies with individuals as the fundamental element. At every level of ecological organization, the underlying structure of the system must be regarded as fluid, not fixed. In the case of individuals, genetic, physiological, and behavioral adjustments to a changing environment can and will occur. These shifts can feed back on population dynamics and even have important ecosystem consequences (e.g., Carpenter et al., this volume). No model will capture all this complexity. Ecological systems will undergo both predictable and unpredictable responses to internal and external change. It is wrong to assume, in moving across levels of organization from individuals to populations to communities and so on, that simple aggregation of constituents at lower levels will explain variability observed at higher levels. For example, an individual-based model probably would not predict the differences in the importance of consumer species in structuring an intertidal food web elucidated by Paine (1992). In this case, two grazers have a disproportionately strong effect on community composition because they can control the development of brown algae, an effect one could probably not predict from physiology, demography, or a static food web model.

Model failure, of course, is important for learning. Models codify what we think we know, and confronting them with predictions from large-scale experiments or observation programs will test how good our knowledge is. One problem with such "learning experiences" is that model development may not be sufficiently rapid to provide forecasts regarding global change. Furthermore, there is no assurance that translation across scales of organization will work, for the reasons given above.

A second key feature of mechanistic models is that much of the effort going into them will emphasize identifying what level of specification is necessary to describe ecological systems. There will not be one simple answer to this question, or even a simple protocol to follow. Instead, the level of detail necessary will vary according to the question being addressed and the scale over which a prediction is needed. For example, consider the kinds of models one might need to forecast the effects of climate warming on a commercially important fish such as the anadromous striped bass, *Morone saxatilis*, whose current range in eastern North America extends from the Gulf of Mexico up the Atlantic seaboard to the Gulf of St. Law-rence. At least two kinds of models might be required. First, the effects of climate warming on a single population may be of interest. In that case, a mechanistically based model of either individuals or age classes would probably be useful. The effects of changing temperature and hydrology might be projected for a specific estuary and consequent responses of the fish population forecast. These models would fail, however, if factors not included in the model, such as disruptions in the community or shifts in habitat availability, have important impacts on the striped bass. A very different model of striped bass would be required to forecast how range limits might extend and contract in response to climate change. Here, a habitat-based model considering the key factors of tolerance to temperature and hypoxia, as well as the availability of resources, would be required (e.g., Coutant, 1990). But a model richly endowed with physiological and habitat requirements might also fail because it ignores the complexity of community interactions such as competition and predation.

HOW COMPARATIVE STUDIES CAN HELP

Comparative studies consider many populations, communities, and/or ecosystems across a specified gradient, region, or larger geographic area (e.g., countries, continents, the globe). The goal of such studies is to describe and answer questions about general patterns. In this way, comparative studies are no different than any other area of scientific endeavor (Peters et al., 1991). Nevertheless, because of its broad scale, the comparative approach is less mechanistically based and depends heavily on statistical inference. Comparative studies are often criticized because the statistical approaches of regression and correlation on which they depend do not lead directly to mechanistic understanding (Lehman, 1986). Identification of large-scale patterns with statistical techniques, however, is a tool that can broaden ecological generalization and hence contribute to understanding. Comparative ecologists often focus on "what works" in developing models instead of worrying about specifying the exact nature of the interactions underlying a pattern. Weaknesses of the approach are that comparative analyses can generate meaningless and misleading relationships, miss important interactions, and confound time or space scales affecting ecological interactions (Carpenter et al., 1991). A necessary part of comparative research is testing patterns against independent data and experiments to root out cases where statistical inference has gone astray. Hence, comparative analysis is not an exclusive alternative to other forms of inquiry, but an additional approach with some distinct advantages.

Comparative studies can make an important contribution to global change research. First, by sampling numerous populations or ecosystems at some point in time, one can develop a metric against which to evaluate future change. Second, studies involving many ecosystems are more likely to document large-scale human impacts than studies focused on a few systems. Finally, comparative studies provide a means for developing probabilistic models that can forecast change at large scales. These models have the desirable feature of being easily understood so that scenarios of change can be evaluated and the results communicated to those who make policy.

As with mechanistic models, models derived from comparative studies may fail, because the models are wrong or because changing environmental circumstances move the models outside their domain of prediction. In the case of comparative studies, model failure is easily recognized because the models are usually quite simple and can be evaluated against independent sets of data. Thus, comparative models and the concepts underlying these models can be readily tested.

Models from comparative studies may also fail to detect important but often subtle ecological interactions, such as those discussed above for the rocky intertidal. On the other hand, there is evidence that large-scale comparisons can detect key interactions. For example, in lakes the presence of large cladoceran grazers modifies the relationship between nutrient loading and algal biomass (Pace, 1984). Although comparative studies do not identify how large cladocerans affect this relationship (presumably by enhanced herbivory), the importance of the interaction can be detected and its significance quantified across a number of ecosystems.

EXAMPLES OF COMPARATIVE STUDIES

The potential contribution of comparative ecological studies to global change research can best be described by a few examples. Concern about the effects of acidic deposition on aquatic ecosystems has led to several large-scale comparative studies. Reckhow et al. (1987) used data from a suite of these studies to examine the relationship between water chemistry and the presence or absence of four species of fish that commonly occur in the Adirondacks region. It was possible to develop empirical models that predict the presence or absence of two species, brook trout (*Salvelinus fontinalis*) and lake trout (*Salvelinus namaycush*), in a given lake as a function of variables related to lake chemistry, such as pH and calcium concentration. These models summarize the degree to which chemical conditions determine trout distribution in the Adirondacks. Not surprisingly, chemistry alone does not provide perfect knowledge, as evidenced by the misclassification of lakes (predicting presence when no trout exist, or vice versa) 20 to 30 percent of the time. Nevertheless, the models provide a management tool that clearly indicates the likely effects of large-scale changes in the acidity of lakes on the distribution of two important fish species. Improvements in lake water quality as a consequence of air pollution regulation should lead to a decrease in the number of acidic lakes in the region and an increase in the number of lakes suitable for supporting trout. The models provide a quantitative basis for making such projections, given different scenarios of change for the chemistry of lakes.

A key problem in global change research is documenting the degrees and types of human impacts at large scales. Again, comparative studies are often a very effective means for accomplishing this goal. We recently conducted a study evaluating concentrations of nitrate in major rivers of the world (Peierls et al., 1991). Data were summarized from 42 rivers, including at least one river from every continent as well as tropical, temperate, and boreal environments. The chemical compositions of rivers are the result of a complex of biogeochemical interactions within their watersheds. For example, differences among drainage basins in relief, area, water discharge, geology, and climate have an important influence on the chemistry of rivers and on the amount of sediment they transport (Berner and Berner, 1987). Despite these complexities, at a global scale over 75 percent of the variability in river nitrate concentrations can be explained by variation in human population density in their watersheds. In other words, the most important factor determining nitrate concentration and export of nitrate from rivers to the ocean is some aspect(s) of human activity. This comparative study illustrates the vivid impact of humans on ecological systems at a global scale. This global pattern is reinforced where long-term data sets on river nitrate concentrations are available as in the case of the Mississippi and Rhine river basins (Cole et al., in press). The global-scale alteration of river chemistry revealed by our comparative analysis indicates that the impact of humans on the biosphere is not simply a far-off prospect based on scenarios of climate warming. Rather, human-accelerated environmental change is an immediate issue (Likens, 1991).

Comparative studies also provide forecasts of expected patterns of large-scale change. Such a forecast has been developed recently by Strayer (1991) in his projection of the likely distribution of the zebra mussel (*Dreissena polymorpha*) in North America. Zebra mussels invaded the Great Lakes of North America in 1985 and are spreading rapidly into other water bodies. Strayer evaluated the distribution and abundance of zebra mussels using data from a large number of European lakes and rivers. He considered a variety of factors, including temperature conditions, water chemistry, and water transparency (a measure of productivity in lakes), that might limit these organisms. His analyses led him to predict that zebra mussels will potentially occupy a range from the southern United States into southern Canada. Indeed, Strayer anticipates that zebra mussels will ultimately occupy all large hardwater ($Ca^{2+} > 1$ meq L^{-1}) streams and many, if not most, hardwater lakes in this range. This comparative study allows a prediction of the ultimate distribution of an invading species decades before its final range is achieved and without detailed studies of the species in each potential lake or stream. The predictions are clear, their basis is easy to understand, and in time we will be able to unambiguously evaluate the model.

RESEARCH RECOMMENDATIONS

These examples illustrate the nature of comparative studies and how they can be used to document human impacts and forecast ecological responses to global environmental change. I believe these examples provide compelling evidence that ecologists need to expand their research approaches to include comparative studies conducted over a number of sites. In addition, our journals are replete with data that might be exploited in literature-based studies that address specific questions (e.g., Peierls et al., 1991).

When considering the impact of global change on ecological systems, there are a number of specific research topics where large-scale comparative studies seem especially promising.

1. Large-scale patterns of species distribution need to be evaluated to determine the limits of prediction by models based simply on physiological tolerance and energy availability.
2. What are the constraints on populations at the margins of their distribution? Are these constraints primarily physiological and energetic, or do other factors interact with these limitations to inhibit species at margins?
3. Comparisons across communities need to be made to determine the impact of climatic variability on the population dynamics, species persistence, and species richness of these communities. From such comparisons, can forecasts be made of how communities will respond to future scenarios of climatic variability?
4. How do shifts in community composition determined by large-scale human impact (e.g., flux of greenhouse gases such as CO_2 and CH_4) influence ecosystem processes? Is there a high level of functional redundancy within ecosystems so that community composition matters little to the rate at which various processes occur? Are processes carried out by only a few organisms (e.g., nitrification) more susceptible to changes in the composition of communities?
5. What species within what types of ecosystems are good indicators of changes in climatic conditions? What species within what types of ecosystems are especially sensitive to deleterious habitat fragmentation? Are there particular kinds of ecosystems that are good indicators of large-scale human impact?
6. What species are likely to prosper under various scenarios of habitat fragmentation? How extensively will those species spread and what impact will they have?
7. Is species richness predictable (e.g., Currie, 1991) and how do such predictions vary as a function of scale? Can models that predict species richness incorporate factors such as climate change and landscape fragmentation to determine how these factors will modify diversity?

This is only a brief listing of important issues where comparative studies will have great value. In addition to these questions, two important feedbacks need to be considered. First, given large-scale ecological changes (e.g., shifts in vegetation), what will be the feedbacks of those changes on climatic and related hydrological systems? Second, how will large-scale ecological changes influence patterns of human exploitation and conservation? Comparative studies will be important in attempting to establish generalizations here as elsewhere.

SUMMARY

Understanding present and pending ecological responses to changes in the global environment requires ecologists to work at large scales. One avenue for this research will be improved models of populations built either from individual physiology or demographic parameterizations. These mechanistic models are likely, however, to fail in making large-scale forecasts of population response to environmental change, because the models will too often neglect unexpected ecological interactions. An additional and useful tool in this effort should be comparative studies across populations, communities, and ecosystems. Comparative studies can provide the basis for evaluating future ecological change by providing large-scale reference data on current conditions. Comparative studies also provide a means for demonstrating large-scale impacts of humans on ecological systems and forecasting future ecological change. Indeed, it is hard to imagine ecologists succeeding in their efforts to predict the consequences of global change without an extensive reliance on large-scale comparative studies.

ACKNOWLEDGMENTS

I thank Peter Kareiva for constructive suggestions and thorough editing of an initial version of this paper.

PART FIVE

LANDSCAPE CHANGE AND HABITAT FRAGMENTATION

CHAPTER 23

SPECIES INVASIONS AND DELETIONS:

Community Effects and Responses to Climate and Habitat Change

David M. Lodge

One of humankind's greatest influences on the earth's biota has been to increase the rate at which species invade non-native habitats, and the rate at which species go extinct at both local and global scales. Biological invasions constitute such a significant ecological problem that several volumes have been devoted to their study in recent years (Table 1). Range contractions and the elimination of species from communities have received less focused attention. One of the major effects of changes in global climate and habitats will be an acceleration of species invasions, extinctions, and range shifts. Indeed, one of our central concerns regarding global change should be how species ranges and communities will change (Mooney, 1991a,b; Ojima et al., 1991). Thus, if we want to manage human activities and mitigate species extinctions and invasions, we must understand how existing communities respond to invasions and extinctions.

In this chapter, I first review current generalizations about invasions in terrestrial and aquatic habitats and evaluate their usefulness in predicting the occurrence and outcome of specific invasions. Second, I suggest an alternative perspective on invasions that stresses the naturally dynamic ranges of organisms; in this context, species invasions and local extinctions are not unusual events in nature. Third, I present three freshwater examples

TABLE 1. Classic and selected recent volumes on species invasions.

Year	Author(s)	Title
1958	Elton	*The Ecology of Invasions by Animals and Plants*
1965	Baker and Stebbins (eds.)	*The Genetics of Colonizing Species*
1983	Parsons	*The Evolutionary Biology of Colonizing Species*
1986*	Groves and Burdon (eds.)	*Ecology of Biological Invasions*
1986*	Kornberg and Williamson (eds.)	*Quantitative Aspects of the Ecology of Biological Invasions*
1986*	Macdonald, Kruger and Ferrar (eds.)	*The Ecology and Management of Biological Invasions in Southern Africa*
1986*	Mooney and Drake (eds.)	*Ecology of Biological Invasions of North America and Hawaii*
1987	Gray, Crawley and Edwards (eds.)	*Colonization, Succession and Stability*
1987*	Joenje, Bakker and Viljm (eds.)	*The Ecology of Biological Invasions*
1989*	Drake, Mooney di Castri, Groves, Kruger, Rejmanek and Williamson (eds.)	*Biological Invasions: A Global Perspective*
1989	Hengeveld	*Dynamics of Biological Invasions*

*An asterisk indicates an official contribution of the International Council of Scientific Unions' (ICSU) Scientific Committee on Problems of the Environment (SCOPE). This list includes neither the many important independently published papers nor the relevant and growing literature on environmental introductions of organisms whose genome has been altered via recombinant DNA techniques.

of species invasions to illustrate how a thorough knowledge of a community can help us anticipate the community rearrangements that may result from global change. One of the three examples includes a modeling approach for exploring the impact of habitat fragmentation on species invasions. Finally, I suggest some questions that might profitably guide future research on the response of ecological communities to global change.

In this chapter, I use the term "colonist" to describe a species that has arrived (with or without human assistance) at a site outside its previous range. "Invader" means an established colonist, i.e., with a self-sustaining population. "Deletion" means local extinction.

CURRENT MODELS OF INVASIONS ARE INADEQUATE FOR PREDICTIONS

A spate of recent books on ecological invasions has built on the foundation of two earlier, classic works by Elton (1958) and by Baker and Stebbins (1965) (see Table 1). The following sections briefly assess the state of the art in verbal and statistical models of invasions. This assessment will establish the context for exploring how communities will respond to changes in climate and habitat.

Verbal models of invasions

The literature is full of case studies of species invasions from which a variety of ecologists have distilled generalizations about invasive species, their impact, and invasible habitats (Table 2). Many of the characteristics describing invasive species and invasible habitats appeal to the intuition of ecologists, but some of the characteristics contradict others. Underscoring these inconsistencies, Pimm (1989) and others suggest that evolutionary trade-offs prevent any organism from combining all the proposed characteristics of invasive species, and that different characteristics may apply to different taxonomic groups. The role of disturbance—probably the most oft-cited habitat characteristic favoring invasion—is controversial. Ramakrishnan and Vitousek (1989) conclude that disturbance may enhance invasibility for plants, but not for animals. In general, there are documented exceptions to most, if not all, of the listed characteristics of species and habitats (see sources in Tables 1 and 2). Thus, these verbal models contain internal inconsistencies and are accurate, at best, in only a probabilistic sense.

Statistical models of invasion and statistical tests of verbal models

Unfortunately, data adequate to test hypotheses or to generate statistical descriptions of invasions are largely lacking. Difficulties with data sets include underrecording of failed colonists, overrepresentation of species likely to be invaders (e.g., biological control studies), narrow taxonomic basis, and overrepresentation of organisms that are easy to observe. Given these serious shortcomings, statistical descriptions must be interpreted cautiously. Nevertheless, I synthesize below the statistical patterns emerging from some of the more complete data sets on invasions (Figure 1).

What percentage of colonists become invaders? In the Netherlands, 38 percent of 220 plant introductions after the year 1550 resulted in the invader "penetrating into natural vegetation" (Weeda, 1987). In Britain, 58 percent of 1058 introductions of a variety of taxa resulted in established populations (Williamson and Brown, 1986); in the Silwood Project on Weed Biocontrol,

TABLE 2. A composite verbal model of species invasions.[a]

CHARACTERISTICS OF SUCCESSFUL COLONISTS

r-Selected; pioneer species; short generation time; high fecundity; high growth rate
Ability to shift between *r*- and *K*-selected strategies
Long-lived
Relatively constant population density
High short-range dispersal rate
Single-parent reproduction
Vegetative reproduction
Much genetic variability
Polyploid (for plants)
C_4 metabolism (for plants)
Phenotypically plastic
Large native range; native of widespread continental habitats
Tolerant of wide range of physical conditions
Abundant in native range
Broad diet
Gregarious
Large body size compared to related species
Small individual size
Human commensal

CHARACTERISTICS OF COLONISTS WITH LARGE EFFECTS ON COMMUNITY[b]

Polyphagous
Predators absent

CHARACTERISTICS OF INVASIBLE HABITATS/COMMUNITIES

Climatically matched with original habitat
Early successional; low cover (for plants)
Vacant niche; low diversity of native species; absence of morphologically similar species
Lack of taxonomic balance, lack of natural enemies
High number of predator species
Absence of predators or grazers in evolutionary history
Absence of fire in evolutionary history
Low-connectance food web
Anthropogenically disturbed

CHARACTERISTICS OF HABITATS LIKELY TO EXHIBIT LARGE INVASION EFFECTS[b]

Simple (e.g., removal of a low number of plant species will cause collapse of food web); anthropogenically disturbed

[a]This list includes the most widely cited generalizations about invasive species and invasible habitats. It is not comprehensive. In order to maintain the emphases of different authors, the list includes some overlapping and some largely contradictory characteristics. The most closely overlapping characteristics are listed on one line, separated by semicolons; the most obviously contradictory characteristics are listed consecutively on different lines. Recent sources proposing and/or evaluating these generalizations include Ashton and Mitchell (1989), Baker (1965), Brown (1989), Crawley (1987), Ehrlich (1989), Johnstone (1986), Loope and Mueller-Dombois (1989), Mack (1989), Noble (1989), Pimm (1989), and Williamson and Brown (1986).
[b]"Large effects" include, e.g., extinction of native species, change in nutrient flow.

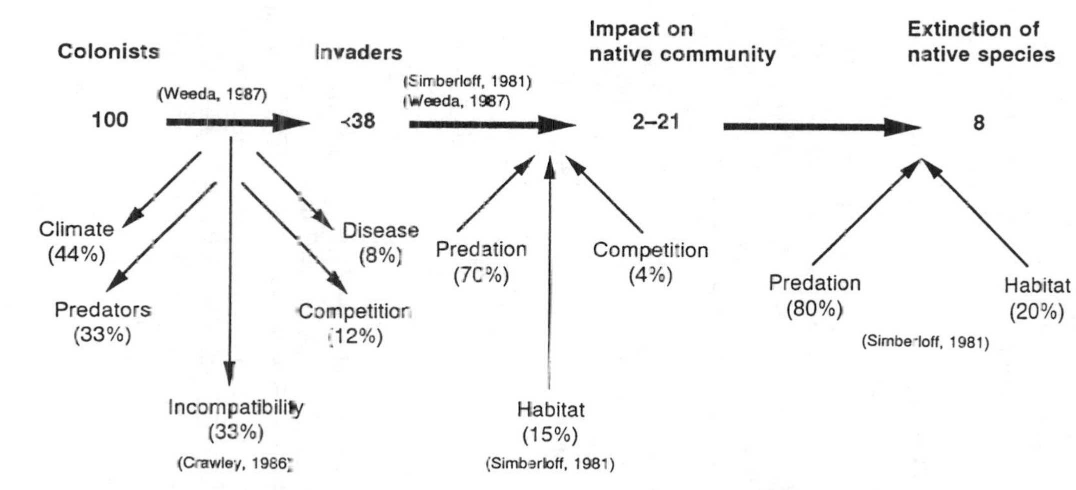

FIGURE 1. Best estimates from the literature of the success and impact of a hypothetical cohort of 100 colonizing species. Mechanisms responsible for failure of colonists, impact on native communities, and extinction of native species are listed with percentages. The percentages for the failure of colonists do not add to 100 because multiple factors contributed to individual failures. These figures are likely to be quite inaccurate, for reasons explained in text.

65 percent of potential control species released became established (Crawley, 1986). Given the concerns in the previous paragraph, even the lowest percentage listed above (38 percent) is certainly biased too high. All we can say with reasonable confidence is that less than 38 percent of colonists are successful (Figure 1).

What causes failure of colonists? In the Silwood Project (Crawley, 1986), in which failed as well as successful introductions were studied, rank order of causes of failure was climate > predators and parasites = incompatibility (with host) > competition > disease (Figure 1). This ordering is consistent with the idea that invasible habitats are climatically matched with native habitat. However, because competition and disease are difficult to recognize, the importance of biotic interactions was probably underestimated. Thus, we should expect species distributions to change as a direct response to climate change, but in ways that will be strongly modified by biotic interactions. The three case studies that I present later in this chapter demonstrate the importance of biological interactions.

What percentage of colonists have an impact on a native community? Different authors used different criteria: 21 percent had "recorded effect," via predation, habitat modification, and competition (Simberloff, 1981); 6 percent became "pests" (Williamson and Brown, 1986); 2 percent "spread aggressively" (Weeda, 1987). Simberloff (1981) concluded that 8 percent of introductions caused extinctions of native species. Debate continues as to whether these are underestimates of significant effects (Herbold and Moyle, 1986), or overestimates because many failed introductions will not have been recorded.

Are components of the verbal model accurate, at least in a probabilistic sense? Probably the most cited and best tested element of the verbal model is that invaders have high r. From a correlative analysis, Lawton and Brown (1986) conclude "that r alone is not obviously the main, or even an important determinant of invasion success" (cf. Crawley, 1986). Perhaps verbal and statistical models that are more taxon- or habitat-specific would be more robust (Mack, 1989; Noble, 1989; Williamson, 1989). At present, however, data are largely inadequate for rigorous testing of general or limited models. Certainly, the general verbal model applies only in a weakly probabilistic sense, if at all.

Can models predict the success and impact of invasions?

Ehrlich (1989) offers one of the best summaries of our predictive abilities regarding invasions: "Ecologists can make some powerful and wide-ranging predictions about invasions. . . . On the other hand, ecologists cannot accurately predict the results of a single invasion or introduction event." Ehrlich goes on to argue that our lack of predictive ability regarding invasions does not suggest a failure of ecology as a science, pointing out that, after all, physicists cannot predict which of two atomic nuclei will decay first; they can only make a probabilistic statement. Unfortunately, Ehrlich's analogy leaves unsolved the practical problem of predicting which of two colonizing species will succeed.

Whereas the question of which of two nuclei will decay first is of no practical importance, the question of which of two colonizing species will succeed may have large implications, including extinction of native species or devastation of crops. Unfortunately, community ecologists operate in a middle-number milieu (Allen and Starr, 1982): we must deal with too many particles to model each individually, but we deal with too few particles to be satisfied with averages or probabilistic models. Statistical treatments of large numbers of invasions are analogous to those appropriate for practical applications in physics. Yet, they leave ecologists unable to solve important and specific practical problems. General, predictive principles are unlikely to emerge from an accumulation of case studies of invasions. Ecological predictions will require intensive study of particular communities and new modeling approaches for predicting community response to environmental change.

AN ALTERNATIVE PERSPECTIVE ON INVASIONS

In this section, I want to propose an alternative to what I perceive as the underlying assumptions in much of the invasions literature reviewed above. These assumptions include: that native communities are equilibrial; that range expansions are largely, if not exclusively, the result of human activity; that the consequences of any species invasion are invariably bad; and that invading species are responsible for an extremely large portion of the changes in native communities. These assumptions are not *entirely* wrong, of course, nor do they underlie all invasions studies. They are, however, pervasive.

During the last century, at least 77 percent of documented bird species introductions into Britain occurred without human assistance (derived from Williamson and Brown, 1986). The naturally dynamic ranges of organisms are amply demonstrated by these and many other data (Hengeveld, 1989). These natural range fluctuations are largely a result of changes in climate, which occur on every ecological and evolutionary time scale (Davis, 1986). The result is that ecological communities are in perpetual disequilibrium (Davis, 1986). Community succession proceeds, therefore, toward a moving target.

Anthropogenic effects clearly increase species invasions, but those introductions occur into communities whose species composition is already naturally dynamic. In this context, as Brown (1989) has argued, not all invading species are "bad." In fact, it becomes difficult to distinguish "natural" and "unnatural" invasions. For the most part, species are not entering static, equilibrial, tightly coevolved communities. The mesic forests of Wisconsin, for example, have only had their current species composition for about 2000 years (Davis, 1986). Human-aided invasions are only one of several causes of changes in native community structure. These observations should not make anyone complacent about anthropogenic effects, the rates of which may be so fast as to be devastating to current communities. My point is that consideration of the natural dynamics of communities, including insights from processes operating on scales of from 10 to 100,000 years, will lead to greater understanding of current and future species range expansions (Davis, 1986).

In addition, because species naturally come *and go* from communities, a singular focus on *invading* species is too narrow. The more general goal —to understand the impact of species interactions in communities—will be furthered by examining the impact of species *deletions* as well as species additions.

The alternative perspective I propose is that species additions and deletions are common and natural phenomena in communities, made much more common by human activities. Whereas the invasion literature demonstrates that generalities encompassing all case studies do not exist, I expect that an examination of both natural and human-imposed species dynamics and their effects on particular communities will provide insight into how communities will respond to global change.

In the next section, I illustrate the dynamics of freshwater communities to both natural and anthropogenic changes. I focus particularly on the effects of adding or removing single species because I believe identifying strongly interacting species (*sensu* Paine, 1980) will be important to predicting and managing the effects of climate and habitat change on communities. I will focus on freshwater communities because: (1) they have received far less attention than other terrestrial and marine habitats in the context of climate and habitat change; (2) availability and quality of water resources will become a central issue under many scenarios of climate change (Ashton and Mitchell, 1989; Ausubel, 1991; Karr, 1991); and (3) freshwater systems are what I know best.

First, I use the known relationship between fish species composition in Wisconsin lakes and lake characteristics to predict the sorts of changes in species composition that could occur during global warming. Second, I describe the major changes in Great Lakes communities that have resulted from intentional and accidental species introductions. Whereas some of these invasions and their effects were predictable, most were not. Third, I describe how detailed studies of an ongoing invasion by crayfishes in Wisconsin lakes can support a model based on percolation theory that predicts both the fate and impact of the crayfish invasion.

AQUATIC CASE STUDIES

Fishes in north temperate lakes

The presence and abundances of fishes in the north temperate lakes of North America are determined by a thoroughly examined set of historical and contemporary ecological factors that operate on different spatial and temporal scales (Tonn and Magnuson, 1982; Rahel, 1984; Tonn et al., 1990; Hinch et al., 1991). These communities therefore offer a good framework for considering the potential impact of large-scale climate and habitat changes (Tonn, 1990). I will limit the following discussion to Wisconsin, but a similar conceptual approach would apply more generally to north temperate lakes (Tonn et al., 1990).

Among small to moderate-size lakes in northern Wisconsin, lake size, depth, and pH are accurate predictors of fish assemblages (Figure 2; Tonn et al., 1990). Physical and chemical variables determine the ability of top predators to survive, and therefore constrain the entire fish assemblage. For example, unless there is a stream to escape into, only species like mudminnows and minnows, with physiological or behavioral adaptations to hypoxia, can survive in small, shallow lakes where winterkill (mortality from hypoxia under ice) is frequent. These small species do not occur in lakes containing piscivores because they are susceptible to predation (Tonn et al., 1990 and references therein). Winterkill offers a refuge from piscivores, which are intolerant of hypoxia.

Among lakes where winterkill is infrequent (Figure 2), seepage (without stream connections) and low-pH lakes are dominated by largemouth bass. Species that co-occur with these piscivores are spiny-rayed, and are therefore resistant to predation. Although other fishes (including pike) occur in

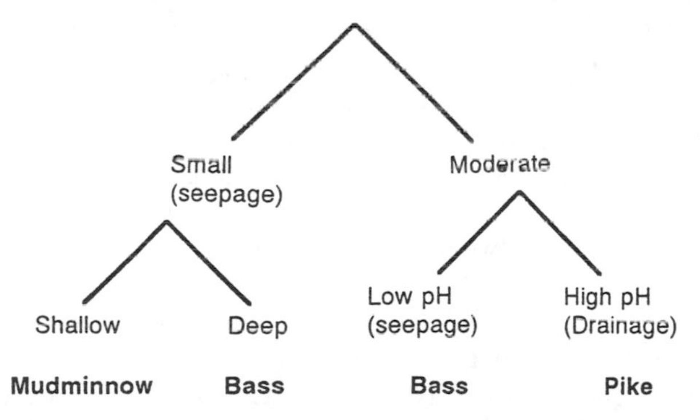

FIGURE 2. Fish assemblages and environmental predictors for lakes in northern Wisconsin. Species listed are the dominant ones in each lake type; except for small, shallow lakes, there is broad overlap in species composition. (After Magnuson et al., 1989.)

these lakes, they are not as tolerant of low pH or mild winter hypoxia as are bass. In lakes of higher pH (typically drainage lakes, i.e., with surface connections), pike dominate. Even if low levels of dissolved oxygen occur during winter in drainage lakes, pike survive by migrating into streams. Within these lakes, predation by pike regulates the abundance of other species (Tonn et al., 1990). Thus, when the physical or chemical factors in a lake change sufficiently to cross the threshold for the support of piscivores, the impact cascades down through the food chain.

These direct and biotically mediated effects of physical and chemical factors on fish assemblages make possible some preliminary hypotheses about the effects of climate and habitat changes. Patterns of fish species composition in lakes near biotic thresholds should vary dramatically from year to year because fluctuations in climatic forces (temperature, rainfall, duration of ice cover, depth of snow cover) will episodically change the frequency of winterkill, stream flow, and ionic concentrations. For example, pike may dominate a drainage lake one year, but be eliminated by winterkill the next. A 20-year data set from the Experimental Lakes Area (ELA) of southern Ontario (Schindler et al., 1990) suggests that the climatic factors driving such natural dynamics are the main factors that will change as a result of global warming. In the ELA, lakes have already become warmer, duration of ice cover has decreased, water renewal rates have decreased, and concentrations of dissolved ions such as calcium (and therefore conductivity) have increased. Decreased precipitation and increased evaporation (Schindler et al., 1990) also imply that lake levels will drop.

In the hypothetical landscape shown on the left side of Figure 3, I have portrayed the current pattern of fish assemblages in northern Wisconsin. On the right side are the same lakes after climate warming has had direct and indirect effects on the fish assemblages. Very shallow seepage lake A disappears. Small, shallow mudminnow lake B becomes fishless as winter water volume declines. Decreased water level and reductions in the winter dissolved oxygen level push smallmouth bass lake C over a threshold, eliminating bass and converting the lake via predation release to a mudminnow lake. Lake D, a moderate-size mudminnow lake, remains within the range of parameters that supports a mudminnow assemblage.

Some of the major effects in this hypothetical lakescape occur via habitat fragmentation—the loss of stream connections between lakes. For example, the surface connection between lakes E and G might disappear, eliminating the stream and lake G as winter refuges for pike. Lake E might therefore cross the pike-to-bass threshold. For local hydrological reasons, the stream between lakes F (a low-pH, low-conductivity bass lake) and G does not disappear, but conductivity in lake F increases over the threshold for pike dominance. Therefore, in the same landscape, opposite transformations can occur, from pike to bass (lake E) and from bass to pike (lake F). Larger lakes in the landscape, like G and H, would undergo changes in tempera-

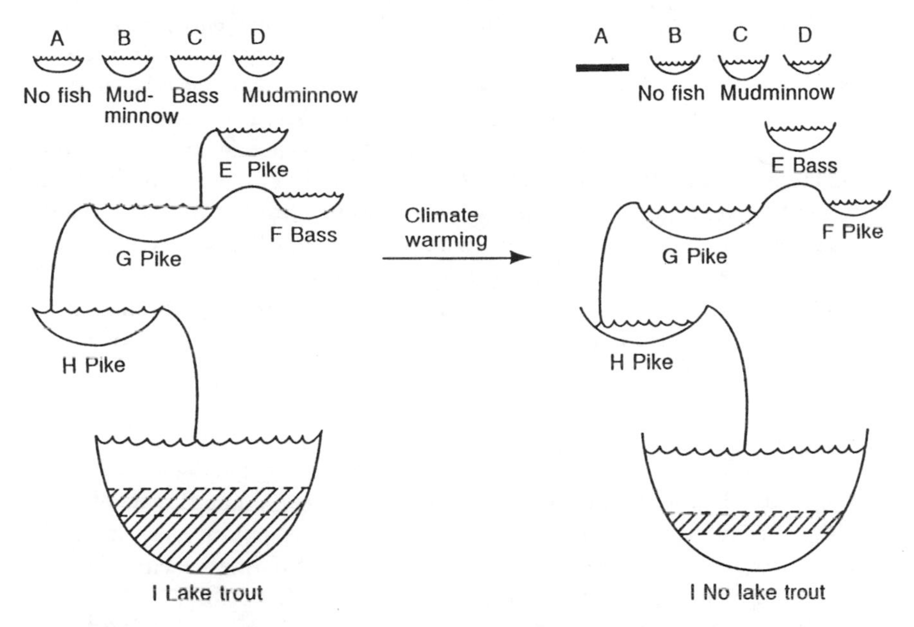

FIGURE 3. Hypothetical fish assemblages in northern Wisconsin lakes before (left) and after (right) climate warming. Lines connecting lake cross sections are streams. Shaded region in lake I represents strata habitable by lake trout. See text for explanations of changes in fish assemblages.

ture, duration of ice cover, and ionic concentrations like those described in Schindler et al. (1990), but they would not be pushed over a fish assemblage threshold. Thus lake size should be related to resistance to some environmental changes.

The largest lake (lake I in my hypothetical landscape), however, is expected to undergo some profound changes. This lake is typical of large, deep lakes in northern Wisconsin and of smaller lakes further north in Canada. The year-round occurrence of well-oxygenated, cool water in such lakes permits the occurrence of lake trout (and other glacial relicts, including opossum shrimp). As noted by Schindler et al. (1990), climate warming leads to the restriction of summer habitat for these species because average water temperature and thermocline depth increase. In addition, hypolimnetic oxygen concentrations would also decline (Blumberg and Di Toro, 1990; Schertzer and Sawchuk, 1990). In some lakes, lake trout and other glacial relicts consequently would become extinct.

In general, then, large changes in northern Wisconsin fish assemblages could easily result from climatic and associated habitat changes. This example illustrates the potential predictive power that emerges from intensive study of particular communities. Furthermore, it illustrates that in order to understand community responses to climate and habitat change, it is not enough to understand the direct physiological consequences of altered physical and chemical conditions. If only physiological consequences of climate change were considered, I would predict that increasing severity of winterkill in a northern Wisconsin lake would have only one major effect—the elimination of bass. If, on the other hand, I include current knowledge of community interactions in the prediction, I would predict that as a result of bass reduction, minnows would increase dramatically, with changes in other trophic levels as a result (see Carpenter et al., this volume). To accurately anticipate community response to climate and habitat change, it is therefore essential to understand the impact of species additions and deletions.

The North American Great Lakes

In the Great Lakes, human intervention has caused dramatic changes in communities of fish and invertebrates over the last century, typified by species introductions and consequent extinctions of native species (Figure 4A). The following account applies most closely to Lake Michigan, and is taken largely from Stewart et al. (1981), Kitchell and Crowder (1986), and Eddy and Underhill (1974).

Native fish assemblages of the upper Great Lakes were dominated by lake trout and coregonids, some of them endemic. The construction of the Welland Canal in 1829 allowed the colonization of the lakes by two marine species, lamprey (in the 1920s) and alewife (in the 1930s). Rainbow smelt were introduced about 1912. Overfishing, combined with predation by lampreys, dramatically reduced native lake trout, coregonids, and other large fishes. With the virtual elimination of the piscivorous lake trout, populations of the exotic planktivores—smelts and alewives—exploded. Native planktivores, especially coregonids, declined further, with some species driven to extinction, probably as a result of competition with exotics. Implementation of effective chemical control of lampreys, beginning around 1955, allowed the reintroduction of lake trout about 1965. To control the nuisance population of alewives, several species of exotic salmonids have been stocked in massive numbers since the 1970s. With increased stocking of salmonids, alewife populations declined in the 1980s; populations of some native planktivores, especially coregonids and perch, increased, apparently in response to decreased competition from alewives.

The net result of these anthropogenically induced changes is a Great Lakes upper food web consisting almost completely of exotic fish species (Figure 4B). As in inland lakes (see Carpenter et al., this volume), changes in the higher trophic levels of Lake Michigan have resulted in large effects on zooplankton, algae, and water clarity (Scavia and Fahnenstiel, 1988).

Currently the lower food web is undergoing direct invasion by at least

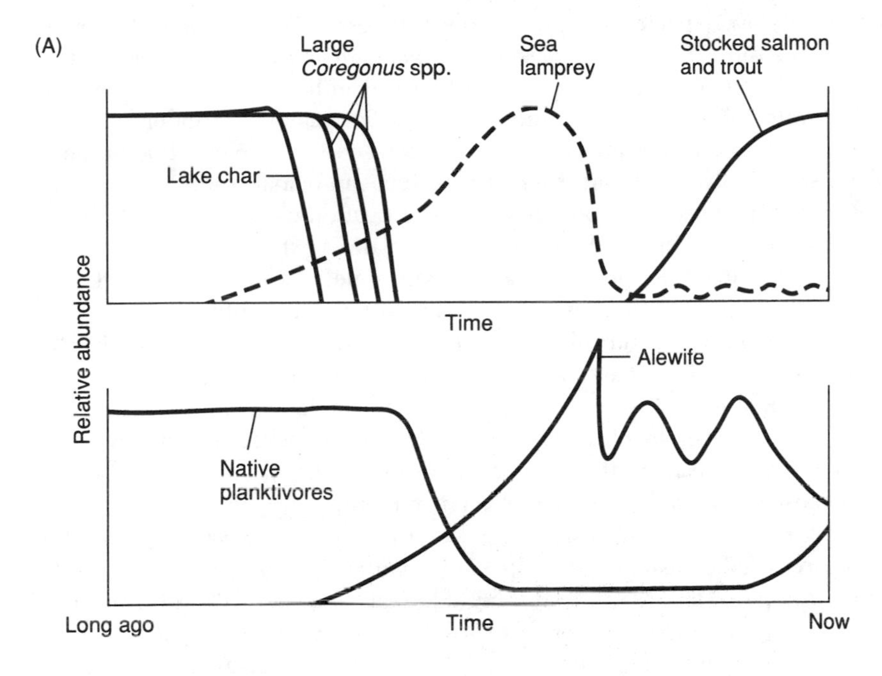

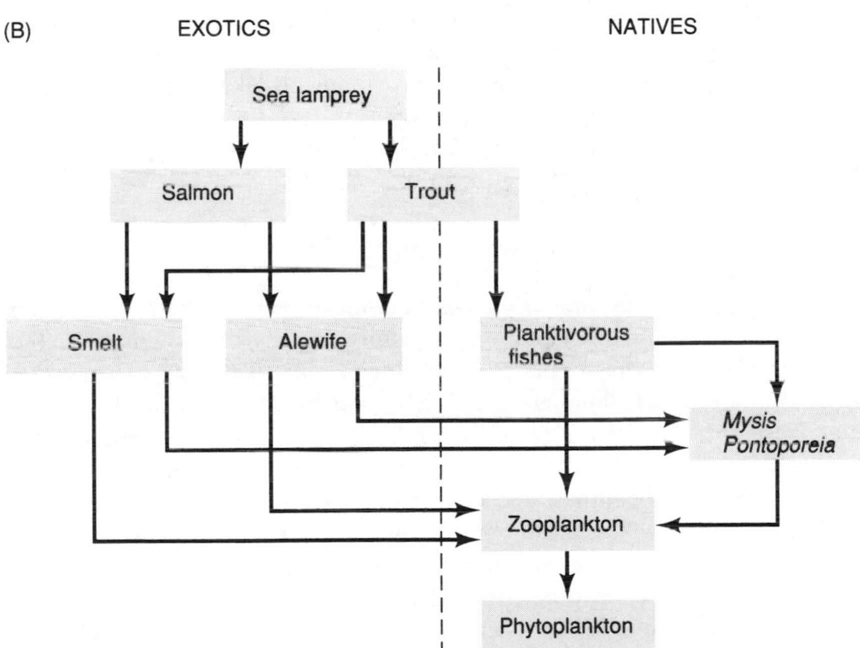

FIGURE 4. (A) Generalized representation of changes in fish populations in the North American Great Lakes. (B) Generalized pelagic food web of the Great Lakes. (From Kitchell and Crowder, 1986.)

two invertebrate species: the predaceous cladoceran *Bythotrephes* (Lehman, 1987), and the zebra mussel, *Dreissena polymorpha* (Hebert et al., 1989; Roberts, 1990). Both are Eurasian and were probably transported in the ballast water of ships, as are hundreds of species each year (Roberts, 1990). The impact of zebra mussels on other species is still unclear, but millions of dollars are already being spent to protect municipal water plants from mussel infestations, with total losses and mitigation costs expected to be in the neighborhood of 4 billion dollars over the next few years (Roberts, 1990). The impact of species invasions and deletions on the Great Lakes has clearly been huge, and will only grow larger in the future.

The outcome of some of the most important species invasions in the Great Lakes could not have been predicted by the verbal models reviewed above. Populations of alewives, lampreys, and smelts, previously anadromous Atlantic Ocean species, now live in freshwater lakes without ever returning to the ocean, thus violating our intuition that the invaded environment must be similar to the native environment.

Under the major global warming scenarios, habitable water volume for most Great Lakes fish species will likely increase (Magnuson et al., 1990; Hill and Magnuson, 1990). Assuming adequate food, growth rates of most fishes would increase as a result of a longer growing season and warmer temperatures (Hill and Magnuson, 1990). However, Hill and Magnuson (1990, and other papers in Regier et al., 1990) emphasize that food web dynamics will be a critical component for projecting the effects of climate change on fish communities. Thus, both the inland lakes and Great Lakes examples emphasize that community responses to global change will depend not only on physiological responses, but also on species interactions, whether the species are native or exotic.

Crayfish dispersal in lakescapes

Over the past 60 years, crayfish species composition in northern Wisconsin lakes has been changing rapidly as a result of species invasions from the southern Midwest (Olsen et al., 1991). Since about 1930, when *Orconectes virilis* was the most abundant crayfish in northern Wisconsin lakes, two congeneric crayfishes have invaded, probably via the bait buckets of fishermen (Olsen et al., 1991 and references therein). *Orconectes propinquus* arrived around mid-century, and *O. rusticus* probably in the 1960s.

Ecologically similar, these three congeners share life cycle patterns, omnivorous food preferences, and rocky habitats in mesotrophic lakes. Morphologically, the species differ primarily in adult size, with *O. propinquus* markedly smaller than its two congeners (Olsen et al., 1991). Hybridization occurs, at least between *O. propinquus* and *O. rusticus* (Capelli and Capelli, 1980; Berrill, 1985).

Data spanning 15 years for nine lakes suggest that *O. propinquus* routinely displaces *O. virilis*. Subsequently, both are replaced, without exception, by *O. rusticus* (Figure 5). A more detailed long-term data set for Trout Lake shows that relative abundances of *O. virilis* and *O. propinquus* can shift, but confirms that *O. rusticus* always displaces both (Lodge et al., 1986).

Establishment of *O. rusticus* results not only in congener extinction, but in reductions in littoral zone macrophytes and invertebrates (Lodge et al., 1985; Lodge and Lorman, 1987; Weber and Lodge, 1990; Lodge et al., unpublished), and probably in changes in nutrient cycling (Carpenter and Lodge, 1986). *O. rusticus* reduces benthos more than its crayfish congeners because it suffers lower size-specific predation rates by fish (DiDonato and Lodge, unpublished), is larger than one congener (decreasing predation rates further and reducing behavioral inhibition by fish predators), and has greater weight-specific feeding rates (Olsen et al., 1991).

The consequences of establishment of *O. rusticus* are now predictable from experience: extinction of two congeneric crayfishes and reductions in benthic invertebrate and macrophyte populations. As with many other examples, however, the invasion models reviewed earlier in this chapter could not have predicted the incredible success and impact of *O. rusticus*. Its native habitat is primarily streams in a small native range, its body size is similar to or smaller than *O. virilis*, and it invades habitats already occupied by two congeners (certainly not a vacant niche). These and other aspects of

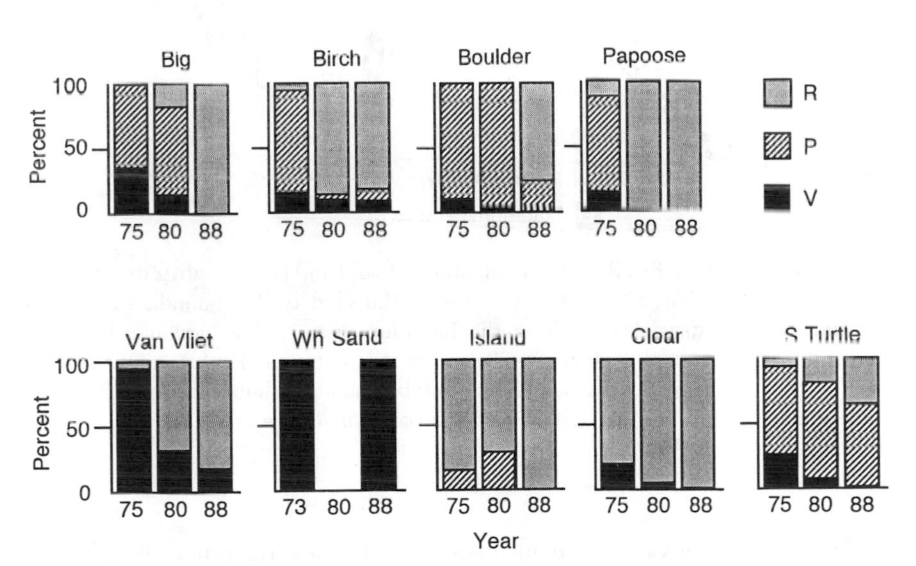

FIGURE 5. Changes in trap index of relative abundance of adult crayfish in nine northern Wisconsin lakes. R, *Orconectes rusticus*; P, *O. propinquus*; V, *O. virilis*. (After Olsen et al., 1991.)

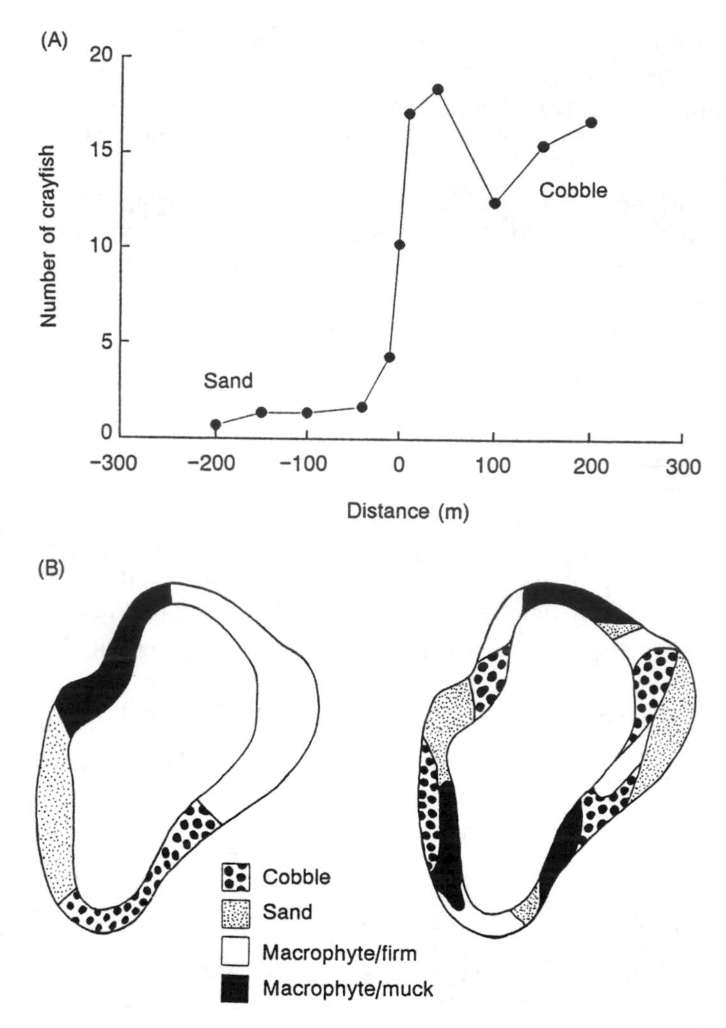

FIGURE 6. (A) Crayfish density in neighboring sand and cobble substrates in Trout Lake, Wisconsin. Negative distances are from the sand–cobble boundary into sand habitat; positive distances are from the boundary into cobble habitat. (B) Hypothetical maps of two lakes showing different spatial patterns of substrates. The lake on the left, with broad expanses of less suitable habitat, should be invaded more slowly than the lake on the right, which has corridors of preferred habitat (cobble). (After Kershner, 1992.)

the *O. rusticus* invasion are inconsistent with several tenets of popular verbal models regarding invasions.

Given the difficulties of predicting which organisms are likely to be invaders, a more realistic but still vitally important goal is to develop invasion-specific models as ongoing invasions proceed. Such efforts will at least allow predictions of subsequent dispersal rates and directions, as well as community impact, and allow mitigation planning. For example, Capelli and Magnuson (1983) discovered some clear statistical patterns in the kinds of lakes invaded by *O. rusticus*. Seepage lakes and lakes isolated from human activity were less likely to be invaded, as were small lakes (Capelli and Magnuson, 1983). Lakes with calcium concentrations of less than 2.5 mgL^{-1} were not invaded at all (Capelli and Magnuson, 1983). Thus, the occurrence of climate-induced increases in calcium concentration in small Canadian lakes (Schindler et al., 1990) suggests that climate warming may increase the number of lakes habitable by crayfish. This, in turn, would induce large changes in within-lake communities. Thus, both the placement of lakes in the landscape and within-lake characteristics are determinants of species dynamics.

Once a lake is colonized by crayfish, features of the lake landscape may help explain the extremely variable rates of population growth and dispersal (Figure 6). Mark Kershner and I are developing a spatially explicit, individual-based simulation model, in which the movement of each crayfish is stochastic, but depends partly on the occupied habitat and the surrounding habitat (Kershner, 1992). Abundant field evidence demonstrates that crayfish respond to habitat structure, strongly preferring rocky (cobble) habitats that provide refuge from fish predators (Figure 6A). The goal of the simulation model is to test the effect of habitat heterogeneity on crayfish dispersal within a lake. For example, we expect that a lake with abundant rocky patches (providing corridors for crayfish movement) will be invaded more quickly and with greater impact than a lake with broad expanses of less suitable crayfish habitat (Figure 6B).

Our approach was inspired by the landscape models of Turner and colleagues (Turner and Bratton, 1987; Turner et al., 1989). These models are being used to test disturbance propagation (Turner et al., 1989), but are underused by ecologists studying invasions. Our model structure includes the simultaneous overlaying of a habitat grid and a crayfish grid (Figure 7). Parameters governing the probability of crayfish movement as a function of habitat type have been derived from field measurements (see Figure 6A). The spread of crayfish in randomly generated habitat grids suggests that crayfish invasions within particular lakes will be very sensitive to the amount of preferred (cobble) habitat (Figure 8). Future generations of the model will incorporate the effect of habitat on population growth, and will thus produce more realistic long-term results, which can be tested using a long-term data set for the crayfish invasions of Trout Lake (Lodge et al., 1986).

It is increasingly important to incorporate heterogeneous environments in models because of habitat fragmentation by humans. The success of some diffusion models (e.g., Andow et al., 1990) suggests, however, that detailed

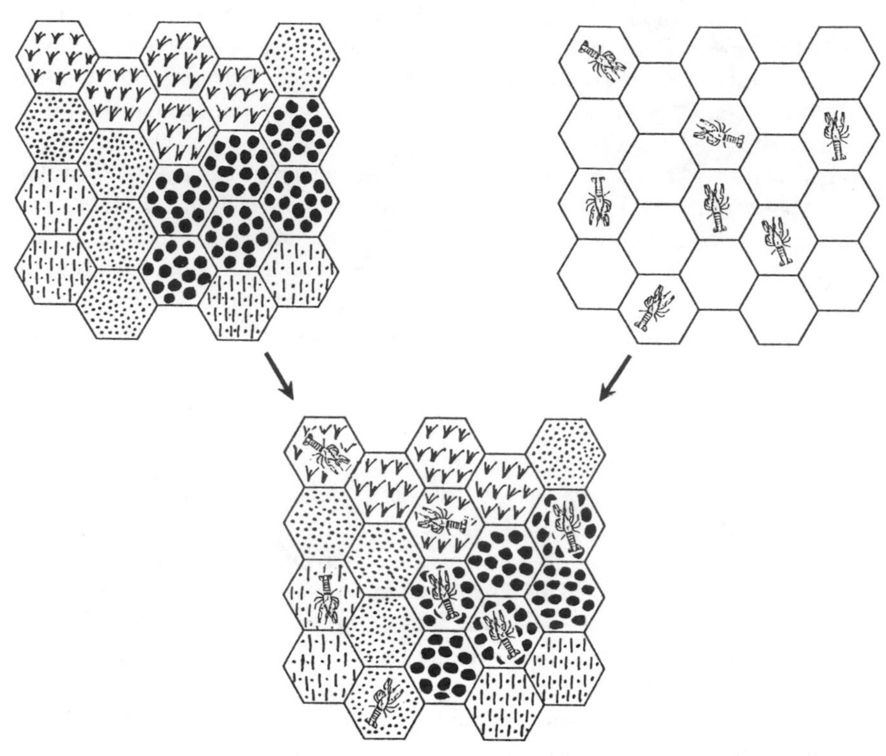

FIGURE 7. Schematic of joint execution (bottom center) of habitat (upper left) and crayfish (upper right) arrays in spatially explicit simulation model of crayfish dispersal. (After Kershner, 1992).

descriptions of heterogeneity may be unnecessary for some scales of observation.

All quantitative models must include accurate parameters to be predictive. Implementation of any models will therefore require a time-consuming and expensive case-by-case approach. Like the previous two case studies, the invasion of northern Wisconsin lakes by *O. rusticus* illustrates the lack of generality and predictive power of the invasion models reviewed above, and the importance of trophic interactions in predicting the impact of invading species. In addition, it emphasizes the need for spatial modeling approaches to predicting the community impact of habitat fragmentation.

CONCLUSIONS AND GOALS FOR FUTURE RESEARCH

General models of invasions such as those reviewed in this chapter do not allow ecologists to make accurate predictions regarding specific species or

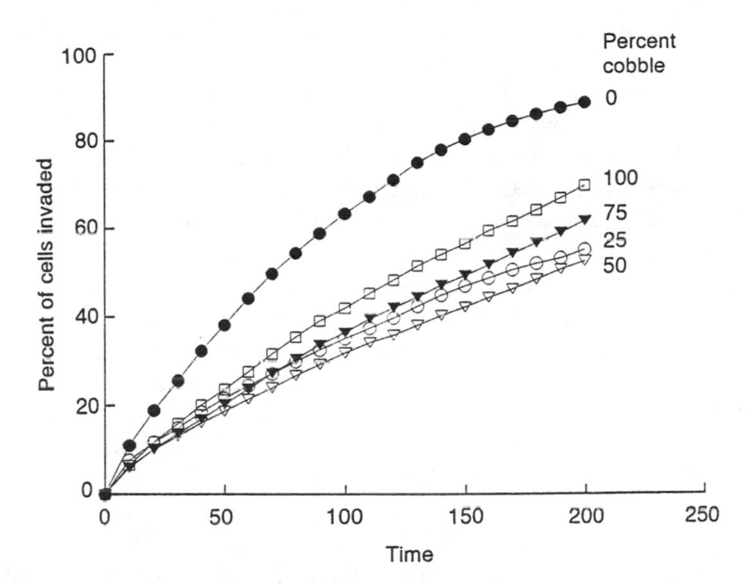

FIGURE 8. Simulation model results showing the percentage of habitat cells colonized by crayfish as a function of time steps and the abundance of cobble habitat (ranging from 0 to 100 percent of the habitat cells). Time has no absolute units. (After Kershner, 1992.)

habitats. Predictive understanding will emerge from focused studies at more specific levels of taxonomy or habitats. The conceptual framework underlying many invasion studies needs to be tightened. For example, it is widely understood that what constitutes a disturbance for one species constitutes an essential part of life for another. Thus, a statement like "disturbance enhances invasion" is unlikely to provide a general basis for predictions.

The impact of changes in climate and habitat on communities cannot be predicted solely on the basis of species responses to physical and chemical conditions. Species interactions, particularly those involving strong interactors (*sensu* Paine, 1980), will often be an important determinant of community response. The case studies in this and other chapters (see Carpenter et al., this volume) have illustrated that the addition or deletion of one species can have profound impacts on community structure and ecosystem function. Understanding the consequences of global change will therefore require a much better understanding of the community and ecosystem roles of individual species. A major challenge will be predicting which species are likely to be strong interactors, and therefore most likely to produce large community and ecosystem effects.

Paleoecological studies indicate that climate is never constant, and consequently most communities are probably disequilibrial (e.g., Davis, 1986).

Thus, species invasions and deletions occur frequently, even without any human influence. However, anthropogenic climate and habitat changes will dramatically increase the frequency of invasions from outside current biogeographic boundaries. The paleoecological perspective comes from trees, which are long-lived and respond more slowly to climate and habitat changes than do animals (Davis, 1986). That fact alone suggests that novel combinations of vegetation and animals will result from environmental change. Several key questions must be addressed if we are to make progress. At what spatial and temporal scales are equilibrial or disequilibrial conceptual approaches more useful? in what sorts of communities? and with what sorts of taxa? Which taxonomic groups will be slow to respond to environmental change and thus be left behind as the ranges of other species shift? Which groups will have ranges that respond most quickly to environmental change and thus become invaders? Multiscale conceptual models like that of Tonn (1990) and Lodge et al. (1987) provide frameworks for considering these issues.

In habitats increasingly fragmented by humans, habitat heterogeneity will probably be a critical factor in understanding the direction and rates of species range changes. Armesto et al. (1991) suggest that the probability of invasion during terrestrial succession is positively related to the heterogeneity of existing vegetation. Patterns of landscape heterogeneity and connectivity among patches (lakes) are particularly obvious and important for colonization patterns of freshwater organisms. Even if warming of northern lakes makes them suitable for more southerly fish, colonization routes are absent for many lakes (Hinch et al., 1991). There is unlikely to be a general relationship between invasibility and habitat heterogeneity that operates across multiple spatial scales. Instead, we should probably expect the relationship between dispersal and habitat heterogeneity to be taxon-specific, size-specific, and dispersal mechanism-specific.

Modeling efforts need to proceed hand-in-hand with empirical work (Kareiva and Andersen, 1988) so that basic demographic parameters can be estimated for models. However, even if accurate parameters were available, most current models would be inadequate to the task of predicting response to climate or habitat change because they do not incorporate temporal changes in the environment or habitat heterogeneity. New modeling approaches are needed, with carefully chosen targets. Given the assessments above, general models need to be modified for application on a case-by-case basis. A major unresolved question concerns the level of realism of climate change and spatial structure that needs to be included for accurate predictions of invasions.

Empirical work and modeling need to focus on threshold responses of organisms to environmental factors affected by climate or habitat change. It is particularly important that we identify the degree of environmental change that is likely to shift a community from dominance by one set of species interactions to dominance by a different set of species interactions.

ACKNOWLEDGMENTS

I thank Peter Kareiva for sharing unpublished materials; Steve Carpenter for helping to sharpen my approach to this topic; Robert McIntosh for providing important references on terrestrial vegetation; and Mark Kershner for discussions, references, and continuing collaboration on the simulation model of crayfish dispersal. Anna Hill, Kareiva, Kershner, McIntosh, and an anonymous reviewer provided constructive reviews of the manuscript.

CHAPTER 24

SPECIES DIVERSITY, SPATIAL SCALE, AND GLOBAL CHANGE

Susan Harrison

An understanding of large-scale dispersal is essential for predicting community responses to global warming and habitat change. Current climate forecasts suggest a mean rise of at least 3°C in global temperature in the next century and a roughly 250 km poleward shift of climatic zones (Peters, 1988). Presumably, present-day assemblages will experience the selective loss of those species that are unable to tolerate the change or to shift their ranges along with moving climatic zones. Insular species will be especially vulnerable to extinction because of the physical barrier they face to shifting their ranges. For the same reason, fragmentation of habitat will exacerbate the loss of species from terrestrial communities (Peters and Darling, 1985). But beyond these qualitative predictions, a great deal remains unknown. One of the most important unanswered questions concerns how well most species are able to reach suitable habitats across distances of tens, hundreds, or thousands of kilometers.

For some taxa, fossil evidence records the responses of species distributions to past climate change. For example, mapping of fossil pollen has shown that many European and North American forest trees moved northward at on the order of 10^2–10^3 meters per year during the post-Pleistocene warming. Controversy exists over whether these rates of movement represent the species' maximum rates of dispersal, or whether species tended to track closely their shifting climatic zones. This is a difficult question to resolve, given that the precise nature of the climatic limits to species distributions is unknown and that the different features of climate did not shift in synchrony. Moreover, the fossil pollen data itself must be used to reconstruct the distributions of past climates. A final problem is the extreme sensitivity of migration rate estimates to the detectability of fossils (see, e.g., Roberts, 1989; Huntley and Webb, 1989; Bennett, 1986).

With these caveats in mind, two important generalizations emerge from the fossil evidence for trees and several other taxa (e.g., beetles, mammals, mollusks). (1) More species appear to have shifted their distributions than to have gone extinct or to have evolved morphologically in situ, suggesting that in many cases, dispersal has been adequate to track change. (2) Species have migrated individualistically, rather than as tight assemblages. Thus it appears that major present-day temperate tree communities have existed only since the Pleistocene. The implications are that broad, first-order predictions about changes in communities can be made on an autecological basis (e.g., Graham and Lundelius, 1984; Peters, 1988; Graham and Grimm, 1990; Cronin and Schneider, 1990).

Here I examine an alternative line of evidence on the question of large-scale dispersal. Analyses of present-day biogeographic patterns are used to ask, at what spatial scale do species ranges tend to be limited by lack of dispersal? If species are often absent from suitable habitats within, say, hundreds of kilometers of their existing ranges, it would suggest generally poor dispersal abilities and gloomy prospects with respect to climate change. However, if within such distances ranges tend to correspond to the availability of habitat, the outlook is somewhat better. Rather than look in detail at the distributions and habitat requirements of a few selected species, I will venture some predictions about broad spatial patterns of diversity that might be generated by dispersal limitation, and test these using distributional data for entire taxa.

BETA DIVERSITY

The focus of this analysis is turnover in species composition among localities, or *beta diversity*. Whittaker (1960) pioneered the concept of partitioning regional (gamma) diversity into its local (alpha) and turnover (beta) components. Whittaker defined beta as

$$(s/\bar{a}) - 1$$

where s is the total number of species in a region (gamma diversity) and $\bar{a}$ is the average number of species at each locality within the region (average alpha diversity). Beta is thus the proportion by which the richness of a region exceeds the average richness of a single locality. It represents the degree of spatial differentiation in communities. (Note that various authors define the terms alpha, beta, and gamma diversity differently, and that "diversity" here refers solely to species richness.)

The causes and interpretation of beta diversity depend on the scale at which "localities" and "regions" are defined. On a very local scale, disturbance and species interactions produce transient differences in composition among sets of equivalent communities (Caswell and Cohen, 1991). At the larger biogeographic scale, the scale with which this paper is concerned, the two general causes of beta diversity are what Shmida and Wilson (1985) term habitat differentiation and ecological equivalency. Habitat differentiation is the change in community composition associated with the variety of habitats in a region. Ecological equivalency refers to species replacement between separate units of similar habitat. Ecological equivalency is of interest here because one of its causes is the absence of large-scale dispersal.

BETA DIVERSITY AND LARGE-SCALE DISPERSAL: PREDICTIONS

Clearly, at large enough distances, isolation and limited dispersal produce species turnover. Consider, for example, the functionally convergent but taxonomically distinct plant communities of the Mediterranean climate regions of Europe, South Africa, South America, and North America. But at what smaller scale does spatial isolation begin to affect community patterns? This is an empirical and largely unaddressed question (for major exceptions, see Shmida and Wilson, 1985; Cody, 1986) whose answer requires analysis of spatial patterns in species distributions.

We may expect the answer to be linked to taxon-specific dispersal abilities. The ranges of species are shaped both by where their habitats are found and by which areas of suitable habitat they have reached in their evolutionary histories. Compared with good dispersers, poorly dispersing species should be absent from more areas of suitable habitat, and thus have more restricted ranges on average. In turn, these smaller ranges produce higher turnover at a given spatial scale. Thus, for a particular set of localities, we expect beta diversity to be higher among sedentary organisms (e.g., soil invertebrates) than motile ones (e.g., birds or butterflies).

This chain of reasoning leads to two testable predictions about beta diversity and its relationship to dispersal limitation.

1. To the extent that dispersal limitation is important in shaping community patterns, distance between localities should be a good correlate of turnover: the composition of a community should become more and more different as spatial separation increases. The interpretation of such a pattern may be confounded by a tendency for habitats to become more dissimilar with distance. However, plotting turnover against distance at least provides a first indication of whether dispersal limitation may be important at a given scale.
2. Among a common set of localities, beta diversity should be higher in poorly dispersing than in better-dispersing taxa.

Here I review some recent work that tests these predictions. The first study analyzes distributional data for 15 taxa on a common set of sites in Britain. Two subsequent analyses use data for varied taxa on different sets of islands or terrestrial habitat isolates. The results suggest that the effects of dispersal limitation on communities are not generally strong until distances of one thousand to several thousand kilometers. At smaller scales, patterns of species distribution appear to be influenced much more by habitat factors. However, fragmentation and barriers (e.g., water) appear to increase the effective isolation produced by distance.

It is often said that the combined effect of anthropogenic habitat destruction, extinctions, and introductions is to homogenize the earth's biota. Work reviewed here quantifies this prediction in terms of the different components of diversity. In the land avifauna of 49 Pacific islands, beta diversity has declined more than local (alpha) or regional (gamma) diversity, evidently because the most narrowly distributed taxa are the most susceptible to local extinction (C. D. Thomas and S. Harrison, in preparation.) We predict that global warming will have a similar impact on many communities, disproportionately reducing the differentiation component of diversity.

BETA DIVERSITY IN THE BRITISH FLORA AND FAUNA

Harrison et al. (1992) compared beta diversity among 15 taxa in Britain, where atlases of distribution maps are available for plants and many vertebrate and invertebrate groups. We chose the largest possible region (transects across Britain), and large localities within it (50 km^2 National Grid squares), in order to maximize the potential influence of distance and dispersal limitation on species turnover. We created two arrays of 50 km^2 grid squares on the longest possible transects, going approximately north–south and west–east, with 17 and 8 squares, respectively. These transects are shown in Figure 1. For 8 animal and 7 plant taxa, we recorded the presence or absence of each species in each square, and calculated measures of beta diversity for each taxon and transect.

To allow comparisons between the two transects of unequal length, we used a modified formula for beta, namely,

$$\frac{(S/a - 1)}{N - 1} \times 100$$

where N is the number of localities (squares) in the region (transect). "Beta-1," as we call this measure, ranges from 0 (complete homogeneity) to 100 (complete turnover). In other words, it represents the percentage by which the total diversity increases with each additional locality.

Alpha diversity varies among localities. Because beta-1 is inversely related to average alpha, its value is strongly influenced by the inclusion of species-poor localities. Moreover, in most British taxa there is a strong tendency for alpha diversity to increase both north to south and west to east, closely paralleling trends in climate. High diversity is associated positively with sunshine and negatively with rainfall. This means that climate-

FIGURE 1. Locations of the two transects of 50 × 50 km grid squares. (From Harrison et al., 1992.)

driven gradients in alpha diversity may produce the misleading appearance of an increase in beta diversity with distance.

To address this problem, we examined another beta measure, which we refer to as "beta-2":

$$\frac{(S/a_{\max} - 1)}{N - 1} \times 100$$

where $\bar{a}_{\max}$ is the maximum value of alpha diversity in the collection of localities. Beta-2 measures the amount by which regional (transect) diversity exceeds the maximum diversity attained within one locality (square). Beta-2, like beta-1, ranges from 0 to 100, and it converges on beta-1 when variability in alpha is small.

We ranked taxa by dispersal ability in three ways:

1. Animals: Flying (bees, breeding birds, butterflies, dragonflies, moths) > nonflying (fishes, molluscs, woodlice).
2. Plants: Small propagules (ferns, mosses, orchids) > medium-sized propagules (composites, umbellifers) > large propagules (Rosaceae, native trees).
3. Animals > plants.

Values of beta-1 and beta-2 for each taxon are shown in Table 1. Beta

TABLE 1. Beta diversity in the British flora and fauna.

Taxon	S[a]	Beta-1[b]	Beta-2[b]	Source
		NORTH-TO-SOUTH TRANSECT		
Bees	24	6.7	1.6	Alford (1973)
Birds	191	3.3	2.5	Sharrock (1976)
Butterflies	59	4.4	0.7	Heath et al. (1984)
Dragonflies	38	12.5	1.2	Hammond (1977)
Moths	90	5.9	0.5	Heath and Skelton (1973)
Fishes	47	8.0	2.6	Maitland (1972)
Mollusks	165	3.7	1.3	Kerney (1976)
Woodlice	24	9.3	3.8	Harding and Sutton (1985)
Mosses	104	3.9	0.9	Smith (1973)
Ferns	59	4.6	1.6	Perring and Walters (1972)
Orchids	37	6.4	1.2	Perring and Walters (1972)
Composites	102	3.9	1.3	Perring and Walters (1972)
Unbellifers	57	5.9	1.2	Perring and Walters (1972)
Rosaceae	53	3.2	1.1	Perring and Walters (1972)
Native trees	48	2.9	1.6	Perring and Walters (1972)
		WEST-TO-EAST TRANSECT		
Bees	22	9.0	2.3	Alford (1973)
Birds	166	5.7	4.3	Sharrock (1976)
Butterflies	57	3.6	0.9	Heath et al. (1984)
Dragonflies	34	12.6	4.4	Hammond (1977)
Moths	87	4.1	1.7	Heath and Skelton (1973)
Fishes	39	7.0	3.7	Maitland (1972)
Mollusks	153	4.7	2.4	Kerney (1976)
Woodlice	26	12.3	6.3	Harding and Sutton (1985)
Mosses	99	10.1	4.6	Smith (1973)
Ferns	37	8.1	4.0	Perring and Walters (1972)
Orchids	36	11.4	5.4	Perring and Walters (1972)
Composites	106	4.0	1.7	Perring and Walters (1972)
Unbellifers	56	6.1	3.1	Perring and Walters (1972)
Rosaceae	49	3.4	2.4	Perring and Walters (1972)
Native trees	39	2.6	0.7	Perring and Walters (1972)

[a]Total number of species in the taxon and transect.
[b]Beta-1 and beta-2 are defined in the text.

TABLE 2. Rankings (highest to lowest) of beta diversity in the British flora and fauna, according to presumed dispersal ability.

	Beta-1	Beta-2
	NORTH-TO-SOUTH TRANSECT	
Animals[a]	FNFFNFNF	NNFFNFF
Plants[b]	SMSMLL	LSM(SM)LS
All[c]	AAAAAPPAPAPAAPP	AAAA(AP)PAP(APP)PPA
	WEST-TO-EAST TRANSECT	
Animals	FNFNFNFF	NFFNNFFF
Plants	SSSMMLL	SSSMLML
All	AAAPPAPAPAAAPAPP	APAPAAPAP(AP)APPAA

[a]Flying (F) versus nonflying (N) animals
[b]Plants with small (S) versus medium (M) versus large (L) propagules
[c]Animals (A) versus plants (P)

diversity appeared to be generally low. Birds, for example, had typical beta-1 values of 3.3 and 5.7. (Recall that this is the average percentage of turnover between localities.) For comparison, the beta-1 value was 38.1 for the native land birds of the six largest Hawaiian Islands (Thomas and Harrison, in preparation).

Rankings of beta diversity by dispersal category, for each of the four possible comparisons (beta-1 and beta-2, N–S and W–E) are shown in Table 2. None of these yielded an ordering of beta diversity as predicted by presumed dispersal ability. Beta values were not higher for plants than animals (or vice versa), for nonflying than flying animals, or for plants with large than medium or small propagules. The one clear-cut ranking (plants, beta-2, W–E) was in the reverse of the predicted direction.

BETA DIVERSITY VERSUS DISTANCE

To estimate the relationship between differentiation and distance, we calculated values of both beta-1 and beta-2 for each pair of localities in a transect (i.e., $N = 2$). For each taxon, we regressed the resulting values on the distances between sites (Table 3). Since site pairs are not independent of one another, we conservatively assumed that the degrees of freedom in these regressions equaled the number of sites minus two, rather than the number of site pairs minus two.

TABLE 3. Regressions of beta diversity in the British flora and fauna on the distance between sites.

Taxon	Beta-1				Beta-2			
	r^2	Slope	Int.	p	r^2	Slope	Int.	p
NORTH-TO-SOUTH TRANSECT (df = 15)								
Bees	.21	2.8	26.4	NS	.01	0.4	12.8	NS
Birds	.90	1.7	5.7	.01	.75	1.5	2.7	.01
Butterflies	.70	3.2	9.6	.01	0	0	0.5	NS
Dragonflies	.50	2.7	25.6	.01	.04	0.6	10.7	NS
Moths	.30	2.9	22.3	.05	.08	−0.5	8.4	NS
Fishes	.57	3.2	20.2	.01	0	−0.1	13.3	NS
Mollusks	.63	2.1	1.8	.01	.14	−0.3	8.3	NS
Woodlice	.49	2.3	21.6	.01	.01	0.1	12.3	NS
Ferns	.59	1.6	13.2	.01	.39	1.3	3.3	.05
Mosses	.25	1.4	22.9	.05	.31	1.4	8.6	.05
Orchids	.64	2.9	13.5	.01	.24	1.4	5.4	.05
Composites	.67	2.3	8.5	.01	.01	0.1	5.3	NS
Umbellifers	.66	2.8	15.4	.01	.06	−0.4	9.7	NS
Rosaceae	.74	1.9	9.8	.01	.45	1.1	3.3	.01
Native trees	.75	1.7	6.7	.01	.59	1.5	1.3	.01
WEST-TO-EAST TRANSECT (df = 6)								
Bees	.35	5.0	15.9	NS	.02	−0.7	11.0	NS
Birds	.66	1.8	6.9	.05	.50	1.6	4.5	.05
Butterflies	.32	1.5	7.3	NS	.04	0.4	2.4	NS
Dragonflies	.15	4.6	28.2	NS	.01	0.9	7.9	NS
Moths	.36	1.2	10.1	NS	.04	−0.5	5.4	NS
Fishes	.30	2.0	16.6	NS	0	−0.3	13.8	NS
Mollusks	.47	2.9	7.4	NS	.43	1.2	3.5	NS
Woodlice	.42	3.4	13.4	NS	.26	2.8	0.1	NS
Ferns	.08	1.1	16.9	NS	.03	0.6	5.7	NS
Mosses	.45	6.1	15.9	NS	.32	2.9	7.1	NS
Orchids	.43	3.8	16.5	NS	.01	0.4	9.8	NS
Umbellifers	.46	4.6	8.3	NS	.03	0.5	7.3	NS
Rosaceae	.29	1.1	7.7	NS	.09	0.6	5.6	NS
Native trees	.51	2.0	4.7	NS	.45	1.6	0.5	NS

In most taxa, beta-1 showed a significant increase with distance. On the N–S transect, differentiation increased by about 4 to 6 percent per 100 kilometers. However, in only a few taxa did beta-2 show a significant increase with distance. The discrepancy between these two measures reflects the overriding role of alpha diversity gradients in creating apparent distance trends in beta-1. In other words, the British flora and fauna are characterized more by a drop-off in diversity from south to north and west to east than by true turnover of species. Reinforcing this conclusion, distance along each gradient is correlated with alpha diversity in most taxa. Distance is also strongly correlated with a climate vector composed of 16 seasonal temperature, sunshine and rainfall values (Harrison et al., 1992). The relationship between climate and alpha diversity in Britain has been explored by Currie and Paquin (1987), Turner et al. (1987, 1988), and Adams and Woodward (1989).

We conclude that beta diversity at this scale is not determined primarily by how well species are able to reach available habitats. Instead, it is probably more a function of the breadths of their environmental tolerances. That is, the taxa with high beta diversity along a common gradient are those with narrowly defined requirements for such factors as temperature, day length, and rainfall, and the associated biotic changes in habitat.

If patterns such as this characterize many biotas at comparable spatial scales, one message for conservationists is that a few species-rich areas may sometimes encompass a large share of a region's total diversity. Northern temperate areas may tend to show this type of pattern, given their recent recovery from glaciation. They may tend to be made up of good migrators, and therefore to be preselected to cope with climate change. The influence of distance and dispersal limitation on regional diversity must presumably be sought on a larger scale (e.g., thousands of kilometers), where major barriers to dispersal exist (e.g., between islands), or perhaps in tropical regions, where ranges are smaller and "mountain passes are higher" (Janzen, 1967; Stevens, 1989). Unfortunately, adequate distributional data are scarce for temperate biotas and almost nonexistent for tropical ones.

BETA DIVERSITY ON ISLANDS AND ISOLATES: BARRIERS TO DISPERSAL

Quinn and Harrison (1988) examined the relationship between areal subdivision and diversity, using data for the distributions of various taxa in archipelagoes or sets of terrestrial isolates (see Table 4). From each of 30 species-by-island (or isolate) data sets collected from the published literature, two cumulative species-area curves were constructed—one with the islands or isolates ranked smallest to largest, and one with them ranked

TABLE 4. Beta diversity in various taxa on islands or terrestrial isolates.

Taxon/Sites	Beta-1	Beta-2	Source
Alpine plants/Adirondacks	5	0.5	Riebesell (1982)
Lizards/Australian reserves	14	0.5	Kitchener et al. (1980b)
Mammals/Australian reserves	10	0.5	Kitchener et al. (1980a)
Reptiles/Bass Strait	31	2	Rawlinson (1974)
Sand dune mammals/Bass Strait	18	0.5	Hope (1973)
Birds/Canary Islands	15	8	Bacallardo (1976)
Ground beetles/Canary Islands	50	12	Machado (1976)
Bats/Caribbean Islands	28	4.3	Baker and Genoways (1978)
Birds/Channel Islands	12	0.2	Power (1972)
Herpetofauna/Channel Islands	34	9	Savage (1967)
Birds/East African parks	10	5	Williams (1967)
Mammals/East African parks	7	3	Williams (1967)
Ground bettles/Faeroes Islands	7	0.8	Bengston (1982)
Plants/Galapagos Islands	23	2	Wiggins and Porter (1971)
Insects/Galapagos Islands	47	9	Linsley and Usinger (1966)
Mammals/Gulf of California	34	4	Case and Cody (1983)
Reptiles/Gulf of California	48	17	Case and Cody (1983)
Birds (extant)/Hawaii	55	22	Shallenberger (1981)
Birds/Great Basin mountaintops	9	2	Brown (1978)
Mammals/Great Basin mountaintops	12	2	Brown (1978)
Mammals/U.S. national parks	11	3	Harris (1984)
Seabirds/Scottish islands	6	1	Bourne and Harris (1979)

largest to smallest. In all but one case, the cumulative species curve rose faster and stayed higher when the localities were ranked smallest to largest. In other words, at all intermediate values of cumulative area, collections of small localities usually supported more species rather than fewer, larger ones. The ratio of the areas under the two curves (termed the "saturation index") measured the strength of this effect, which is closely linked to beta diversity.

Beta diversity values were calculated using the same data sets (Table 3). Comparison of the results showed that beta diversity was not significantly higher for nonflying than for flying organisms. Quinn and Harrison (1988) similarly concluded that values of the saturation index showed no clear effect of dispersal ability. However, beta was significantly higher among true islands than sets of terrestrial isolates (Mann–Whitney test; $N = 13, 8$; for beta-1, $U = 123$, $p < 0.001$; for beta-2, $U = 83$, $p < 0.05$). Presumably, water barriers in conjunction with distance produce more isolation and higher turnover than distance alone.

However, distance effects were still not strong. The typical rates of increase of beta-1 with distance were on the order of 2 to 10 percent per 100 kilometers, comparable to the rates seen in the British biota, and the regression parameters revealed no obvious differences between islands and terrestrial isolates or between flying and nonflying organisms.

BETA DIVERSITY IN THE PACIFIC AVIFAUNA

With C. D. Thomas (in preparation), I have calculated beta diversity values for 392 species of nonmigrant land birds on 49 Pacific islands (or clusters of tiny islands), including New Zealand. Data for extant or historically extinct species came from the field guides of Pratt et al. (1987) and Falla et al. (1978), and fossil information from Olson and James (1982), Savidge (1987), Freed et al. (1987), and Holdaway (1989). Beta diversity was substantially higher than in any of the previously examined data sets: beta-1 was 33.7, when now-extinct species are included, and 29.7 when only surviving native species are included (see below). These high values presumably reflect the effect of thousands (rather than hundreds) of kilometers in separation between islands, in conjunction with the water barrier.

To estimate the relationship of turnover to distance, values of beta-1 for pairs of islands were plotted against interisland distance for a subset of 12 islands selected to span the entire region (Figure 2). Also shown on Figure 2 are pairwise beta-1 values for two smaller groups: the 6 largest Hawaiian islands and 6 Fijian islands.

Figure 2 shows that beta-1 increases strongly with distance over the range 1000–3000 km, after which it asymptotes at slightly less than 100. Although habitat dissimilarity may contribute to this pattern, we know of no reason to expect strong distance gradients in the habitat attributes of these islands. We conclude that the pattern reflects the tendency for very isolated representatives of the same type of habitat to support different sets of species ("ecological equivalency" in the terminology of Shmida and Wilson, 1985).

HUMAN IMPACT ON BETA DIVERSITY

Thomas and Harrison (in preparation) also compared the components of Pacific island bird diversity before and after anthropogenic extinctions and introductions. While gamma (Pacific regional) bird diversity declined slightly, and average alpha (island) diversity has actually increased, beta diversity has declined substantially (Table 5). Introductions reduced beta

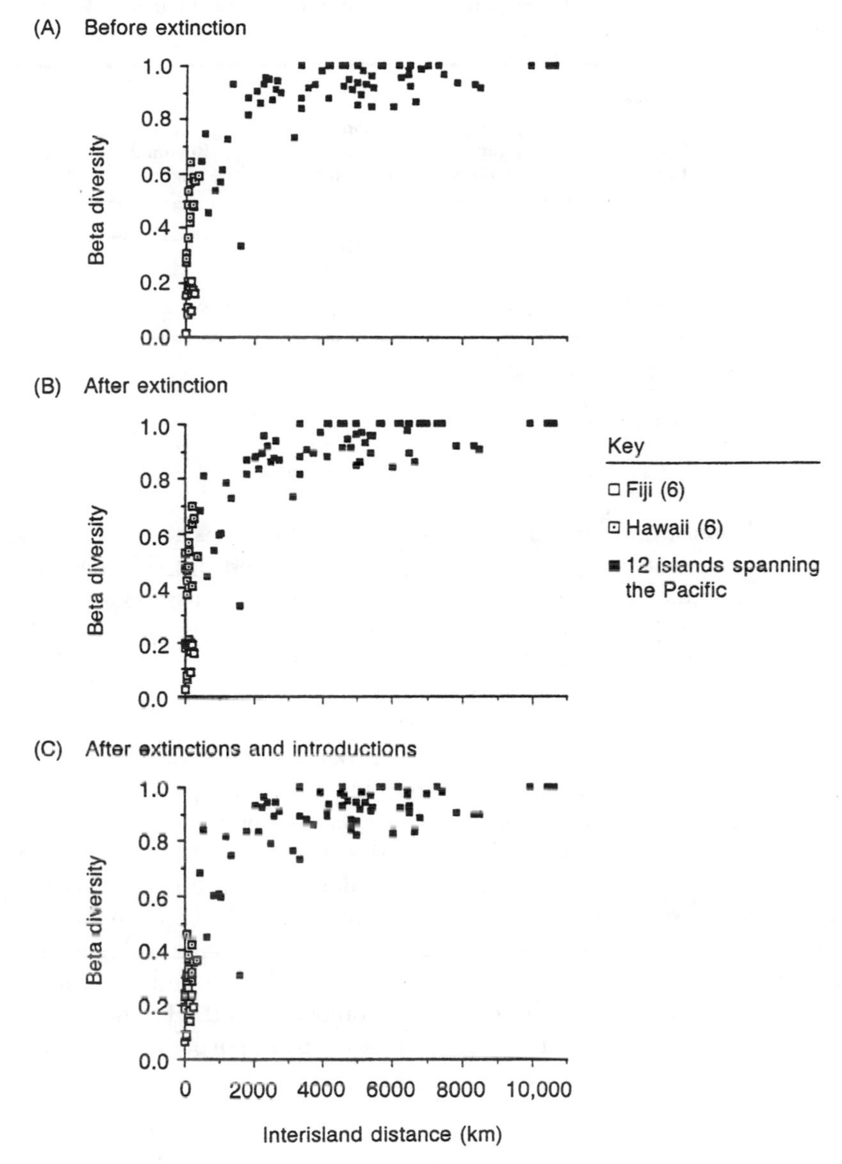

FIGURE 2. Beta diversity versus interisland distance for the nonmigrant land avifauna of 12 Pacific islands; pairwise values for the 6 largest Hawaiian islands and 6 Fijian islands are also shown. (From C. D. Thomas and S. Harrison, in preparation.)

TABLE 5. Components of diversity in the nonmigrant land bird fauna of 49 Pacific islands.

	Native species before extinctions	Species after extinctions	Species after introductions	Regional change	Global change
Alpha[a]	17.98	15.76	22.65	—	—
Beta[b]	33.7	29.7	27.9	—	—
Gamma[c]	346	241	326	−20	−102

[a]Average number of species per island
[b]Beta-1, as defined in text
[c]Total number of species on all islands combined

diversity because introduced species now occupy more islands on average than surviving native species do (averages = 3.7 and 3.2, respectively). Extinctions reduced beta diversity because they tended disproportionately to strike species that occurred on fewer islands. Sixty-five (32.5%) of 200 species originally confined to one island became extinct, while only 44 (6.4%) of 681 occurrences of the 109 species found on multiple islands did so.

Why were single-island species more prone to extinction than populations of multiple-island species? Two nonexclusive possible explanations are (1) recolonization or "rescue" effects, and (2) taxonomic heterogeneity, such that single-island species have characteristics such as flightlessness, ground nesting, lack of disease resistance, etc., rendering them especially vulnerable. We found some evidence for heterogeneity in proneness to extinction. Probabilities of extinction were higher in species belonging to higher taxa endemic to the Pacific than in non-endemic taxa (Thomas and Harrison, in preparation). Thus, humans have not only homogenized the Pacific avifauna in terms of species diversity, but have also disproportionately reduced higher-level taxonomic diversity.

SUMMARY AND CONCLUSIONS

Over long periods of time, how capable are organisms of long-distance dispersal? Little evidence of limited dispersal was found for the contemporary plant and animal taxa analyzed here. Within distances of a few hundred kilometers, on continents in the temperate zone, patterns of species distribution appear not to be shaped primarily by isolation. Dispersal limitation appears to become a major influence on species distributions over scales of thousands of kilometers, and where water barriers add to the isolation produced by distance.

Similar conclusions were reached by Cody (1986) in studies of the causes and spatial scale of species turnover in the shrub flora and bird fauna of Mediterranean climate regions. Also in qualitative agreement are the studies of the fossil record cited earlier, which suggest that species distributions have often kept pace with past changes in climate.

Compared to the other possible outcomes, these results permit a certain degree of optimism with respect to the effects of global warming. However, it is undeniable that greenhouse warming will be much more rapid than any natural climate changes in the past. This presents the major impediment to extrapolating from any biogeographical pattern, fossil or contemporary, to the prediction of coming changes. Moreover, the situation is almost certainly different in tropical biotas, which contain a much higher proportion of narrowly distributed species (Janzen, 1967; Stevens, 1989). This could be related to the fact that tropical biotas lack a history of recent postglacial colonization, and thus may consist of poorer migrators on average. Tropical communities may therefore prove more vulnerable to climate change than temperate ones.

Global climate change will inevitably bring extinctions, and these will differentially affect the more narrowly restricted and poorly dispersing species within any region. Thus, one of its more pronounced effects on communities will be to reduce beta diversity—in other words, to further homogenize the biota of the earth.

ACKNOWLEDGMENTS

Discussions and collaborative work with John Lawton, Thomas Lewinsohn, James Quinn, and Chris Thomas formed much of the basis for this paper. Peter Kareiva, Donald Strong, and participants in the NSF Workshop on Global Change provided useful and entertaining comments.

CHAPTER 25

EFFECTS OF GLOBAL CHANGE ON THE DYNAMICS OF INSECT HOST–PARASITOID INTERACTIONS

M. P. Hassell, H. C. J. Godfray, and H. N. Comins

Models of world climate change predict increases of 1.5°C–4°C in global mean temperature over the next few decades (MacDonald, 1990; Schneider, this volume). Precise predictions of temperature change, however, or how these will affect local patterns of rainfall, are uncertain (Schneider, this volume), and we remain "completely unsure as to how changes in global mean temperature will affect the structure and function of terrestrial ecosystems" (Crawley, 1990). Perhaps even more dramatic than climate change is the habitat destruction attributable to human activities. However, even with trends as undeniable as tropical deforestation, there is enormous uncertainty about the accuracy of estimates for habitat loss reported by particular countries (Groom and Schumaker, this volume). These uncertainties should in no way prevent population biologists from becoming actively involved in anticipating direct or indirect outcomes resulting from climate shifts and habitat fragmentation. Two approaches are needed: detailed studies of how environmental change is likely to affect particular ecological systems, and more general tactical studies on how environmental change may affect the dynamics of interacting species. The former requires the development of complex models parameterized from detailed field studies, while the more general models can progress more readily with what we already know about various species associations.

In this chapter we apply the second of these approaches to two different problems, both involving questions of how the population dynamics of interacting insect hosts and their parasitoids may be affected by environmental change. The considerable importance of such interactions to global species diversity and their effects on so many agricultural systems makes them important subjects for study. But in addition we hope that the conclusions we draw will point to consequences of environmental change that may apply in many ecological systems. First, we assume that shifts in climate differentially affect the development rates of hosts and parasitoids, and ask how this may alter the dynamics of their interaction. We will show that asynchronies between the populations creates a refuge effect that, depending on its severity, may lead either to extinctions or to enhanced population stability. We then turn to the question of how habitat fragmentation (as a consequence of climate change or human exploitation of habitats) may influence host–parasitoid interactions, and show that the answer depends sensitively on understanding the characteristic spatial dynamics of the system.

DIFFERENTIAL DEVELOPMENT RATES

Many insect parasitoids and their hosts have discrete generations, often with only a single generation each year. Successful parasitism requires temporal coincidence between the life history stage of the parasitoid that attacks the host and the stage of the host that is susceptible to parasitism. No single species of parasitoid attacks a host at all stages of its life cycle. Parasitism is invariably concentrated on particular stages of the host's life cycle, typically the egg, larval or pupal stages, although there are some parasitoid species that do attack adult hosts. In a few parasitoid species, the first instar parasitoid larva is responsible for host location. But in the great majority of cases, it is the adult female parasitoid that is the searching stage, and usually for only part of the adult stage, since the first few days or weeks after adult emergence are required for egg maturation.

In this section we explore the dynamic consequences of different patterns of host and parasitoid phenology, a subject that has received surprisingly little attention in the literature (but see Griffiths, 1969). We imagine that shifting average temperatures in different seasons of the year may differentially affect host and parasitoid development rates. Species that are normally well synchronized in time may become less so, thereby creating greater variance in the levels of parasitism among the host population. Most global climate projections emphasize that the greatest warming in temperate regions is likely to come in the winter months. The most crucial time for host–parasitoid synchronization is at the end of this period; thus even modest changes in winter climate may have particularly important consequences. Of course, if hosts and parasitoids respond in exactly the same way to climate change, synchrony may be preserved. However, the means by which hosts and parasitoids regulate diapause vary widely, and we think it is unlikely that hosts and parasitoids will respond in concert to perturbations in climate. Indeed, in many cases hosts and parasitoids rely on entirely different physical cues for entering or exiting diapause; for example, the host may regulate diapause using thermal cues whereas the parasitoid uses day length. There is a large literature on diapause in parasitoids and their hosts, which has been extensively reviewed by Tauber et al. (1983, 1986). Our aim in this study is to make a first attempt to analyze the dynamic consequences of some of these phenological disruptions.

The model

Let us consider a host population that is attacked by a specialist species of parasitoid. At the beginning of the season, all hosts are in a nonsusceptible stage and all parasitoids are in a nonsearching stage. For narrative simplicity, we shall assume that at this time (1) hosts are eggs and only become susceptible to parasitism when they hatch, and (2) parasitoids are pupae and begin to search for hosts as soon as the adult females emerge. Hosts are susceptible to attack until they pupate and parasitoids search until they die. The model, however, is fully general, and the susceptible host stage and the parasitoid searching stage may occupy any part of the life cycles of the two species. Our aim is to produce the simplest description of the host–parasitoid system that includes the essential features of variable synchrony between the two populations. We thus deliberately omit much of the biology in order to focus primarily on this phenomenon. Only the outlines of the model are presented here; full details are given in H. C. J. Godfray, M. P. Hassell, and R. D. Holt, unpublished. (A comparable simulation model has been discussed by Munster-Svendsen and Nachman, 1978.)

The model consists of two parts: a within-generation component that describes the dynamics of emergence, maturation, and parasitism; and a between-generation component that relates the numbers of surviving and parasitized hosts at the end of one season to the numbers at the beginning of the next season. The between-generation component is particularly simple. We assume that each parasitized host gives rise to one adult female parasitoid in the following season, and that each surviving host has a net finite reproductive rate, λ, of two. Thus, at equilibrium, parasitoids must destroy 50 percent of the host population.

The within-generation component is more complex. Host and parasitoid phenology is determined by three factors: (1) the dates at which hosts begin to hatch and parasitoids begin to emerge, (2) the degree to which host hatching and parasitoid emergence is synchronized within each population, and (3) host maturation out of the susceptible stage and parasitoid death. Here we explore the dynamic consequences of changing the relative dates of the first appearance of hosts and parasitoids. We assume that the hatching and emergence pattern of hosts and parasitoids is as described in Figure 1. Finally, we assume that hosts spend a fixed period (10 days) in the susceptible stage and that parasitoid mortality is negligible over the period when hosts are available and can thus be ignored.

In order to describe parasitism, we define H_t and P_t as the numbers of susceptible hosts and searching parasitoids, respectively, that are alive at time t. The rate of loss of hosts to parasitism is $H_t f[P_t, H_t]$, where $f[P_t, H_t]$ describes the instantaneous risk of parasitism. The simplest form of $f[P_t, H_t]$ is a function purely of the number of parasitoids: $f[\cdot] = aP_t$, where a is the per capita attack rate. This form of the function is used in both the models of Lotka (1925) and Volterra (1926) and (implicitly) those of Nicholson and Bailey (1935), and implies that the risk of parasitism increases linearly with parasitoid density (Figure 2). The instability in these models (neutral cycles or divergent oscillations, respectively) is due to the lack of any density dependence in the risk function $f[P_t, H_t]$. In this study, we assume that the risk of parasitism increases with the number of searching parasitoids, but at a decelerating rate (Figure 2). Such density dependence in parasitism might arise in a number of ways: some hosts may be in a physical or

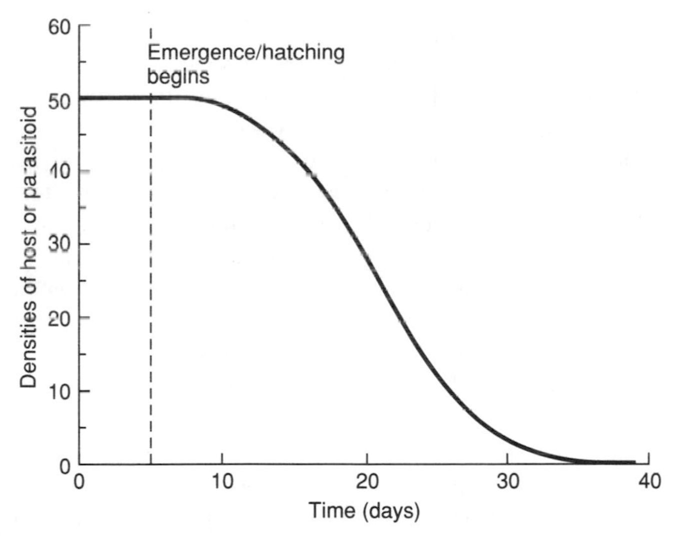

FIGURE 1. The shape of the distributions governing host hatching and parasitoid emergence. The curve shows the change in density of either host eggs or parasitoid pupae.

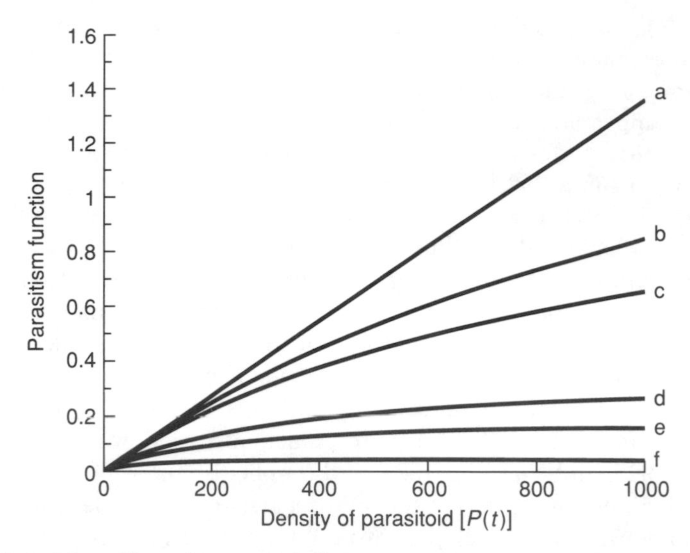

FIGURE 2. The effect of parasitoid density on the parasitism function $f\ [P_t, H_t]$. The extent of parasitoid dependence is inversely proportional to the parameter k. (A) $k \rightarrow \infty$. (B) $k = 1$. (C) $k = 0.5$. (D) $k = 0.1$. (E) $k = 0.05$. (F) $k = 0.01$.

probabilistic refuge, a fraction of hosts may be able to defend themselves physiologically from parasitism, or there may be interference among parasitoids at high density (Hassell and Pacala, 1990). The amount of density dependence is indexed by a parameter k, and the model converges to $f[\cdot] = aP_t$ (no density dependence) as $k \rightarrow \infty$. This particular function has the advantage of converging under certain conditions to a familiar, discrete-generation model whose dynamics are very well understood.

The properties of the model are illustrated using a set of biologically defensible parameter values that describe a typical host–parasitoid interaction where both species have a single generation per year. Keeping all other parameters constant, we change the date of first parasitoid emergence relative to that of the host, and ask how much density dependence is required to stabilize such interactions. Such changes are the sorts of perturbations to host–parasitoid interactions one would expect as a result of global warming. The degree of intrinsic density dependence in the system is described using the parameter k, which is used as a measure of the stabilizing effect of temporal refuges. An example of within-generation dynamics using our chosen parameters is shown in Figure 3. The within-generation model is described by a set of differential equations, some containing time lags. We make use of the lumped age-class techniques developed by W. Gurney, R. Nisbet and colleagues (Nisbet and Gurney, 1983; Gurney et al., 1983; Gurney and Nisbet, 1985) and previously em-

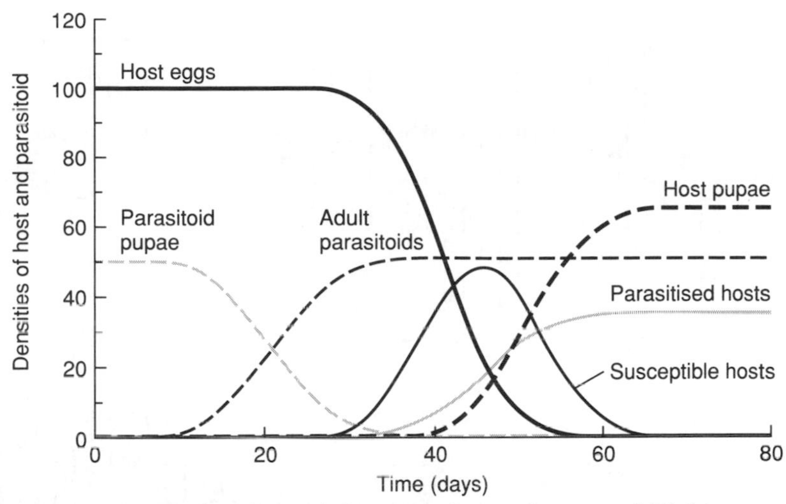

FIGURE 3. An example of typical changes in host and parasitoid life history stages within one generation. The figure represents a system that is not at equilibrium: the following year there are both fewer hosts and fewer parasitoids.

ployed in host–parasitoid studies (Murdoch et al., 1987; Godfray and Hassell, 1989). The equations are numerically integrated to obtain the number of parasitized and unparasitized hosts at the end of the season.

Phenological disruptions may change fundamentally the outcome of host–parasitoid interactions

The results of the simulations are shown in Figure 4 in terms of the level of density dependence (k) and the time of first parasitoid emergence relative to the susceptible host stage. We consider first the stability properties of the system when parasitoids begin to emerge at day 5. As density dependence increases (k gets smaller), the system is stable above a critical threshold, but then becomes unstable again when density dependence gets very strong. These results are not due to phenology because parasitoid emergence in this case is completed before any hosts hatch, and because the number of parasitoids searching during the period of host susceptibility is almost constant. The stability boundaries are thus simply a result of the intrinsic density dependence and handling time, a fact that can be confirmed for certain limiting cases where our model with complex within-generation dynamics collapses to standard, discrete-generation models.

The stability boundaries at the left of Figure 4 thus represent the behavior of the model in the absence of phenological effects. As parasitoid emergence is delayed, the stability properties of the system change in a

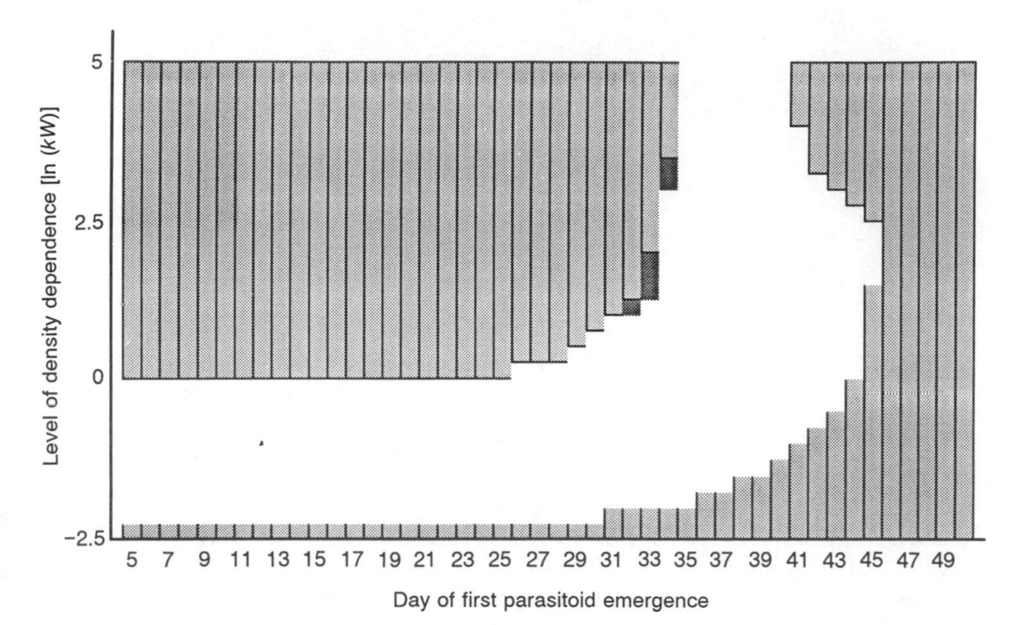

FIGURE 4. Stability boundaries for the model described in the text in a space with axes representing the date of parasitoid emergence relative to that of the host and the amount of density dependence (k). Unfilled areas represent regions where the system is asymptotically locally stable; light-shaded areas represent regions of instability. The small dark-shaded regions represent model runs where the system showed cycling behavior after 1000 generations.

number of ways. First, if parasitoids begin to emerge between days 26 and 34, the system requires less density dependence in the parasitism attack rate in order to be stable; when the first emergence occurs between days 35 and 40, the system is stable even when there is no density dependence in the parasitism term; and finally, when the first emergence occurs between days 41 and 45, the system again can be stable as long as the parasitism term contains enough density dependence. Second, the maximum density dependence that the system can tolerate and still be stable decreases as parasitoid emergence is delayed. Third, the interaction can never be stable, whatever the density dependence in the parasitism term, if the parasitoids do not begin to emerge until day 46. Fourth, apparently persistent population cycles are observed in a narrow region of parameter space.

These results arise because a delay in parasitoid emergence increases the heterogeneity in the risk of parasitism. Those hosts that emerge early in the season experience a reduction in, or even the absence of, parasitoid attack; they are in an absolute or partial temporal refuge. The presence of a temporal refuge introduces density dependence in the parasitism function;

as parasitoid numbers increase, the risk of parasitism increases less than linearly since some hosts are always able to escape parasitism by virtue of their phenology. As with spatial refuges, an increase in the size of a temporal refuge results in larger host population densities (Figure 5). When handling time is non-zero, very high host population densities cannot be regulated by parasitoids, and this accounts for the reduction in the maximum amount of density dependence that the system can tolerate and for the instability of systems with little or no density dependence when parasitoid emergence begins between days 41 and 45. Finally, when parasitoid emergence does not begin until day 46, so many hosts are in a refuge that no matter how effective the parasitoid is, it can never reduce the host sufficiently for regulation to occur.

As long-term changes in global climate become increasingly apparent, the year-to-year variance in various climatic variables is also expected to increase. In this context, we should also examine the dynamic consequences of the emergence patterns of hosts and parasitoids varying from generation to generation. We have done this with a series of simulations in which years of late parasitoid emergence alternate with years in which they emerge early. We have used the same parameter combinations as in the first set of simulations and assumed $k = 0.05$ ($\ln(10k) = 1.6$); in years when parasitoids emerged early, the date of first emergence was day 25, while in late years, the date of first emergence was day 37 (see Figure 4). These parameters

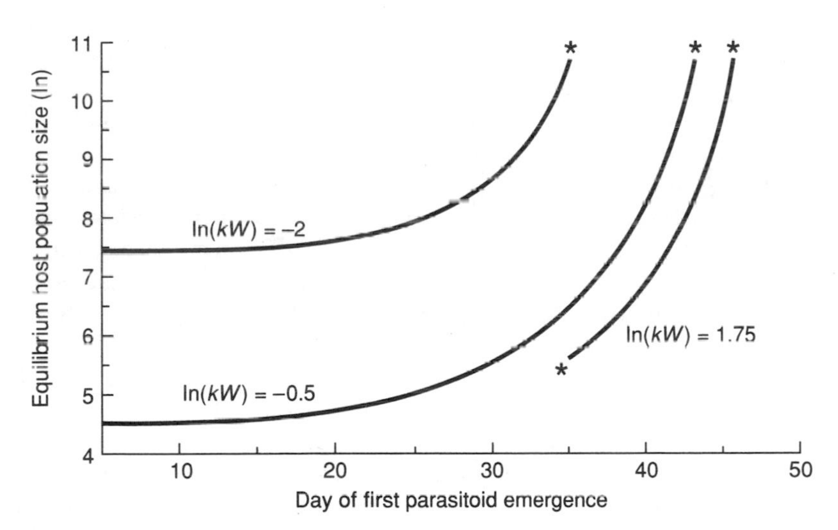

FIGURE 5. Equilibrium host densities (before reproduction) in stable regions of the parameter space of Figure 4. Host density increases as the density dependence grows stronger (smaller k) or the temporal refuges grow larger (later parasitoid emergence). The asterisks indicate points where the equilibria become unstable.

were chosen to make the system stable if the parasitoids always emerged late, and unstable if they always emerged early. With varying sequences of early and late years, the interactions were deemed to persist if the densities of the component populations remained within the bounds 10^{-3}–10^{5} for 500 generations.

Interestingly, occasional early years have a marked destabilizing effect on the interaction. No simulations persisted if early years occurred at a frequency greater than 1 in 20. If early years occurred exactly every 20 years, the system did persist, but this was generally not found if early years occurred randomly with an average frequency of 1 in 20. Only when early years occurred at a frequency of 1 in 25 or less was persistence the most frequent outcome. The reason for occasional years of early parasitoid emergence being so destabilizing appears to lie in the relatively high equilibrium parasitoid densities achieved in runs of late years. If all the parasitoids then emerge early, the complete host population is exposed to very high levels of parasitism. The few hosts that survive to reproduce produce only a small generation the following year, some of which will be in temporal refuges. The parasitoid population thus crashes to very low levels, leaving the host population to increase rapidly in the absence of regulation.

HABITAT FRAGMENTATION AND SPATIAL DYNAMICS

Much work over the past two decades has pointed to the potential importance of spatially patchy environments in influencing the dynamics of species interactions (e.g., DeJong, 1979; Atkinson and Shorrocks, 1981; Hanski, 1981,, 1983; Comins and Hassell, 1987, Crowley, 1981; Reeve, 1988; Taylor, 1988). Of particular recent interest has been the relationship between the dynamics of populations at the local and regional (metapopulation) scales, considering especially the extent to which natural populations persist as a set of linked subpopulations each of which is prone to be unstable (see Gilpin and Hanski, 1991, for an excellent review). Such persisting metapopulations will be sensitive to the number of subpopulations and the ease of movement between them. Any reduction of habitat size or fragmentation of the habitat thus runs the risk of disrupting the whole system, either by reducing the number of subpopulations below some critical level required for the combined metapopulation to persist, or by interfering with the dispersal required to link the locally unstable subpopulations.

In this section we examine this general problem in the context of an insect host–parasitoid interaction in which subpopulations occur in discrete patches linked by dispersal. We use our model framework to emphasize that an understanding of spatial dynamics can be crucial for predicting how changes in habitat area can affect species persistence. More particularly, we shall show that quite small changes in parameters that have a marked affect on spatial dynamics (but not necessarily on temporal trends in average population sizes) can markedly influence the degree to which the habitat can be fragmented and yet maintain resident populations.

The model

Let us consider a habitat that is divided into discrete patches of food plants on which an insect species oviposits and the resulting larvae feed. The immature stages of these insects are hosts for a specialist parasitoid species whose life cycle is synchronized with that of its host. Such a patchy environment can be modeled in a very simple way: as a two-dimensional arena in which the populations are distributed amongst a square grid of "cells" or patches. Prior to parasitism in each generation, there is a dispersal phase in which a fraction of adult hosts, μ_N, and a fraction of adult female parasitoids, μ_P, leave the patch from which they emerged, while the remainder stay behind to reproduce in their patch of origin. In most previous studies, these dispersing individuals are distributed over all other patches according to some specified rule. In this study, however, rather than entering a "pool" for such global dispersal, the dispersing μ_N hosts and μ_P parasitoids diffuse outward to colonize equally the eight neighboring patches nearest to the patch from which they emerge. Letting $N_{i,t}$ and $P_{i,t}$ be the predispersal host and parasitoid population densities in patch i at time t and $N'_{i,t}$ and $P'_{i,t}$ be the densities after dispersal, we have

$$N'_{i,t} = (1 - \mu_N)N_{i,t} + \mu_N\overline{N}_{i,t} \tag{1a}$$

$$P'_{i,t} = (1 - \mu_P)P_{i,t} + \mu_P\overline{P}_{i,t} \tag{1b}$$

where $\overline{N}_{i,t}$ and $\overline{P}_{i,t}$ are the average host and parasitoid populations over the eight nearest neighboring patches. For patches along the boundary of the arena, the definitions of $\overline{N}_{i,t}$ and $\overline{P}_{i,t}$ are slightly different, depending on whether one assumes cyclic, absorbing, or reflective boundary conditions. Unless otherwise specified, we use reflective boundary conditions; this has little effect on the outcome except that simulations with cyclic boundaries tend to produce more symmetrical patterns (because of their more symmetrical mathematical properties).

To lay bare the effect of this diffusive dispersal alone, we assume that all patches are identical in "quality" and that parasitism is random in each patch, following the usual Nicholson and Bailey (1935) formulation:

$$N_{i,t+1} = \lambda N'_{i,t}\exp(-aP'_{i,t}) \tag{2a}$$

$$P_{i,t+1} = cN'_{i,t}[1 - \exp(-aP'_{i,t})] \tag{2b}$$

Here λ is the finite rate of increase of the host population, a is the per capita parasitoid attack rate, and c is the conversion efficiency of parasitized hosts into adult female parasitoids of the next generation. Because of the uniform environment, all three parameters are spatially homogeneous. The assumption of random parasitism within patches has the advantage of not contributing itself to the persistence of the overall populations, but this can easily be changed to accommodate aggregated attacks within patches. In particular, a clumped distribution of attacks can be elegantly represented by replacing the zero term of the Poisson distribution in Equations 2a and 2b with the zero term of the negative binomial distribution (May, 1978), in which case aggregation of attacks increases as the clumping parameter, k, tends to zero. Further details of this model are given in Hassell et al. (1991) and Comins et al. (in press).

Complex spatial dynamics emerge as a result of local dispersal

In a homogeneous environment, host–parasitoid interactions with discrete generations will not persist unless the parasitoids attack hosts sufficiently nonrandomly, or unless host or parasitoid populations are affected by some form of density-dependent effect. Previous work has shown that in a subdivided environment, such associations can persist with parasitoids searching randomly within patches, provided that there is sufficient variability in the risks of parasitism of individual hosts (Chesson and Murdoch, 1986; Pacala et al., 1990; Hassell et al., 1991). The model described above goes further in demonstrating that purely deterministic processes, whereby some constant fractions of hosts and parasitoids (μ_N and μ_P, respectively) move to immediately neighboring patches in each generation, can also lead to the populations persisting, even when parasitoids search randomly within patches and there are no explicit density-dependent mechanisms present.

The assumption of diffusive rather than global dispersal is crucial to the properties of the model outlined above. With global dispersal (which means each patch receives the same fraction of the pool of dispersing hosts and parasitoids), and with the within-patch dynamics unstable, metapopulation persistence is impossible. However, the change to diffusive dispersal introduces a wide range of spatial dynamics in which the combined population can readily persist. After rescaling the model, four parameters must be considered in illustrating these properties: the host rate of increase, λ, the dispersal fractions μ_N and μ_P, and the width of the square grid of patches, n. We shall be concerned in particular with the effects of changing arena size (we assume a maximum n of 30 patches) and other kinds of subdivisions of the habitat on the persistence of the interacting populations; but first we outline some basic properties of the system. All simulations are started with random initial densities of host and parasitoid populations in a single patch and with all other patches empty.

In simulations with small arenas ($n \leq 10$) the populations always become extinct within a few hundred generations (except for the "crystal lattice" cases described below). Both the metapopulation and its constituent patches show divergent oscillations of hosts and parasitoids leading to global extinction (involving numeric underflows in the simulations). Markedly different results are obtained when the arena width is increased to between 20 and 30 patches. Three general types of behavior are now found, which we describe as "spatial chaos," "spirals," and "crystal lattices" (Figure 6), all of which permit long-term persistence of the host and parasitoid populations within a relatively narrow range of average population densities per gen-

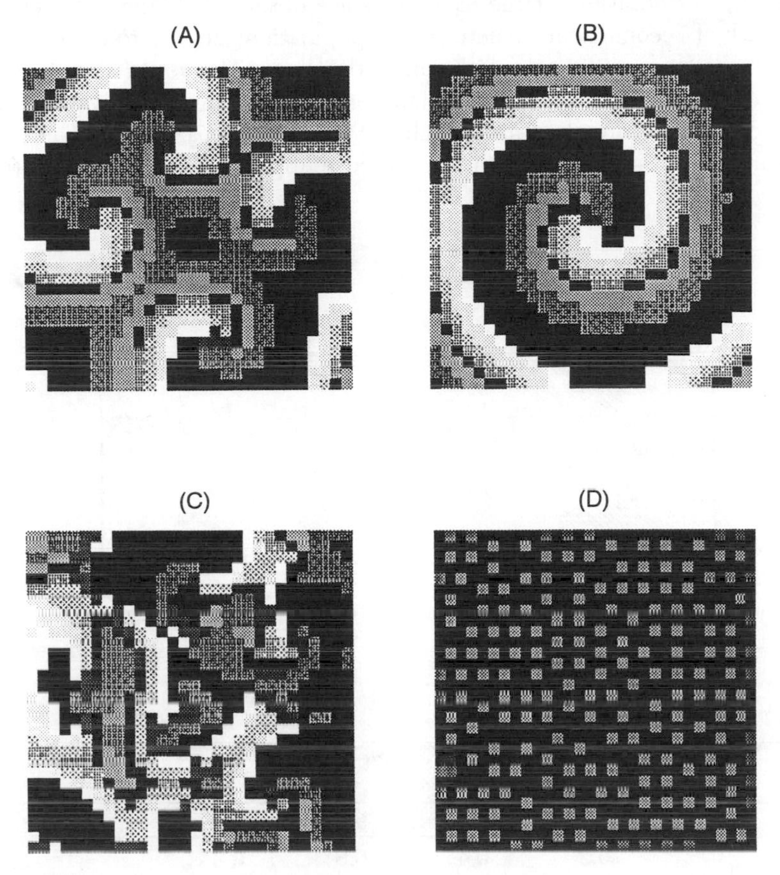

FIGURE 6. Examples of the different patterns of spatial dynamics obtained from the model described in the text. Each picture is a "snapshot" in time, with the nine different shadings representing different relative abundances of hosts and parasitoids within a patch (see Comins et al., in press, for further details). (A,B) Typical small and large spiral waves, respectively, obtained in the spiral region of Figure 7. (C) Spatially chaotic patterns obtained in the chaos region of Figure 3. (D) The crystal lattice pattern obtained for $\mu_P \rightarrow 1$ and small μ_N (top left of Figure 7); although host and parasitoid populations continue to show some fluctuation within patches, the pattern of relative densities appears as a static mosaic of high- and low-density patches.

eration. Figure 7 shows the approximate boundaries for these different patterns in relation to the host and parasitoid dispersal fractions (μ_N and μ_P, respectively) for a chosen value of λ and an arena width of $n = 30$.

The *spatial chaos* pattern is characterized by the host and parasitoid populations fluctuating from patch to patch with no long-term spatial organization. Randomly oriented wave fronts are observed, but each persists only briefly. The total metapopulations generally remain within narrow bounds, but occasional large excursions are observed (Figure 8A). Despite the lack of recognizable structure, the populations appear to coexist indefinitely (as long as the arena is sufficiently large). The *spiral structures* are characterized by the local population densities forming spiral waves, which rotate in either direction around almost immobile focal points. The phase-space dynamics of each patch form a close approximation to a fixed track,

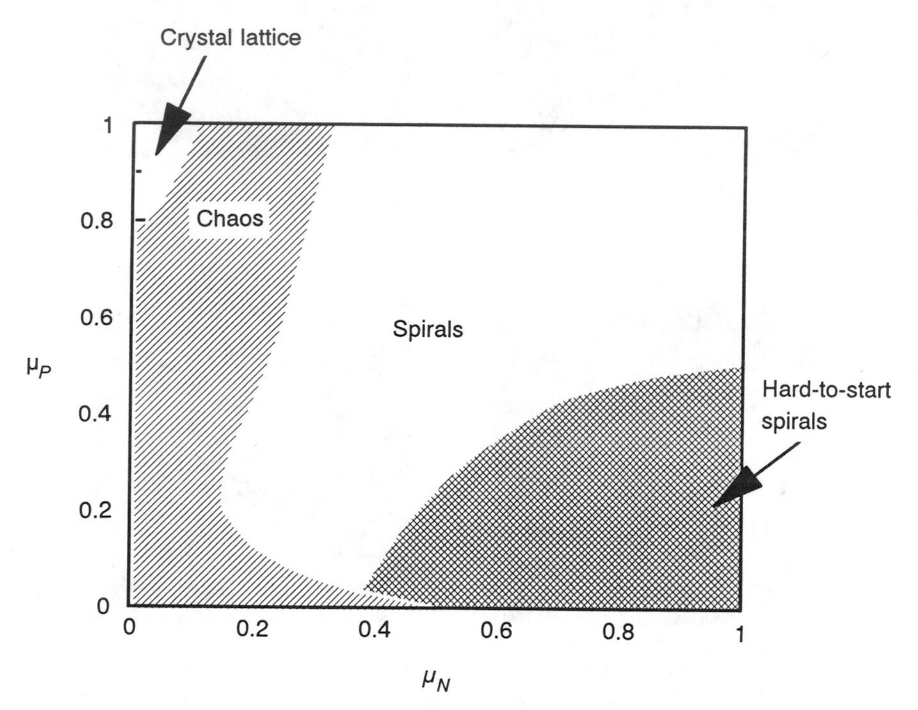

FIGURE 7. Dependence of the observed persistent spatial pattern on μ_N and μ_P, for arena width 30 and $\lambda = 2$. The boundaries are obtained by simulation and are approximate. The single hatched area indicates the region in which the spatial pattern is chaotic; the cross-hatched area represents parameter combinations for which the persistent spiral pattern is unlikely to be established by starting the simulation with a single nonempty patch. Spirals may be established in these cases by starting with a lower μ_N and increasing it after 50 to 100 generations. Nonpersistence occurs for some combinations with very small μ_N or μ_P; this area is imperceptible in the figures. (After Hassell et al., 1991.)

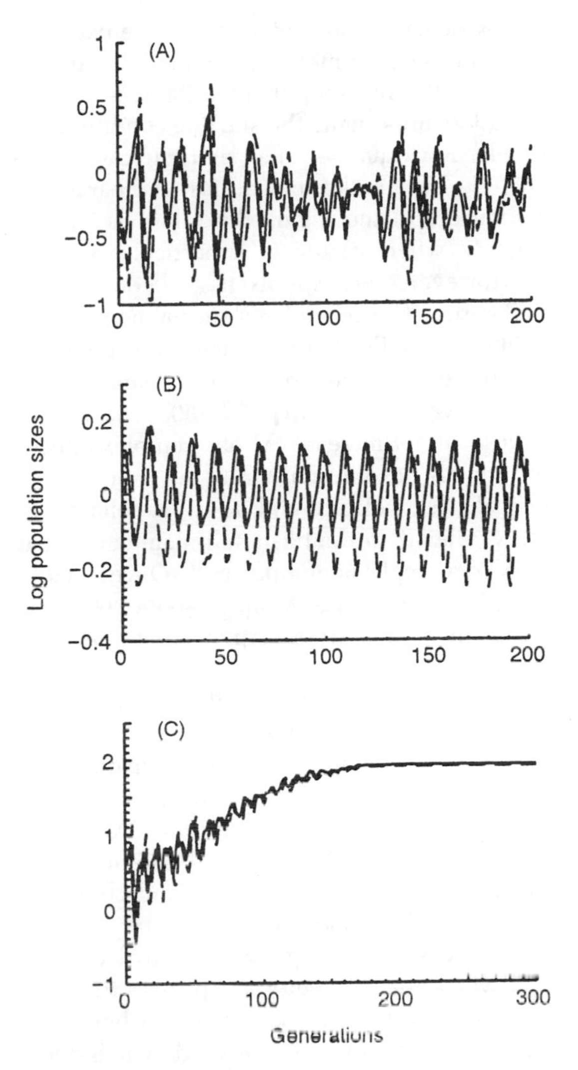

FIGURE 8. Typical time series of average population size corresponding to the three classes of spatial behavior shown in Figure 6. (A) Chaotic population dynamics, where spatial chaos occurs. (B) Stable-limit cycles, where spatial spirals occur. (C) A stable equilibrium, where the crystal lattice spatial pattern is observed.

even though no exact repetition occurs—although they are still apparently chaotic, since the position and number of focal points varies slowly with time in nonrepeating patterns. The combined metapopulation exhibits what appear to be stable limit cycles (Figure 8B). Finally, the rather extreme combination of very low host dispersal and very high parasitoid dispersal gives persistent *crystal lattice* structures in which relatively high-density patches occur at a spacing of two grid units. These patterns are stable even when $n = 2$ and lead to stable metapopulation dynamics (Figure 8C).

Nonpersistence of the metapopulations for $n = 30$ may occur for two reasons. First, and less important, the starting conditions for the simulation may not be suitable. For example, in an extreme case, persistence is never observed if all patches start with uniform host and parasitoid densities. This is also seen in Figure 7 in the region described as "hard-to-start spirals," which do not persist when started in a particular manner from a single nonempty cell. However, once spirals have been initiated, they permit persistence in all cases. The second reason for nonpersistence is intrinsic dynamic instability. It is difficult to delineate the nonpersistence area exactly, but it is restricted to parameter combinations in which either μ_N or μ_P is very low (e.g., $\mu_N < 0.02$ or $\mu_P < 0.02$).

Varying the host rate of increase (λ) has complex effects on this model, but persistence remains possible for λ ranging from close to unity to very large. The principal effects of changing λ are as follows. First, for large λ ($\gtrsim 10$), it becomes difficult to establish self-maintaining patterns from our starting condition of a single nonempty cell. Once established, however, spirals may be stable for very large λ (up to about 200). Second, increasing λ tends to reduce the spatial scale of persisting spirals. Conversely, the spatial scale of spirals is increased for small λ. There is thus a minimum value of λ for which a persistent pattern will fit into a given sized arena (for $n = 30$, spirals can be established for $\lambda > 1.2$). Third, increasing λ appears to favor the formation of spirals (rather than spatial chaos) at low host dispersal rates. This may be because large λ increases the effectiveness of host dispersal (diffusion over several generations from occupied patches is effectively faster if aided by rapid reproduction in newly infected patches).

Parasitism within patches in Equations 2a and 2b is random and therefore inherently destabilizing. The effects of stabilizing parasitism can be explored by using May's (1978) negative binomial distribution of attacks within patches, in which the parameter k expresses the degree of contagion in the distribution of parasitoid attacks (random when $k \rightarrow \infty$ and increasingly clumped as $k \rightarrow 0$). With $k < 1$, the single-patch dynamics are stable, as is the total population over all patches. Only if the within-patch dynamics are oscillatorily unstable do we obtain the range of patterns shown in Figure 7 (e.g., $k \geq 1.2$ is needed in the case $\mu_N = 0.8$, $\mu_P = 0.89$, $\lambda = 2$ and $n = 30$).

The effects of habitat fragmentation depend on characteristic spatial dynamics

Habitat shrinking Habitat fragmentation will affect population dynamics either by reducing habitat size or by dividing the environment and providing only limited opportunities for dispersal between the remaining fragments. We now examine how both of these possibilities can influence the dynamics of the model outlined above. In the first place, and not unexpectedly, it is clear that the probability of long-term persistence decreases as the number of patches is reduced (Figure 9). More interestingly, the extent of this effect appears to depend very much on the characteristic spatial scale of the dynamics. Thus, with parameter combinations producing crystal lattice patterns (Figure 7), the overall populations persist in a stable interaction even with $n = 2$. At the other extreme, interactions in which large-scale spirals occur are especially vulnerable to shrinking arena sizes, while interactions producing chaotic spatial patterns are intermediate between the two. This trend is also clear from Figure 9, which represents a slice across Figure 7 in which $\mu_P = 0.89$. Within the region of chaos, all interactions are seen to persist for side lengths of 15 and above; whereas within the region producing spirals, the probability of extinction increases, so that with the relatively large spirals generated with $\mu_N = 0.8$, some extinctions occurred

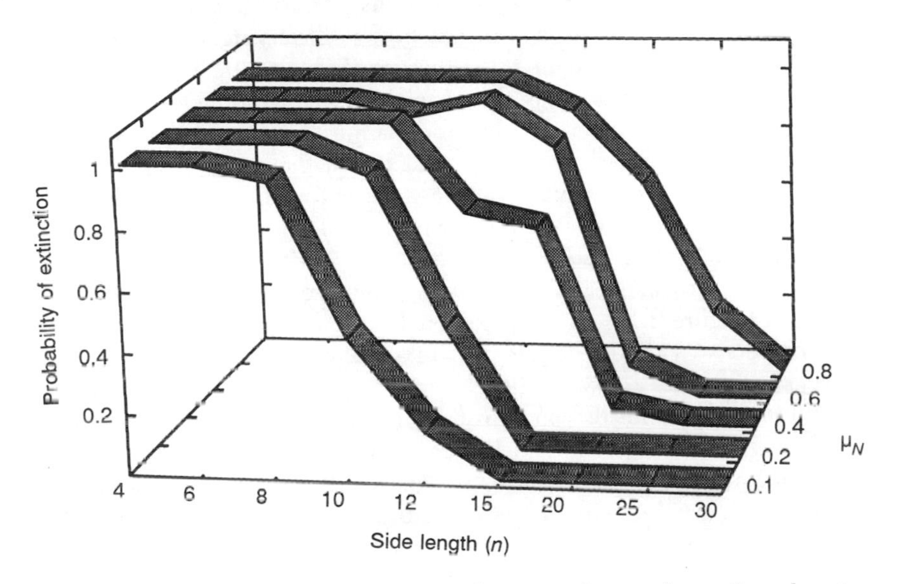

FIGURE 9. Extinction probabilities in relation to the numbers of patches in a square grid of side length n, and the fractions of hosts dispersing to neighboring patches (μ_N) ($\mu_P = 0.89$, $\lambda = 2$). Extinction is measured as the proportion of 50 replicates failing to persist over 2000 generations. Each replicate is started by setting non-zero population densities in only the third patch from the left in the top row. The same 50 pairs of initial host and parasitoid densities are used for all the parameter combinations. Local extinction occurs by numeric underflow (densities less than about 10^{-45}); however, the results are robust when local extinction thresholds for both hosts and parasitoids are modeled explicitly. (From Hassell et al., 1991.)

for all simulations with $n < 30$, and there was no persistence at all with $n < 15$. Intuitively, therefore, it appears that failure to persist in small arenas is due to the insufficient space in which to fit a self-maintaining pattern. These general trends remain true for different values of λ, and also if the dispersal distance is increased (Comins et al., in press).

Habitat subdivision Habitats may fragment in ways other than the simple shrinking of the total area. They may also become partially subdivided by inhospitable corridors which, however, still permit some movement across the barriers. Mankind's ever-increasing network of roads must continually be having this effect on the habitats through which they plunge. By pushing ecological communities closer to the limits of their ranges, climate change is also likely to produce similar effects as habitats shrink in a patchy way, leaving pockets with limited connections for the species within them. The importance of spatial dynamics to the persistence of such fragmented communities can be illustrated in a phenomenological way using the host–parasitoid model described above.

Our approach is the very simple one of using barriers of one patch width to subdivide the 30 × 30 grid of patches with varying numbers of gaps allowing movement between the subareas (Figure 10). Two conclusions stand out. The first reinforces the conclusion of the previous section that interactions with characteristic spatial dynamics on a large scale are by far the most easily disrupted. Thus, an interaction with a single persisting spiral filling the 30 × 30 arena always becomes extinct when a barrier bisects the habitat, even when this takes the form of a "sieve," as shown in Figures 10A and 10B. On the other hand, interactions with small-scale spirals or chaotic spatial dynamics can persist much more easily as the habitat is disrupted (Figure 10C) and, as expected, interactions with crystal lattice patterns persist with any degree of habitat subdivision as long as the units are no smaller than a 2 × 2 grid.

The second conclusion concerns the amplitude of fluctuations in average population sizes. In general, such fluctuations can be greatly increased by habitat fragmentation. An example is illustrated in Figures 11A and 11B.

In short, partial subdivision of habitats will affect species persistence in a way strongly influenced by the characteristic scale of the spatial dynamics, and will tend to lead to more frequent population outbreaks.

Environmental variation

A recurring theme from our analyses so far has been the importance of understanding spatial dynamics in order to predict how habitat fragmentation may affect population dynamics. The same remains true in considering how dynamics are influenced by changing patterns of environmental variables that affect demographic parameters. We shall look at the sensitivity of

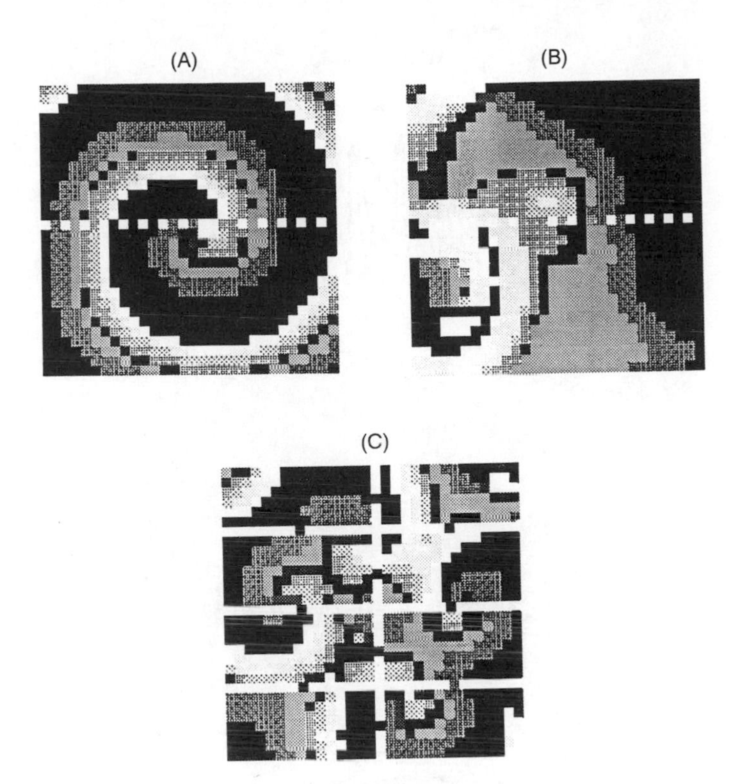

FIGURE 10. The effect of barriers on population persistence. (A) An arena with a large spiral bisected by a "sieve" of alternating barriers and gaps. (B) The same arena after 50 generations, showing the breakdown of the spiral pattern. This is rapidly followed by extinction, defined as for Figure 9. (C) An arena with small spirals subdivided into eight regions connected by single-cell gaps. Such systems persist despite the habitat disruption.

our spatial model, and in particular the resistance of the different spatial patterns, to superimposed random variation from patch to patch.

The patch-to-patch variation is introduced by multiplying the host rate of increase (λ) by a uniform random multiplier. Interestingly, the extent to which this tends to disrupt the different spatial patterns is once again linked to the relative scales of the "disturbance" on the one hand and the spatial dynamics on the other. Thus, the crystal lattice patterns are easily destroyed by quite modest random patch-to-patch variation. Once spirals or chaotic patterns are established, however, even quite appreciable levels of random variation have little effect on their evolution, since the patch-to-patch scale of variation is now much smaller than the spatial scale of the spirals. Larger-scale spatial patterns become disrupted once the scale of the random variation is made much larger than individual dispersal distances; for instance,

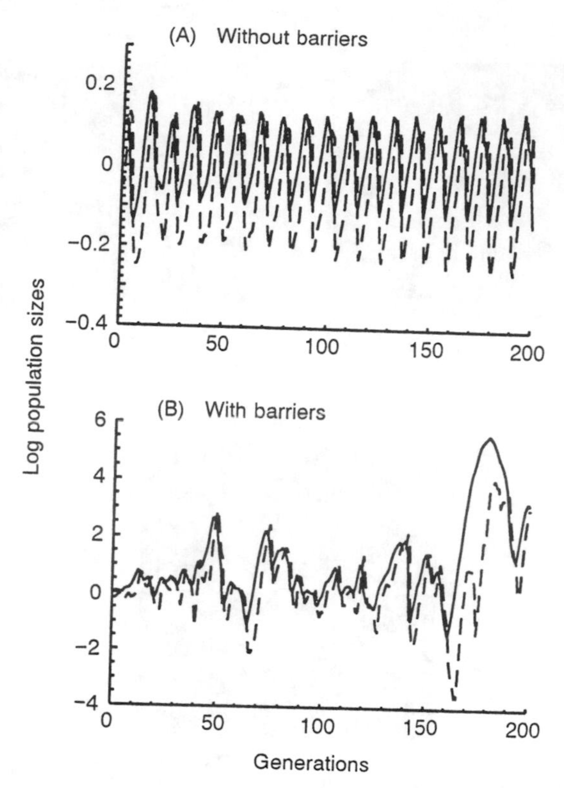

FIGURE 11. Examples showing the considerable increase in the range of population fluctuations arising from imposing the barriers illustrated in Figure 10C ($\lambda = 2$, $\mu_N = \mu_P = 0.6$).

by defining a parameter, R, so that the number of independent uniform random numbers in each direction of the array of patches is $n/(R + 1)$. The intervening multipliers of λ are obtained by a two-dimensional linear interpolation.

Random variation on a patch-to-patch scale has the additional interesting effect of increasing the probability that self-maintaining patterns will be established. For example, in simulations with $\lambda = 2$, $n = 30$, spatial chaos and spirals may be generated from completely uniform initial population densities if λ is chosen uniformly from the range 0.25 to 3.

CONCLUSIONS

It is perhaps surprising, in view of the considerable interest in the dynamic effects of spatial heterogeneity, that so little attention has been paid to the role of temporal asynchrony as a form of heterogeneity between interacting populations (but see Griffiths, 1969). Many discrete-generation models of host–parasitoid interactions implicitly assume that although the risk of parasitism varies among host individuals, each host experiences a constant risk of attack throughout the period it is susceptible to parasitism. This will often be untrue. Although natural selection will tend to increase the temporal synchrony of hosts and parasitoids, asynchrony may occur if there are constraints on the parasitoid's biology, if the system has been disrupted by humans, or if host and parasitoid respond differentially to changes in weather patterns during development. Some parasitoids may not be able to search for hosts at times of the year when hosts are present because the climate at that time of year is too wet, dry, hot, or cold. Asynchrony may also occur if parasitoids introduced into new areas for biological control are imperfectly adapted to their new environment.

In this study we have analyzed a host–parasitoid model incorporating variable degrees of synchrony between hosts and searching parasitoids. In effect, such asynchrony introduces a partial refuge effect in which, by virtue of their timing, some hosts are less at risk from parasitism than others. Indeed, the familiar proportional refuge model (Hassell, 1978) can be derived as a limiting case of the present model if parasitoid emergence is fully synchronous, but host hatching occurs over a long period of time relative to the length of the window of susceptibility to parasitism. When only a small fraction of the host population is protected, the proportional refuge model predicts limit cycles in the populations (Hassell and May, 1988). Such limit cycles are also found in the present model, but only for very narrow regions of parameter space. A further limiting case occurs when there are constant numbers of searching parasitoids during the susceptible host stage and when handling time is zero; the model now corresponds exactly to the discrete-generation model of May (1978).

We conclude that a lack of temporal coincidence between searching parasitoids and susceptible hosts can contribute to the stability of host–parasitoid systems. In certain cases, temporal heterogeneity can be sufficient to stabilize a model that otherwise has unstable, Nicholson–Bailey dynamics. Although we have only analyzed a small fraction of the possible types of host–parasitoid interactions, the major patterns we discern seem relatively robust. However, changes in climate can have many effects other than promoting consistent patterns of asynchrony. For example, occasional and irregular early springs may prompt early emergences of parasitoids, which in turn expose the entire population of hosts to parasitism. Interestingly, our simulations indicate that even if events such as this are extremely rare (e.g., once every 20 years), they could totally undermine the stability of host–parasitoid dynamics. This phenomenon emphasizes that if ecologists want to anticipate the consequences of global change, they need to consider rare weather events as well as long-term averages.

In a homogeneous environment, host–parasitoid interactions with discrete generations fail to persist unless parasitism is nonrandom, or unless host or parasitoid populations are subjected to some form of explicit density-dependent effect. In a subdivided environment, such associations can persist if there is sufficient variability in the probability of individual hosts being parasitized. In this paper, we show that purely deterministic processes, whereby some constant fractions of hosts and parasitoids (μ_N and μ_P, respectively) move to immediately neighboring patches in each generation, can also lead to persistence of the overall population, even when parasitoids search randomly within patches and no explicit density-dependent mechanism is present. The densities of host and parasitoid subpopulations in a two-dimensional array of patches can exhibit spiral waves, or spatial chaos, or crystal lattice patterns, depending on the magnitude of the parameters, μ_N and μ_P, and the host rate of increase, and provided the habitat is large enough. These complex patterns of spatial variation in population density arise even though the environment in all the patches is the same. The patterning is thus deterministically generated by the interplay of local dispersal and local dynamics.

Qualitatively similar results are obtained using a quite different approach—a cellular automaton model—in which there are simple rules of movement instead of the explicit expressions for the dynamics in each patch and the movement to neighboring patches (Hassell et al., 1991). The automaton has a finite number of possible cell states, as opposed to the continuous set of host and parasitoid population densities in a patch in the mathematically explicit model. In each generation, the state of every cell in a square arena may change. Diffusive effects are paralleled by the fact that the new state depends not only on the current state of the given cell, but on the states of the nearest eight cells. Full details are given in Comins et al. (in press). Interestingly, the same qualitative behaviors of spatial chaos and spiral waves are obtained. The value of this approach of having only qualitative categories of patch densities and a very simple set of "transition rules" is in emphasizing those general properties of mathematically explicit models that are detail-independent.

Habitat size is critical for the persistence of the above patterns. With the exception of the crystal lattice pattern, the probability of extinction rises rapidly as the arena width becomes smaller. The failure to persist in small arenas appears related to there not being enough space in which to fit the characteristic pattern of the spatial dynamics. Thus, the large spiral in Figure 6B is much more sensitive to reduction in arena size than is the more irregular chaotic pattern in Figure 6C, while the crystal lattice (Figure 6D) can persist even in arenas of width two. In a similar way, large-scale patterns are much more easily disrupted, with the total populations then becoming extinct, by partial subdivision of the environment with only limited dispersal between the divided segments.

The extent to which the underlying deterministic patterns are sensitive to superimposed random variation at the patch-to-patch level also depends on the scale of the spatial variation. Hence, random variation in the host growth rate, λ, does little to disrupt the pronounced spiral patterns, even with quite appreciable levels of random variation, but can dramatically disrupt a fine-scale crystal lattice pattern. Thus only if the scale of random variation approaches, or is greater than, the characteristic scale of the spatial dynamics does it have an important disruptive effect.

While this model includes a minimum of biological detail and thus certainly cannot be applied directly to natural interactions, it does point to how important the characteristic spatial dynamics can be in determining how habitat fragmentation may affect the dynamics of real systems. Hence, the same system, but with different basic demographic parameters affecting its spatial patterns, could respond in completely different ways to the same level of habitat disruption. This is thus yet another study emphasizing how important it is for field biologists to examine within-generation patterns of heterogeneity, in addition to the time series of average population trends over several generations. Instead of expecting to be able to predict some generic consequences of habitat fragmentation, ecologists will have to settle for discovering a suite of possible consequences that depend on the scale of fragmentation relative to the scale of an organism's population dynamics.

CHAPTER 26

CONSERVATION PLANNING FOR SPECIES OCCUPYING FRAGMENTED LANDSCAPES: The Case of the Northern Spotted Owl

Kevin McKelvey, Barry R. Noon, and Roland H. Lamberson

Landscape fragmentation can be a major threat to the persistence of species with very specific habitat and area requirements (e.g., Wilcove et al., 1986). Such a species is the northern spotted owl (*Strix occidentalis caurina*), whose populations are experiencing fragmentation at a number of spatial scales. The analyses we discuss here focus on the geographic scale (Lord and Norton, 1990) and are concerned with the fragmented pattern of old-growth forests that results from timber harvest. In the case of the northern spotted owl, concerns over the fragmentation of habitat have been translated into a specific plan for the owl's conservation (Thomas et al., 1990). In this chapter we report some of the models that were used to guide decisions and thinking about the size and geometry of the habitat reserves that form the core of this conservation plan. We would like to emphasize that we are reporting only a small fraction of the studies and analyses that have been pursued to help maintain spotted owl populations, and that our results represent the synthesis of a massive team effort. Before discussing the specific details of the models, we briefly discuss the general issues surrounding the relationship between spotted owls and old-growth forests, as well as some basic biology and demography of spotted owls.

THE NORTHERN SPOTTED OWL

More than any threatened or endangered species before it, the northern spotted owl epitomizes the struggle between groups representing disparate value systems in a land of dwindling resources. The dilemma is often portrayed as a choice between resource utilization, employment, and economic vitality versus diverse, healthy ecosystems that provide habitat for a large array of species and populations. The confrontation has been particularly intense and prolonged because northern spotted owls are largely restricted to coniferous forests in mature and old-growth age classes, which have high economic value (see review in Thomas et al., 1990). During the last 3–4 decades, these forests have been greatly reduced and fragmented by timber harvest and there is concern that continued habitat reduction will severely threaten the species' viability. For more than a decade, debate has focused on the degree of threat to the species and on the amount and distribution of mature and old-growth forests required for the species' conservation.

Throughout its range, the northern spotted owl is closely associated with old, dense, large-diameter forest stands (i.e. old growth) that provide forage, cover, and suitable nest sites. Nearly all observations of successful nesting, breeding, and fledging of young are from old-growth forests, or from those younger forests that include the structural characteristics of old-growth forest (Thomas et al., 1990, Appendix F).

Timber harvest, fire, and clearing for agricultural and urban development have reduced spotted owl habitats to less than 10 percent of their original area. Most of this habitat loss has occurred since 1950, a precipitous trend that is projected to continue into the future unless strong conservation policy is instituted (Murphy and Noon, 1992). This broad regional decline in old-growth forests and the associated decline in northern spotted owl numbers prompted a petition for federal intervention under the Endangered Species Act, and the species was given threatened status in 1990 (Federal Register 55(123): 26114–26194, 26 June 1990).

Reproductive success in northern spotted owls can vary greatly from year to year. In some years, most pairs nest and produce offspring, whereas in other years most pairs do not even attempt to nest. This variation may be related to fluctuations in the owl's primary prey, mostly small mammals. A key aspect of the species' life history is that almost all juvenile birds are forced to disperse from their natal areas, sometimes over long distances, in search of both a suitable site and a mate (Allen and Brewer, 1985; Forsman et al., 1984, Meslow, 1985; Gutierrez et al., 1985; Miller, 1989).

This chapter presents the results of our efforts to characterize the life history structure of the spotted owl, to determine the demographic parameters that most influence rates of population change, and to integrate, in the form of computer simulation models, the species' demography with variation in habitat amount, distribution, and quality. The computer simulation models were developed specifically to aid the development of a conservation plan. The models therefore parallel the evolution of our thoughts, starting with simple models designed to answer the most basic questions (e.g., are owl populations declining?) to complex models designed to test the efficacy of particular reserve designs on the landscape.

METHODS

First we employed standard Lotka–Leslie methods to make inferences about the effects of variation in certain life history parameters on population growth rate (Keyfitz, 1968; Mertz, 1971; Nichols et al., 1980; Caswell, 1989). We examined the effect of changes in one demographic character at a time with all others held constant. We assumed (1) reproduction was characteristic of a birth-pulse population (Caughley, 1977, p. 6), (2) no density dependence, (3) a 1:1 sex ratio, and (4) formulated our models exclusively in terms of females.

Parameter estimates and model structure were based on information from the study of marked populations. Owls were captured during the breeding season, aged, sexed, and individually marked with both colored and U.S. Fish and Wildlife Service leg bands. Reproductive status was determined for all individuals using methods outlined by Forsman (1983). Young were counted after fledging had occurred. Individuals were initially placed into four age classes (juvenile, first-year subadult, second-year subadult, adult) according to criteria developed by Moen et al. (in press). Because of similarity in survival rates and breeding status, second-year subadults were considered in the adult stage. Annual survival rates for banded juveniles were estimated empirically (number observed in year $t + 1$/number banded in year t). Estimates for banded subadults and adults were calculated using Jolly–Seber capture–recapture models for open populations (Jolly, 1965; Seber, 1965) using program JOLLY (Brownie et al., 1986). The method assumed that an owl was dead when it was replaced on its territory by another owl and subsequently went undetected for two or more years (Franklin et al., 1990).

The estimate of fecundity combines two important components: the number of females fledged per nest and the proportion of females 2 years of age or older that breed. (Age at first breeding is variable, but for most birds it occurs at age two years.) Fecundity was estimated each year as the mean number of young fledged per pair, and assumed a 1:1 sex ratio at birth. We believe our estimates for most demographic parameters (e.g.

Franklin et al., 1990) are quite accurate because the data represent a large number of marked birds studied for 5 to 8 years.

LIFE HISTORY STRUCTURE

The basic demographic life history of the spotted owl is presented in Table 1. The parameter b_x denotes the expected number of female fledglings produced by a female of age x. Annual survival probabilities were assumed distinct for the three age classes: juvenile (s_0), subadult (s_1), and adult (s). The parameter l_x represents the probability of survival to age x (by definition $l_0 = 1.0$). For $x \leq 2$, probability of survival to age x is given by $l_x - s_0 s_1 s^{x-2}$ (Table 1). Adult refers to ages subsequent to the second year of life. Within the adult age class, survival rate was assumed to be constant. We used parameter estimates from Franklin et al. (1990) (Table 2), except where otherwise noted. For our projection models, we divided the population into three stages: juveniles, subadults, and adults. Time was expressed on an interbirth interval of 1 year, and we assumed an age at first reproduction of 2 years.

With long-lived species, estimates of age-specific parameters are impossible to attain or have large sampling variances because of small sample sizes. A useful approximation to an age-structured model is a stage-structured model (Lefkovitch, 1965; Caswell, 1989). Estimating the population dynamics of spotted owls with a stage model would yield misleading conclusions only if the species experienced reproductive senescence prior to

TABLE 1. Spotted owl life history.

x[a]	l_x[b]	b_x[c]	$l_x b_x$[d]
0	1.0	0	0
1	s_0	0	0
2[e]	$s_0 s_1$	b	$s_0 s_1 b$
3	$s_0 s_1 s$	b	$s_0 s_1 s b$
4	$s_0 s_1 s^2$	b	$s_0 s_1 s^2 b$
⋮	⋮	⋮	⋮
x	$s_0 s_1 s^{x-2}$	b	$s_0 s_1 s^{x-2} b$

[a] Age expressed in years.
[b] Probability that an individual aged 0 (a newly hatched bird) will survive to enter age class x.
[c] Expected number of female fledglings produced by a surviving female of age x.
[d] Net maternity function.
[e] Age at first reproduction is 2 years.

TABLE 2. Demographic parameters for the northern spotted owl.

Parameter[a]	Estimate	Sample size	Standard error
Juvenile survival (s_0)	0.159	63	0.046
Subadult survival (s_1)	0.868	34	0.008
Adult survival (s)	0.868	80	0.008
Fecundity (b)	0.382	197	0.190
Rate of population change (λ)	0.925	—	0.032

(Data from Franklin et al., 1990; Franklin, in press.)
[a] $H_0 = \lambda \geq 1.0$; $H_A = \lambda < 1.0$
$Z = (0.925 - 1)/0.032 = 2.343$
$P < 0.01$; Decision: reject H_0

about 15 years (Noon and Biles, 1990). Given the current high estimates for s (Table 2), early senescence seems unlikely.

RESULTS: DETERMINISTIC ANALYSES

Finite rate of increase

In general for the spotted owl, λ can be defined as the single, positive, real root of the equation (see Table 2)

$$1 = \sum_{x=1}^{\infty} \lambda^{-x} l_x b_x = b s_0 s_1/(\lambda^2 - \lambda s),\ 0 < s < 1,\ \lambda > s \tag{1}$$

which can be rewritten as

$$\lambda^2 - s\lambda - b s_0 s_1 = 0 \tag{2}$$

and solved for λ as

$$\lambda = [s + (s^2 + 4 s_0 s_1 b)^{1/2}]/2 \tag{3}$$

Alternatively, λ can be computed as the dominant eigenvalue of the stage projection matrix.

The solution to Equation 3, based on the parameter estimates we used (Table 2), yields the estimate $\lambda = 0.925$. This is equivalent to a 7.5 percent annual decline in population size with a population half-life of 9 years. However, λ merely estimates population change over the period of study, and the projection of future population size is based on the unrealistic assumption of constant parameter values.

Sensitivity analyses

If simplifying assumptions are made about adult survival and fecundity, implicit differentiation of the characteristic Equation 2 can be used to evaluate the effects of changes in the vital rates. Survival and fecundity rates are measured on different scales, however, which makes direct comparison of their sensitivities difficult. Elasticities, a measure of the sensitivity of λ to proportional change in the vital rates, solve this problem (Caswell et al., 1984). For the spotted owl, population change is most sensitive to variation in adult survival rate and relatively insensitive to variation in prereproductive survival or fecundity. In addition, Noon and Biles (1990) found λ to be relatively insensitive to age at first breeding.

Interval estimates of λ

A simple deterministic calculation of λ neglects the fact that birth and survival rates are estimated with some sampling error. For this reason, a formal statistical test of the $H_0:\lambda \geq 1.0$ is relevant to any species whose populations are believed to be declining. Methods to estimate the sampling variance of λ are provided by Lande (1988, p. 206) and Caswell (1989, p. 185). When we applied these methods we found that spotted owls are apparently declining in northern California (Table 2; Franklin et al., 1990; Franklin, in press). Similar findings of significant population decline have been reported for the Oregon Coast Range (E. D. Forsman, pers. comm.; Thomas et al., 1990, Appendix L).

Senescence

The effects of a senescent decline in fecundity are explored by incorporating a maximum age, $w - 1$, beyond which no reproduction occurs (i.e., $b_x = 0$, $x > w$) (Mertz, 1971; Nichols et al., 1980). The life span of spotted owls, and whether or not fecundity remains constant, are unknown. However, it is unlikely that one could obtain survival estimates for all adult owls as high as 87 percent if spotted owls lived less than 15 years. Thus we feel confident that adults do readily live to age 15 without significant senescence. This is an important point because mathematical analysis reveals that any senescence occurring after age 15 has little effect on λ, sensitivities, or elasticities.

Given evidence of declining populations in two geographically separated portions of their range, coupled with the precipitous loss of their preferred habitat, the logical conclusion was to attribute the owl's decline to habitat loss. We were then led to investigate how the amount, distribution, and quality of habitat might affect the demography of owl populations.

DYNAMIC PROJECTION MODELS

No natural population is exposed to purely deterministic forces. Thus the population dynamics of the spotted owl should be examined in light of both demographic and environmental stochasticity. Further, estimating the sensitivity of population growth to variation in individual demographic parameters can be misleading. Demographic parameters do not vary independently and thus should be considered simultaneously. Unfortunately, the effects of stochastic factors and the covariance of life history attributes are usually not amenable to analytic models. These factors, and their effects on rates of population change, however, can be explored in simulation models.

A conservation strategy for any species must ultimately be described by a map that integrates information on the species distribution, the distribution of current and potentially suitable habitat, and economic, political, and legal constraints. To develop a conservation strategy for the northern spotted owl we needed insights into how the arrangement of owls and their habitat across the landscape affected their population dynamics. Our task was to test and refine various iterations of the conservation map until the arrangement of conservation areas described a reserve system with a high likelihood of ensuring the owl's persistence for 100 years (Thomas et al., 1990; Murphy and Noon, 1992).

An initial map showing the distribution of owls and their habitats indicated that the strategy for persistence of the northern spotted owl would need to be evaluated in the context of a metapopulation (Levins, 1970; Shaffer, 1985). The distribution of habitat polygons was discontinuous across the landscape, reflecting natural discontinuities in vegetation structure and composition, in topographic conditions, and the effects of extensive human-induced habitat disturbance and fragmentation.

To provide a framework to guide reserve design for the northern spotted owl, we developed a number of simulation models incorporating various degrees of spatial information. Our approach, which investigated the effects of variable dispersal efficiencies on the population dynamics of territorial animals occupying heterogeneous landscapes, was based on previous work by Lande (1987). Lande's model, based on monogamous territorial breeders with obligate juvenile dispersal, was directly applicable to the life history structure of the spotted owl.

NONEXPLICIT SPATIAL MODELS

The individual territory model

The individual territory model assumes that all newly fledged juveniles disperse, and that adult birds who experience loss of their territory (e.g., due to timber harvest) also disperse (Lamberson et al., in press). The model assumes a landscape which is of fixed spatial extent and contains a fixed number of potential home range territories (or "sites"). However, only a fraction of these sites are capable of being occupied, namely, those that contain suitable habitat for spotted owls.

The state variables in the model, updated annually, are (1) the number T of currently suitable sites (which may be decreasing over time due to timber harvest); (2) the number P of suitable sites that are occupied by nesting pairs; and (3) the number S of suitable sites that are occupied by single males. The key equation which describes search success in this model (see Lande, 1987) is

$$P\text{ (success)} = 1 - [1 - (\text{unoccupied suitable sites/total sites})]^m \quad (4)$$

where m is the number of sites that can be searched prior to mortality (additional equations describing the model dynamics are given in Lamberson et al., in press).

A nesting pair will annually produce young (according to a deterministic or stochastic likelihood), and these will disperse at the end of the season, the males seeking an unoccupied site and the females seeking a site occupied by a solitary male. Dispersal success is density dependent, calculated by assuming random search of accessible sites. Search capabilities, together with the occupancy ratio of searched sites, determine the bird's potential for successful dispersal (consistent with Lande, 1987).

Due to the dynamic and nonequilibrium nature of this model, there were several new findings including:

1. The probability p of population survival for 250 years as a function of initial population size (expressed as a percentage of suitable sites occupied) is affected by assumptions about environmental stochasticity. When there is no stochasticity, p is a stair-step function, rising abruptly from 0 to 1 at the unstable equilibrium threshold point. Adding environmental stochasticity smooths out the stair-step; that is, the extinction threshold becomes "blurred."
2. In a similar fashion, survival probability p, as a function of the percentage of suitable habitat, is affected by environmental variation. Once again the deterministic case shows a stair-step function, with the jump from 0 to 1 occurring at the threshold point (Figure 1). As before, adding environmental variance makes the extinction threshold less abrupt.
3. The crowding of older owls into the remaining suitable habitat as timber harvest continues temporarily produces very high occupancy rates—much higher than would be expected under long-term stable conditions (Figure 2). Thus, predicting long-term population status from short-term occupancy data can be very misleading.

The territory cluster model

In the individual territory model, search is based on simple sampling with replacement: all cells are equally likely to be searched. In reality, dispersing

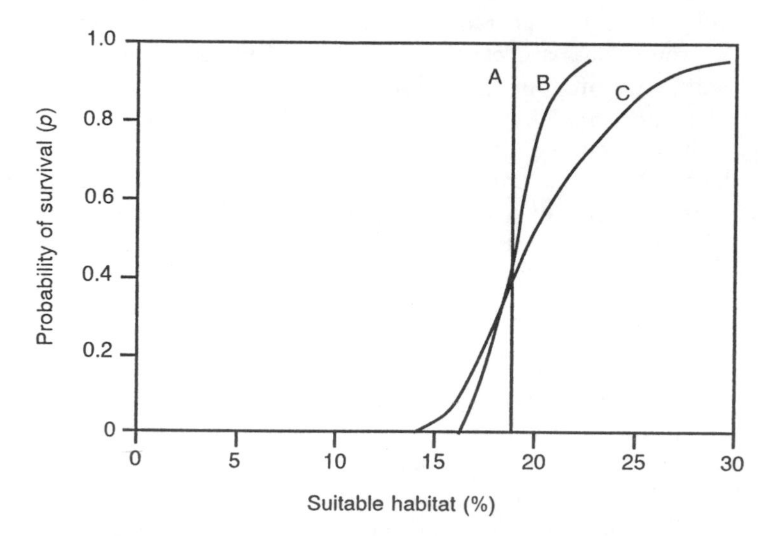

FIGURE 1. The 250-year survival probability as a function of the percentage of sites that are suitable owl habitat in the landscape. Curves A, B, and C represent the conditions of low, medium, and high environmental variability.

juveniles have a high probability of searching the areas close to their nest sites and a low probability of searching distant areas. Their success will therefore be based on the local habitat availability. If suitable habitat is unevenly distributed, then the search efficiency in areas where the habitat is clustered may be significantly higher than the global proportions would suggest.

The basis of the territory cluster model (Thomas et al., 1990, Appendix M; R. H. Lamberson and B. R. Noon, unpublished) was a rectangular array of circular clusters containing a variable number of owl sites (territories). All sites within a cluster were assumed to be of identical size, and a site was either suitable or unsuitable. A suitable site was capable of supporting survival, mate attraction, and reproduction. Clusters could be either totally or partially suitable—the carrying capacity of a cluster equaled the number of suitable sites. As in the individual territory model and Lande (1987), the response variable was the proportion of suitable sites occupied at any point in time.

The matrix between clusters was assumed to be entirely unsuitable as owl habitat. Assuming a constant percentage of the habitat as potentially suitable and restricting the suitable habitat to clusters had two important consequences: (1) as cluster size increased, the distance between adjacent clusters increased predictably; and (2) the probability of choosing a dispersal azimuth that intersected an adjacent cluster was constant.

The life history was a three-stage, female projection model. Fecundity

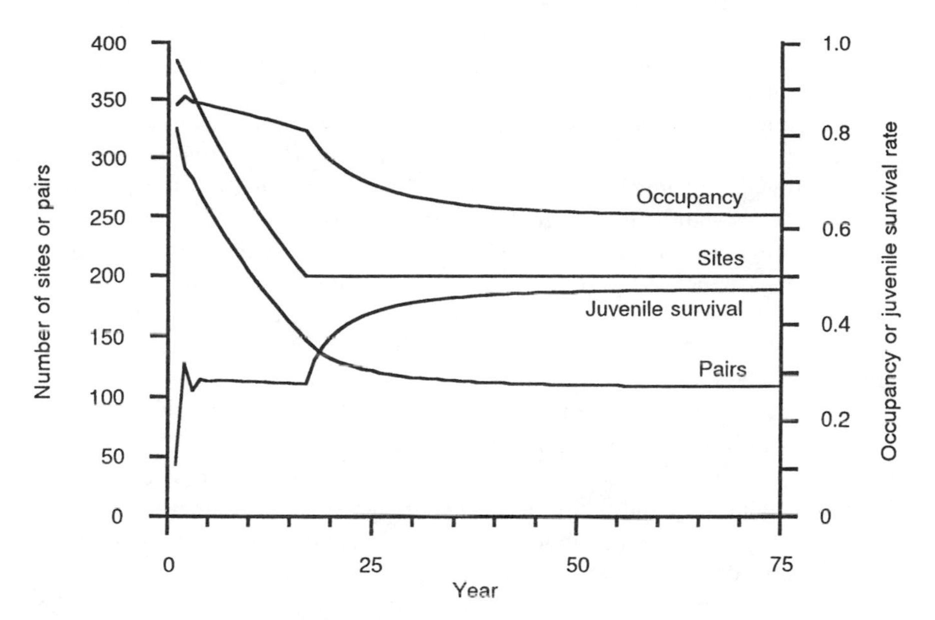

FIGURE 2. Trends in the number of pairs of spotted owls, number of sites, site occupancy by pairs, and juvenile survival rate based on a 75-year simulation. The simulation assumed that 4 percent of the suitable habitat was lost per year until 20 percent of the landscape remained suitable, and that juveniles could search 20 sites in the landscape. Note that juvenile survival increases as adult occupancy decreases.

was treated stochastically, with expected value equal to field data (see Table 2). Parameter estimates were identical to Franklin et al. (1989, 1990), except that the value of adult survival (s) was increased to provide the possibility of a stable population. (Thus, the model assumes that habitat conditions within the clusters would be better than those in regions where previous demographic studies of owls have been conducted.)

Dispersal dynamics The model distinguished between dispersal within and among clusters (cf. Doak, 1989). Within-cluster dispersal was identical to the individual territory model, with each dispersing owl allowed to sample with replacement a given number, m, of sites within the cluster (Equation 4). If a dispersing juvenile did not succeed in finding a suitable site within its natal cluster (based on a fixed number of searches), it was forced to disperse between clusters. Between-cluster dispersal was modeled as a straight-line path moving away from the natal cluster at a random azimuth. Two sources of mortality existed: first, if the direction chosen did not intersect a neighboring cluster, the bird died; second, if a correct direction was chosen, the risk per unit of distance traveled was assumed to be constant. This assumption means that as distance between the clusters increases, the probability of reaching an adjacent cluster declines exponentially.

The number of possible combinations of model parameters is immense and a large number of sensitivity analyses were performed (Table 3). The most significant results, however, were based on the relationship between mean occupancy and cluster size. The number of sites per cluster was varied across a range of values and the 100-year occupancy trend plotted. For a simulation in which 60 percent of the sites within a cluster were suitable, occupancy did not stabilize until clusters held at least 10 sites. Clusters of 10 sites stabilized at about 70 percent occupancy of the suitable sites, while clusters of 15 stabilized at about 85 percent, a figure representing nearly full occupancy given the adult survival rates. Further increases in cluster size had little effect on occupancy (Figure 3). In general, mean occupancy increased as the percentage of sites that were suitable in each cluster increased. The marginal difference in occupancy among clusters of different sizes, however, was less pronounced beyond cluster sizes of 30 (Figure 4). Another approach to choosing an optimal cluster size is to estimate the number of owls occupying a fixed amount of suitable habitat. Based on the mean occupancies at 100 years from Figure 3, the efficiency of land use (owl pairs per unit of area) is clearly higher for larger cluster sizes.

In summary, the results suggest that providing for clusters of territories should increase the persistence likelihood of spotted owls, primarily by facilitating juvenile dispersal (cf. Doak, 1989). Given the assumptions of this model, clusters containing 15–20 sites with at least 60 percent being suitable at any point in time, and with moderate connectivity among clusters, should provide for a stable reserve system.

TABLE 3. Qualitative results of sensitivity analyses for the territory cluster model.

Factor varied	Sensitivity[a]
Within-cluster search efficiency	Low for large clusters; high for small clusters
Resistance to between-cluster dispersal	Low for large clusters; high for small clusters
Search time outside of natal cluster	Low–moderate for large clusters; high for small clusters
Initial population size	If population was in large clusters, low; small clusters, high

[a]Measured in terms of reduction in mean pair occupancy.

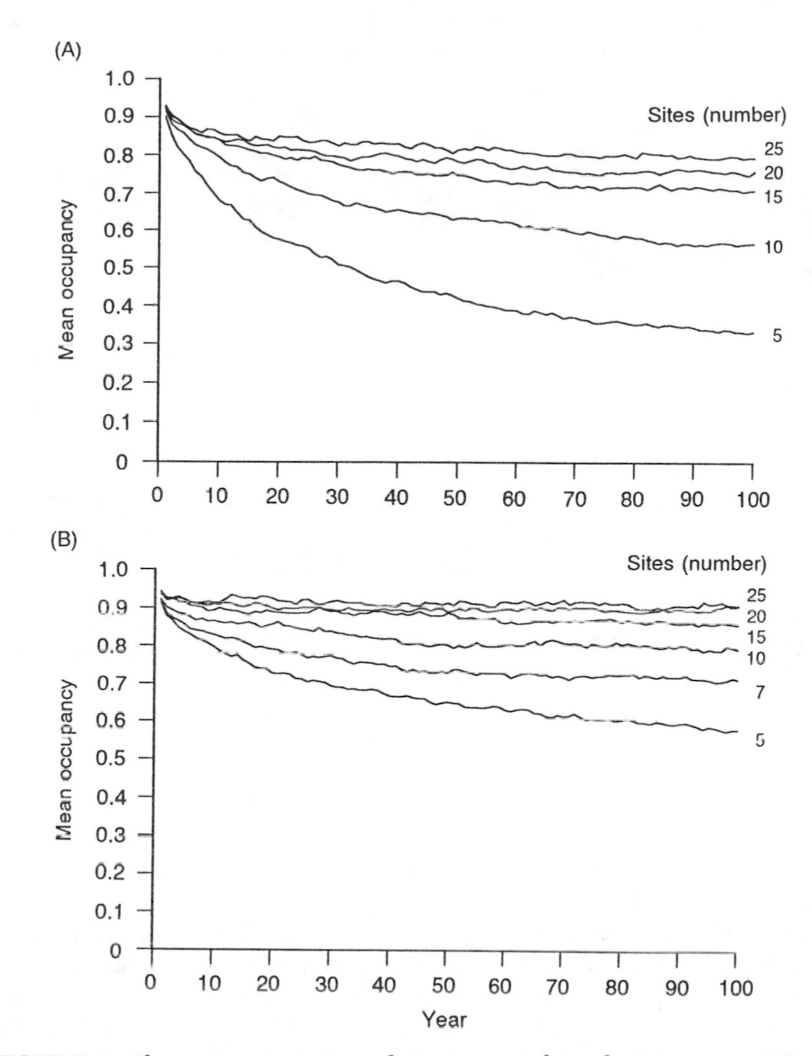

FIGURE 3. The mean proportion of sites occupied in clusters over a 100-year simulation. The number of sites per cluster varied from 5 to 25, with 35 percent of the landscape in clusters. Dispersal exponential coefficient was 0.0354. (A) 60 percent of the sites in the clusters were assumed suitable. (B) 100 percent of the sites in the clusters were assumed suitable.

A spatially explicit landscape model

The territory cluster model provides basic rules for achieving reserve design stability, but the landscapes that it can model are limited (evenly distributed

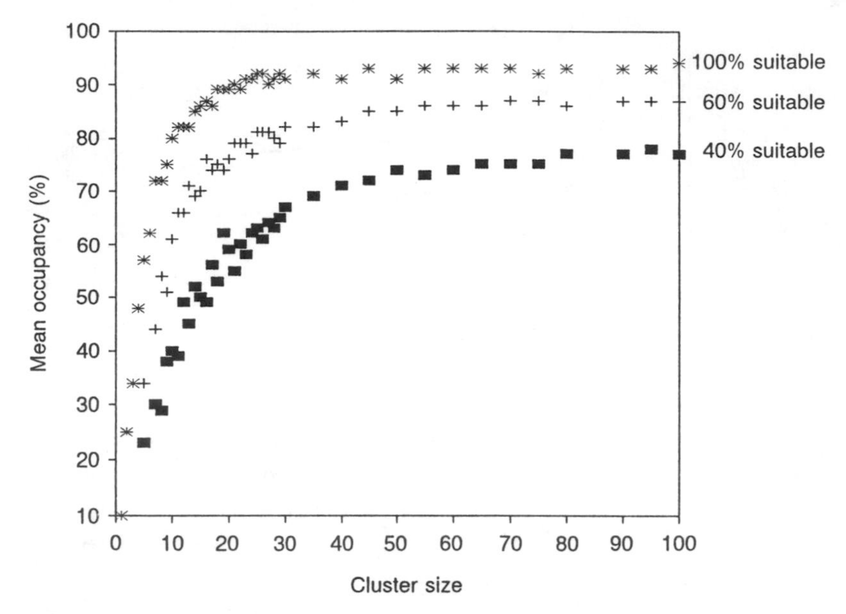

FIGURE 4. Simulated steady state mean occupancy rate as a function of cluster size. The curves represent 40, 60, and 100 percent of the sites within each cluster as suitable habitat.

circular clusters and wraparound boundaries). The landscape model was designed to provide greater spatial flexibility and to allow suitability to be defined along a gradient of qualities, rather than defining a site as simply suitable or unsuitable.

The landscape model links an organism's survival and reproduction explicitly to its current habitat location. As a consequence, a population's rates of survival and fecundity vary based on landscape configuration. In addition, the model allows for habitat areas that vary in their degree of suitability for nesting. Lastly, the model assumes that each organism must search the landscape to find new territories and mates. The model is a two-sex, single-organism simulator. Each organism is born, moves, attempts to find a mate and breed, and dies. This format allows the behavior of each individual to be simulated by following a series of probabilistic rules.

Male and female behavior in this model is similar to that in the territory cluster model. Males search for territories to occupy. If they find a suitable nest site, they stop moving and become territorial. The likelihood of settling in a given site is a function of the habitat quality of that site. Males remain on their selected site until they die or the site becomes unsuitable for nesting. If the site becomes unsuitable, the males become nonterritorial and reinitiate search (Figure 5). Females are born and disperse from the

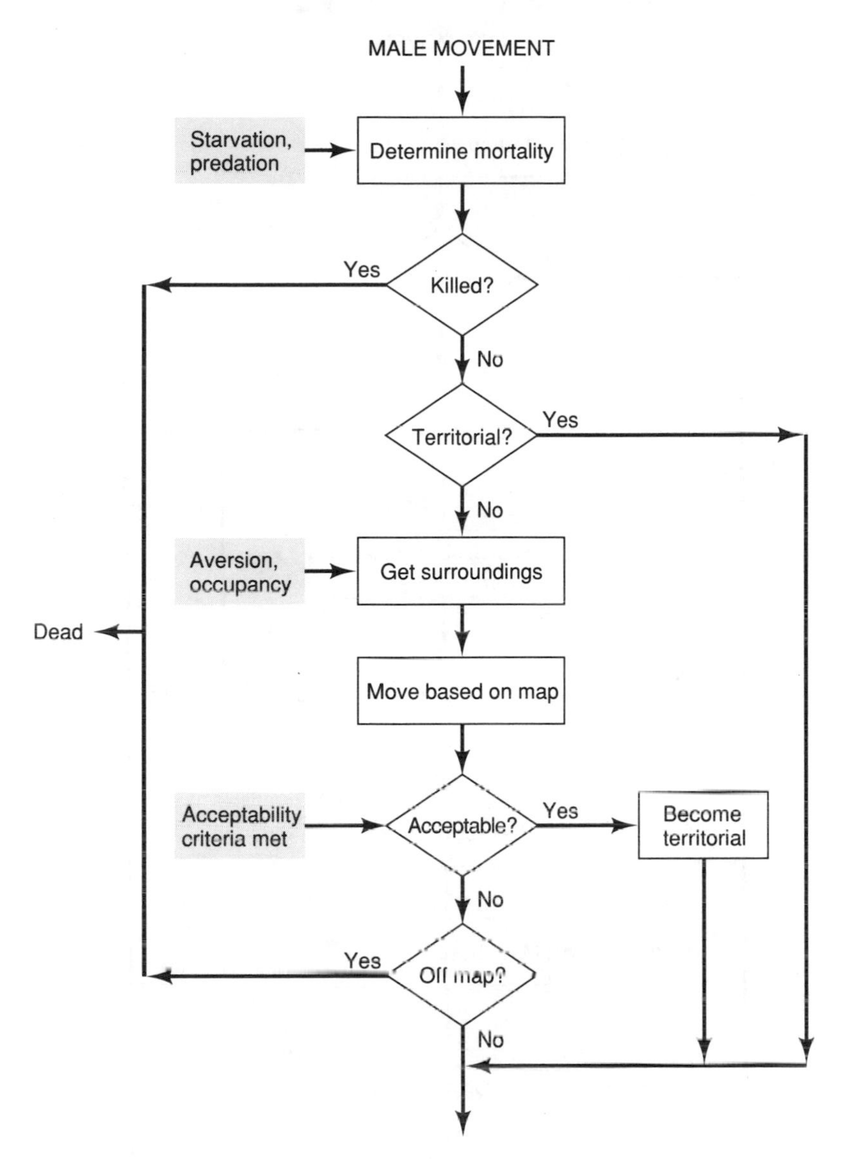

FIGURE 5. Flow diagram representing the process for determining male behavior at each time step in the landscape model.

natal site looking for unpaired, territorial males. When they find a territorial male, they obligately pair (Figure 6). Once paired, females remain on site until they die or the site becomes unsuitable for nesting. Paired individuals split up when one member of the pair dies or the site becomes unsuitable for nesting. If the female dies, the male remains territorial and stays on the

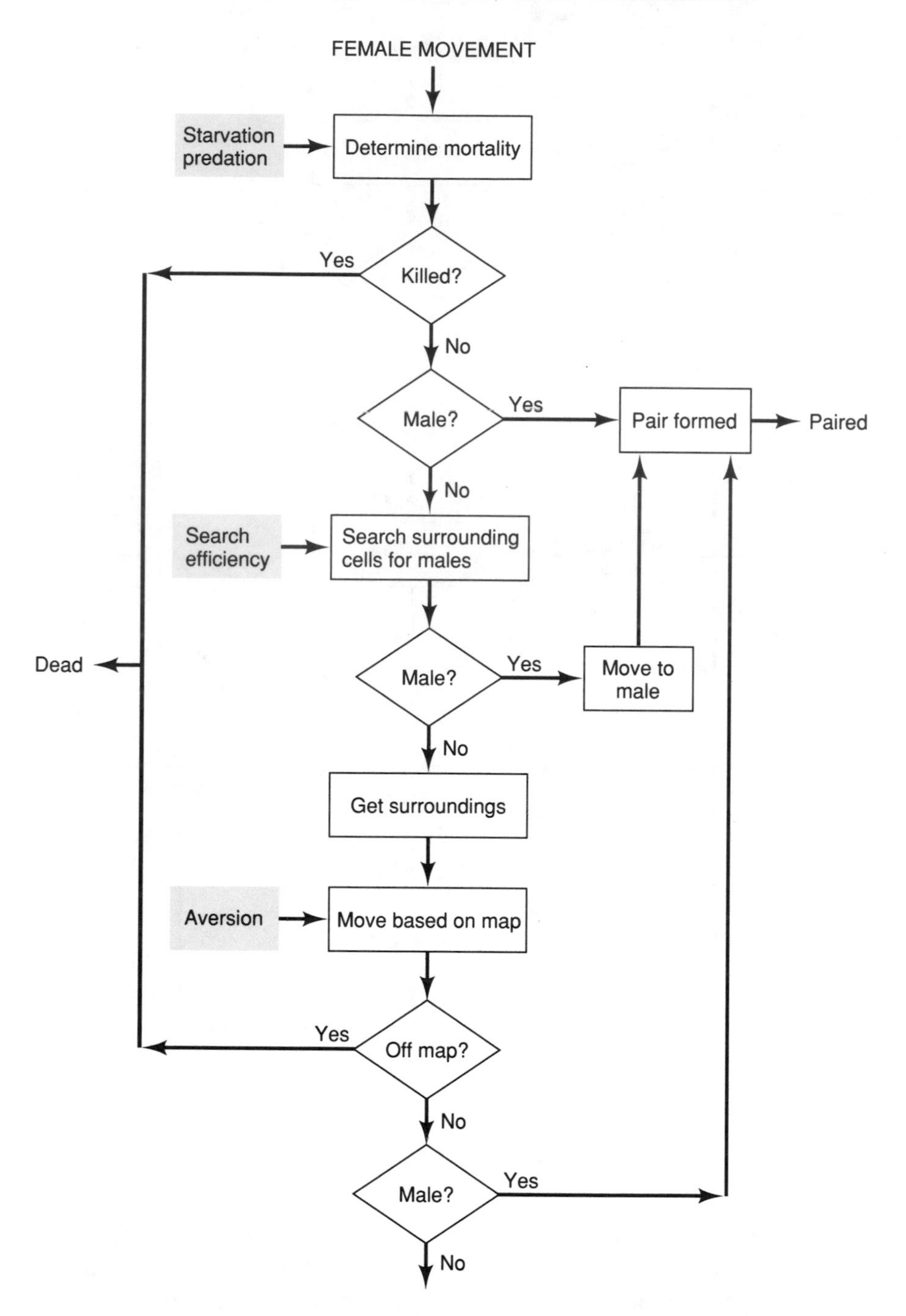

FIGURE 6. Flow diagram representing the process for determining female behavior at each time step in the landscape model.

site. If the male dies, the female has no site fidelity and will initiate search for a new mate (Figure 7). If the site becomes unsuitable for nesting, both members search independently for a new site.

Demographics All demographic parameters are linked to site quality. Individual mortality and fecundity are determined by the quality of the site occupied at the beginning of each time step. In keeping with the stage-structured approach, risks are assumed to be constant within a stage over the course of a year. The year is broken up into i time steps, and the risk per step for an owl in stage class j occupying habitat type k is defined as one minus the ith root of the yearly survival for class j in habitat type k.

Movement The map is divided into a fixed array of grid cells, with each cell representing one territory-sized unit. The grid is hexagonal to allow more realistic movement. The rate of movement is dependent on the size of the grid cell and the number of time steps per year. Individual moves at each time step are restricted to adjacent cells. All of the mobile classes of owls (nonterritorial males and females) have the opportunity to move at each time step. To ensure that certain birds or areas of the map are not given

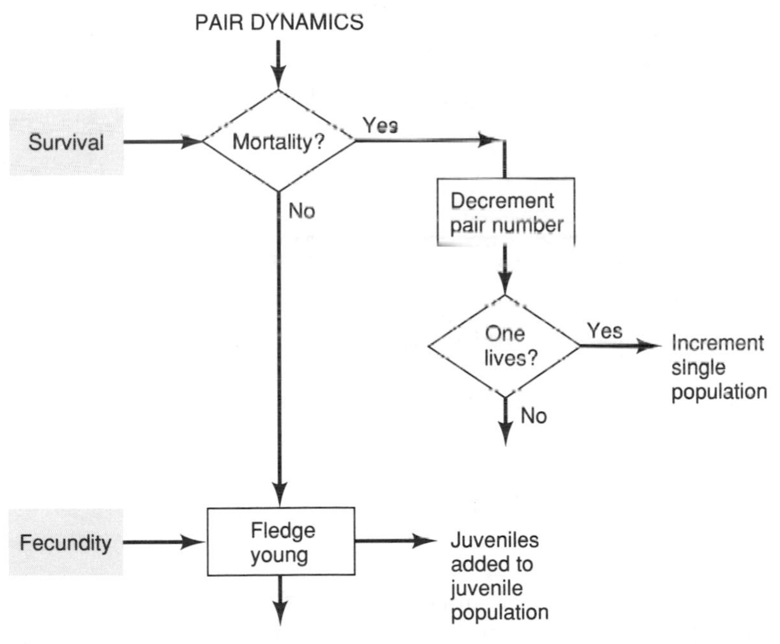

FIGURE 7. Flow diagram representing pair dynamics on a yearly basis in the landscape model. Pairs remain reproductive only if both members survive.

preferential access to open territories or mates, the order of movement is fully randomized at each time step.

In its simplest implementation, movement is a random walk. The model, however, allows owls to search with "intelligence"; that is, they may favor movement through good habitat and avoid poor habitat. Similarly, females move obligately to known territorial males, and nonterritorial males may be averse to crossing defended territories. This intelligent behavior is modeled by giving the owls absolute knowledge of the quality of the cell they occupy and partial knowledge of adjacent cells. They have no knowledge of more distant parts of the landscape. This knowledge takes the form of a series of switches and weighing factors that condition the probability of movement (Table 4).

Three boundary conditions can be specified at the map edges: absorbing, reflecting, and wraparound. In addition, internal reflecting zones can be created by specifying a land type for which the owls show complete aversion.

Fecundity Fledglings are newborns that survive to disperse. It is assumed that there are good and bad years for fledging. If it is a good year, the pair produces fledglings according to a beta-distributed random variable, scaled to range from zero to the maximum clutch size. The number of fledglings in a clutch is the beta value rounded to the nearest integer. There are therefore two levels at which variability can affect the number of fledglings. If the area under the beta distribution is concentrated around the mean clutch size, the population will pulse based on the frequency of good years. When a good year occurs, all pairs will produce about the mean number of

TABLE 4. A summary of factors that can affect an individual's movement in the model.

Factor	Based on	Sex	Form
Become territorial	Habitat quality/ occupancy	M	Probabilistic switch
Aversion	Habitat quality	M/F	Weighting
Site fidelity	Habitat quality	M/F	Weighting
Linear propensity	Behavior	M/F	Weighting
Territorial aversion	Occupancy	M	Weighting
Female finds male (current cell)	Occupancy	F	Absolute switch
Female finds male (adjacent cell)	Occupancy	F	Probabilistic switch
Global boundary	—	M/F	—

fledglings. If the probability of a good year is 1.0, variability in the number of fledglings will occur on an individual territory basis and will depend on the parameters of the beta distribution. Both parameters are linked explicitly to habitat quality and the stage class of the pair.

GIS interface An automated link exists between the model and vector-based geographic information systems (GIS). Using this link, a hexagonal grid with a cell size equal to the mean home range size is intersected with a vegetation map, and the habitat quality of the area contained within each grid cell is analyzed. Habitat quality is an index based on the relationship between the attributes of the cell and the demographics of the organism. The number of fledglings, for instance, is correlated with the proportion of old growth in a home-range-sized area around the nest site (J. Bart, pers. comm.).

Maps can also be created by the model and transformed into vector-based GIS maps. The ability to move information freely between the GIS and the model allows a dynamic interaction between land management decisions and the potential impact of those decisions. Vegetation changes can be simulated at the stand level within the GIS, and the impact of those changes can be estimated. Model output, such as mean pair occupancy for each cell (output as a map), can be overlaid on the stand-level map to determine which stand types correlate with areas displaying high or low occupancy rates.

Effects of spatial configuration We have used the model to project population trends for five hypothetical landscapes with an identical number of suitable sites (Figures 8–12). Other than habitat configuration, there were no differences in the initial values of any model parameters. The map boundaries were wraparound, so the exact location of the habitat within the map frame was unimportant. The demographic parameters (Table 5) were set to yield a $\lambda = 1.0$ and were not modified by habitat quality. In these simulations only two habitat qualities, suitable and unsuitable for nesting, were specified. The movement parameters deviated only slightly from a random walk: birds were twice as likely to choose suitable habitats; males treated occupied habitats identically to unsuitable habitat; and birds were twice as likely to continue to move in the same direction as to choose a different direction.

Model results parallel the territory cluster model, showing that a clustering of suitable habitat is both more efficient in terms of mean population level and more stable in terms of lowered extinction probabilities than is a random structure. The shape of reserve areas also has an important effect on their stability properties. A cluster with a low ratio of edge to area (Figure 10) is more stable than continuous clusters of identical area but with varying degrees of irregularity (Figures 8, 9, 11, and 12). A high edge-to-

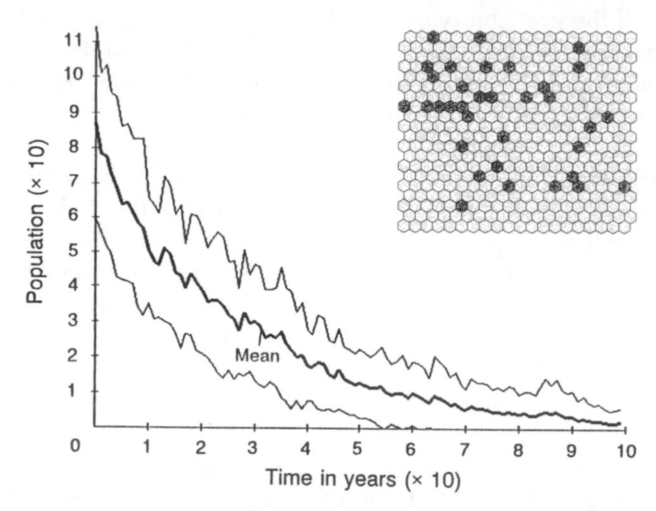

FIGURE 8. Landscape simulation with suitable habitat randomly scattered. The results are based on 30 simulations. The heavy line represents the mean population, the thin lines are one standard deviation from the mean.

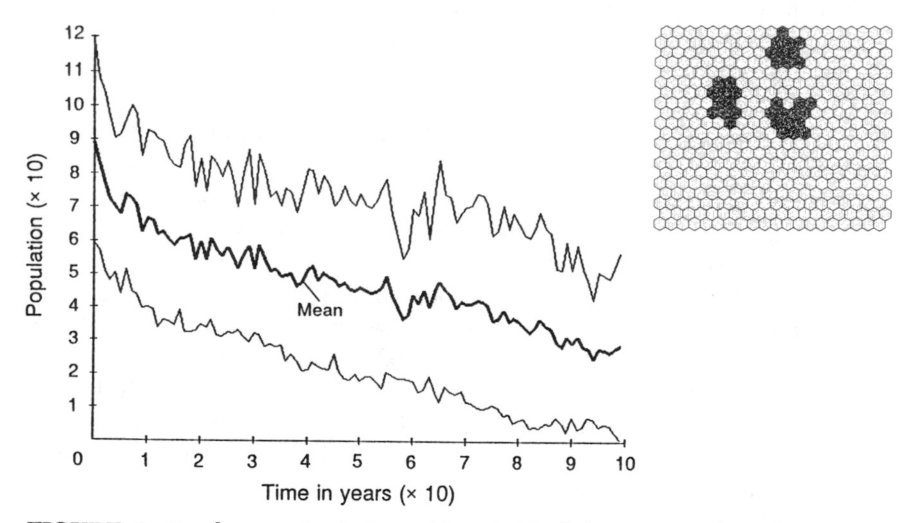

FIGURE 9. Landscape simulation with suitable habitat arrayed in three small blocks. The results are based on 30 simulations. The heavy line represents the mean population, the thin lines are one standard deviation from the mean.

area ratio has a negative impact on demographic stability: the rate of decline in a large, highly irregular cluster (Figure 12) is similar to the decline rate of the dispersed cluster system (Figure 8).

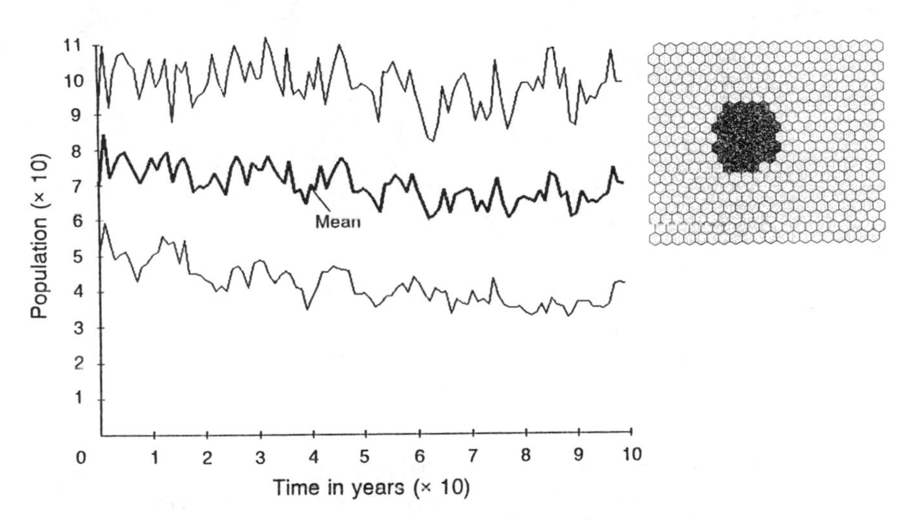

FIGURE 10. Landscape simulation with suitable habitat arrayed in one large regular block. The results are based on 30 simulations. The heavy line represents the mean population, the thin lines are one standard deviation from the mean.

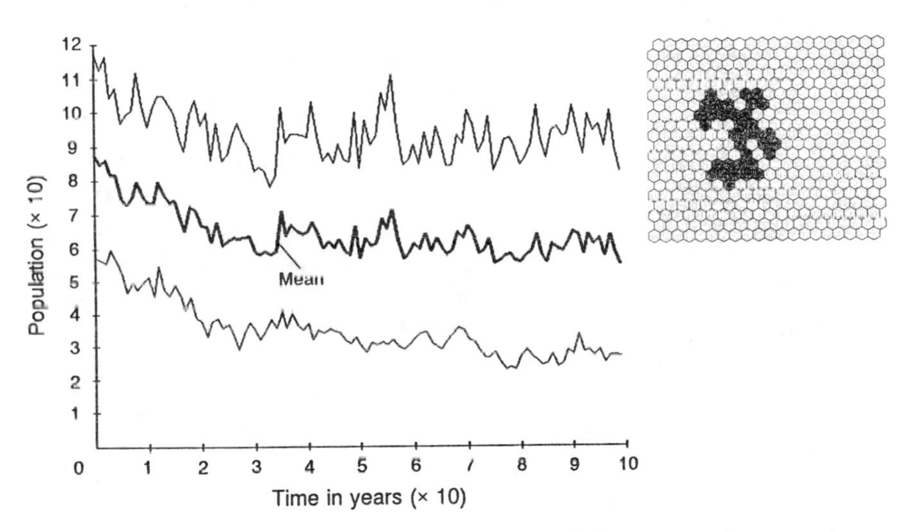

FIGURE 11. Landscape simulation with suitable habitat arrayed in one large, irregular block. The results are based on 30 simulations. The heavy line represents the mean population, the thin lines are one standard deviation from the mean.

Source–sink relationships The previous results demonstrate the impact of reserve shape when each landscape cell is either suitable or unsuitable for breeding. In that case, the breeding population is limited entirely to the

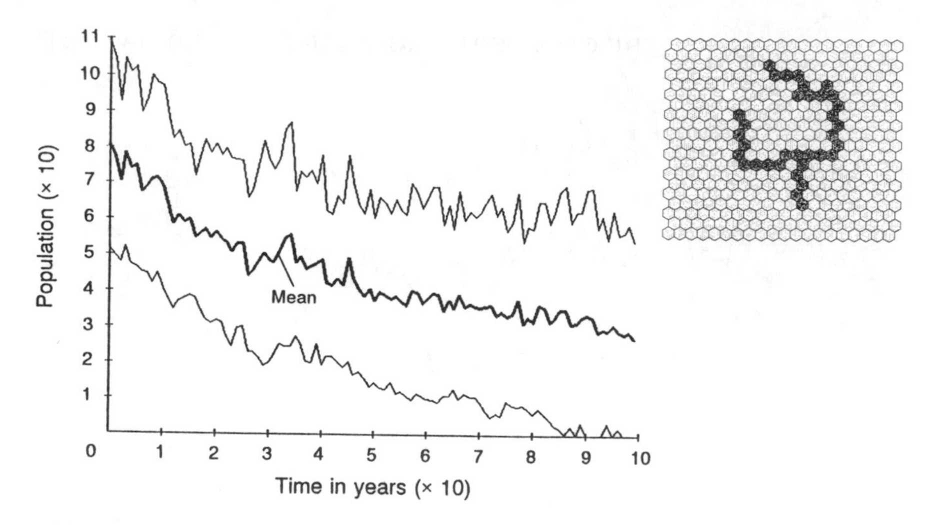

FIGURE 12. Landscape simulation with suitable habitat arrayed in one very irregular block. This is similar to the form of reserves expected in riparian corridors. The results are based on 30 simulations. The heavy line represents the mean population, the thin lines are one standard deviation from the mean.

TABLE 5. List of the parameter values used for landscape model simulations.

Parameter[a]	Value	Source
Juvenile survival	0.29	Franklin et al., 1990
Subadult survival	0.935	Thomas et al., 1990
Adult survival	0.935	Thomas et al., 1990
Birth rate	0.335	Franklin et al., 1990
Aversion	1.0, 0.5[b]	
Boundary	Wraparound	
Linear propensity	2.0	
Site fidelity	0.5	
Territorial aversion	0.5	
Female finds male	0.5	
Time steps	40.0	
Runs	30.0	

[a]Parameters were chosen to produce as optimistic an estimate of owl survival as could be supported by these data. Parameters varied with site quality only where explicitly stated.

[b]Only two habitat types were placed in the map, one representing the best habitat and the other representing the worst. This split corresponds to suitable/unsuitable designations found in the individual territory and territory cluster models.

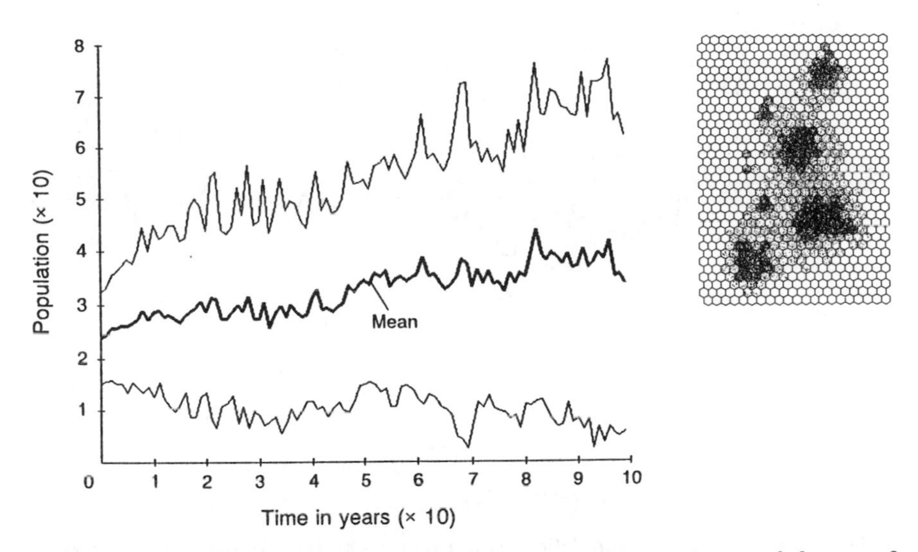

FIGURE 13. Landscape simulation with a reserve system consisting of clusters of suitable habitat surrounded by marginal habitat. The population was initialized with owl pairs occupying all suitable sites. The results are based on 30 simulations. Shading is based on mean pair occupancy.

suitable habitat. In a landscape containing a gradient of habitat qualities, there will be source locations that produce an excess of young, and sink locations into which some of these juveniles travel. When the possibility of nesting in the sink locations exists, populations will occur exterior to the designated reserve network, even though the vital rates from these areas do not provide for a self-supporting population (compare Figures 13 and 14). These results suggest that the presence of sink areas adjacent to habitat clusters may have a negative impact on the stability of the reserve. Even though the mean population size of the entire landscape is higher in the source–sink system, the mean occupancy of the reserve clusters is lower and the variability of the system increases with time.

DISCUSSION

Inferences from the deterministic analyses

The analyses of Lande (1988) and Noon and Biles (1990) demonstrate that λ shows an extreme sensitivity to variation in adult survival rate and a relative insensitivity to fecundity or preadult survival rates. Two types of information, however, are relevant to a species' population growth rate. One is the sensitivity of λ to variation in life history characters; the other concerns life history attributes showing the most natural variation. Varia-

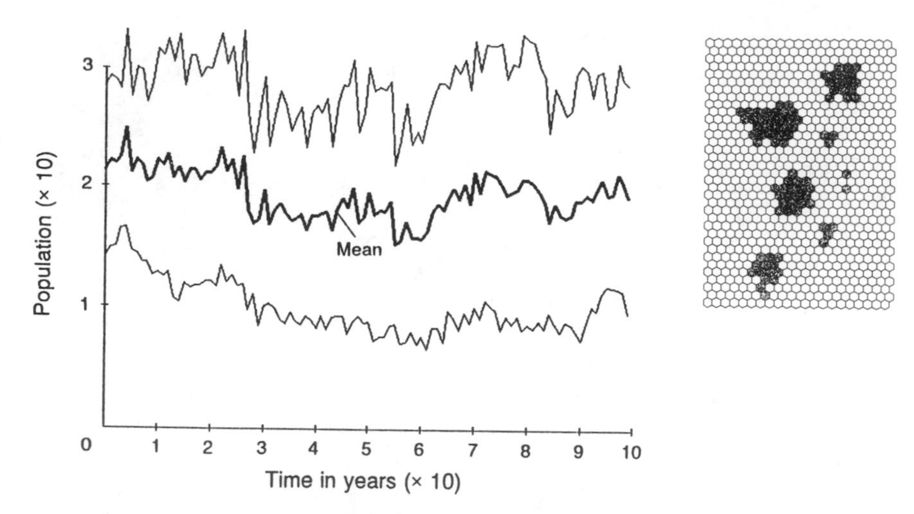

FIGURE 14. Landscape simulation with a reserve system consisting of clusters of suitable habitat surrounded by unsuitable habitat. The population was initialized with owl pairs occupying all suitable sites. The results are based on 30 simulations. Shading is based on mean pair occupancy.

tions in growth rate (λ) may be associated more closely with attributes showing extensive natural variation. In the case of the northern spotted owl, the effects of landscape pattern on survival during dispersal may induce high levels of both spatial and temporal variation in first-year survival rate. As habitat fragmentation continues, the uncertainty of successful dispersal will become progressively more relevant to the likelihood of persistence. The effects of landscape pattern on demographics are difficult to track analytically, but can be addressed through computer simulation.

Shifts in life history parameters

Because the static stage-projection model is a single-sex, nonspatial model, its extrapolation to population dynamics in real landscapes is limited. The simulation models diverge from this structure by including the dynamics of both sexes and the effects of search efficiency—as, for example, in the individual territory model (Figure 15). In this model, juvenile survival, dependent on finding a suitable site, is largely defined by search efficiency. Similarly, search efficiency regulates the movement between a pool of singles and the reproductive adult class. Pair survival occurs only when both members survive with probability s^2 rather than s, for a single-sex model (Figure 15). Therefore, if the probability of pair formation is low, the number of reproductive females will decline more quickly than indicated by the stage matrix model.

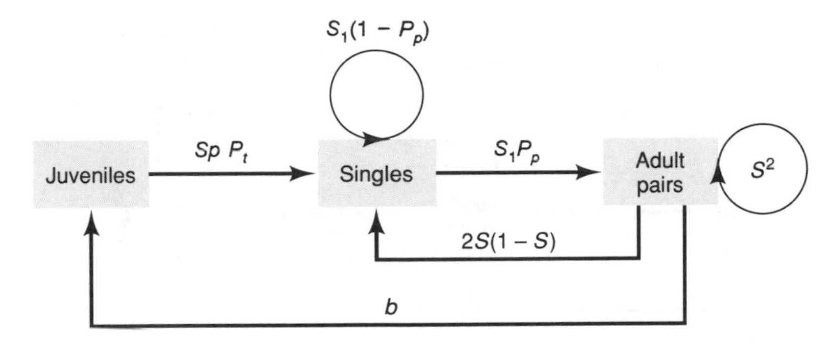

FIGURE 15. A flow diagram of the life history structure used in the individual territory model. P_t is the probability of finding a territory, P_p is the probability of pairing.

The territory cluster model has a simpler life history structure than the individual territory model, but more complex spatial dynamics. The only demographic parameter affected by search is s_0 ($= s_p s_d$). Being a single-sex model, it underestimates pair turnover rates, particularly in small isolated clusters, and contains no Allee effect. These two features lead to optimistic predictions regarding the fate of owl populations.

The landscape model is a two-sex model that maintains a strict stage structure but allows for movement between reproductive and nonreproductive classes (Figure 16). In this model, survival rates are differentially affected by search to the extent they are explicitly linked to properties of a landscape map. For example, survival may decrease in poor-quality habitat. In contrast to the previous two models, in the landscape model juveniles that fail to locate a suitable site do not die. As a consequence, search efficiency has less of a direct effect on s_0, but has a strong effect on the dynamic movement between the reproductive and non-reproductive stages To the extent that search inefficiency prevents pair formation, fecundity (b) declines and λ decreases accordingly.

The role of clusters

In a spatial model with search, a system composed of clusters is more stable than a diffuse system because clusters produce regions where search efficiency is maximized. In a cluster of suitable territories, a population can recover from low occupancy because search efficiency will remain high. Dispersing males have a high probability of finding habitats immediately adjacent to existing pairs. Dispersing females will more easily find territorial males because of their adjacency. The key to stable populations within clusters is that they be large enough to avoid frequent local extinction due

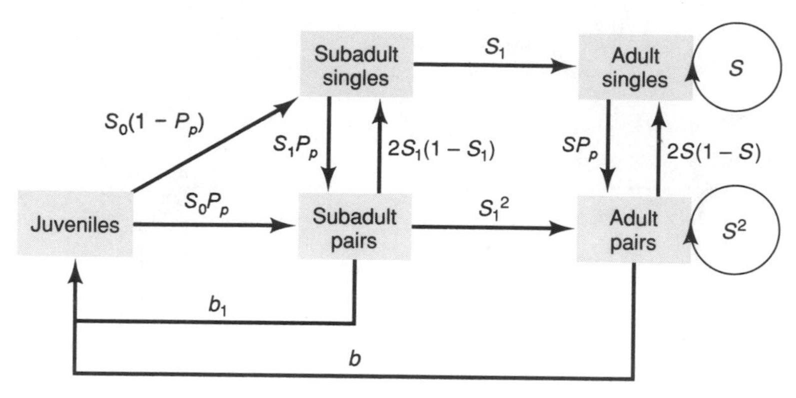

FIGURE 16. A flow diagram of the life history structure used in the landscape model. P_p is the probability of pairing.

to stochastic demographic events. Based primarily on demographic uncertainty, this suggests a minimum cluster population size of about 20 breeding pairs (see also Richter-Dyn and Goel, 1972). Even in a diffuse system with a large number of suitable territories, pair occupancy will be low because of low recolonization rates. As a result, in a highly fragmented system, positive density-dependent growth rates usually associated with low population densities are unlikely. As the density of reproductive pairs decreases, the probability of new pair formation at isolated sites also decreases due to a decreased density of dispersing owls.

The effects of shape

The negative effects of cluster irregularity can also be explained through the effect of shape on search efficiency. Circular clusters are the most stable because the density of suitable habitat is locally maximized. All other geometric forms will have reduced search efficiency when compared with a circular cluster. The effects of cluster irregularity are, in fact, probably more pronounced than is indicated by Figures 10–12. In these simulations, the risks associated with crossing poor-quality habitat were identical to those encountered in suitable habitat. If greater risks were encountered in the poor habitat, the effects of irregular cluster shape would be accentuated.

Source–sink dynamics

The increase in the variance of population size observed in landscapes in which breeding occurs in sink locations outside of the clusters (Figure 13) is primarily due to increased variance in adult survival (s). If breeding is limited to the clusters, the population will equilibrate with a relatively constant proportion of the population within the clusters and dispersing through the matrix. Changes in λ will be bounded by the levels of demographic and environmental stochasticity, and population variability will remain relatively constant over time. In contrast, when breeding occurs exterior to the clusters, the variability in λ will depend not only on environmental stochasticity but also on the proportion of the population within the clusters. The bounds on λ will be determined by the proportion of the population found outside the clusters $\pm$ the impact of environmental stochasticity. Because the simulations reported here were initialized with reproductive pairs restricted to clusters, the divergence from this state increases with time (Figure 13).

The decline in mean occupancy within the clusters in the source sink system (Figure 13) is due to decreased interaction between the clusters. Dispersing juveniles that settle exterior to the clusters would, if this option were not available, continue searching. Some of this search would return owls to clusters, increasing the recolonization rates. Thus a reserve design may be more stable if the boundaries between the reserve and the surrounding landscape are very distinct, since sink areas exterior to the reserve system may lead to suboptimal choices on the part of the dispersing juveniles. In particular, because juveniles settle outside of clusters, reserve systems with adjacent marginal habitat may be more prone to extinction than reserves surrounded by totally hostile habitats (cf. Figure 13 with Figure 14). This effect can arise even though the total population (number in the reserve plus adjacent habitats) is enhanced by marginal habitats.

CONCLUSIONS

The negative consequences associated with unsuccessful search are qualitatively similar in all three models but, due to differences in the mechanics of search, different inferences can be drawn from each model. All of these inferences yield insights into the probable efficacy of particular reserve designs.

The primary inferences drawn from the individual territory model are that the introduction of environmental stochasticity does not change the inflection point for the extinction threshold and that, in landscapes experiencing a high rate of habitat loss, both occupancy and demographic rates may underestimate the risk of extinction. The territory cluster model demonstrated that, for a given area, large clusters spaced farther apart on the landscape were more stable than small clusters that were closer together, and, assuming low to moderate dispersal rates among adjacent clusters, 15–20 suitable territories per cluster represents a likely stability threshold. The landscape model demonstrated that the shape of clusters is nearly as important as their size. Further, it raised the possibility that, in a landscape consisting of a gradient of habitat qualities, the presence of marginal habitat adjacent to a reserve may not help the owl population, and may even increase its risk of extinction. In addition, the landscape model integrates landscape-level vegetation characteristics into a spatially explicit distribution of habitat qualities. This allows one to project the population consequences of current and future landscape patterns.

The models we have described in this chapter are much more than academic exercises. In particular, explicit links between vegetation characteristics and the expected value of the vital rates at the scale of individual territories have been developed for the northern spotted owl in Washington and Oregon (J. Bart pers. comm.). Concurrently, vegetation maps are being drawn based on Forest Service and Bureau of Land Management data which describe all federal lands in western Oregon and Washington in terms of habitat suitability for spotted owls. Merging our landscape models with such detailed habitat maps will allow us to better design old-growth reserves that favor the persistence of the northern spotted owl. In addition, the strategy we develop should be applicable to a wide range of threatened species with narrow habitat requirements.

CHAPTER 27

HABITAT FRAGMENTATION AND GLOBAL CHANGE

James F. Quinn and James R. Karr

Systematic reduction in the earth's capacity to support living systems is perhaps the most alarming consequence of anthropogenic impacts. Habitat destruction plays the dominant role in humanity's negative influence on the biota, and is tied to growing human populations, technological advances, and rising levels of economic activity. The pressure for development is relentless throughout the world. Although the idealized goal of conservationists may be halting habitat loss, ecologically robust compromise solutions are required to minimize further biotic impoverishment.

Ecologists must identify conservation strategies that will succeed despite a limited percentage of the countryside that is preserved in a natural state. Equally important is treatment of lands and waters between reserves. Intervening spaces influence the viability of reserves, and they can be managed for important biological values even though they do not represent pristine wilderness.

Few countries have protected more than the 3 percent of land area accorded national park or statutory wilderness status in the United States. Although the detailed calculations are debatable (Lugo, 1988), the relationships between species richness and area suggest that less than half of the original species are likely to be found in such a small fraction of the original wilderness (Simberloff, 1986). Further, many species are unlikely to persist in a park system fragmented into isolated reserves (Soulé et al., 1979).

CONSERVATION IN FRAGMENTED LANDSCAPES

Strategies for establishing nature reserves have been the subject of a sometimes acrimonious debate (Simberloff and Abele, 1982; Wilcox and Murphy, 1985). Nevertheless, a consensus, perhaps best represented by the northern spotted owl recovery plan (see McKelvey et al., this volume), seems to be emerging among ecologists regarding strategies for setting aside natural areas to minimize extinction risks. In short, the consensus advocates multiple protected areas, each large enough to plausibly maintain local populations of those species most at risk of extinction due to "demographic stochasticity" (Shaffer and Samson, 1985) or severe inbreeding (Frankel, 1974). Typically, the "charismatic megafauna" of large, rare, often predatory, vertebrates is emphasized, at the expense of plant, invertebrate, or aquatic species. Although the long-term success of conservation efforts depends on protecting an array of major taxa and a landscape perspective that includes all habitats (e.g., forests, streams, wetlands), legal and institutional mechanisms for identifying and protecting threatened habitat types and ecological processes are virtually nonexistent.

Accepted models of extinction processes (e.g., Richter-Dyn and Goel, 1972) suggest minimum viable local populations of 20–50 individuals for long-lived terrestrial vertebrates with relatively constant year-to-year mortality and fecundity schedules, such as the northern spotted owl (see McKelvey et al., this volume). On the other hand, local populations numbering many thousands may be at risk in annual species, or in species subjected to intense environmentally driven variation in life history parameters (Leigh, 1981). On both theoretical (Leigh, 1981) and empirical (Karr, 1982b, 1990) grounds, extreme variability in population size suggests considerable risks of extinction in even locally abundant species.

Even under pristine conditions, many species occur in metapopulations with ephemeral local populations. Recovery occurs by reintroduction from surviving subpopulations, leading to a regional extinction–immigration balance. Thus, subdivision with some interpopulation migration protects the metapopulation from local catastrophes such as fires or floods (Quinn and Hastings, 1987). Conservation biologists generally prescribe maintaining "corridors for migration" to enhance the reestablishment of ephemeral local populations.

Patterns of regional diversity support the proposition that effective long-term protection of biological diversity will require the preservation of diverse habitats dispersed over a large landscape. In particular, existing biogeographic surveys suggest that collections of geographically dispersed natural habitat fragments (islands, mountaintops) harbor a greater fraction of their regional biotas than do the same number of square kilometers in large, contiguous blocks of habitat (Quinn and Harrison, 1988). Apparently, similar habitat patches often have little overlap in their species lists, leading to high between-habitat diversity (see Harrison, this volume). Under such circumstances, single megareserves, no matter how pristine and well-buffered, cannot protect the nonoverlapping biotas of outlying areas. Skeptics of subdivided reserve systems (e.g., Wilcox and Murphy, 1985) argue that geographically distributed small reserves are subject to high extinction risks, and therefore will be ineffective in contributing to long-term conservation. Obviously, some species (particularly charismatic megavertebrates) require extensive contiguous areas to maintain a reliable source population for recolonizing outlying reserves. On the other hand, a conservation strategy that does not place conservation lands where unique species occur is doomed to failure.

In essence, conservation biologists have settled on a conceptual approach that relies on probabilistic conditions for the persistence of fragmented populations of selected rare and usually atypical species. Thus, contemporary conservation strategies hinge on identifying and setting aside enough sufficiently large habitat fragments to preserve existing colonization–extinction balances for a few charismatic, extinction-prone species, thereby putting all our eggs in a few well-watched baskets. We suppose, without overwhelming evidence, that these species are reasonable "miners' canaries" for a complex of species dependent on the same general kinds of environmental conditions. The potential pitfalls of this approach are obvious. For example, it will not be successful for taxa restricted to small, specialized habitats (e.g., bogs, streams, rocky outcrops) that are not distributed homogeneously in the major regional habitats (Sharpe and Zhao, 1990; Zimmerman and Bierregaard, 1986). Thus, any successful reserve system requires a mix of core megareserves for the megafauna and multiple, perhaps small, reserves to capture unique species, habitats, and metapopulation functions.

ROSY SCENARIOS

If environmental rules do not change, standard population models suggest that conditions sufficient to protect fragmented populations may not be especially stringent. Target species must have a few "source areas" in which individuals tend to more than replace themselves under average survivorship and fecundity schedules. [Species that do not replace themselves locally, as suggested by life history data for the northern spotted owl (Thomas et al., 1990; McKelvey et al., this volume), are problematical.] Source areas are not necessarily where the bulk of the population is found (see Groom and Schumaker, this volume). For most vertebrates, the year-to-year extinction risk for source subpopulations larger than 20–40 individuals is likely to be no more than a few percent. A few independently varying subpopulations of this kind can insure that target species will outlast the average ecologist, much less the usual wildlife recovery plan. For example, the model of Quinn and Hastings (1987) predicts that a population broken into ten independent protected subpopulations, each with a 1 percent per year probability of extinction, would still have about a 99 percent chance of avoiding total extinction in the next century.

The simplest metapopulation models (Levins, 1970; Levin, 1974; Slatkin, 1974), which assume occasional long-distance migration between subpopulations, also suggest optimism. If migration is panmictic, the original models take on a logistic form, and per-capita colonizations increase most rapidly when species are rare, due to the increase in opportunities for new colonization as habitable patches are emptied by local extinction. Thus, the most cited metapopulation models generally predict long-term regional persistence of fragmented populations, even in the face of severe and spatially widespread environmental fluctuations (Harrison and Quinn, 1989). In short, conventional wisdom does not predict sudden dynamic surprises, particularly population collapses, in fragmented populations facing gradual environmental change. However, as illustrated by more detailed models for the spotted owl (McKelvey et al., this volume), confidence in the global stability of metapopulations may be misleading.

ECOLOGICAL SURPRISES

The first element of prudent management is to anticipate and avoid surprises, such as abrupt collapse of previously stable populations and species associations. Collapses may be more likely when habitats vary dramatically in quality. In particular, a small area of high-quality habitat (a "source") may produce a net outflow of migrants that sustains populations in large marginal areas ("sinks"). Anthropogenic fragmentation that eliminates or isolates the source area, or long-term environmental change that decreases reproduction in the source, could easily lead to rapid population collapse over a much larger area (Pulliam, 1988).

Dispersal limitation also may lead to ecological surprises. Hassell and colleagues (this volume) have shown that complex behavior, including "checkerboard" distributions, cycles in both space and time, and various forms of chaos can result from even the simplest models with dispersal limited to adjacent cells. Barriers to dispersal (caricaturing human disruption of a natural habitat matrix) lead to abrupt changes in dynamics. Perhaps the ubiquitous patchiness of species distributions reflects analogous dynamics in natural landscapes that are similarly susceptible to qualitative change or collapse in population dynamics.

Epidemiological models yield similar concerns. To persist, a parasite must on average infect one new susceptible host before killing or being lost from its present host. As hosts become more sparse or less susceptible to infection, the parasite becomes unable to successfully replace itself with at least one offspring established in a new host. As a result, parasite–host systems often exhibit abrupt host density thresholds, above which the parasites become epidemic, below which they decline toward extinction (see Dobson and May, 1986, for a review). The same logic applies to parasite or predator species that exploit hosts or prey with patchy, ephemeral distributions, and to species limited to a particular kind of habitat patch, particularly those with strong territoriality (Lande, 1987).

Dynamically, in today's fragmented forest, a northern spotted owl could be viewed as a benign parasite on a patch of old-growth trees. As the density of old growth stands is reduced, we can expect the owl–old growth system to cross the "host" density threshold below which the owls rapidly disappear. Depending upon uncertain assumptions (e.g., dispersal behavior of juveniles) and parameter estimates (e.g., mortality rates), models suggest that northern spotted owls require 20–35 percent of the region to be composed of suitable stands (Thomas et al., 1990; McKelvey et al., this volume). Most endangered species are probably both more specific in their host habitat needs and poorer dispersers than northern spotted owls, and may require greater densities of pristine patches to remain above the epidemiological threshold. In a recent review, Hastings (1991) suggests that an analogous density threshold is a general consequence of fragmented distributions.

Regardless of population size, tightly coupled species may rapidly collapse in gradually changing environments. The dependence of many freshwater mussels on larval dispersal by specific fish species (Strayer, 1990) and the reciprocal dependence of figs and forest vertebrates as food and as dispersal agents, respectively, are vivid examples of associations made precarious by highly specialized interdependencies (Terborgh, 1983; Gautier-Hion and Michaloud, 1989).

CLIMATE CHANGE: CHANGING THE RULES

Whether one subscribes to the rosy scenario that a few well-protected core populations assure persistence, or the more pessimistic view that many sudden, dynamic collapses are likely, global climate change complicates the picture. In general, climate change makes management strategies for maintaining biological diversity more stringent than those inferred from models of a static environment. Coupling climate change with other anthropogenic influences (e.g., habitat fragmentation and chemical contamination, especially contamination associated with reproductive dysfunction, Colborn et al., 1990), makes effective conservation programs much more difficult.

Discussions of climate change have emphasized long-distance shifts in vegetation and associated faunas (see Peters, 1988, and Root, this volume, for reviews). Judging from post-Pleistocene patterns, warming in the next half century could require range shifts of hundreds of kilometers for many species. In the presumably less fragmented environment following the last glaciation, range shifts appear to have averaged a small fraction of a kilometer per year, one to two orders of magnitude less than the shift needed to adjust to the rapid anthropogenic changes expected during the next half century. Furthermore, because suitable intervening habitats have often been eliminated by agriculture or urbanization, the rates of future range shifts may be severely reduced relative to the rates detected in fossil communities.

Climate change also may influence the probability of ecological surprise on a local scale. Under static conditions, the persistence of a few permanent source areas may insure the regional stability of susceptible populations, as long as the intervening matrix remains passable to dispersing individuals. Long-term environmental change can threaten source areas by decreasing the survival and fecundity of the target organisms, either through direct physiological effects or through secondary effects on associated species. For example, annual movements of migratory birds may be sufficient to transfer plant propagules, but the establishment and reproductive maturation of long-lived plants may be too slow to meet the needs of birds that can no longer depend on populations decimated by changing climate. Thus, both mutualistic interactions and the relative time scales of the life histories of the mutualists may influence population persistence.

If the location of physiologically suitable source areas shifts over time, long-term persistence depends upon colonizing new "source" habitat patches. Limitations in dispersal ability may prevent this from happening rapidly enough to offset current losses, resulting in population collapse. Thus, without changing population size, the dynamics of some populations may change from a globally stable source–sink configuration to an unstable metapopulation lying below a viability threshold of the kind seen in epidemiology (Dobson and May, 1986).

RESEARCH AGENDA

Many of the information needs most critical to adaptive management of a fragmented landscape have been developed in preceding chapters. We discuss those issues we found most compelling.

Biological inventories and "choosing canaries"

Distributional information for most species, even many of considerable conservation interest, are woefully incomplete. For example, most of the large mammal populations reported to have disappeared from U.S. National Parks (Newmark, 1987) apparently represent unrecorded extant species rather than actual disappearances (Quinn et al., in press). Similarly, our knowledge of the loss of bird species from the best-studied tropical forest in the world (Barro Colorado Island, Panama) is based primarily on knowledge of what now occurs on the nearby mainland, instead of on recorded disappearances of species known to have occurred on the island (Karr, 1982a,b, 1990).

Worse yet, systematic inventories (species lists and range maps), which themselves are rare, provide no foundation for predicting the influence of habitat fragmentation or climate change. Studies complete enough to provide a reasonable sample of interannual climatic variability, population censuses, mechanistic analyses of climatic impacts on life history parameters, and experimental investigations of interspecific interactions, can be counted on one's fingers (Dunham, this volume). Yet this is probably the quality of information needed to predict the effects of environmental change on any species. Such studies are unattainable for more than a few species with the available workforce and resources.

Thus, the identification of appropriate indicators of change in ecological communities—particularly those that might help use to anticipate ecological surprises—must become a top priority. The lack of simple yet coherent suites of ecological indicators hampers our ability to evaluate the status of ecological resources and to communicate the consequences of biotic impoverishment to policy makers or to the public at large (Karr, in press a). Furthermore, divergent views among practicing ecologists obstruct a useful synthesis. Population and community ecologists are guided by an evolutionary paradigm based on natural selection, whereas ecosystem ecologists are guided by a thermodynamic perspective (Karr, in press b). It is hard to imagine ecologists with these divergent viewpoints agreeing on the criteria that should be used for identifying ecological health (e.g., indicator species, species richness, production, trophic structure, individual health). A suite of indicators provides the best approach, just as a battery of medical tests is often needed to indicate individual health (Karr, 1991).

Dispersal limitation and the maintenance of local uniqueness

Harrison (this volume) raises the complex issue of the role of historical accident in determining ecological patterns. If dispersal abilities do not limit the invasion of species into suitable habitats, as suggested by the insensitivity of between-habitat diversity to distance, then rapid range adjustments in response to both fragmentation and climatic change might be expected. In contrast, the same data may be interpreted as an indication of strong community divergence, even on a very local scale—a pattern consistent with severe dispersal limitation. The between-habitat diversity values found by Harrison, over a wide range of taxa and geography, suggest a typical species overlap of less than 30 percent (often much less) among sites, regardless of distance between them. Comparable patterns are commonplace in systems of nature reserves. High between-habitat diversity, however, does not prove a lack of dispersal. It may reflect the existence of alternative community states, a matrix of successional stages, or a matrix of distinct habitats of sufficient geographic area and age to support the evolution of distinct biotas. Regardless of what causes variation in divergence, landscapes dominated by historical patterns cannot be counted upon to retain their structure in the face of rapid anthropogenic change.

Managing ecological invasions

In landscapes whose structure is disrupted by fragmentation, many native species will be unable to keep pace with the shift in climate zones. Human actions to foster their movements seem inevitable, although the secondary consequences of aiding migration of desirable species may be detrimental. Worse yet, we may unwittingly enhance the success of undesirable exotics —the so-called weedy species—resulting in the homogenization of the world (Culotta, 1991). We must ask ecological theory, however inadequate, to provide guidance for "managed introductions," which might also be viewed as ecological experimentation on a meaningful geographic scale.

Only a minority of native species are likely to be candidates for purposeful introduction into suitable but unoccupied habitats, whether newly available due to climatic change or unoccupied because of historic extinction. For other species, the maintenance of corridors for natural migration has become an accepted prescription. The desirability of corridors bears more scrutiny than it has received (Simberloff and Cox, 1987). Often diseases, pests, and exotic invaders may move more readily along corridors than the target native species, thereby serving to degrade and homogenize otherwise unique fragments of the original landscape.

More generally, the capital needed to purchase corridors and protect their margins from encroachment may often be better dedicated to acquiring new reserve areas. The ecological connectedness of conservation lands may be less important for increasing dispersal than for preventing the loss of animals into dangerous interspersed sinks. For example, dispersed juvenile northern spotted owls are apparently subject to increased great horned owl predation in new clearcuts (Thomas et al., 1990). In such cases, managing the source–sink balance, particularly by managing the worst sinks, is undoubtedly more important than maintaining ribbons of pristine habitat between protected reserves.

Identifying and protecting unoccupied patches

Metapopulations persist because the rate of colonization of unoccupied patches balances the rate of loss from occupied patches. If wetlands temporarily unoccupied by a protected bog plant are fair game for conversion into cornfields or condos, the plant will be unable to maintain the balance, and its regional distribution will contract over time. Criteria for identifying unoccupied patches are probably best verified by experimental introductions to test habitability (see Harrison, 1989, as an isolated example where this has been done on a meaningful geographic scale).

Unfortunately, without records of past local distributions, it is difficult to make a compelling case under endangered species laws for preserving unoccupied patches, because absence could suggest that the patch is unsuitable. In any case, relying on legal endangered status runs a severe risk of delaying management until the "surprise" threshold has long been exceeded. These issues show the weakness of basing resource decisions on endangered species legislation as opposed to establishing societal goals based on efforts to protect ecological integrity (Karr, 1991, in press a).

The advent of rapid climatic change makes the problem of preserving "empty patches" even more pressing. For species whose ranges are set by physiological tolerances, provisions may be made for adaptive range shifts (Peters, 1988; Root, this volume). Other species, particularly those whose dynamics are dominated by interspecific associations, are likely to shift from a stable source–sink configuration, where several well-protected sources can insure persistence, to more of a shifting mosaic of transient populations, in which a fair density of unoccupied but invadable "host" patches are needed for the metapopulation to rise above the "surprise" threshold.

Mechanisms of extinction

Mechanistic literature on the dynamics of extinction in small populations concentrates on the sampling properties of population propagation in an implicitly constant environment ("demographic stochasticity" in numbers and genetic drift in allelic composition). These models predict a distinct threshold (often labeled the "minimum viable population") above which persistence is virtually certain. A naive corollary has been distressingly widely accepted and applied in policy making, holding that populations kept above a threshold (often 50 or 500) are acceptably safe and that those below are doomed to extinction.

The role of the genetic structure of fragmented populations in the risk of extinction has provoked considerable discussion (see Lande, 1988 for a review). It is conceivable that declining heterozygosity and inbreeding depression in local populations could reduce reproductive output enough to provoke demographic collapse (Gilpin, 1991). For this reason, some authors argue that corridors or other mechanisms promoting gene flow are essential for metapopulation persistence (e.g., Frankel, 1974, 1981). As with the demographic role of corridors and migration, the appropriateness of this prescription depends upon the biology of the target species. For most populations, genetic effects on risks of extinction *per se* will be dwarfed by environmental effects, at least until the population is on its last legs (Lande, 1988).

Several models show that environmentally driven variation in vital rates, even with a net predicted average growth in population size, can lead to a substantial short-term risk of extinction (Lewontin and Cohen, 1969; Leigh, 1981; Ludwig, 1976; Shaffer and Samson, 1985; Harrison et al., 1991). Most such models assume that environmental conditions are chosen from a continuous distribution of "white noise." Unfortunately, all but the simplest cases have proven mathematically intractable, so even such fundamental issues as the relationship between extinction risk and population size or habitat area are not easily predicted. Under realistic conditions, even relatively large populations may run a substantial chance of "random" extinction, suggesting that the search for a safe threshold for viability may be illusory. Many paleontologists believe that past extinction events resulted mainly from rare extreme environmental fluctuations (see Raup, 1986 for a review). However, the theory of catastrophe-driven extinction is in its infancy (Hanson and Tuckwell, 1981; Gripenberg, 1985; Ewens et al., 1987). In this context, it is disturbing that most climate models predict not only general warming, but more intense climatic disturbances and greater variability as a result of increased greenhouse gases (Schneider, this volume).

Climatic extremes often may cause extinction by modifying interspecific interactions rather than through direct physiological effects, i.e., "the confusion of fundamental and realized niche" (Pacala and Hurtt, this volume). Communities that have been particularly susceptible to local extinction in modern times include oceanic islands (Vitousek, 1988) and freshwater lakes and streams (Lodge, this volume). Introduction of exotic predators, diseases, and habitat alteration are the primary causes of these extinctions. The effects of further environmental degradation will continue to be most pronounced in habitats with strong interspecific interactions, particularly "trophic cascades" among trophic levels (Paine, 1980). Often the most vulnerable communities may be depauperate (Strong, in press), rather than the high-diversity settings typically targeted for protection.

INSTITUTIONAL CONSIDERATIONS

The diversity of the contributions to this volume reflects both the strengths and missed opportunities of research on the ecology of long-term environmental change. Ecology differs from climatology (see Schneider, this volume) and many other sciences (physical and chemical oceanography, geophysics, etc.) in that most research is initiated by individual investigators and is typically supported by funding lasting only a few years. The strengths of investigator-initiated research include the flexibility to innovate and to implement data acquisition protocols tailored to local circumstances. However, individuality and flexibility have inevitable consequences. Individual investigators tend to conduct univariate experiments and analyses, reflecting limitations of individual expertise and equipment. Both personalized protocols and study-to-study differences in the variables measured make comparisons among studies difficult. Short funding horizons discourage the multiyear data sets needed to test for the effects of gradual environmental change, or for the effects of rare extreme events.

Many ecologists believe that competition for investigator-initiated research funding has prevented environmental biologists from presenting a unified front calling for public support of research on biotic impoverishment. The paltry and declining support available for describing and studying biological diversity is extremely disappointing when compared with the public funds astronomers garner to photograph no more spectacular and considerably less useful celestial sights.

On the theoretical front, the bewildering variety of ecological models described in this volume makes a compelling case that conceptual approaches in ecology are not narrowly constrained by any single monolithic paradigm. This provides a striking contrast with the state of theory regarding the effects of greenhouse warming on the physical environment, where much of our understanding is based on the work of a few centers conducting large, cross-disciplinary modeling studies (Schneider, this volume).

While there is disagreement among ecologists on the utility of centralized modeling studies, there is almost universal agreement that a lack of simple monitoring data on the biological environment precludes the analysis needed to plan for long-term environmental change. For example, much of the data identified as essential by contributors to this volume (e.g., rates of geographic range shifts, local extinction and immigration risks, compositional differences among communities, and effects of exotics) can only be derived from long-term distributional and population records — data more akin to that gathered by U.S. Weather Bureau stations and U.S. Geological Survey stream-gauging stations than to the short term grant- and publication-driven activities of most ecologists.

Reasons for the lack of a national biological survey in the United States to define and monitor changes in our biota are varied, but undoubtedly include the fragmentation of responsibility among government agencies (e.g., Environmental Protection Agency; U.S. Geological Survey; National Oceanographic and Atmospheric Administration; Fish and Wildlife Service), the lack of a user constituency with substantial economic clout, and the lack of understanding of the extent to which we as a society depend on the integrity of biological support systems (Karr, in press a). Without widespread ecological literacy, society is unlikely to develop a policy that advocates maintenance of ecological integrity.

One of a growing number of initiatives, which include National Institutes for the Environment, various national and state biodiversity bills, and the Sustainable Biosphere Initiative, may provide an institutional platform to address these problems. Unfortunately, the information needed to address long-term environmental degradation is still not well specified. Ideally, one would like to obtain frequent maps of the population status of each species and of selected assemblage attributes, coupled with as yet unspecified physical environmental data for each sample site. Given available budgets and biologists, such data are unattainable for most species and places, so appropriate "canaries" still must be sought. Standardization of biotic inventory strategies among competing government and private land-use agencies may be as important as the detailed choice of information gathered. Unfortunately, progress toward information standards seems nonexistent except in aquatic systems, where biological criteria (indicators of ecological condition compared with some least-disturbed reference community) are being widely used to protect the biological integrity of water resources (USEPA, 1990; Karr, 1991, in press a; Dudley, 1991).

Describing the biological effects of climate change is not merely an academic exercise. Biology drives global change in the physical environment as much as global weather drives biology (Ojima et al., 1991; see Schimel, this volume). For example, changes in albedo, runoff, and evapotranspiration (all of which influence atmospheric dynamics) largely result from the same land uses that fragment stands of natural habitat. Often, recovery must be derived from inhabitants of the remaining fragments. A renewed commitment is needed if we are to identify and protect the species and habitats at greatest risk, and to counteract ecological surprises.

Also, we see a need for increased risk-taking in ecological analysis. Most ecologists promote parallel objectives of maximizing diversity, minimizing ecological meddling, and maintaining intact communities and biological processes. Global change is likely to make attaining all of these objectives difficult, if not impossible. Post-Pleistocene communities facing much slower warming did not track the changes through range shifts of intact communities; existing associations were continually disassembled and novel ones formed. Thus, an insistence on maintaining historically transitory preindustrial associations seems pointless. We must accept that change is inevitable, and reject the assumption underlying much public policy that a static "pristine" state, represented by a snapshot immediately before the Industrial Revolution, can be maintained indefinitely. To manage that change, we must shed our squeamishness about purposefully manipulating natural communities.

In the next few decades, the physical environment is expected to change an order of magnitude faster than at any time in the recent past. Large-scale geographic shifts are blocked by fragmented landscapes, and many species are already under stress from habitat destruction, chemical contamination, and exploitation. For many species, rapid adaptation and geographic accommodation may be the only hope. Under these circumstances, ecologists should reconsider their blanket antipathy to purposeful introductions of locally non-native species, selective breeding, game farming, market mechanisms to promote management of "unnatural" but diverse associations in harvested ecosystems, and other proactive ecological intrusions. Ecological understanding is not up to predicting what will happen, and monitoring and adaptive management will be essential to successfully mimic adjustments that might have occurred more slowly in our absence. We will undoubtedly make serious mistakes. However, we will make even larger mistakes if we manage for stasis in the declining fragments of pristine habitat and accept our current level of ignorance regarding the processes that govern biodiversity.

SUMMARY

Conversion of natural habitats is undoubtedly the most important process threatening the long-term persistence of most of the earth's species and the biological processes upon which we and those species depend. Most conservation will have to be concentrated on fragmented lands comprising a small proportion of the habitat originally available. Our ability to protect the species in those refuges depends on dynamics within patches, among patches within the larger landscape, and in the interstitial spaces between patches. Global climate change threatens our ability to succeed because it changes all three of those components in ways that are difficult to predict. Some species will be lost from their present habitats as physiological conditions change; their persistence will depend upon dispersal across a fragmented landscape to currently unoccupied locales. Persistence of other species is threatened by conversion from a stable source sink configuration to dispersal-limited metapopulation dynamics, subject to sudden catastrophic collapse. Further, the threat from weedy species, those ubiquitous commensals that disperse with human society, increases as landscapes are fragmented and as climate change increases the selective pressures for rapid dispersal ability.

Past climate change has disrupted persistent species associations, and rapid changes expected in the next century will undoubtedly lead to increased invasion of "warm weather" exotics, dramatic shifts in community structure, and local extinction of many affected species. Managing the changes will require better information to anticipate sudden ecological surprises, greater appreciation of the importance of unoccupied patches, and improved understanding of the dynamics of dispersal and extinction. Managing the change also will require a willingness to engage in purposeful manipulation of threatened species, a more informed public, and coordination of governmental efforts to reduce the undesirable consequences of global climate change.

PART SIX

THE CONTEXT FOR POPULATION AND COMMUNITY RESEARCH ON GLOBAL CHANGE

CHAPTER 28

POLICY IMPLICATIONS OF GLOBAL CLIMATE CHANGE

Gordon H. Orians

Human-induced changes in the earth's climate are being added to an already long list of threats to the future of biodiversity on the planet. Climate change is a problem that has the potential to undermine many of the remedial actions taken under the assumption that today's climate will persist into the foreseeable future. Awareness of the global implications of human activities has only recently become widespread, however, because the changes they cause are the cumulative effects of many small-scale actions, most of whose individual contributions are miniscule. It is difficult to perceive or conceive that human activities already coopt about 40 percent of global terrestrial production (Vitousek et al., 1986). The fact that a multitude of small-scale actions contribute to global change makes dealing with global change problems especially difficult. The cumulative effects of repeated environmental perturbations are both more substantial than and often qualitatively different from those caused by the individual events. Therefore, traditional methods of assessing the significance of development on a project-by-project basis may fail to predict, and hence fail to help us manage, cumulative effects.

Devising wise policies to deal with global climate change is made difficult by the great uncertainties surrounding predictions of future changes at all scales, but especially at local and regional scales. Because existing climate models treat clouds and oceanic circulation crudely, surprises may lurk in the outputs of better models, not to mention in reality (Schneider, this volume). There are also great uncertainties in the responses of human institutions to climate change, especially given the great range of economic, political, and social resources available to different countries to deal with the changes thrust upon them. The likely responses of natural and human-modified ecosystems to climate change are equally problematic. It is extremely difficult to predict the consequences of climate change on the structure, functioning, and distribution of ecological systems, both because predictions of climate change are imprecise and because our poor state of understanding of the dynamics of ecological systems would prevent accurate predictions even if future climates were known.

DEVISING A STRATEGY FOR DEALING WITH CLIMATE CHANGE

It is likely that there will be winners and losers among both species and human societies if global climate warms. However, the losers are likely to dominate the arena, especially if change is rapid or unpredictable. To improve our ability to assess the consequences of climate change, we must direct our attention to assessing the kinds and amounts of change that might occur and identifying the major processes that drive those changes; determining how best to offset the effects of the driving variables so as to reduce the amount of climate change; and assessing how we might best adapt to whatever changes are forthcoming. There is much debate over how much and what kind of anticipatory and preventative actions should be taken.

Emphasis on the predicted increase in global mean temperature has at times obscured the fact that the temporal and spatial scales of climate changes are of critical importance biologically (Schneider, this volume). All climate models predict complicated regional patterns of change. Some regions are expected to experience little temperature change, others considerable warming. Changes in precipitation are likely to be positive in some places and negative in others. The spatial and temporal scales of biological responses and of responses of human institutions are key to assessing the efficacy of potential response options. If climate warms gradually, so that global mean temperature increases 3–5°C over the next century, this rate of change will be rapid relative to the population dynamics of long-lived species, such as many trees, and the rates at which their ranges can shift (Pacala and Hurtt, this volume). Rates of evolutionary change are even slower (Travis and Futuyma, this volume).

Considerable uncertainty surrounds the best ways to assess the likely consequences of climate change and the desirability of different policy options. Cost–benefit approaches have been very popular in the professional policy arena, but they raise both philosophical and technical problems. The use of such models is based on the assumption that humans have exclusive rights to use and manipulate the biosphere in whatever ways we choose.